The Dancing Chain
History and Development of the Derailleur Bicycle

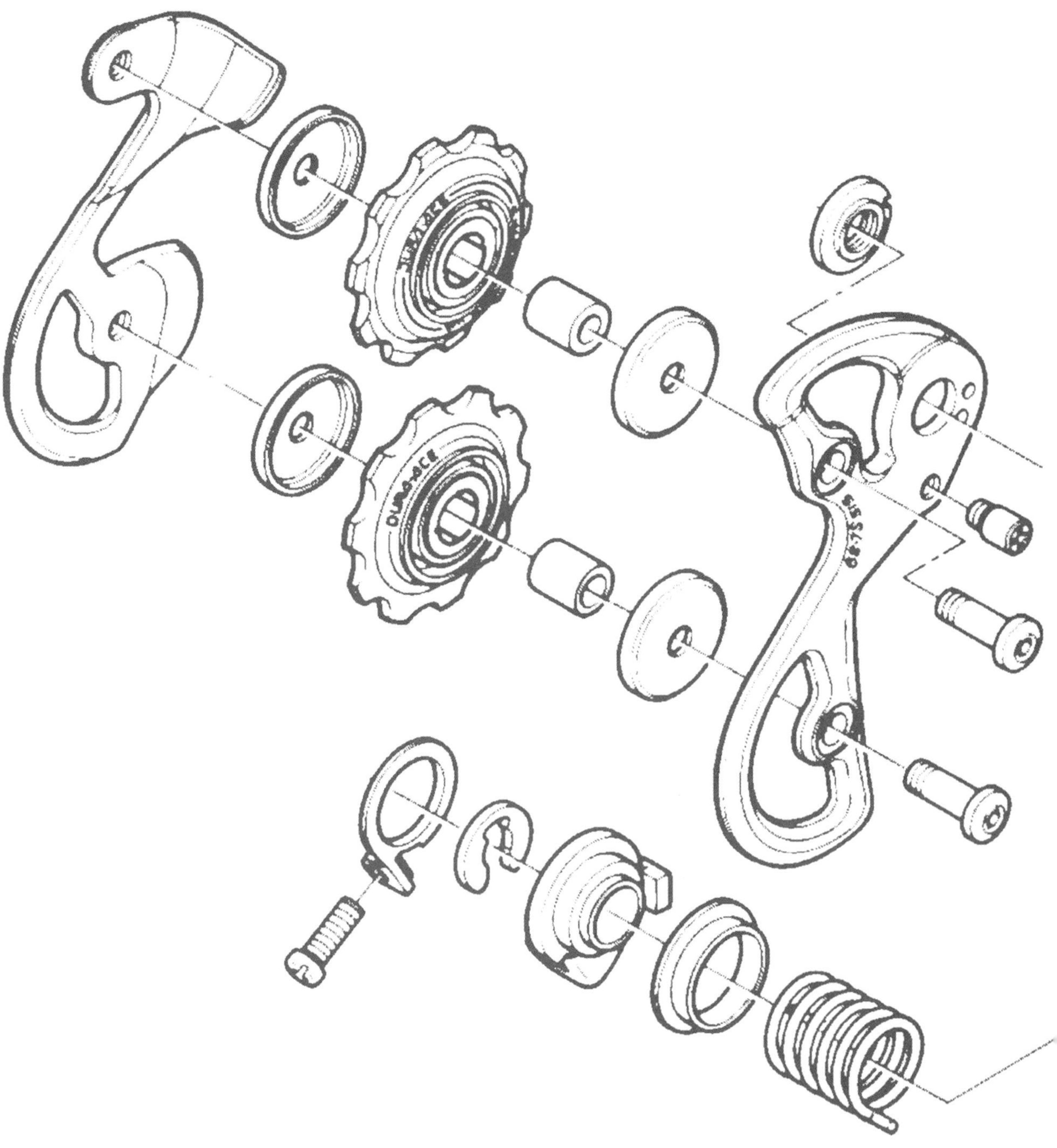

Cycle Publishing / Van der Plas Publications, San Francisco

# The Dancing Chain

## History and Development of the Derailleur Bicycle

Fourth edition, revised, updated, and expanded

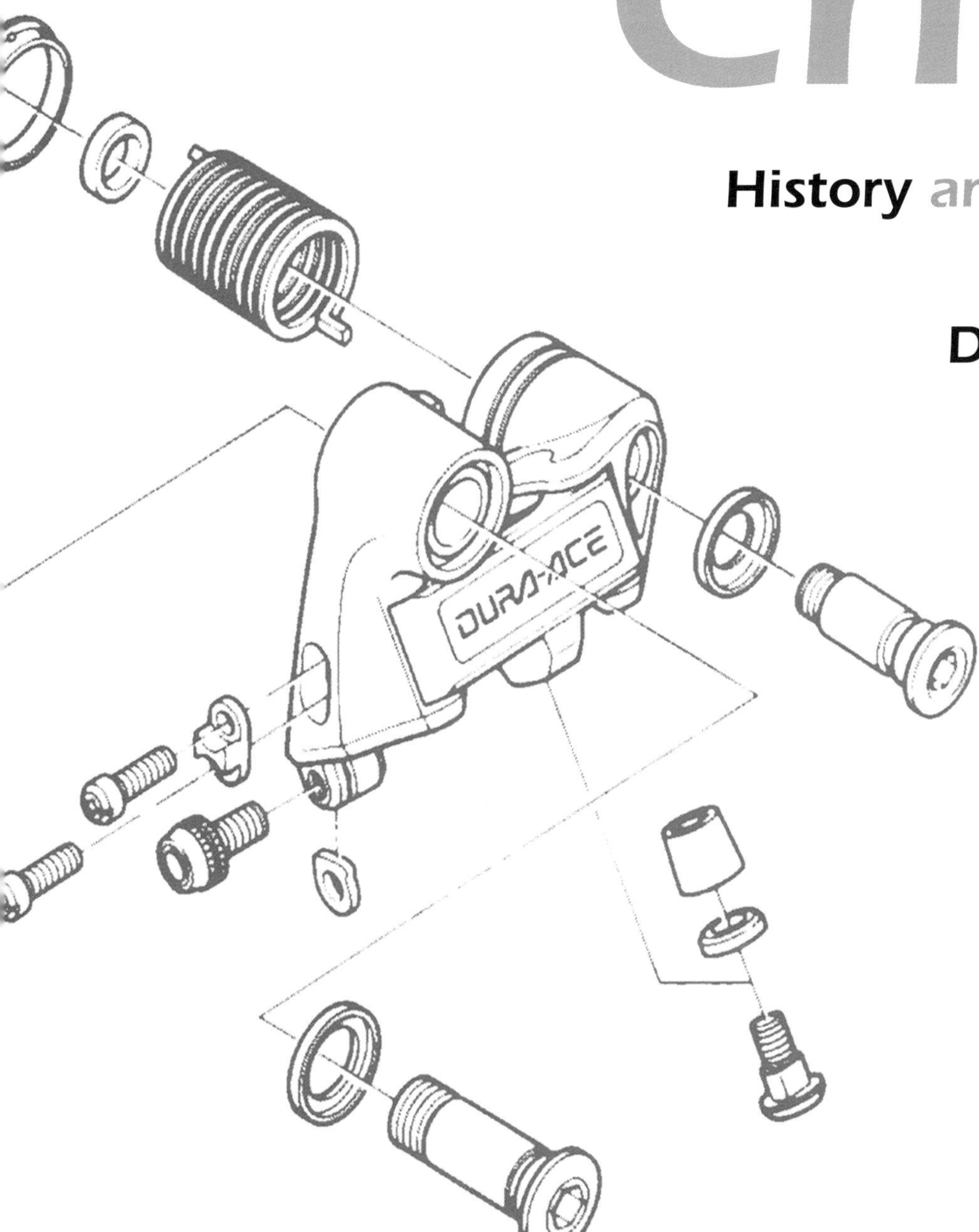


**Frank Berto**

With contributions by

Tony Hadland
Jan Heine
Raymond Henry
Gordon Selby
Ron Shepherd
Walter Ulreich

Fourth edition, revised, updated, and expanded, 2012
Printed in Hong Kong

Published by:
Cycle Publishing / Van der Plas Publications
1282 7th Avenue
San Francisco, CA 94122, USA
Tel: (415) 665-8214
Fax: (415) 753-8572
E-mail: con.tact@cyclepublishing.com
Web site: http://www.cyclepublishing.com

Distributed or represented to the book trade by:
USA: Midpoint Trade Books, Kansas City, KS
Great Britain: Orca Book Services / Chris Lloyd Sales and Marketing Services, Poole, Dorset
Australia: Woodslane, Warriewood, NSW

Cover design: Kent Lytle, Lytle Design, Alameda, CA

Frontispiece: Exploded view of a Shimano Dura Ace RD 7400 rear derailleur, a design that is still as "modern" today as it was when it was first introduced in 1985

*Publisher's cataloging in Publication Data*
Berto, Frank, 1929—
The Dancing Chain: History and Development of the Derailleur Bicycle
p: 400. 28.0 cm. Includes bibliographic information and index
1. Bicycles and bicycling—manuals and handbooks
2. History of technology
I. Authorship: Ron Shepherd, coauthor; Tony Hadland, contributor; Jan Heine, contributor; Raymond Henry, contributor; Gordon Selby, contributor; Walter Ulreich, contributor
II. Title: History and Development of the Derailleur Bicycle
ISBN 978-1-892495-69-3 (hardcover)
Library of Congress Control Number 2008925815

Dedicated to the memory of
**Ron Shepherd** († 2004),
who got the ball rolling, and without whose enthusiasm *The Dancing Chain* might never have come to be.
Photo: Ron Shepherd, in wheelchair, with Frank Berto, at the 10th International Cycling History Conference, 1999.

## Acknowledgments


The following people helped make this third edition more accurate and complete than the preceding editions:
**Tony Hadland** provided corrected and expanded material on cycling in Britain, and on Raleigh, Sturmey-Archer, and Moulton.
**Gordon Selby** provided information that allowed me to write an enhanced history of British Cyclo, Osgear, and Simplex. He also corrected many of my misconceptions about cycling in Britain in the 1950s and 1960s.
**Raymond Henry** provided additional information on, and illustrations of, early French derailleurs.
**Michael Zellman**, of SRAM, provided historical material on Fichtel & Sachs.
**André Ghestem**, of Shimano France, provided the 1980s Bertin-Milremo catalogs with their lovely Rebour line drawings.
**Hillary Stone** pointed out errors in the first edition.
**Bruce Epperson** and **Carl Burgwardt** provided details on old Columbia bicycles.
**Dr. Chester Kyle** provided background on speed records and human-powered vehicles.
**Derek Roberts**, **George Bickerstaffe**, **John Southgate**, and **Michael Barber** provided corrections and suggested changes to the first edition.
**Jan Heine**, provided additional illustrations, test results, and corrections for this edition.
My wife, **Connie Berto**, proofread the entire manuscript.
And about 50 additional people who wrote to me pointing out errors or omissions in earlier editions. Thank you to everyone who wrote to me.

# Introduction to This Edition

I personally financed publication of the first edition of *The Dancing Chain* in 1999. Ron Shepherd was responsible for the first five chapters, covering the pre-World-War-II years, and I was responsible for last seven chapters covering the post-World-War-II period.

When the last copies were sold in 2002, Rob van der Plas of Cycle Publishing agreed to publish a second edition. I finished rewriting the entire book in 2004, and it was published in early 2005.

In 2004, I wrote a revised section "Bicycle" for *Encyclopaedia Britannica*. This became an E-mail joint-venture with eight historians from the International Cycling History Conferences. This research allowed me to rewrite the period from Von Drais and Michaux to the introduction of the first derailleurs.

The last copies of the second printing of the second edition were sold in 2008. The mechanics of publishing and printing have changed in the interval, and Rob asked me to prepare an enlarged third edition. I agreed to add the gear-train and derailleur developments for the years 2006 through 2009 and any other relevant new material.

In the interim, I had become friends with Jan Heine, who had started *Vintage Bicycle Quarterly* (VBQ, later renamed *Bicycle Quarterly*, or BQ) in 2002. This publication concentrates on the "Golden Years" of the classic French touring bicycles—from the 1930s to the 1960s. I read every issue from cover to cover, because my expertise starts from the 1970s, when I began writing technical articles for *Bicycling* magazine.

In 2005, Jan wrote and published *The Golden Age of Handbuilt Bicycles*, featuring gorgeous color photographs by Jean-Pierre Pradères of 50 classic hand-built French bicycles. The oldest bicycle is a 1909 La Gauloise and the newest is a 2003 Alex Singer, but most are from the "Golden Age"—the years 1930 to 1960.

Jan rode and evaluated all 50 bicycles, and he kept complete notes. His notes covering 21 different derailleurs were published in his newsletter. There are more bicycles than derailleurs, because most of the classic French touring bicycles used the same Cyclo Standard, the classic touring derailleur.

Jan agreed that I could include his derailleur test results and many of his photographs in this edition of *The Dancing Chain*.

I was also able to include several additional Daniel Rebour illustrations, many of them obtained from André Ghestem, of Shimano France, and never seen in print before.

## About the Author

Frank Berto was born in Vancouver, BC in 1929. When he was about five years old, he was given a small Meccano set (the British equivalent of what Americans know as an Erector set). It was his favorite toy, and building Meccano models was his inoculation for mechanical engineering. He rode a single-speed Hercules roadster everywhere until he got a driver's license.

He graduated from the University of British Columbia in Mechanical Engineering in 1952 and went to work for an international oil company in Aruba, Netherlands Antilles, where he met Connie and got married. He obtained a Masters Degree in Mechanical Engineering from California Institute of Technology in 1958 and started work for Chevron in San Francisco, specializing in instrumentation and oil measurement.

During the 1970s bicycle boom, he, with his wife Connie and their five children, had seven 3-and 10-speed bicycles. Being involved in Boy Scouts, he found himself walking up hills while the troop pedaled past. That started him thinking about bigger sprockets and smaller chainwheels.

Thanks to a helpful bike shop and SunTour derailleurs and freewheels, he installed lower gearing on his bike. The gearing on the other bikes in his stable were similarly converted in a process of trial-and-error improvements. In the early 1970s, there was very little in print about bicycle gearing, and he wrote his first gearing articles for *Bike World* and *Bicycling!*

He wrote that the Japanese SunTour and Shimano derailleurs shifted better and cost less than the derailleurs made by Campagnolo, Huret, and Simplex. This was heresy to the other bicycle experts of the day. His second sermon was about bicycle gearing. He wrote that riders could have whatever gear range they wanted —especially a low enough low gear.

By 1977, he was riding half a dozen centuries (organized 100-mile rides) a year. In 1984, he road-tested loaded-touring bicycles and bought his first custom-built touring bicycle, complete with "half-step plus granny" gearing. He went on several long bicycle tours, including what's known in Britain as the End-to-End: some 700 miles, from Land's End, at the south-western tip of England, to John O'Groats, at the north-eastern tip of Scotland.

Between 1975 and 1990, he wrote more than 100 articles for *Bicycling*, which at the time billed itself as the world's largest-circulation bicycling magazine. When he retired from the oil business in 1986, he became their technical editor, and wrote an article for almost every issue. This was an educational period, and his writing was a "learn as

Author Frank Berto, second from right, flanked by three of his contributing authors for the first edition of *the Dancing Chain*, from left to right, Walter Ulreich, Raymond Henry, and Ron Shepherd, during a presentation at the 1996 International Cycling History Conference in Buffalo, New York.

you earn" process. He made factory visits and met the experts. *Bicycling* provided him with all the latest components as soon as they were available.

In 1988, he put all the collected information and expertise together and wrote *Bicycling Magazine's Complete Guide to Upgrading Your Bike*. This book dealt only with issues relevant to road bikes, but it appeared just as the mountain bike took over the market.

The mountain bike was developed on Mount Tamalpais, a few miles from his home. He knew everyone involved and thought at first that it would be just another California fad that would soon pass.

He ceased writing for *Bicycling* in 1991, when they de-emphasized technical content. Since then, he has written for a dozen different periodicals and has been involved as an expert witness in bicycle patent lawsuits.

Frank met Ron Shepherd in the mid-1970s. They were the two apostles of low gears, and corresponded regularly. Ron got him interested in bicycle history and in Moulton bicycles. He attended the 7th International Cycle History Conference (ICHC) in 1996 and presented a paper on rear derailleurs. Ron presented a paper on derailleur-pulley resistance. They shared a room, and Ron convinced Frank to join his project to write the "definitive history of the derailleur."

Between 1996 and 2007, Frank gave 8 papers at the History Conferences. His 1998 paper on the invention of the mountain bike led to publication of his second book, *The Birth of Dirt*, which is the source for much of the information in Chapter 11. The facts and ideas and presented in several of his other papers, such as "Sunset for SunTour," History Repeats," "Hub Gear Efficiency," and "The End of the Great American Bicycle Boom," have been incorporated in this edition.

His international oil-industry experience has convinced him that we will have a serious oil shortage in this decade. In fact, he's pretty sure it has already started. "Peak Oil" is the buzzword. His survival plan includes a Toyota Prius hybrid car and a Moulton NS bicycle.

**Publisher's Note Regarding Units**

We have adhered to the following convention for quoting units of measure and their conversion: Dimensions, speeds, and distances are generally quoted in the unit system in which they were originally recorded (e.g., Imperial units for early British and U.S. figures, and metric units for current figures, except some U.S. measures. Conversions into the other system are given in parentheses following the original figures.

Metric units are quoted in their standard unit designations (e.g., m for meters, km/h for kilometers per hour, and g for grams. For English measures, we have followed U.S. conventions instead of British conventions (specifically, abbreviations are followed by a period, or full stop).

# Table of Contents

# 1 The First Bicycles: 1817–1869

THE BICYCLE is a two-wheeled, rider-propelled machine. Historians disagree about the precise details of the invention of the bicycle, and many dates are questionable, but I am convinced that Karl von Drais' 1817 *Draisienne* was the first bicycle. The pedal-driven bicycle initially became popular in the late 1860s, but it took a quarter of a century to evolve into the standard pneumatic-tired, chain-driven design we know today. This chapter will look at the genesis of the bicycle and the developments and inventions that led up to bicycles suitable for the use of multispeed gearing and derailleurs.

## Early Myths and Recent Frauds

Numerous claims of earlier inventors of the bicycle have been circulated over the years, and all have been proven to be either fakes or hoaxes. Some of these fakes will be discussed in this section.

Leonardo da Vinci has been credited with having sketched a bicycle in 1492 in the Codex Atlanticus. However, that drawing has been proven to be a forgery, added at the time of the volume's restoration in the 1960s.

The (otherwise well-respected) French author Louis Baudry de Saunier erred when he said that a certain Compte de Sivrac rode *vélocifères*, or *célérifères*, in the 1790s. The Compte de Sivrac never existed, and the terms *vélocifères* and *célérifères* referred to fast horse-drawn carriages, not precursors of the bicycle.

The basis for the claim that velocipedes had appeared in Mi-

Above: Fig. 1.1. Baron Karl von Drais, born in 1775, invented the first bicycle in 1817.

lan in 1811, and their use on public roads declared illegal in a police ordinance against riding *velocipedi* in Milan was a forged document produced in the 1950s.

### Terminology

Bicycle historians have had a problem with nomenclature for more than a century. The name for the historically current and popular bicycle has changed (in chronological order) from *Draisienne* to *hobbyhorse*, to *vélocipède de pédale*, to *velocipede*, and finally, to *bicycle*. Whenever a substantially revised new model took over, the old model got a new name.

Around 1870, the term *bicycle* came into general use. When bicycle wheels began to be built with wire spokes and solid rubber tires in the early 1870s, the old wooden-wheeled velocipedes were called *boneshakers*, and the new models were called *bicycles* or *high bicycles*. When the safety bicycle appeared in the 1880s, the old high bicycles were called *ordinaries* or *ordinary bicycles*, to distinguish them from the new safety bicycles.

By the 1890s, the pejorative term *penny-farthing* came into use, comparing the front wheel to the large British penny and the rear wheel to the much smaller farthing (quarter-penny).

Something similar took place in the 1980s: At the beginning of the decade, most bicycles sold in the U.S. were drop-handlebar "ten-speeds," but gradually mountain bikes became more popular. By the early 1990s, 90% of the bicycles sold were mountain bikes, so they became *bicycles*, and the drop-handlebar machines became *road bikes*.

## Karl von Drais, Inventor of the Bicycle

I have a three-part definition of "inventor." An inventor must have an original idea, and not copy someone else's prior idea. He must make the first prototype, which cannot be a copy of someone else's prototype. Finally, he must actively participate in the subsequent developments that led to the utilization of the invention.

There is indisputable evidence that Charles Friedrich Christian Louis Baron Drais zu Sauerbronn ("Karl von Drais"), of Baden, Germany, invented the first bicycle—the first two-wheeled, rider-propelled machine. Von Drais called his bicycle a *Laufmaschine*, (running machine), but *Draisienne* and *velocipede* (fast foot) became the popular names.

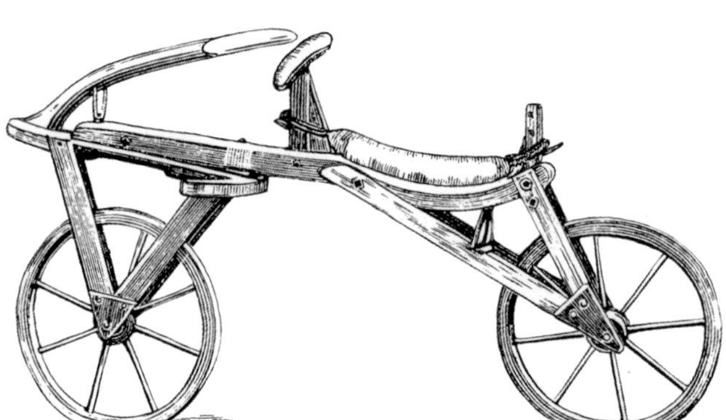

Above: Fig. 1.2. Karl von Drais' *Laufmaschine*, or Draisienne, was the first bicycle, even though it was propelled by kicking instead of pedaling. It was made largely of wood and weighed about 50 lbs. (23 kg). It was built in 1817 to Drais' design by the Karlsruhe coach-builder Johann Frey.

Right: Fig. 1.3. Denis Johnson's 1818/1819 Pedestrian Curricle used more metal than Von Drais' machines and had lighter wheels. Surviving examples weigh less than 50 lbs. (23 kg). These machines soon became known as "hobby-horses." This is a photo of No. 25, the oldest extant Johnson machine, on display at the Mark Hall Museum in Harlow, England.

Von Drais invented his machine in 1817, and he reports covering 9 miles in less than an hour. The following year, he exhibited it in Paris. The machine was made of wood, and the seated rider propelled himself by kicking off against the ground. A balance board supported the rider's arms.

In spite of von Drais' patents, copies were soon being produced in other countries, including Great Britain, Austria, Italy, and the United States. Von Drais' breakthrough was establishing that a two-wheeled vehicle could be balanced by steering the front wheel in the direction of lean. It has been suggested that he may have picked up the idea from ice-skating.

Denis Johnson, a London coach builder, purchased a Draisienne and patented an improved model in 1818 as the *pedestrian curricle*. The following year he produced more than 300. They were commonly called *hobby-horses*. They were very expensive, and many buyers were members of the nobility. Caricaturists called the devices *dandy horses*, and riders were sometimes jeered in public. The design raised health concerns, and riding was only practical on smooth roads. Johnson's production lasted less than two years.

In 1821, Lewis Gompertz added a hand crank, connected to a rack-and-pinion drive, to a Johnson hobby-horse. Gompertz obtained a patent on his "Addition." He built and sold a small number. His goal was to increase the distance traveled between leg kicks. Gompertz' machine required a "ratch-wheel," his name for a freewheel. From our perspective, one wonders why he did not put pedals on the front wheel rather than his cumbersome hand-powered lever.

The brief Draisienne–hobby-horse fad did not lead to the sustained development of two-wheeled vehicles. However, von Drais and Johnson clearly established that steerable two-wheeled machines could remain balanced while in motion. The estimated 500 or so early hobby-horses built were easy to copy, and they remained in circulation to provide inspiration for the designers and builders of subsequent machines.

## Tricycles and Quadricycles

The search for practical human-powered vehicles did not end with the hobby-horse, but was pursued in a variety of different directions. Between 1820 and 1860, numerous innovative machines were produced in small numbers, but they rarely went beyond the initial prototype, because they were too heavy, too complicated, or too expensive.

Right: Fig. 1.4. The Duke of Leinster bought this hobby-horse in Ireland in 1819. Sir John Powers had a similar machine.

Left: Fig. 1.5. Lewis Gompertz' 1821 "Addition" to Johnson's hobby-horse had a rather complicated hand-operated drive mechanism to the front wheel.

In 1819, B. Smythe, a surveyor, invented the British Facilitator or Travelling Car. It probably never got past the patent drawing stage, but the design had two speeds—one for climbing and one for descending.

Willard Sawyer started as a carpenter in Dover and ended up as the most significant manufacturer of four-wheeled velocipedes, although that was probably never his main business. He made velocipedes from 1838 until 1878. Most were driven by foot treadles attached to the front wheels, but some were propelled by hand.

The machines were steered by the rear wheels. Sawyer's outstanding workmanship reduced the weight, which helped hill-climbing. Sawyer velocipedes were made in three quality levels, and the top-quality models were sold to royalty. In 1851, Sawyer drove his velocipede 80 miles (135 km) from Dover to London for the Great Exhibition.

## Early Treadle-Driven Velocipedes

A small number of two-wheeled machines with rear treadle-drives were built in southwestern Scotland during the early 1840s. Kirkpatrick Macmillan, a blacksmith from Dumfriesshire, is most often associated with them. He is said to have traveled 40 miles to Glasgow in 1842. However, documentation is problematic, and in spite of the numerous "replicas," we do not know for sure whether Macmillan's machine had two, three, or four wheels, or whether he powered it with his arms or his legs.

Gavin Dalzell (pronounced "day-ell"), of Lesmahagow, Scotland, built a similar two-wheeled machine in the mid-1840s and is said to have operated it for many years. It had wooden wheels with iron rims. The rider's feet swung treadles backward and forward, moving a pair of rods connected to cranks on the rear wheels. Thomas McCall, another Scotsman, built similar machines in the late 1860s.

There is some documentation that suggests that Alexandre Lefèbvre of St.-Denis, France, may have built a two-wheeled velocipede powered by treadles connected to cranks on the rear wheel as early as 1842. Lefèbvre may have ridden his bicycle on many trips, including one from St.-Denis to Le Havre. He may have taken this velocipede with

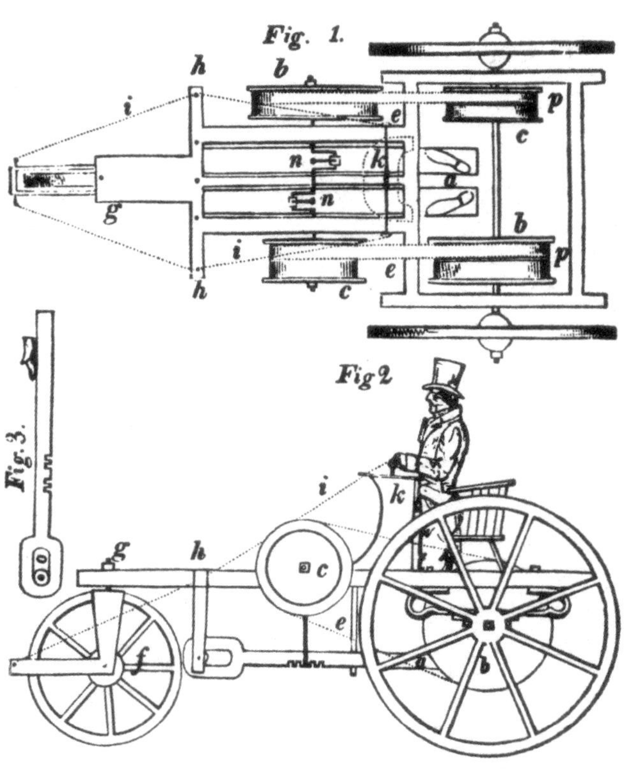

Above: Fig. 1.6. Willard Sawyer on one of his 1851 velocipedes, which survives in the Science Museum in London.

Left: Fig. 1.7. B. Smythe's 1819 British Facilitator, or Travelling Car, probably was the first English 2-speed human-powered vehicle.

him when he emigrated to California in 1861.

[Editor's Note: This claim too is disputed, since the Lefèbvre machine at the local history museum in San Jose, California, has not been dated, and the few experts who have had a chance to see it agree it is likely to be of later provenance. It may have been built when Lefèbvre lived in the United States, i.e. after 1861.]

Viewed with hindsight, driving the rear wheel with treadles, levers, and cranks was a sounder concept than attaching pedals to the front wheel. However, neither the Scottish machines nor Lefèbvre's velocipede were commercially exploited. There is no evidence that they contributed to subsequent development.

## Early Pedal-Driven Velocipedes

The bicycles that we ride today can be traced back to the first velocipedes powered with pedals attached to the front wheel that were built in Paris during the early 1860s. There is no conclusive evidence to prove who first conceived the idea of applying pedals to the front wheel or who actually did so. Either way, these machines were originally known as *vélocipède de pédale*. The word *bicycle* came into use in Europe in 1868 to replace that cumbersome phrase.

### Pierre Lallement

There is evidence that Pierre Lallement, a French mechanic, built and demonstrated a two-wheeled velocipede in Paris in mid-1863. At the time, he was working for M. Strohmayer, a Parisian maker of carriages for children and invalids. Lallement took the parts for an improved velocipede with him when he went to the United States in 1865, and he completed his new *veloce* in Ansonia, Connecticut.

A U.S. patent was issued in 1866, but a manufacturer could not be enlisted, and Lallement returned to France sometime in 1868. In that same year, French Michaux velocipedes started a craze in America. Lallement was

Above: Fig. 1.8. Treadle-driven velocipede built by Thomas McCall in 1860. Macmillan's machine, if indeed it was a two-wheeler, is presumed to have looked something like this.

Right: Fig. 1.9. Gavin Dalzell's restored velocipede, presumed to date from 1846, as pictured in 1906. Both the McCall and the Dalzell machines are now on display in the Glasgow Museum of Transport. The claimed dates are to be taken with a pinch of salt.

able to sell his patent to American entrepreneur Calvin Witty for $2,000, and this patent would soon have consequences for the American bicycle industry.

### Pierre and Ernest Michaux

Michaux et Cie. was a company that made carriage locks in Paris. Pierre Michaux and his son Ernest presented their pedal-driven velocipede to the public in the mid-1860s. The best evidence indicates that they built it in early 1864 (not 1861 or 1855, as claimed in many histories), and that they built a few more in 1865 and 1866. Some had malleable cast-iron frames, appar-

Left: Fig. 1.10. Alexandre Lefèbvre's treadle-drive velocipede may have preceded the Michaux front-drivers by 20 years, although here too, the claimed date is not based on any contemporary record. This undated machine, first identified by Andrew Ritchie in his 1973 book *King of the Road,* is in the "History San Jose" museum in San Jose, California.

Bottom left: Fig. 1.11. The 1866 U.S. patent drawing for Pierre Lallement's velocipede.

Bottom right: Fig. 1.12. An 1869 photograph of Pierre Lallement in Paris on a Michaux velocipede.

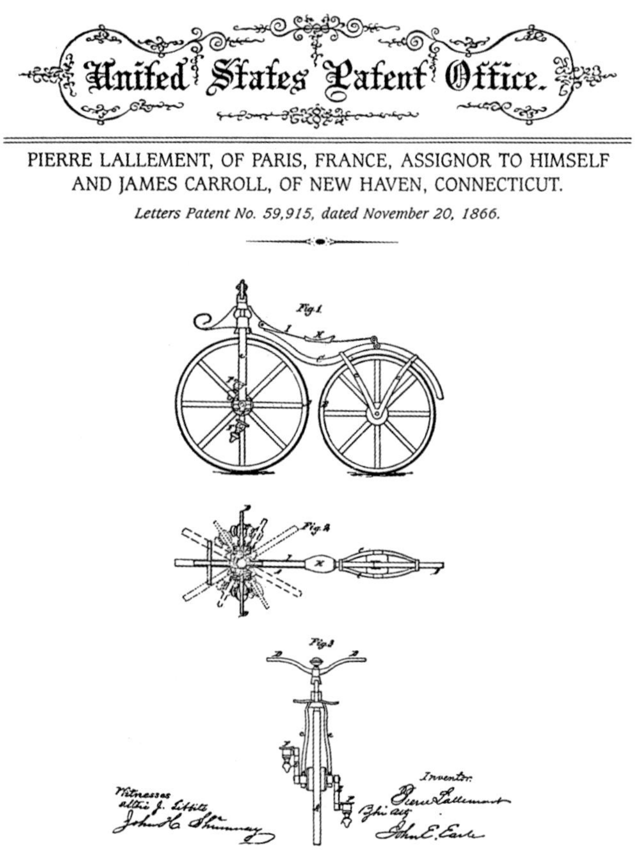

ently in anticipation of large-scale production. Cranks and pedals were attached to the front wheel, which was 34 to 36 in. (85–90 cm) in diameter. The rear wheel was slightly smaller. Neither Pierre nor Ernest Michaux ever filed patent claim to the basic design, but they did patent a number of improvements in April 1868. The 1866 model weighed around 45 kg (100 lbs), so it wasn't a great hill-climber.

The Michaux' role as the pioneer manufacturer of pedal bicycles is inextricably linked with that of the Olivier brothers, René and Aimé. In 1865, these two wealthy young men rode their Michaux velocipedes more than 800 km (500 miles) from Paris to Marseilles. Their subsequent enthusiasm for the new sport helped it to become a worldwide craze for the young, fit, and well-to-do. The Olivier brothers paid 50,000 francs for a 69% equity in Michaux et Cie, which then moved to a much larger factory.

The first models had a serpentine-shaped malleable iron frame. Shortly thereafter, the company switched to a diagonal frame made of wrought iron, which quickly became the industry standard. Major production commenced in 1867, the year of the Paris Exposition, when many visitors saw velocipedes on the streets. Its popularity spread, and the number of makers multiplied, and by the autumn of 1868, the new velocipedes were a familiar sight across France, sales reaching new heights despite relatively high prices.

In 1869, the Olivier brothers took full control of the Michaux company and changed the name to Compagnie Parisienne des Vélocipèdes. Production reached about 200 velocipedes per month. By that time, more than 100 French companies were making velocipedes.

Many major bicycle improvements were invented or developed in 1869. Numerous "véloce" clubs were formed, and they sponsored bicycle races. Four hundred riders entered the 122 km (76-mile) Paris–Rouen race, which was won by James Moore in 10¾ hours, riding a Suriray bicycle. Jules-Pierre Suriray had invented the ball bearing. He was in charge of the Paris prisons and used convict labor to grind and polish the balls.

Eugène Meyer manufactured the first steel-rimmed wheels with tensioned wire spokes in 1869. They were much lighter than the wooden-spoked wheels in use up to that time, and enabled bicycle manufacturers to further increase the size of the driving wheel. The rider now pushed down almost vertically on the pedals, making steering easier. The new wheels were had rubber tires, which reduced noise and vibration and gripped the road better when going uphill.

Above and right: Figs. 1.13 and 1.14. First and second generation Michaux machines, respectively with "serpentine" and "direct" (or "diagonal") frame designs. The direct drive to the front wheel remained unchanged.

William Van Anden patented the freewheel (which allows coasting) in 1869 in the United States. Several months later, Antoine Boeuf and François Nicolet patented the freewheel in France.

The hard ride of wood-spoked wheels and iron rims gave early velocipedes the sobriquet of *boneshaker*. Wire-spoked wheels with solid rubber tires were stronger and lighter and helped soften the ride. In 1870, as the boneshaker was in the process of developing into the practical bicycle that would become the *ordinary*, the Franco-Prussian War set back the French velocipede industry. Bicycle manufacture survived in France, but large-scale bicycle production moved across the Channel to Britain, and most of the subsequent developments took place in Britain.

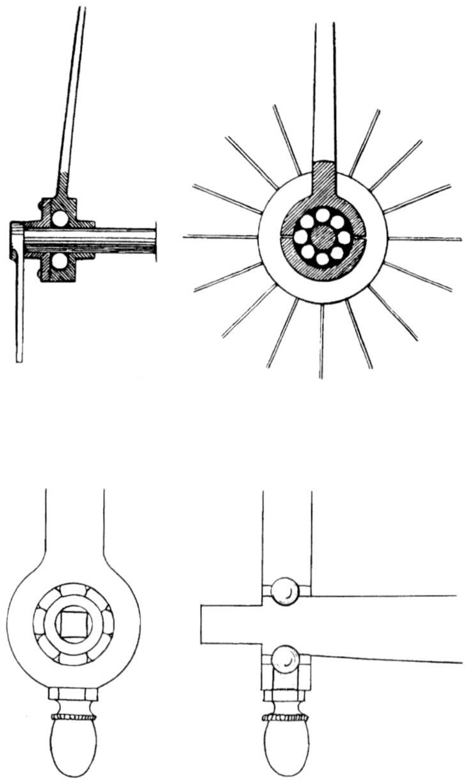

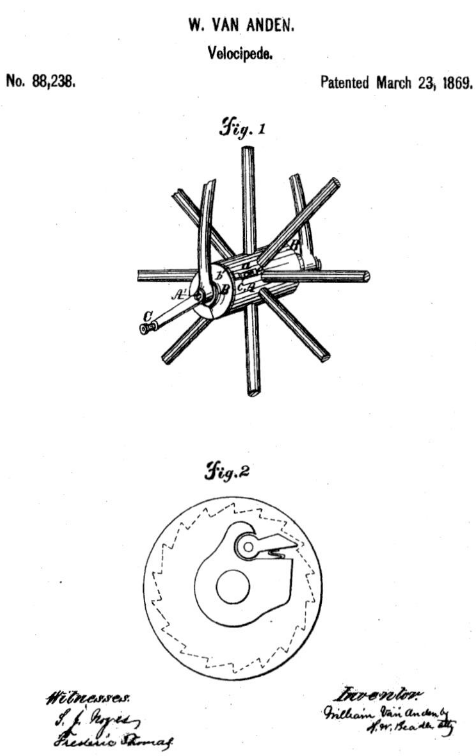


Above left: Fig. 1.15. Jules-Pierre Suriray's 1869 patent for the first ball bearings.

Bottom left: Fig. 1.16. William Van Anden's 1869 freewheel patent. Gompertz had used a similar "ratch-wheel" on his Addition of 1821, but he did not get a patent.

Right: Fig. 1.17. Champion velocipede racer James Moore, depicted in 1869 on his Suriray machine.

Left: Fig. 1.18. The Reynold & Mays Phantom of 1869 was progressive in many respects. It was the first British velocipede to have rubber tires, steel tension spokes, and front and rear suspension. It had a hinge in the middle of the frame, which resulted in lively—some say erratic—handling.

### The First Velocipedes in Britain

Major bicycle production in Britain began in 1868, when Rowley B. Turner brought a Michaux bicycle to Britain and showed it to his uncle, Josiah Turner, the manager of the Coventry Sewing Machine Company. Rowley Turner ordered 500 machines, 250 for sale in Britain and 250 for France. The French sales were lost because of the Franco-Prussian war, but the British market easily absorbed the entire batch, where this type of machine became known as "boneshaker."

By the early 1870s, bicycle technology and usage was securely established. The elegant high bicycle had replaced the crude boneshaker, which had been based on wooden carriage technology. Hollow steel tubular frames and forks, quality ball bearings, tension-spoked wheels, steel rims, solid rubber tires, and standardized parts became the norm.

James Starley's 1871 Ariel set the design standard for the ordinary bicycle. The Ariel had a 48-inch (1.20 m) front wheel and a 30-inch (76 cm) rear wheel. Starley's prolific improvements for bicycles and tricycles over the next decade earned him the title

Right: Fig. 1.21. The Smith, Starley & Co. Ariel bicycle of 1872 was the first British all-metal high bicycle to be manufactured in quantity. Note the lever, which twisted the wheel to tension the spokes and to transmit the pedaling torque.

Above: Fig. 1.19. Monument erected to the memory of James Starley in Coventry. For years, the nose of Starley's image on the monument was broken, but we are pleased to report it was restored to its old glory in 2008.

Right: Fig. 1.20. William Van Anden's 1869 Dexter Velocipede was an advanced design with a suspended saddle and a freewheel to allow coasting downhill. The rear brake was needed because this bike had a freewheel.

"Father of the Cycle Trade." By 1874, the focus of the bicycle industry had clearly shifted from Paris to Coventry, and England led the technological development of the bicycle into the 20th Century.

### The First Velocipedes in America

The first American bicycle craze was inspired by news from Paris. It began in late 1868 and quickly spread to the major East Coast cities, and even to California. New York had the world's first cycling periodical, *The Velocipedist*, published by the bicycle manufacturer Pickering & Davis.

Small American manufacturers sprang up, and more than 250 patents were issued in two years. Bicycles were promoted at indoor riding academies, which were often rinks from the earlier roller-skating craze. Enthusiasm quickly flagged when long-distance travel on poor roads proved impractical.

Calvin Witty used the patent he had purchased from Lallement to charge a $10 royalty for every bicycle sold. This hurried the demise of the velocipede, although most makers ignored the patent. Interest had died out by 1871, not to be revived until after the Philadelphia Centennial Exposition in 1876.

Two British companies exhibited bicycles at the 1876 Philadelphia Centennial Exposition. Albert A. Pope, a Boston industrialist, liked what he saw and began to import British ordinaries. By 1880, the Pope Manufacturing Co. was making the Columbia, a copy of the British Duplex Excelsior. This was the beginning of the American bicycle industry.

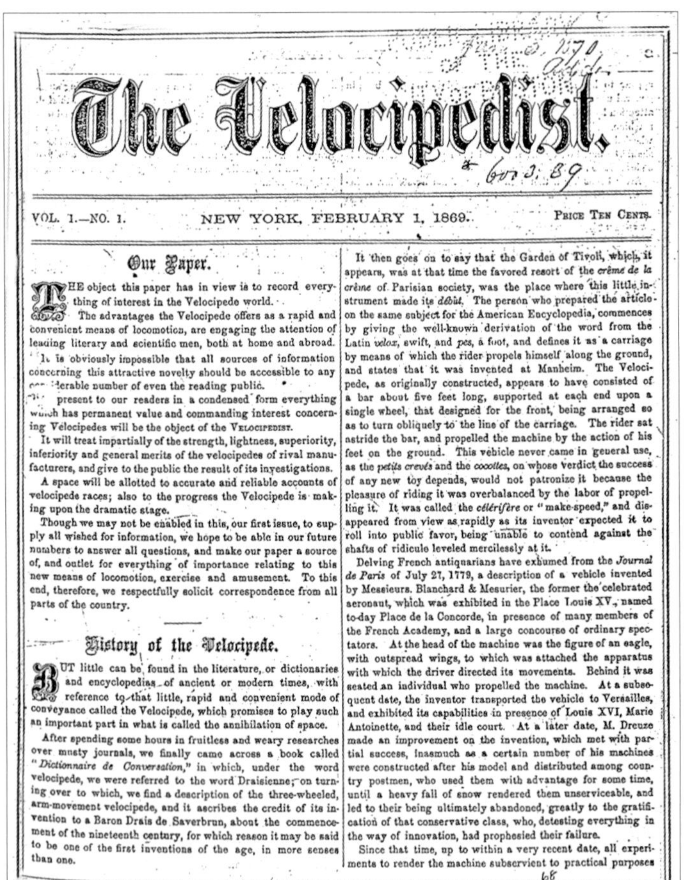

The Velocipedist.

VOL. 1.—NO. 1. NEW YORK, FEBRUARY 1, 1869. PRICE TEN CENTS.

Our Paper.

THE object this paper has in view is to record everything of interest in the Velocipede world.

The advantages the Velocipede offers as a rapid and convenient means of locomotion, are engaging the attention of leading literary and scientific men, both at home and abroad.

It is obviously impossible that all sources of information concerning this attractive novelty should be accessible to any considerable number of even the reading public.

To present to our readers in a condensed form everything which has permanent value and commanding interest concerning Velocipedes will be the object of the VELOCIPEDIST.

It will treat impartially of the strength, lightness, superiority, inferiority and general merits of the velocipedes of rival manufacturers, and give to the public the result of its investigations.

A space will be allotted to accurate and reliable accounts of velocipede races; also to the progress the Velocipede is making upon the dramatic stage.

Though we may not be enabled in this, our first issue, to supply all wished for information, we hope to be able in our future numbers to answer all questions, and make our paper a source of, and outlet for everything of importance relating to this new means of locomotion, exercise and amusement. To this end, therefore, we respectfully solicit correspondence from all parts of the country.

History of the Velocipede.

BUT little can be found in the literature, or dictionaries and encyclopedias of ancient or modern times, with reference to that little, rapid and convenient mode of conveyance called the Velocipede, which promises to play such an important part in what is called the annihilation of space.

After spending some hours in fruitless and weary researches over musty journals, we finally came across a book called "*Dictionnaire de Conversation,*" in which, under the word velocipede, we were referred to the word Draisienne; on turning over to which, we find a description of the three-wheeled, arm-movement velocipede, and it ascribes the credit of its invention to a Baron Drais de Saverbrun, about the commencement of the nineteenth century, for which reason it may be said to be one of the first inventions of the age, in more senses than one.

It then goes on to say that the Garden of Tivoli, which, it appears, was at that time the favored resort of the *crême de la crême* of Parisian society, was the place where this little instrument made its *début.* The person who prepared the article on the same subject for the American Encyclopedia, commences by giving the well-known derivation of the word from the Latin *velox*, swift, and *pes*, a foot, and defines it as a carriage by means of which the rider propels himself along the ground, and states that it was invented at Manheim. The Velocipede, as originally constructed, appears to have consisted of a bar about five feet long, supported at each end upon a single wheel, that designed for the front, being arranged so as to turn obliquely to the line of the carriage. The rider sat astride the bar, and propelled the machine by the action of his feet on the ground. This vehicle never came in general use, as the *petits crevés* and the *cocottes*, on whose verdict the success of any new toy depends, would not patronize it because the pleasure of riding it was overbalanced by the labor of propelling it. It was called the *célérifère* or "make-speed," and disappeared from view as rapidly as its inventor expected it to roll into public favor, being unable to contend against the shafts of ridicule leveled mercilessly at it.

Delving French antiquarians have exhumed from the *Journal de Paris* of July 27, 1779, a description of a vehicle invented by Messieurs. Blanchard & Mesurier, the former the celebrated aeronaut, which was exhibited in the Place Louis XV., named to-day Place de la Concorde, in presence of many members of the French Academy, and a large concourse of ordinary spectators. At the head of the machine was the figure of an eagle, with outspread wings, to which was attached the apparatus with which the driver directed its movements. Behind it was seated an individual who propelled the machine. At a subsequent date, the inventor transported the vehicle to Versailles, and exhibited its capabilities in presence of Louis XVI, Marie Antoinette, and their idle court. At a later date, M. Dreuze made an improvement on the invention, which met with partial success, inasmuch as a certain number of his machines were constructed after his model and distributed among country postmen, who used them with advantage for some time, until a heavy fall of snow rendered them unserviceable, and led to their being ultimately abandoned, greatly to the gratification of that conservative class, who, detesting everything in the way of innovation, had prophesied their failure.

Since that time, up to within a very recent date, all experiments to render the machine subservient to practical purposes

68

Left: Fig. 1.22. America's (and probably the world's) first cycling periodical, *The Velocipedist*, was first published in February 1869. The editor promised, amongst other things, to "treat impartially of the strength, lightness, superiority, inferiority and general merits of the velocipedes of rival manufacturers..."

Below: Fig. 1.23. Two months later—and a bit snazzier looking—the first French cycling periodical appeared: *Le Vélocipède Illustré*.

# 2 The Search for Speed: 1870–1890

BICYCLES first became popular in the late 1860s, but it took a quarter of a century for them to evolve to the standard pneumatic-tired, chain-driven design we know today. This chapter will describe the multitude of ingenious proposals for multispeed gearing from the early days of pedal-driven bicycles to about 1890, when the safety bicycle replaced the direct-drive bicycle. First, I will describe the bicycles and tricycles with pedals and cranks turning the front wheel, and then the solid-tired machines with rear-wheel drive.

## Front Wheel Diameter

The velocipede's cranks were directly connected to the front wheel, so that its speed was limited by the pedaling speed ("cadence") and the wheel diameter. Larger front wheels went faster and handled better on bad roads.

Tension spoking, using thin metal-wire spokes under tension, allowed building larger front wheels, all the way up to 60 in. (1.50 m) in diameter, limited by the rider's leg length. This development led to what would later become known as the "high-wheel," "penny-farthing," or "ordinary" bicycle. A typical ordinary weighed about 40 lbs. (18 kg), but track-racing models could weigh as little as 16 lbs. (7 kg).

The ordinary was inherently unsafe. Mounting and dismount-

Right: Fig. 2.1. Michaux velocipede. With the pedals connected directly to the front wheel, the speed was limited by the rider's ability to pedal fast and the size of the front wheel.

ing was an acquired skill. The rider sat almost directly over the large front wheel. From that position, he could be pitched forward onto his head if the front wheel hit an obstacle on the road. The only way to stop or slow down the machine was by applying reverse force on the pedals or by means of a lever-operated spoon brake that applied friction to the outside of the wheel. Severe braking, or even hard back-pedaling, could also pitch the rider forward over the handlebars. Finally, the ordinary was expensive, and most riders were athletic young men from the upper-and middle-classes.

## The Front-Drivers

The early front-drive bicycles were awkward to ride. To keep the front wheel pointed straight ahead, the rider had to pull and shove on the handlebars at the same time as he pushed on the pedals. On level roads, very rapid pedaling was needed to go at the speed of a runner. On hills, the iron rims tended to lose traction, especially on wet surfaces. The following comment was published in *Le Guetteur* in 1868:

> If the Earth were a huge level floor; if it had neither climbs nor descents, nor hills nor valleys, then at best the velocipede could be used. But the Earth is not a huge floor, such that when the road steepens a little, the rider is not only obliged to put foot to ground, but also to carry his mount. The man becomes the horse!

In 1876, *Bicycling News* recalled the first pedal-driven bicycles in Britain:

> Shortly after the introduction of bicycles into this country, riders of those noisy and imperfect instruments entertained a wholesome dread of ascents. The low-wheeled, long-cranked, iron-tyred velocipede employed by the early British bicyclists would rattle down a hill admirably, but the most skillful riders found comparatively slight ascents troublesome, and invariably walked up those which were severe. The horror of hills was made worse by the tendency of iron tires to revolve upon a slope without ascending it.

Nevertheless, people everywhere adopted cycling as a new mode of travel and devised improvements in the design of bicycles.

The idea of the rider's feet turning at a different speed from the wheel was first mentioned in a patent lodged in London in 1868 by Ferdinand Villepigue:

> The application of a differential movement to the fore [driven] wheel, whereby communicating a difference of velocity, or a dif-

Fig. 2.2. The first velocipedes had front wheels 32–36 in. (81–91 cm) in diameter. Over the next few years, the wheel diameter was increased to provide greater speed.

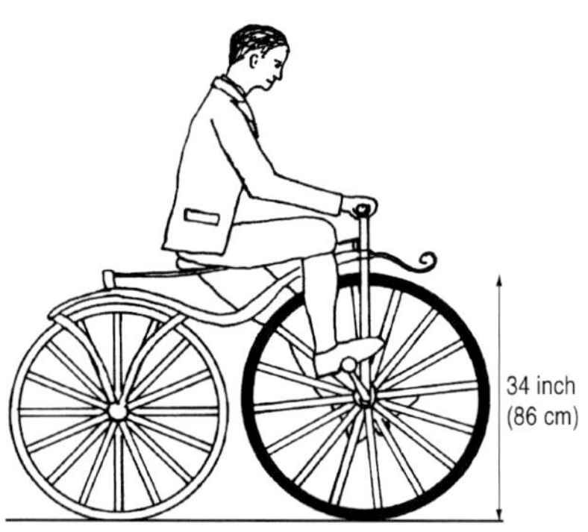


Fig. 2.3. James Moore's velocipede had a front wheel with a diameter of 1.00 m (39 in.). Moore won the 1869 Paris to Rouen long-distance race at an average speed of 11 km/h (7 mph), which was faster than the speed of the stagecoach.

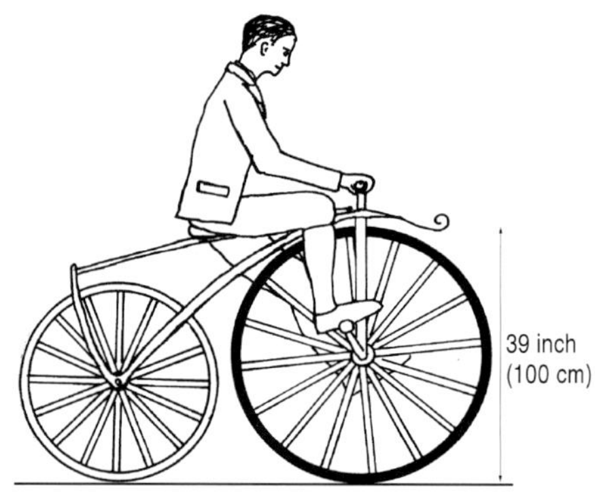


Fig. 2.4. By 1869, Messrs. Snoxel & Spencer advertised velocipedes with front wheels that had a diameter of up to 48 in. (1.22 m).

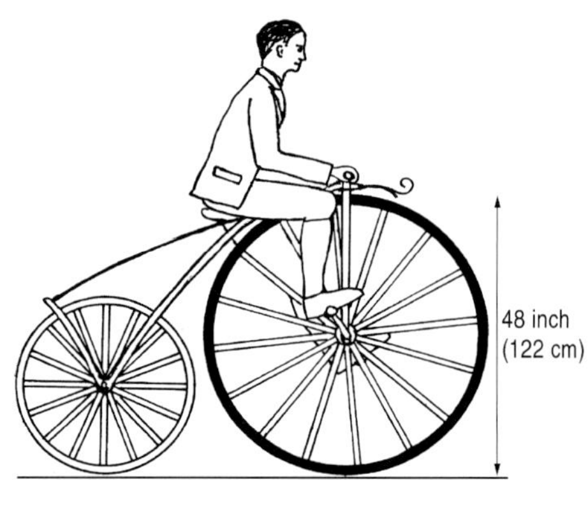

ference in the number of revolutions made respectively by the crank and the fore wheel; that is to say, that according to the slope or other conditions of the road you are travelling over, or the speed which you wish to attain in relation to the force expended, you can make the wheel describe more or fewer revolutions than the crank. Thus, for example, if the road be good the wheel may perform two revolutions, while the crank revolves once; or contrariwise the wheel may be made to make only half a turn while the crank is making a whole one if the ground be bad and the inclination steep.

## France: Fertile Ground for Bicycle Gears

In this section, contributing author Raymond Henry explains why France was the logical country for bicycle gears to be developed. In 1868, several variable-gearing devices were proposed in France to make velocipedes go faster. It was not so much that it was the first country to develop a bicycle industry, but the combination of climate, scenery, social conditions, and roads were all favorable to bicycle touring. In France, cycling is possible most of the year. It rarely gets too hot in summer, and spring and fall are mild without persistent rain.

By the middle of the 19th century, France's road network was the densest in Europe, and it still is today. Counting national, regional, and local roads, there is on average 1.4 km of road for every sq. km (2¼ miles of road for every square mile). In addition, a myriad of rural lanes crisscross the countryside. Towns and villages are closely linked, so one can readily find alternative routes.

The roads of France have always carried people and goods. In the 19th Century, there were mail services and stagecoaches. Many craftsmen and apprentices, traveled from place to place on their "Tour de France" to learn and ply their trade.

The spread of local administration and compulsory education created a uniformity of language, culture, and regulations, so that travelers felt at home everywhere, despite regional varia-

Above: Fig. 2.5. James Starley's 1871 Geared Ariel was probably the first geared bicycle. This is all that remains in the Herbert Art Gallery and Museum, in Coventry. Note the Starley spoking.

Below: Fig. 2.6. Scene in rural France, drawn by an early touring cyclist. The French countryside is richly dotted with historic towns, picturesque villages, and monuments. There is excellent food and accommodation at inns and hotels everywhere. Many of the early cycling clubs rode to hotels known to provide elaborate Sunday meals.

tions. When cyclists took to the roads, they joined in this pattern of comfortable mobility. As early as 1865, the Olivier brothers took a pleasure trip on their velocipedes from Paris to Avignon, a distance of 700 km (430 miles).

## The First Variable Gears for Bicycles

In 1868 and 1869, three patents were lodged in France for variable gears on front-drive velocipedes by Florentin Constant Hazard, Charles Dubos, and by Barberon & Meunier.

In 1870, a Dr. Matthieu wrote in *Le Cycliste* that he had used a 2-speed bicycle. It must have been the model invented by partners Barberon & Meunier, in the small town of Nérondes, in central France. This is a hilly region, which explains the need for variable gearing. It appears that Barberon & Meunier not only patented, but also built a 3-speed system, but it is not known whether it was marketed. This

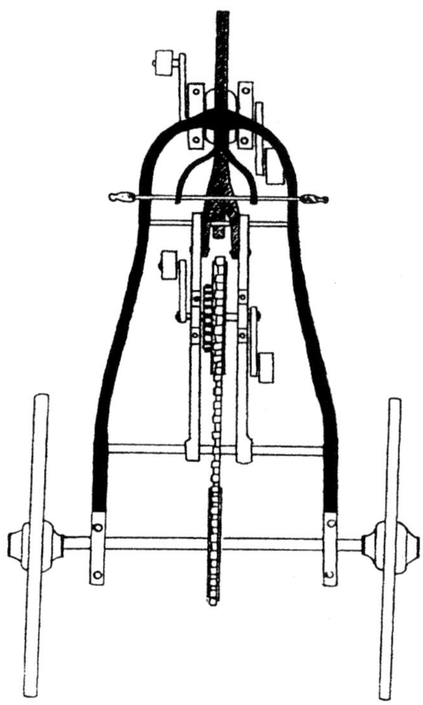

Above: Fig. 2.7. Charles Dubos' 1868 multi-geared tandem tricycle patent shifted the chain across a double chainwheel by hand. Dubos noted that more than two chainwheels were also possible.

Below: Fig. 2.8 A and B The 1869 Barberon & Meunier patent for a "bicycle with double speed" was the first application of *retro-direct* gearing. Pedaling forward, the direct drive to the 1.00 m (39 in.) diameter front wheel was used for climbing hills. Pedaling backward drove the wheel through a belt and pulleys to give a 78-inch gear for level roads.

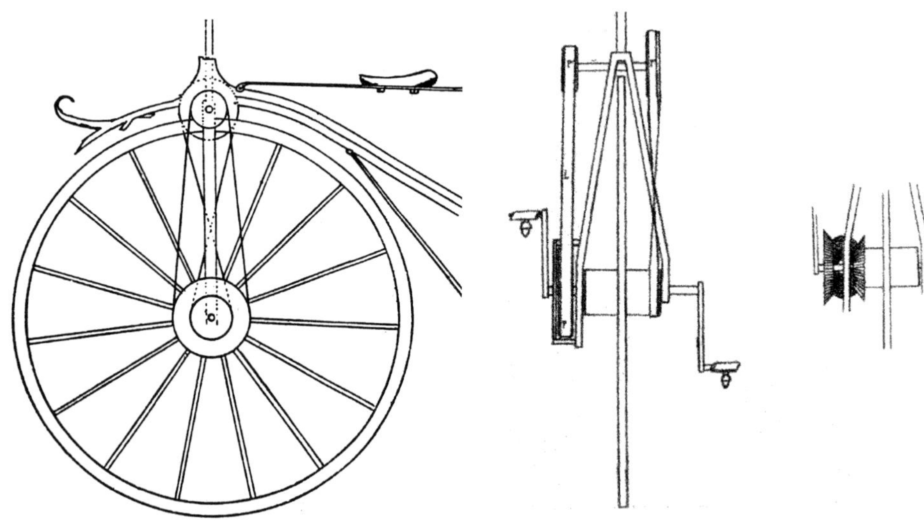

Left: Fig. 2.9. Florentin Constant Hazard's 1868 patent had gear wheels and sliding rods connected to the cranks to increase the speed at which the front wheel turned. Alternatively, by pressing a button, the cranks could be locked to the wheel so that they drove it directly.

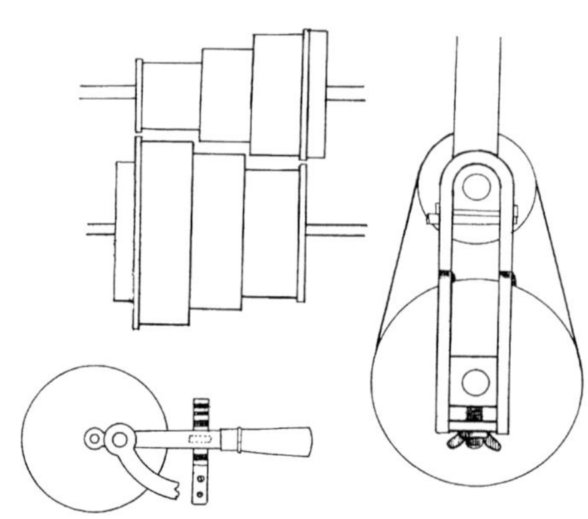

Above: Fig. 2.10. Barberon & Meunier's 1869 patent for 3-speed variable gearing.

was the ancestor of the modern derailleur.

The Barberon & Meunier patent included ratchets, like primitive freewheels, so that one gear ran free when the other was engaged. Although ratchets were common at the time, their application in bicycle transmission was new.

Subsequently, Barberon & Meunier made three remarkable additions to their original patent:

1. A mechanism for shifting a belt or chain between two sets of three discs, either smooth or toothed. The patent explained that when one activates a lever on the handlebars, the chain is pushed from one disc to another, changing the gear. Pushing the lever to the left increases the gear; pushing it to the right decreases it.

2. A primitive gear hub, dated October 1869.

3. A different retro-direct system, dated January 1870. This reversed the drive by a system of cogs and a short chain. This was more efficient than the belts, which slipped if they were not tight enough.

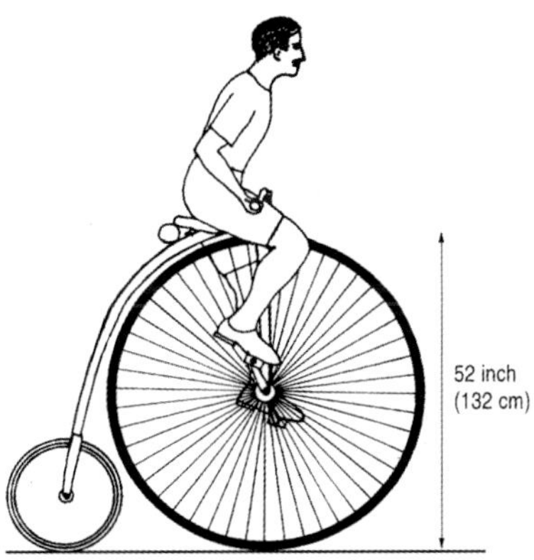


Above: Fig. 2.11. Typical high bicycle with a 52-in. (1.32 m) front wheel. Manufacturers made the front wheel as large as the rider could straddle, thus increasing the distance the bicycle would travel for each turn of the cranks.

Below: Fig. 2.12. The wheel diameter on a high bicycle could be 60 in. (1.52 m) or more. Long-legged people had a natural speed advantage. Someone who could straddle a large wheel was known as "a big wheel."

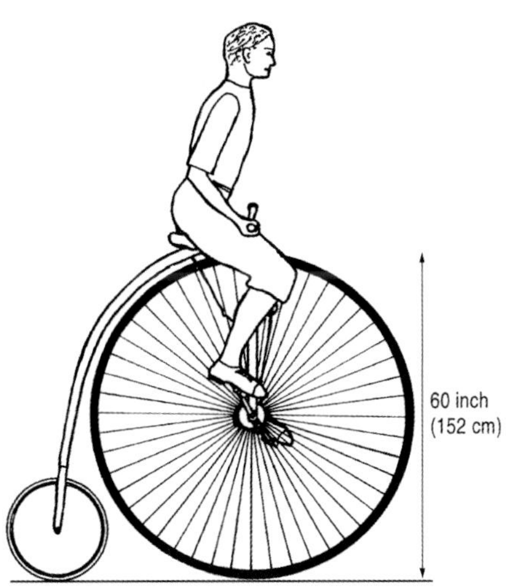


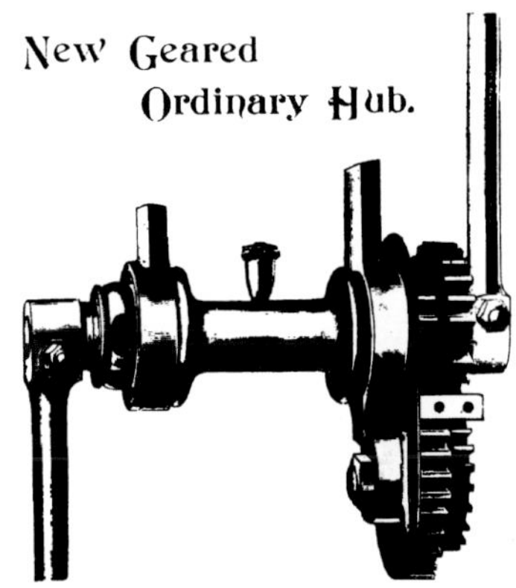


Above: Fig. 2.13. The Geared Ordinary used two sets of gears to allow a smaller front wheel diameter. This complication never became popular.

THE SIGNIFICANCE OF "SINGLE GEARING."
The "gear" of a bicycle is equivalent to the diameter of a driving wheel of an "ordinary" bicycle. How the rider of a 48in. ordinary with a safety geared to 90in. would look on an ordinary geared the same (viz., with 90in. driving wheel).

Above: Fig. 2.14. The safety bicycle on the right has a 27-in.-diameter rear wheel, geared up to 90 in. It has a gear of 90 in. The ordinary bicycle in the middle has a 48-in.-diameter front wheel, so it has a one-to-one gear of 48 in. The ordinary bicycle on the left has a 90-in.-diameter front wheel, so it has a gear of 90 in., like the safety bicycle on the right—if only the rider could reach the pedals.

Barberon & Meunier also applied for a patent in London in 1869, and gained provisional acceptance. Unfortunately, their company disappeared in the turmoil of the Franco-Prussian war.

### Riding High

Eugene Meyer's steel-rimmed wheels with tensioned wire spokes enabled bicycle manufacturers to further increase the size of the driving wheel.

The size of the front wheel of the high bicycle was a compromise that provided a gear that was approximately equal to the lowest gear on a modern racing bicycle. Fast riding on a high bicycle required rapid pedaling. In 1882, British cyclist Herbert L. Cortis became the first person to cover 20 miles (32 km) in an hour. To achieve this feat, Cortis had to pedal at a rate of 120 revolutions per minute.

Although they were expensive, high bicycles became very popular with the athletic young men who could afford them. High bicycles were the fastest things on the road. By the 1880s, more than 500 bicycle clubs had been formed throughout Britain. The nationwide Bicycle Touring Club, later to become the Cyclists' Touring Club, was formed in 1878, as was the Bicycle Union, which catered primarily to racers.

High bicycles offered an entrancing method of traveling long distances and exploring the countryside. Britain was still a rural paradise. There was a real sense of adventure, because there were no road maps, and riders literally ventured off into the unknown. In *Some Notes About the Anerley Bicycle Club,* G.H. Smith recalled:

> The loneliness of those roads is past all belief to those who never cycled over them and know only the whirl of traffic that congests the highways today. Perfect quiet reigned out in the country, miles would be

Above: Fig. 2.15. Mounting a high bicycle takes skill. With the left foot on the step, you push forward with the right leg, straighten the left leg, slide onto the saddle, and then quickly pick up the rotating pedals.

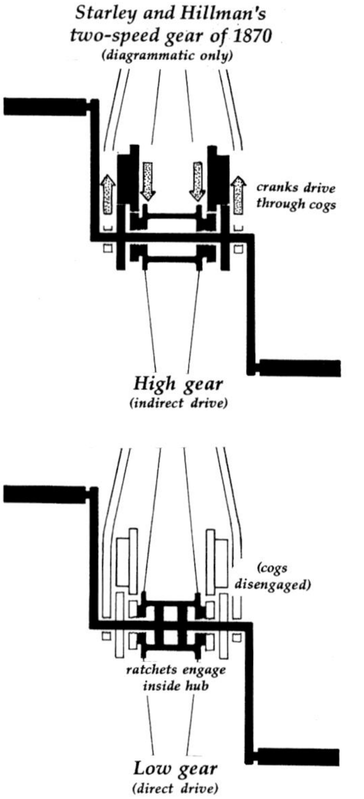


Above: Fig. 2.16 A and B. Starley & Hillman's 2-speed gear, patented in London in 1870 (diagrammatic only). This was an exception in an era when the fashion was for simple and elegant machines without complicated additions such as variable gears.

Right: Fig. 2.17. The 1885 Coventry Machinists' Company's Club Racer was their best high bicycle. It weighed just 23 lbs. (10 kg). The note about tangent spoking suggests that CMC had become aware of the problem with radial spoking on a driving wheel.

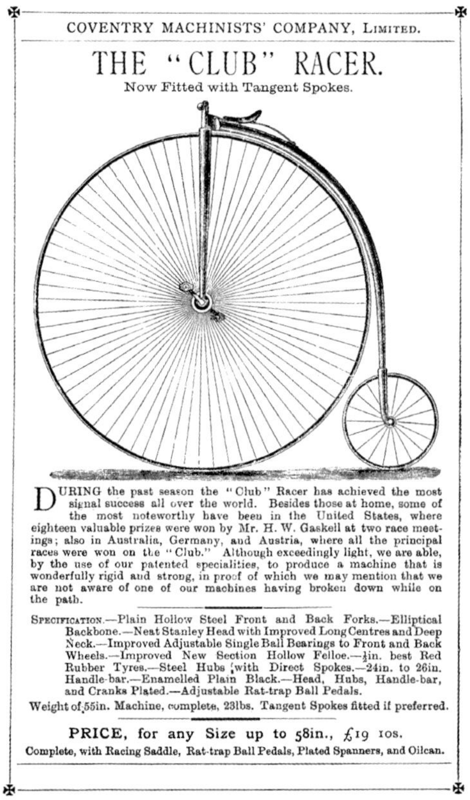

covered without meeting a vehicle, and those that were met were mostly farmer's wagons slowly drawn by heavy horses.

High bicycles were exported from Britain to many other countries, including the U.S., France, the Netherlands, Germany, and as far as Russia and Australia.

France was a prosperous country, and quickly paid off the war levy imposed by the Prussians. The industrial revolution had created a well-to-do middle class with money to spend on leisure. It was mainly this middle class which took up cycling. In those days, bicycles were handmade and expensive; they were regarded as status symbols.

*Le grand bi*, as the high bicycle was known, was exhibited in the Jardin des Tuileries in Paris in 1875. Gentlemen showed off their bicycles in fashionable streets, such as the Avenue de la Grande Armée, or parks, such as the Bois de Boulogne, or participated in amateur events under the gaze of the ladies. Of course, some took up cycling more seriously and went off touring in the countryside.

In 1876, high bicycles were exhibited in the U.S., and the next year the Pope Manufacturing Company was making them locally. Their popularity spread, particularly in New York State and California.

## The Rear-Drivers

Shortly after the appearance of front-drive bicycles, rear-drive machines were also manufactured. Although rear-drive is more complicated than simply attaching cranks to the front wheel, it has many advantages. It is much safer. The rider can sit in a comfortable position to pedal, and the machine can be steered without having to constantly compensate for the pedaling action. Also, pedaling can be "geared up" so that the machine travels any desired distance for each turn of the cranks.

### Driving Belts

As early as 1819, the British Facilitator, or Travelling Car, used a 2-speed belt and pulley-drive. In 1869, Frederick W. Shearing, of Norfolk, proposed a bicycle with a continuous-belt drive to the rear wheel. In the same year, Barberon & Meunier patented a belt-driven velocipede with two speeds.

Belt-drives certainly were not a new idea. Arkwright's 1771 cotton spinning mill used belt-drives to allow a single water wheel to drive long line-shafts with numerous pulleys driving the spinning and carding machines. As recently as the 1940s, one could

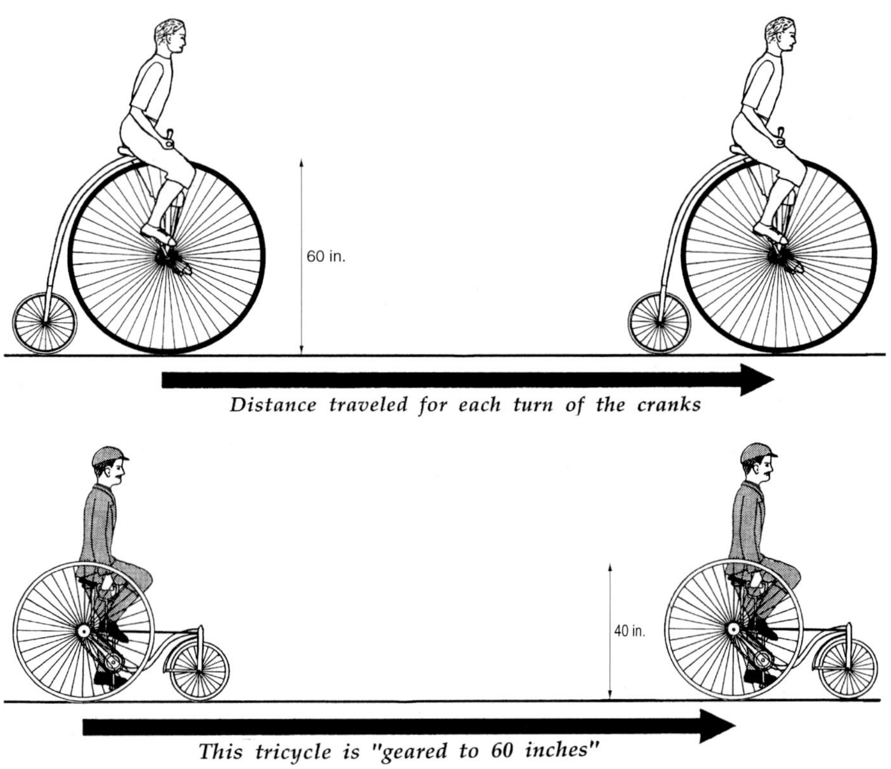


Right: Fig. 2.18 A and B. Gearing was used on tricycles before it was used on the bicycle. This illustrates how a geared tricycle compared with a high bicycle that would travel the same distance for each turn of the cranks. For example, Mr. Percy Furnivall won the British one-mile amateur tricycle championship of 1885 on a Beeston Humber with 40-in. (1.02 m) wheels that had a 15-tooth sprocket on the crank axle and a 10-tooth sprocket on the driving axle. Thus, the tricycle was "geared up" to 60 in.

find a single electric motor with line shafts and step pulleys driving the lathes in machine shops.

From the 1850s onward, numerous patents had been lodged for varying speed by shifting belts across sets of pulleys of differing diameters. Multi-speed belt-drives worked with tricycles, but there were serious problems accommodating wide belts on a bicycle.

## Chain-Drives

Chain drives appeared at about the same time as pedal bicycles, and the two ideas were quickly combined. Today, almost all bicycles are driven by chains. Chain drives are reliable and astonishingly efficient. Chains engage positively without slipping. A driving belt is less efficient and requires tension on both runs to create friction to drive the pulleys. Water or dirt on the belt, or any stretching of the belt itself, can cause it to slip. A driving chain, on the other hand, positively engages teeth on the driving and driven wheels, and needs no tension in the non-driving run of chain.

A sketch of a bicycle rear-driven by rotary cranks and an endless chain appeared in *The English Mechanic* in 1866. That same year, a boat was propelled on a lake near Paris by an electric motor, driven by pin chains.

Meyer and Guilmet built a chain-driven bicycle in Paris in 1868 or 1869. André Guilmet

Left: Fig. 2.19. This 1894 British cotton weaving mill with 300 looms was driven by belts and line shafts. It was still operating a century later, powered by the original tandem-compound steam engine.

Above: Fig. 2.20. Many different mechanisms have been used to link the cranks to the rear wheel: connecting rods, gears, continuous belts and chains, oscillating belts and chains, shaft-drives, and even hydraulic drives. Continuous drives—first by leather belts, but soon superceded by chains—were the most common.

Below: Fig. 2.21. Leonardo da Vinci's 1490 sketches of various chains. Leonardo, or his pupils, may have drawn driving chains, but they could not build one. That had to wait nearly four hundred years for the industrial revolution, when precision machines could stamp out links of exactly the same length.

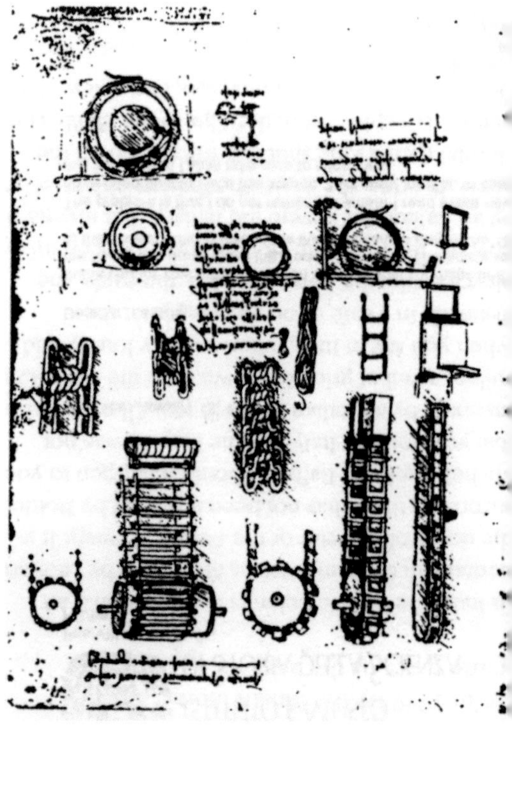

was a clockmaker, while Meyer was the engineer who developed Guilmet's ideas. The dating is controversial because this bicycle was so far ahead of its time. It had a chain-driven rear wheel, steel rims, tension-wire spokes, and a tubular frame.

It probably did not work very well, because the simple pin-chain would not have been able to take the strain of hard pedaling. The chain pitch was just 12 mm (just under ½ in.). This bicycle disappeared in the chaos of the Franco-Prussian war. It's interesting to speculate on the path

Fig. 2. Plate or card chain

Fig. 1. Pin or stud chain

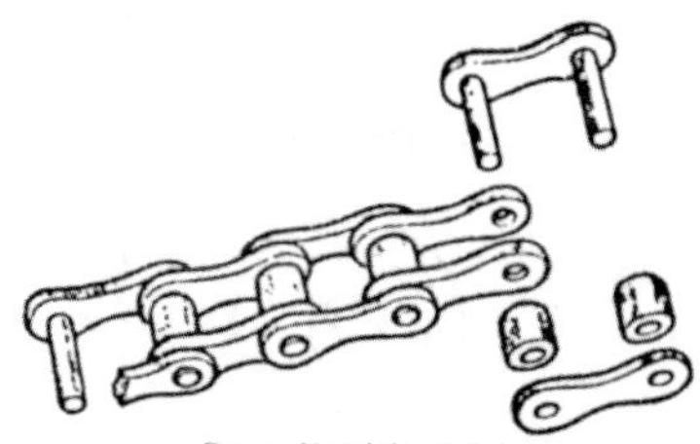

Fig. 5. Slater's bowl chain

Above, top to bottom:
Fig. 2.24. Plate or card chains were used to synchronize weaving machines. They were not designed for power transmission.

Fig. 2.25. Pin or stud chains were also used for textile machines. They were not suitable for heavy loads either.

Fig. 2.26. 1864. Bowl chain (exploded view). James Slater, of Manchester, patented a chain that had "bowls," or bushes, covering the pins. The bushes revolved on the pins, reducing wear on the pins and the sprocket teeth.

MORGAN'S PATENT CHAIN

Above: Fig. 2.27. Morgan's Patent Chain was used on geared ordinaries and tricycles in the mid-1880s. It must have worn rapidly.

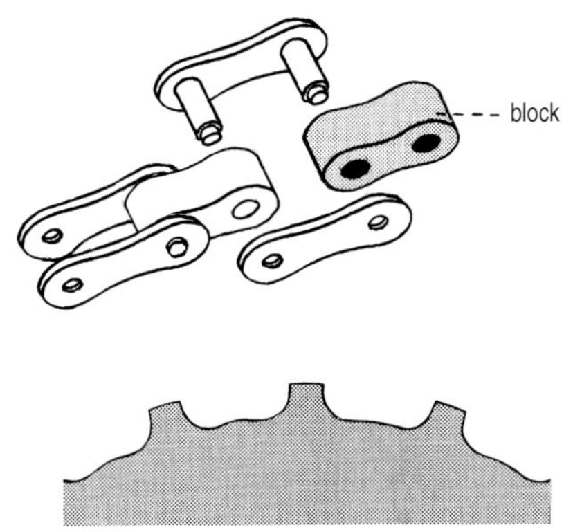


Above: Fig. 2.28. Block chain, also called Humber chain. The lower detail shows the tooth profile of a sprocket for use with a block chain.

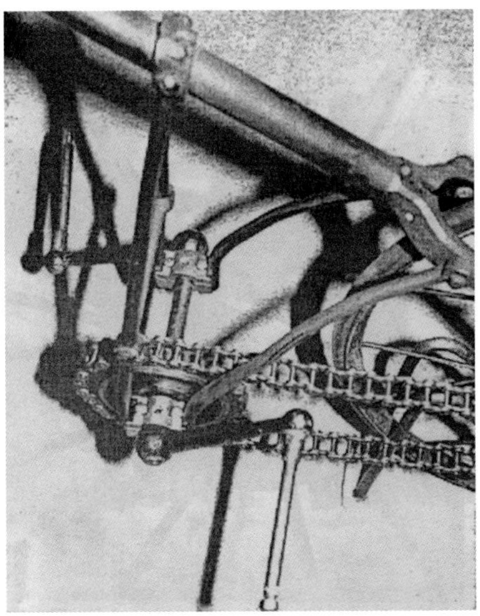

Above: Fig. 2.22. The pin chain that was used on the Meyer-Guilmet bicycle.

Right: Fig. 2.23. The 1869 Meyer-Guilmet chain-driven bicycle is the "Missing Link" between the boneshakers of the 1860s and the safety bicycles of the 1880s; or rather, it *was* the first safety bicycle.

Above: Fig. 2.29. The double-roller chain (lower detail) used the same 1-inch-pitch sprockets as the conventional block chain (upper detail).

of bicycle development if Rowley Turner had brought a Meyer-Guilmet back to England instead of a Michaux.

Numerous kinds of chains were available in the 1860s, but they were not particularly suitable for bicycle drives.

During the 1880s, the simple block chain was the most popular bicycle chain even though roller chains were available. In the U.S., block chains continued to be used right through to the 1950s for track racing. Block chains had a pitch of 1 in. (25.4 mm) and used special chain-wheels and sprockets with a thickness of ¼ in. (6 mm) in the 19th century, and 3/16 in. (5 mm) later.

In 1880, Hans Renold patented the bush-roller chain. Renold had purchased Slater's chain works, and he manufactured "long-link" bush-roller chains with pitches of 1.0, 1.02, 1.19, or 1.3 in. (25.4, 26, 30, or 33 mm) to the customer's order. Renold did not make pitches smaller than 1 in. (25.4 mm), so Renold chains were less suitable for bicycles. After Renold's patent expired in 1894, Perry and other manufacturers made ½-in. (12.7 mm) pitch "short-link" bush-roller chain. Eventually, in 1899, Renold also started making ½-in. (12.7 mm) pitch chain.

Long-link chains remained popular for several years. Archibald Sharp's famous 1896 textbook *Bicycles and Tricycles* referred only to 1-in. pitch chains. Eventually, ½-in. pitch bush-roller chain took over the bicycle market. With proper lubrication, the rollers rotate freely, reducing friction and wear. Renold's bush-roller chain worked so well that the design has been used ever since with little modification.

In France in the 1890s, the company of Rafer Fils Frères, in St.-Chamond, adopted the British 1-in. (25.4 mm) pitch, but they also made chains with pitches of 15 mm, 5/8 in. (15.87 mm), and 18 mm.

There is the tale of a British cyclist on his first Continental tour. While in Normandy, his chain broke. The cyclist wheeled his machine to the nearest bike shop and anxiously demanded "*une chaîne anglaise*" (an Eng-

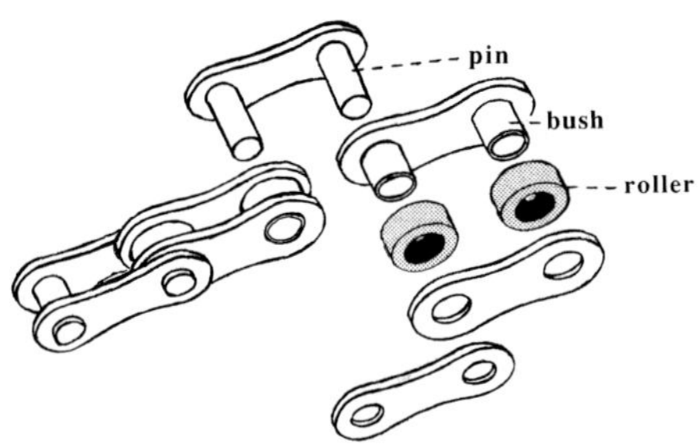


Above: Fig. 2.30. The bush-roller chain (exploded view) has free-turning rollers around the bushes.

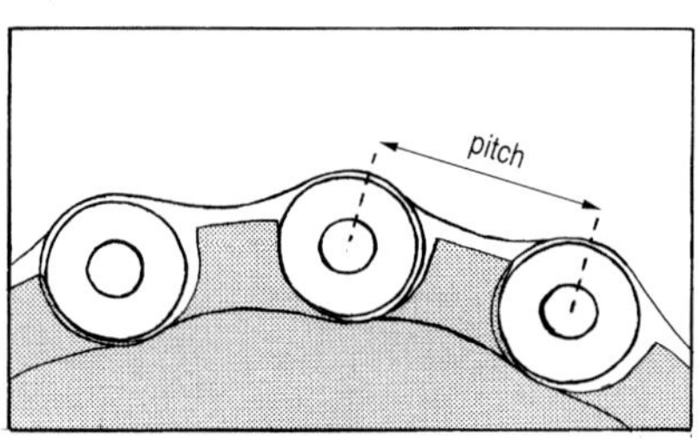


Above: Fig. 2.31. The pitch of a chain is the distance from the center of one link pin to the center of the next one. (However, for a block chain, the pitch is from the center of one link pin to the center of the second one along.)

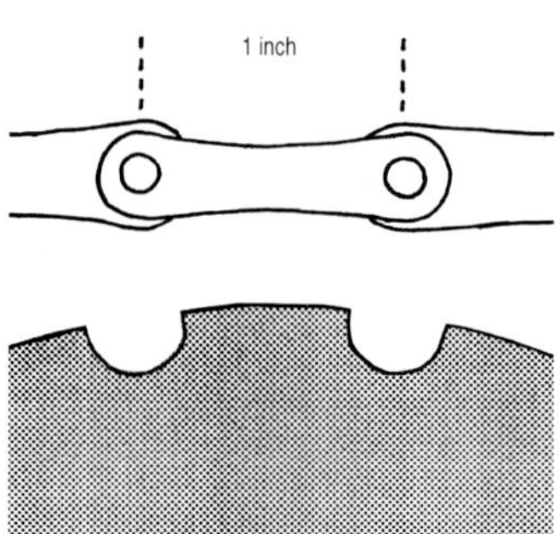


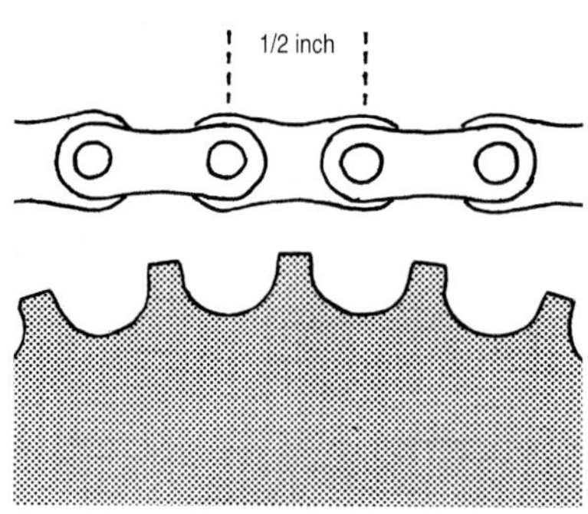


Above: Fig. 2.32. Short-link roller chain with ½-inch pitch. The lower detail shows the tooth profile for short-link chain. The ½-inch-pitch size increased in popularity, and from 1900 onward was the standard pitch around the world.

Left: Fig. 2.33. Long-link roller chain with 1-inch pitch. The lower detail shows the tooth profile for use with the long-link chain.

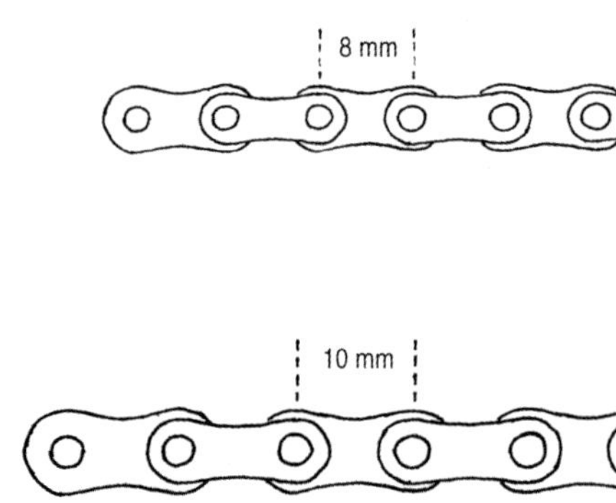


Above: Fig. 2.34. Two attempts to produce shorter-pitch chains. The 8-mm-pitch Chainette of 1910 (above) and Shimano's 10-mm-pitch chain of the 1970s (below). Both were short-lived.

lish chain). The shopkeeper shrugged his shoulders and replied, "*Monsieur, c'est toute la même chose!*" (Sir, they're all the same!). Unfortunately, it's no longer "*toute la même chose,*" and today's tourist must hope that the shopkeeper will have at least five different chain widths in stock.

### Chain Width

A bicycle chain is labeled not by its outside width but by the nominal width of the sprockets on which it fits. Ed Berry's bike (described in Chapter 3) has ¼-in. (6 mm) wide sprockets. The width of the chain is a little over ¼-in. between the inside plates.

Sprockets that are 3/16 in. (5 mm) wide and matching chains were the next standard. Then, ⅛-in. (3 mm) wide sprockets and chains became the standard, and these continue to be used throughout the world for single-speed and hub-geared bicycles. A ⅛-in. chain can flex sideways enough to shift across 4 sprockets. In 1935, 3/32-in. (2.4 mm) sprockets and chains were introduced. These narrower chains were more flexible and could shift across 5 or 6 sprockets. The slimming process has not ended, and today's10- and 11-speeds require sprockets that are only 1/16 in. (1.6 mm) thick. and chains on which the pins don't protrude.

### Chain Efficiency

A clean, well-lubricated roller chain is amazingly efficient. The 1897 *American Machinist* reported tests by Professor R.C. Carpenter of Cornell University, who found that "the friction losses from the best chain varied from one-half to three-quarters of one percent."

In 1930, the National Physics Laboratory, at Teddington, England, measured the efficiency of a Renold bush-roller chain as being between 98.1 and 98.9%.

Measurements using a differential dynamometer at the University of Melbourne, by Dell'Oro and Malone in 1995, showed the power transmission loss of a bicycle chain drive to be only 0.5%, consistent with Carpenter's measurements a century earlier. Chester Kyle and I measured chain efficiency in 2002 (see Chapter 15). Although our tests indicated somewhat higher losses, the more significant finding was that large sprockets are more efficient than small sprockets.

## Tricycle Drivetrain Developments

Tricycles with cranks on the front axle were part of the velocipede

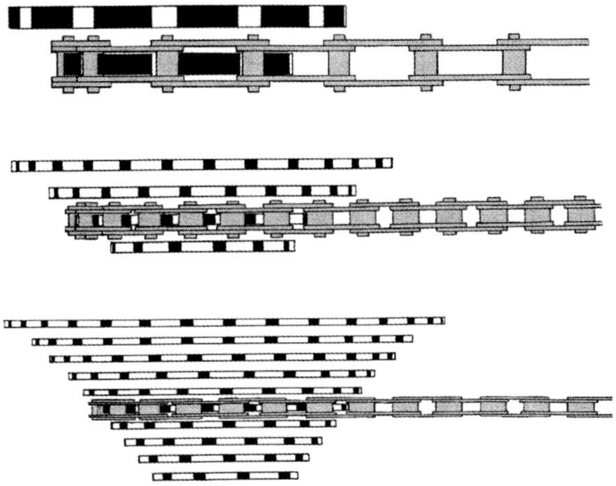

Left: Fig. 2.35. Chain width and number of sprockets compared. Top: ¼-in. wide chain and two sprockets. Middle: ⅛-in. wide chain and four sprockets. Bottom: 1/16-in. wide chain and ten sprockets.

Above: Fig. 2.36. James Starley on the 1877 Salvo tricycle, which he manufactured in Coventry. This was the first practical use of driving chains. Starley probably used Slater's bowl chain.

craze of the 1860s. By the 1870s, rear-driven tricycles had large wheels to increase their speed and to cope with poor roads. Tricycles were more stable than bicycles, and riders could stop without falling off. They were convenient for women wearing long skirts, the elderly, and the non-athletic. Tricycles were also better suited for carrying luggage. On the down side, they were heavy, complicated, and expensive, and they took up a lot of storage space. Tricycles were preferred by well-to-do people who wished to ride in a dignified, leisurely manner.

Above: Fig. 2.37. The Crypto-Dynamic Two-Speed Gear provided direct drive and a 15% reduction in low gear.

Below: Fig. 2.38. Disassembled view of the Crypto-Dynamic Two-Speed Gear, showing the epicyclic gears.

Mud was a serious hazard for the first driving chains. G.H. Smith reported:

> In wet weather, the macadam, and to a lesser degree, the chalk, became very slippery and accounted for many croppers, whilst the sand would be lifted up by the splash of the water into pedal and other bearings; any chain-driven machine suffered intensely, the block chains simply sucking in the grit until they became so gorged with it they were literally incapable of bending and were little short of bars of solid metal.

As road surfaces were improved, chain drives became more practical. Initially, the chain went around a chainwheel and sprocket of equal size. But soon it was realized that chains could be "geared up," so that small-wheeled tricycles could go just as fast as large-wheeled machines.

### Variable Gears for Tricycles

Tricycles could be pedaled slowly uphill, pausing for a rest if necessary. Riders often had to dismount and push their tricycles, or even pull them with ropes. It is not surprising that there were many people who proposed variable gearing for tricycles. The first known such proposal was

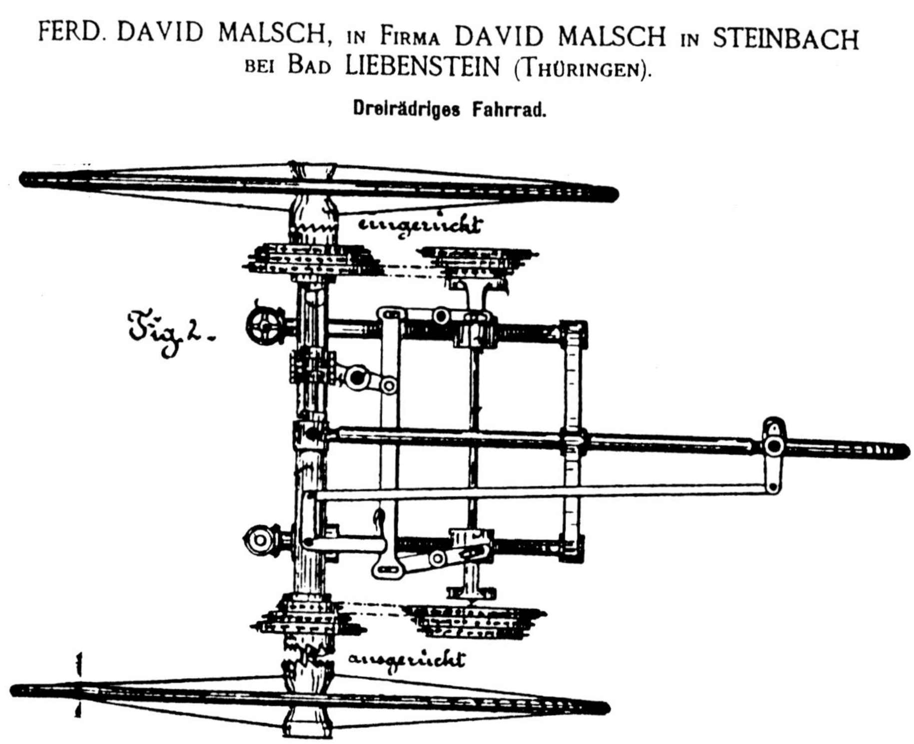


Above: Fig. 2.39. This complex German 1886 Malsch 6-speed tricycle transmission probably died in the prototype phase.

Charles Dubos' 1868 chain drive. In 1881, J.E. Howard and W.W. Stall, of Boston, patented a "speed attachment," which allowed either of two different-sized driving cogs to be locked onto the drive shaft with what they described as a "feather." Some tricycles used two chain-drives with different gear ratios, so that one or the other could be engaged.

In 1882, William Thomas Shaw and William Sydenham patented the Crypto-Dynamic Two-Speed epicyclic gear in England. The 1884 Crypto-Dynamic Two-Speed Gear was the first epicyclic gear that was widely used in tricycles. It replaced the chainwheel and weighed about 5 lbs. (2.2 kg) more. It was lever-actuated and provided an intermediate freewheel position for coasting downhill.

In 1886, Ferd. David Malsch, of Thüringen (Germany), patented a tricycle with three chainwheels and three sprockets on either side. By turning a hand wheel, the drive on either the left or right side could be engaged, with the chain running on any one of six chainwheel and sprocket combinations. Another handwheel was used to wind the pedal shaft forward or back to take up the chain slack.

In 1888, Karl Heinrich Schürer patented a tricycle drive that consisted of two chainwheels and two sprockets. The single chain drove around either a large

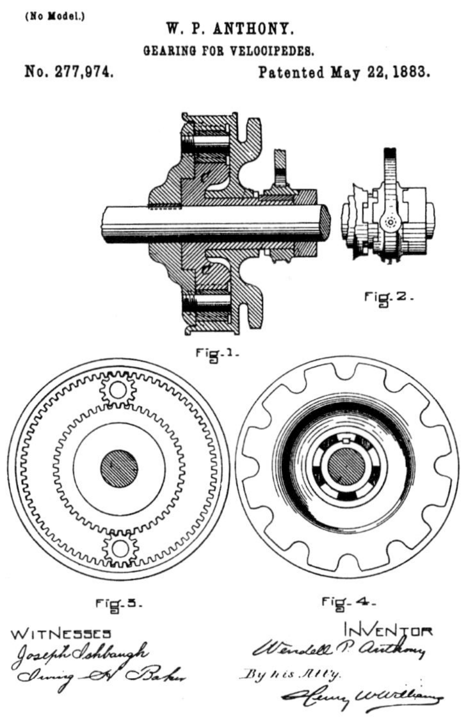


Above: Fig. 2.42. In 1883, Wendell P. Anthony, of Rhode Island, obtained a patent for a 2-speed epicyclic drive that stated: "The object of which is to enable the rider to produce either of two different speeds..., in other words, to gear the tricycle for 'power' or for 'speed' at the will of the rider, according to the conditions of the road-bed."

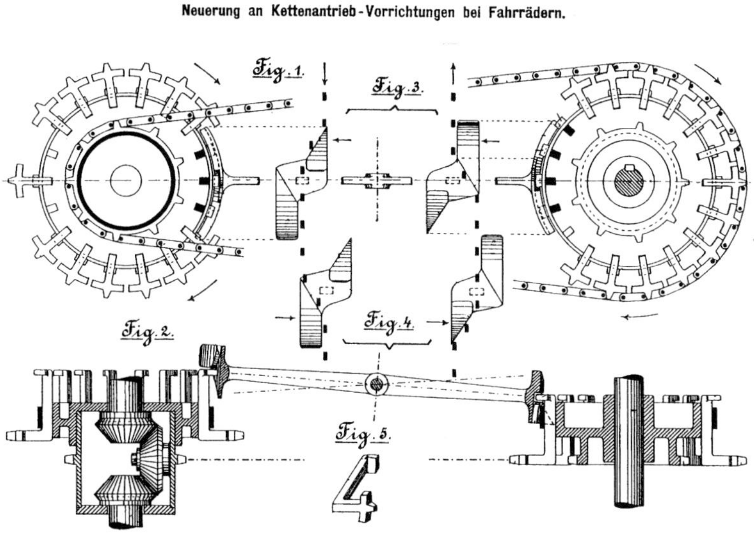


Above: Fig. 2.40. The 1888 Schürer 2-speed tricycle transmission was not a commercial success.

The inside secret - how an epicyclic gear works

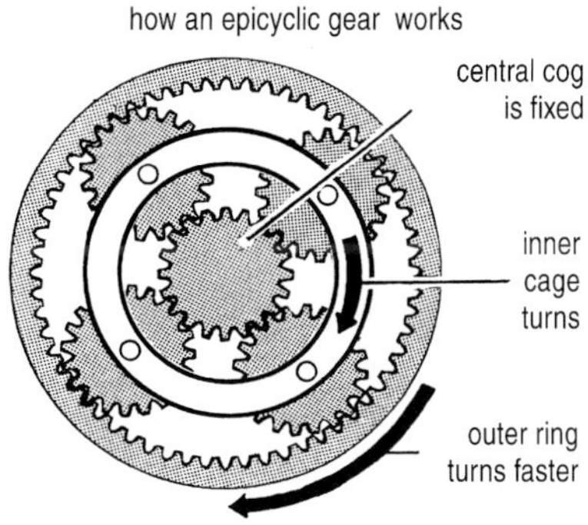


In high gear the sprocket connects to the inner cage, and the outer ring drives the hub

In low gear the sprocket connects to the outer ring, and the inner cage drives the hub

Left: Fig. 2.41. Epicyclic gears have a cage containing a set of cogs ("planets") which roll around a fixed central cog ("sun"). There is also an outer toothed ring, which rolls faster around the outside of the set of cogs. By connecting the sprocket to the cage, and the hub to the outer ring, the driven wheel turns faster, i.e. is in a higher gear. If the connections are made the other way round, a lower gear results.

Below: Fig. 2.43. The same year, Robert Jay, of London, patented a 3-speed tricycle drive.

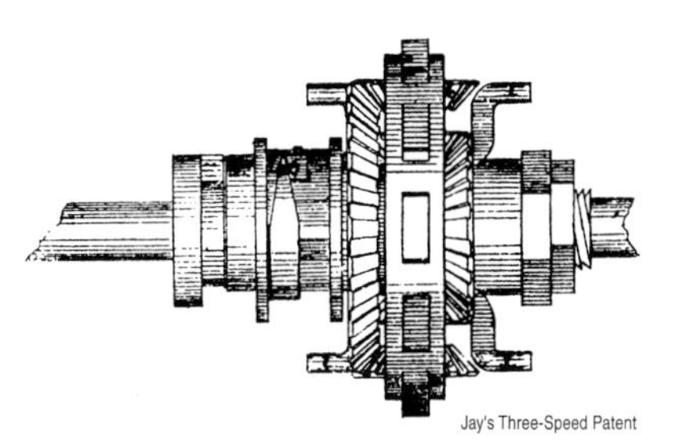

Jay's Three-Speed Patent

chainwheel and a small sprocket for high gear, or a small chainwheel and a large sprocket for low gear. To shift gears, a stout lever scooped the sliding teeth of the large chainwheel in under the chain. Simultaneously, it slid the sprocket teeth clear of the chain line. In either gear, the total number of teeth on the chainwheel and the sprocket combined was the same.

Schürer used an 18-tooth chainwheel with a 9-tooth sprocket for a total of 27 teeth, or a 9-tooth chainwheel with an 18-tooth sprocket to again give a total of 27 teeth. Thus, the same length chain was required for either high or low gear, and there was no need for any device to take up the chain slack. This constant-chain length trick was used right through to the 1920s, when French cycle tourists used chainwheel-sprocket pairs with the same total number of teeth.

### Epicyclic Gearing

Over the next few years, numerous inventors proposed the use of epicyclic gearing to provide gear shifting while in motion. The first British patent for epicyclic gears dates from 1878. Other early patents for tricycle 2-speed epicyclics included:

- 1882, Thomas Shaw and William Sydenham's Crypto Dynamic (England)
- 1882, Charles Harvey and William Paddock (U.S.)
- 1883, Albert E. Wallace (U.S.)
- 1884, Thomas Millward and Charles Leni (England)
- 1885, Charles E. Pratt and Albert E. Wallace (U.S.)
- 1885, Curtis H. Veeder and John D. Veeder (U.S.)

### Other Variable Gear Systems

Epicyclic gearing was popular mainly because it was a neat, tidy, and weatherproof solution, but numerous other methods of varying the gearing were also patented.

Rather than detach and replace the chainwheel, inventors came up with the idea of expanding or shrinking the chainwheel while riding. In Britain, two early proposals were made for using a length of special chain on an expanding chainwheel. As the chainwheel expanded or contracted, the special chain flexed to form a continuous surface for the normal driving chain.

In 1884, Charles Linley, John Biggs, and George Tandy proposed a chainwheel that consisted of two discs mounted face

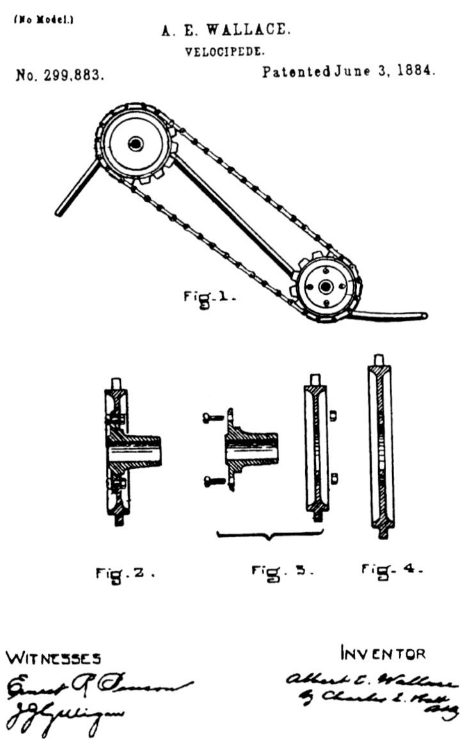


Above: Fig. 2.44. In 1884, Albert E. Wallace, of Hartford, Connecticut, patented a system that allows different-sized chainwheels to be put on a tricycle to change the gearing.

Right: Fig. 2.45. Johan Malsch's 2-speed double-chain bicycle, or *bichain*.

to face, with spiral grooves on their inside surfaces. A special chain had pins that fit into the grooves. Deep inside the spirals, the special chain formed a small-diameter chainwheel. As the chain pins slid out around the grooves, the special chain formed a larger chainwheel, and chain slack was taken up with an idler pulley. Linley was later to produce many other bicycle inventions, including one of the first workable derailleur gears.

### The First Bichain

Another way of changing gears was to have two separate chain transmissions, either of which could be made to drive. The 1887 Sparkbrook tricycle may have been the first machine with this arrangement, known as a *bichain*. In his 1887 *Cycling Handbook*, Lord Bury reported:

> In the 2-speed arrangement fitted by the Sparkbrook Company, two pairs of pulleys and two chains are employed, one set gearing up for speed, the other gearing down for power. The upper pulleys are fixed side by side, centrally on the main axle, and the chains are carried down to two pulleys running loose on the crank-spindle within the lower bracket. The clutch proper is fixed to the spindle, and is capable of engaging with either of the two pulleys, according as a little lever, pivoted to the central tube, is depressed by the left or right foot.

There were English bichain tricycles in France. Vélocio had one in 1891 and remembered having seen others in St.-Étienne. These machines had two chains and a system of clutches to allow either chain to be engaged.

Johan Malch produced such a bicycle with two separate chains, either of which could be made to drive. The German author Wilhelm Wolf described it in his 1890 book *Fahrrad und Radfahrer* as follows:

> Even more practical, it seems to us, is an arrangement with two different drive ratios on a single bicycle or tricycle, invented by Malch in Munich and patented in Germany and England. This uses a splined shell which couples two chainwheels or sprockets with different diameters on the crank or rear hub, so that one of the two chainwheels is transmitting

Left: Fig. 2.46. Much like the 1869 Meyer-Guilmet, Lawson's 1879 Bicyclette used a chain to drive the rear wheel. It had many of the same features as the later safety bicycle, but it was not a commercial success.

Right: Fig. 2.47. The 1885 Singer Xtraordinary was an elegant lever-driven model introduced toward the end of the high-wheel period. It weighed 50 lbs. (23 kg).

power while the other runs free, or optionally both run free, allowing the drive mechanism to be reconfigured for climbing and descending.

Above: Fig. 2.48. The 1884 American Star was a "backside-front" high-wheel-safety. Invented by George W. Pressley in 1880, and built by H.B. Smith Machine Co., it featured a reciprocating treadle-drive.

## The Search for Safety

The high bicycle was truly unsafe at any speed, and it is surprising that it remained popular as long as it did. The transition from front-drive high bicycles to rear-drive safety bicycle took place gradually from 1870 to 1890 in both Britain and the U.S.

There were two main kinds of front-drive safeties. One used levers to drive the cranks, lowering the pedals and allowing the rider to sit lower and further behind the front wheel. The up-and-

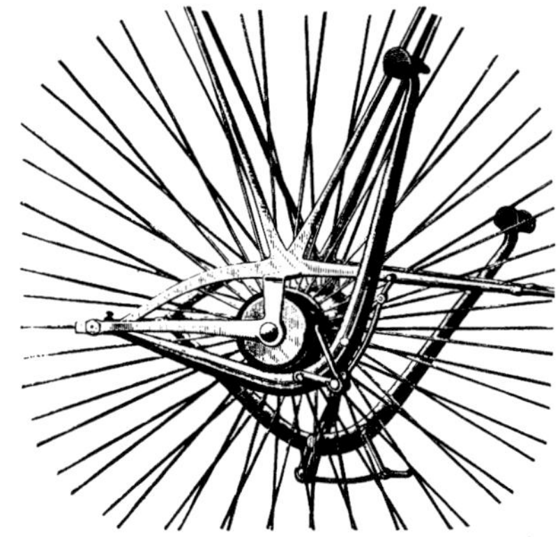

Changeable Power Attachment

Above: Fig. 2.50. The optional "Power Attachment" for the Star changed the anchoring point of the strap and lowered the gear by 20% for climbing hills.

Left: Fig. 2.51. Kangaroo was the name of the geared ordinary, or dwarf safety, introduced by Hillman, Herbert & Cooper in 1884. The 36-in. (91 cm) diameter front wheel was geared to 56 inches using a pair of Morgan's Patent chains. Many races were won on this style of bicycle, but there were problems with the front-wheel bearings that made them less suitable for everyday use.

Above: Fig. 2.49. The Eagle was the only American high-wheel-safety bicycle with rotary cranks (as opposed to the reciprocating lever-driven Star). It was based on L.B. Gaylor's patent and was built from 1889 until 1892.

Right: Fig. 2.52. The 1887 Geared Facile added sun-and-planet gearing to the lever drive. It was lost in the competition with rear-driven safeties.

down pedal action was good for hill climbing. The other used step-up chain gearing to allow a smaller front wheel. These smaller-wheeled geared ordinaries were referred to as *dwarf safeties* or *kangaroos*. They were fast and a bit tricky to handle.

Henry Sturmey's *Indispensable Handbook to the Safety Bicycle* was published in 1885. It described 7 lever front-drive bicycles, 44 geared front-drive bicycles, and 9 chain driven rear-drive bicycles.

The Star was an American design that retained the ordinary's high wheel and rider position but inverted the wheel arrangement, with the large wheel in the back and the small wheel in the front.

The H.B. Smith Machine Co., of Smithville, New Jersey, made

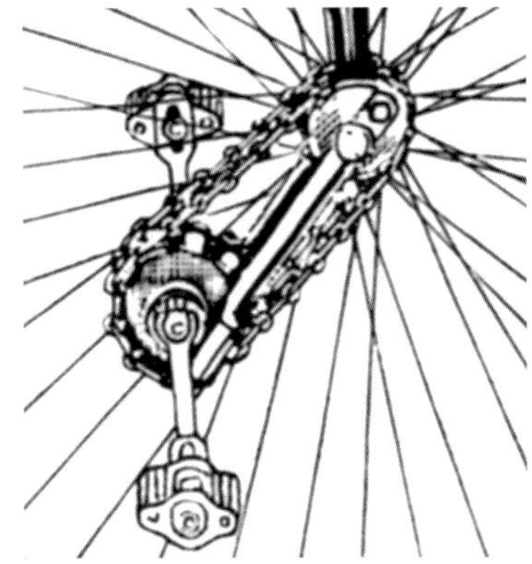

Above: Fig. 2.53. Close-up of the Kangaroo chain-drive.

THE KING OF THE ROAD.

RUDGE SAFETY. (See page 24.)

RUDGE BICYCLETTE (See page 20.)

SEE THIS TRICYCLE BEFORE BUYING.
Unequalled for Ladies' use.

Top left: Fig. 2.54. The Rudge Light Roadster came in wheel sizes from 48 to 60 in. (1.22 to 1.52 m). The 50-in. (1.27 m) model weighed 36½ lbs. (17 kg), including tool kit, and cost $132.

Bottom left: Fig. 2.55. The Rudge Bicyclette had 30-in. (76 cm) wheels, geared to 60 in.

Second column top: Fig. 2.56. The Rudge Safety came in wheel sizes from 36 to 40 in. (91 cm to 1.02 m). The 40-in. (1.02 m) model was geared to 56 in.

Second column bottom: Fig. 2.57. The Rudge Crescent Tricycle had a 26-in. (66 cm) front wheel and 40-in. (1.02 m) rear wheels. It cost £175. The manufacturer claimed it had "automatic steering," presumably meaning it was self-centering.

Below: Fig. 2.58. "Last gasp" of the treadle-drive. The 1892 Rover Star was a refined version of the American Star, with a smaller rear wheel for increased safety. Only 25 of these machines were ever built.

the Star bicycle from 1880 to about 1888. They sold well and set numerous speed records. They had a unique drive-train, in which the pedals moved up-and-down on long levers. Each lever pulled a leather strap wrapped around a drum, which turned the rear wheel through a freewheel mechanism. A return spring raised the pedal to the top position and rewound the strap.

### 1887, the Pivotal Year

By 1887, in both Britain and the U.S., buyers could buy an ordinary bicycle or choose from safer alternatives. Most bicycle makers made three or four different kinds of bicycles (or tricycles) and let the market pick the winner. By 1890, the verdict was in, and ordinaries were no longer in demand. The last catalog year for ordinaries in England was 1892.

The Coventry manufacturer D. Rudge & Co. was a major exporter of cycles to the U.S. Their Boston agent's catalog showed four main options, which are depicted in the accompanying illustrations Figs. 2.52–2.55.

## Chain-Driven Safety Bicycles

Following the rear-drive bicycles of Meyer-Guilmet and Lawson mentioned earlier, BSA, Humber, and Rover produced chain-driven models in 1884 and 1885. These solid-tired machines were the forerunners of the modern bicycle.

In the early 1880s, the term "safety" was used to describe any bicycle that was not high-wheel. Today's bicycle historians have narrowed the definition: The distinguishing features of the safety bicycle were a chain-driven rear wheel, equal-sized wheels, and a triangulated diamond frame. There is also general agreement

Right: Fig. 2.59. The first version of Starley & Sutton's Rover safety bicycle of 1884 had indirect steering, a 36-inch (90 cm) diameter front wheel, and a 30-inch (76 cm) diameter rear wheel, geared to 55 in.

Left: Fig. 2.60. The 1886 Rover was the definitive model. It had 30-inch (76 cm) diameter wheels, geared to 56 in.

Above: Fig. 2.61. Peugeot's 1888 two-speed safety bicycle apparently had a spur-gear gearbox in the bottom bracket.

that J.K. Starley's Rover was the first true safety bicycle, even though the early models lacked the last two features. The Rover safety bicycle took two years and three models to evolve into the definitive design of 1886.

Above and below: Figs. 2.62 and 2.63. The Crypto Bantam was a geared-up front driver that was made into the 1890s. The lower illustration shows what made it tick: the planet gear system at the front wheel hub.

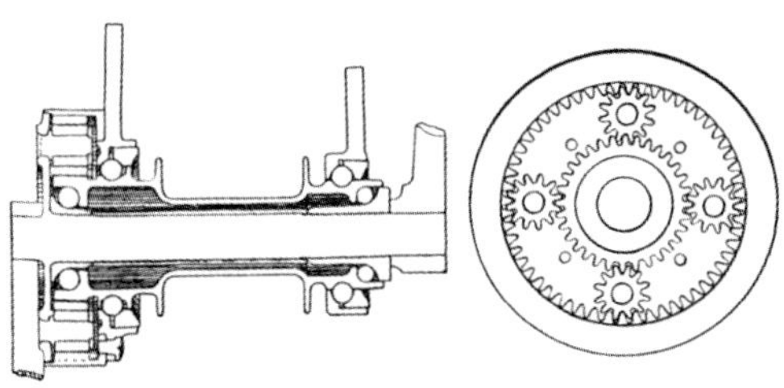

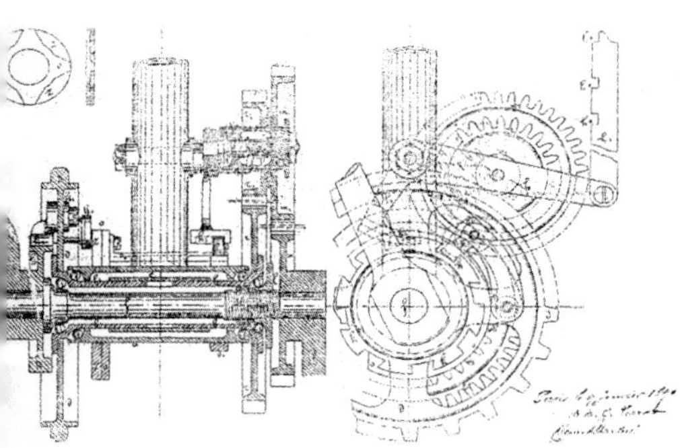

Above: Fig. 2.64. Terrot's1890 patent for a 2-speed bottom-bracket gear using spur gears.

## Early Two-Speed Bicycles

The single-speed chain-driven bicycle remained predominant for the next eighty years, but right from the start, some manufacturers provided variable gearing.

In 1888, Peugeot marketed a 2-speed safety bicycle in France. The advertising for it said:

> The mechanism is fitted on the pedal shaft. To operate it, one only has to press with the foot on one of the small steps placed on the right and left of the fork carrying the chainwheel and the pedals.

This system was in their catalog for only one year. It increased the price of a bicycle considerably, which may be why it was abandoned.

In 1890, Terrot was granted a patent on a 2-speed bottom-bracket gear using spur-gears. It may never have been sold.

Variable gear experiments continued haphazardly, but first the basic bicycle had to be improved and the mass market developed. That happened after 1890, when the variable gear problem was tackled more seriously.

## Bicycle Racing: the Hour Record

In 1870, J.T. Johnson set the first record at14.2 miles (22.79 km) on an ordinary bicycle. In 1882, H.L. Cortis raised the hour record for the ordinary bicycle to 20.2 miles (32.48 km). In 1888, H.E. Laurie raised the hour record to 21.1 miles (33.91 km) on a chain-driven track bicycle.

Since then, the distance covered has steadily increased, until Chris Boardman rode 56.375 km (a record that was retroactively disqualified by the ICU). Interestingly, since neither hill-climbing nor speed changes are relevant, all these records were set on track bikes, i.e. directly-driven bicycles without gearing.

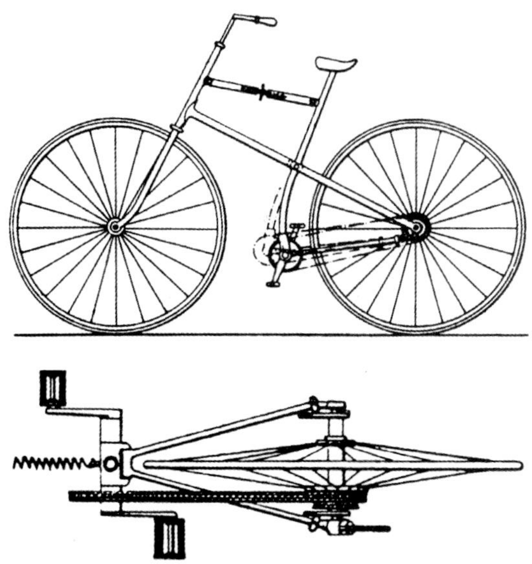

Above: Fig. 2.65. In 1889, Haase & Stamm, of Germany, patented a design on which the chain could run on different-sized rear sprockets. To take up the chain slack, the seat tube could be swivelled forward or back against the tension of a spring. It could be adapted for a 2-speed bicycle as well as for a 3-speed tricycle.

# 3. The Bicycle Boom: 1891–1899

Above: Fig. 3.1. John Boyd Dunlop, inventor of the pneumatic tire, out riding his pneumatic-tired safety bicycle. Note that it has relatively narrow tires for the time.

THE PNEUMATIC tire was probably the most significant bicycle development of the 1890s. The early safety bicycles had solid rubber tires. In 1888, John Boyd Dunlop, a Scottish veterinarian living in Belfast, introduced the first pneumatic cycle tire. This provided a more comfortable ride and greatly reduced rolling resistance.

By 1893, virtually all bicycles sold in Britain, France, and the U.S. had pneumatic tires, which greatly increased their popularity. By 1898, W.W. Hamilton had raised the unpaced hour record to 25.3 miles (40.78 km) on a pneumatic- tired track bicycle, and even higher speeds had been achieved in paced races.

## Boom Years

The pneumatic tire, the tension-spoked wheel, and the chain drive did more than the crank and pedal to establish the bicycle as a serious alternative to the horse. A new standardized bicycle design spread throughout the world, which led to a revolution in personal mobility.

In the early 1890s, bicycles were expensive, and only the rich and fashionable adopted them. Although many of them soon wearied of the new invention, their patronage had stimulated improved designs and lower production costs.

By the middle of the decade, middle-class workers could afford them, especially slightly outdated models that were available on the second-hand market.

Bicycles were now being used for everyday travel, recreation, racing, and touring—both locally and on epic journeys around the world. Large numbers of women also took up cycling, thereby greatly expanding the market. In fact, the bicycle became an icon of the women's movement.

The British bicycle boom began in 1891. By 1896, 200 companies were making more than 800,000 bicycles a year in Britain. They saturated the market, and the British bicycle boom ended by 1898.

The American bicycle boom began in 1894. By 1899, 300 bicycle companies made more than 1.1 million bicycles a year. Prices dropped, and the large bicycle companies established the American Bicycle Trust to cut production and maintain prices. They failed in their efforts, bicycle prices had fallen dramatically by the turn of the century, and there was one bicycle for every ten Americans. By 1904, the U.S. boom was over. In the U.S. bicycle was a major part of the "Gay Nineties," and bicycles were used by millions of people in cities across the country, both for travel and for recreation.

In both Britain and the U.S. the bicycle boom ended and sales plummeted after just a few years. This resulted in numerous bankruptcies of bicycle companies and a significant decrease in bicycle prices.

The end of the bicycle boom is often erroneously blamed on the automobile. However there were not enough cars at that time to have a significant effect. A more likely reason was the dynamic growth of early mass-transit systems, such as electric streetcars and electric inter-urban railways. These provided an attractive alternative to bicycle travel—especially in adverse weather. By 1902 there

*Table 3.1. Estimated U.S. bicycle production, 1877–1914*

| Year | Conventional estimate | Revised estimate* |
|---|---|---|
| 1878 | 50 | 50 |
| 1879 | 3,682 | 900 |
| 1880 | 7,314 | 2,000 |
| 1881 | 10,900 | 3,500 |
| 1882 | 14,600 | 5,000 |
| 1883 | 18,200 | 6,000 |
| 1884 | 21,200 | 7,500 |
| 1885 | 25,500 | 10,350 |
| 1886 | 29,100 | 11,500 |
| 1887 | 32,700 | 13,000 |
| 1888 | 36,400 | 16,750 |
| 1889 | 40,000 | 20,500 |
| 1890 | 70,000 | 29,890 |
| 1891 | 110,000 | 46,990 |
| 1892 | 150,000 | 60,700 |
| 1893 | 200,000 | 54,650 |
| 1894 | 300,000 | 109,300 |
| 1895 | 500,000 | 218,600 |
| 1896 | 1,200,000 | 437,200 |
| 1897 | 1,700,000 | 673,200 |
| 1898 | 2,000,000 | 928,250 |
| 1899 | 1,1282,250 | 1,182,250 |
| 1900 | 997,000 | 997,000 |
| 1901 | 811,000 | 811,000 |
| 1902 | 620,000 | 620,000 |
| 1903 | 400,000 | 662,500 |
| 1904 | 252,923 | 252,923 |
| 1905 | 240,000 | 374,789 |
| 1906 | 225,000 | 367,117 |
| 1907 | 210,000 | 357,132 |
| 1908 | 199,000 | 176,455 |
| 1909 | 233,707 | 233,707 |
| 1910 | 250,000 | 388,148 |
| 1911 | 290,000 | 393,894 |
| 1912 | 350,000 | 426,561 |
| 1913 | 300,000 | 464,080 |
| 1914 | 398,899 | 398,899 |

* Revised estimates are based on Bruce Epperson's research, as published in the *Proceedings of the 11th International Cycling History Conference* (see Bibliography).

Courtesy of Bass Photo Collection
Indiana Historical Society

Above: Fig. 3.2. By the end of the 19th century, the spread of electric trolley service was perceived as a threat to cycling.

were 22,000 miles of street railways in the U.S., nearly all electrified. These trolleys were carrying 5 billion passengers a year, and were a major impetus for urban sprawl, which in turn made cycling less practical.

### Speed and Distance Records

Remarkable performances were achieved on the single-speed bicycles of this era. By 1894, the British 24-hour record for a safety bicycle had reached 426 miles (685 km) compared with the 312 miles (502 km) for a high bicycle.

In 1893 Henri Desgrange raised the hour record to 35.33 km (21.96 miles). In 1898 W.W. Hamilton raised it to 40.78 km (25.34 miles).

In 1899 Charles M. Murphy rode one mile behind a Long Island Rail Road locomotive in 57.8 seconds. He averaged 62.7 mph (100.8 km/h).

### Bicycle Touring

European and American adventurers traveled by bicycle across Europe, America, Siberia, and even around the world. For lack of adequate gearing, they had to dismount and wheel their machines up every mountain road,

Above: Fig. 3.3. In 1899, Charles Murphy set a new 1-mile paced speed record at more than 60 mph (96 km/h) behind a train car. Although only a marginal improvement over E.E. Anderson's 1896 record of 1 min. 3/5 sec., it established a milestone, and Murphy would forever after be called "Mile-a-Minute" Murphy.

Right: Fig. 3.4. Touring cyclists in France at the end of the 19th century. The Touring Club de France was founded in 1890. It was an immediate success, and its membership reached tens of thousands within a few years' time. Members ventured into all parts of the country, including the hills, and soon felt the need for variable gears.

including Europe's highest, the 9,055-foot (2,760 m) Stelvio Pass between Italy and Austria.

In France, cycling also satisfied a longing for freedom of travel. "The bicycle is a social godsend," wrote journalist Pierre Giffard. He dubbed it "*la petite reine*" (the little queen) to express the fondness with which the bicycle was regarded.

France is a country of varied relief. There are high mountain roads and hills in most regions. Where the country is flat, such as in the marshy Camargue or in the wheatfields of the Beauce, there are often strong winds. A single-speed bicycle is uncomfortable to ride under these conditions. The gear is either too high or too low. Soon there was a strong demand for variable gearing, and the bicycle makers tried different ways to meet this demand.

## A Plethora of Patents

Inventors identified with bicycles. With the enormous popularity of cycling, tens of thousands of inventors registered patents on bicycle improvements from 1875 to 1900. At one time, the U.S. Patent Office was housed in two buildings in Washington, DC—the larger building held the bicycle patents, and the smaller building held everything else.

Many inventions addressed the issue of variable gearing to provide speed for easy terrain and less effort for climbing. Variable gearing systems were patented in Britain, France, Germany, the U.S., and many other countries. In his book *Variable Gears*, published in London in 1908, "Logos" listed more than 750 patents for variable gearing for the period 1877–1906.

Tony Hadland (Britain), David Herlihy (U.S.), and Walter Ulreich (Germany, Austria, and Switzerland) searched out the patents cited in this and the following chapters.

Despite the remarkable number of patents, few practical gearing systems were actually installed on bicycles. In most instances the patent drawings are the only evidence we have. On the few occasions when prototypes were built, they failed to attract much response from the cycling public.

Most people were quite satisfied with the basic single-speed bicycle, though it was still relatively expensive. Widespread acceptance of variable gearing had to wait until the 20th century. Nevertheless, it is fascinating to realize that most of the ideas for variable gearing had already been thought out before the 19th century was over. In this chapter, we recall the fertile imagination of the inventors who foresaw the need for variable gearing.

It will be helpful at this point to make a note about some of the terminology and definitions used in this book. *Gear* is a very broad term. First, it can mean equipment, such as camping gear. It can also mean the mechanical advantage of the bicycle geartrain, e.g. a gear of 60 in. British bicycle writers usually call the mechanism that provides variable speeds on a bicycle a mechanism (or a "mech" for short).

To American bicycle writers, a gear usually means a toothed

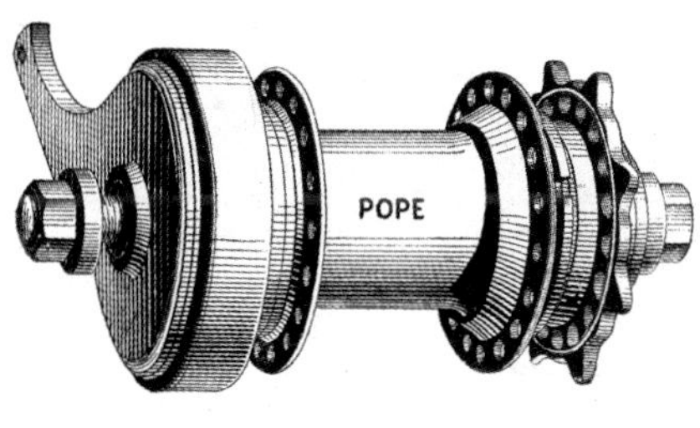


Above: Fig. 3.5. The back-pedal (or coaster) brake became widely available on U.S. bicycles in the mid-1890s. It was the last significant development in gear-trains in America.

Right: Fig. 3.6. The Pope Manufacturing Company made the Columbia Century bicycle in 1893. Riders were going on "Century" rides (100 miles in one day). Note the elliptical chainwheel, which did not catch on at the time, but continues to be reincarnated to this day.

wheel. Little gears are called *pinions*. To both British and Americans, the big gear in front in a chain drive is a *chainwheel*, and the little gear in the back is a *sprocket*. *Cog* is a broad term for either a pinion or a sprocket. *Hub gears* and *bracket gears* are the names for internal gear mechanisms that mean the same thing to both British and Americans.

To make sense of the different proposals for variable gearing, it is convenient to classify chain drives into either *direct-drive* or *indirect-drive* systems.

*Direct drive* means that the cranks are fixed to the chainwheel, and the sprocket is fixed to the rear wheel, so the rider's effort is transmitted from:

- Pedals to cranks, to chainwheel, to chain, to sprocket, to rear wheel.

However, the size of the chainwheel or the sprocket may be varied. Derailleurs are direct-drive systems.

*Indirect drive* means that a set of gears is added to the drivetrain. In the case of bracket gears, the gears in the bottom bracket make the chainwheel move at a different speed from the cranks. The rider's effort is transmitted from:

- Pedals to cranks, to SET OF GEARS, to chainwheel, to chain, to sprocket, to rear wheel

In the case of hub gears, the gears in the rear hub make the rear wheel move at a different speed from the sprocket. The rider's effort is transmitted from:

- Pedals to cranks, to chainwheel, to chain, to sprocket, to SET OF GEARS, to rear wheel

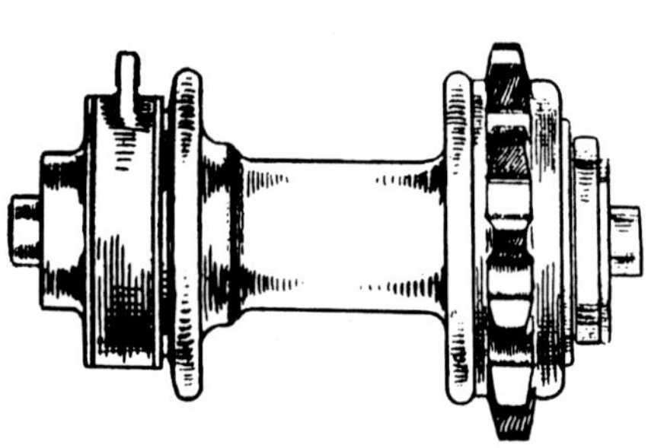

Above: Fig. 3.7. The 1891 Crypto Dynamic Two Speed Hub was the first successful hub gear.

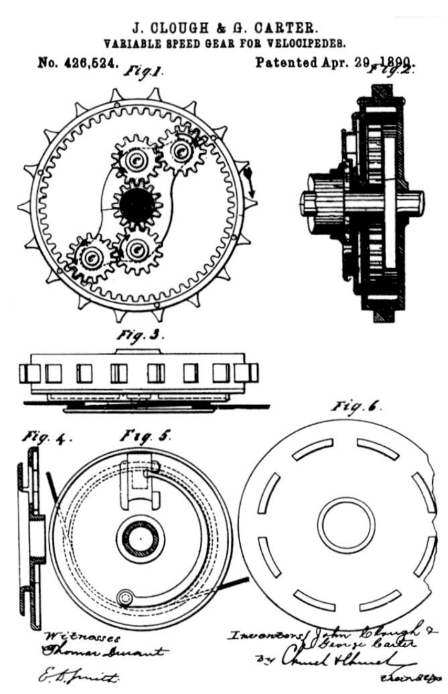


Right: Fig. 3.8. J. Clough and G. Carter, of England, obtained a U.S. patent for a 2-speed epicyclic gear in the bottom bracket in 1890. It was apparently never manufactured.

## Indirect-Drive Systems

Before examining the direct-drive derailleur type, we will look at the wide range of indirect drives, which operated through gears. Many indirect drives used epicyclic gearing, which was described in Chapter 2, and we'll encounter a lot more examples in Chapter 4.

In 1891, the Crypto Dynamic Two Speed Hub gear won praise from *Cycling* for its neat and tidy appearance:

> It is quite unnecessary nowadays to point out the manifold advantages of high and low speed gears, and the main obstacle to their more general adoption so far has, we believe, lain in their complication, and unwieldy nature. But to the "Crypto" gear no such objection can be raised, and as will be seen from the annexed illustration it is extremely neat in appearance, the diameter being little in excess of that of an ordi-

Above: Fig. 3.9. The 1892 Cycle Gear Co.'s Two-Speed Change Gear gave gears of 56 in. and 70 in. respectively.

nary hub, and all the working parts run on ball bearings.

In 1892, the Cycle Gear Co. in Coventry used Clough and Carter's patent to produce an epicyclic chainwheel that could be installed on either bicycles or tricycles. Advertisements for their Two-Speed Change Gear claimed that the gear gave fast riders "one or two miles an hour over the highest speed they can do without it." It was used to win races over distances of 50 miles (80 km) and 68 miles (109 km). For "ordinary" cyclists, it was claimed that, "Our gear gives delightful speed along level and down hill, and perfectly easy hill-climbing." However, the Two-Speed Change Gear did not appear at the annual Stanley Show in London, and nothing further was heard of it.

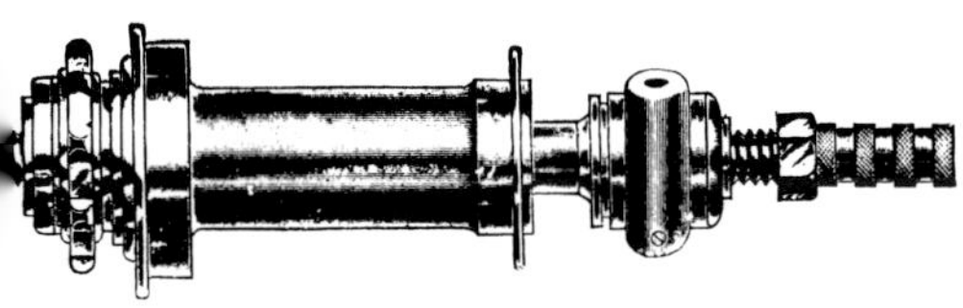

Above: Fig. 3.10. The 1895 J&R two-speed hub gear. It was presumably based on a British patent by Indiana machinist Seward Johnson for a compact two-speed hub gear.

Below: Fig. 3.11. The Hub 2-speed gear was patented in Britain by William Reilly in 1896 and put into production the same year. The Hub was neat, light in weight, and could be installed on any bicycle. It was the first hub gear to achieve widespread popularity.

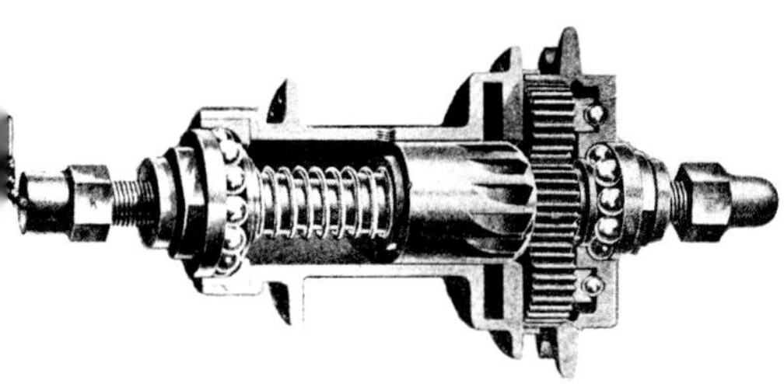

### Other Indirect Drives

Numerous other types of gear-driven mechanisms were patented. In 1889, Bostel and Loosley patented a 2-speed shaft-drive in London.

Other early 2-speed bracket gears included the following products:

- Wood and Moss' bracket gear, patented in London in 1890, was apparently similar to the Collier gear.
- Charles Terrot's bracket gear, patented in1890 in Germany, France, and the U.S., which could drive the chainwheel either directly or through a countershaft.
- Augustus Pursell's 2-speed bracket gear, patented in the U.S. in 1891.

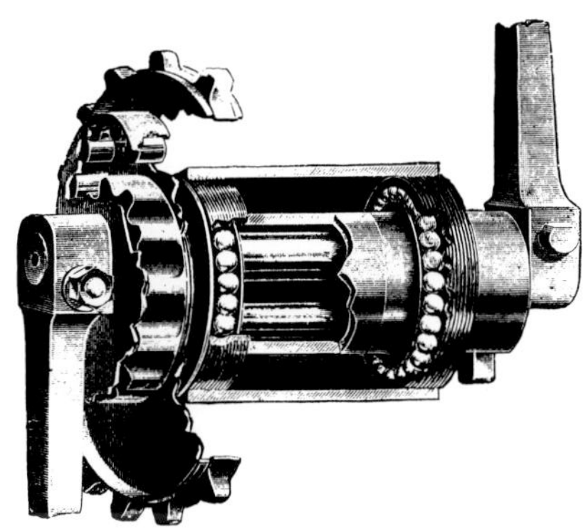

Above: Fig. 3.12. The 1889 British Collier 2-speed bottom-bracket gear consisted of an eccentric 12-tooth cog inside the chainwheel. When released, the cog made the chainwheel turn faster and gave a higher gear. The Collier worked well, but it weighed 4½ lbs. (2 kg) and could only be installed on a frame with a special bottom-bracket shell.

At first, variable gears did not find favor in the U.S. In 1894, two leading cycle manufacturers, A.H. Overman and Albert A. Pope (Columbia), both wrote letters to Frank Leslie's *Illustrated Weekly*, saying there was no advantage in devices for hill-climbing due to their added weight and complication and increased stress on the

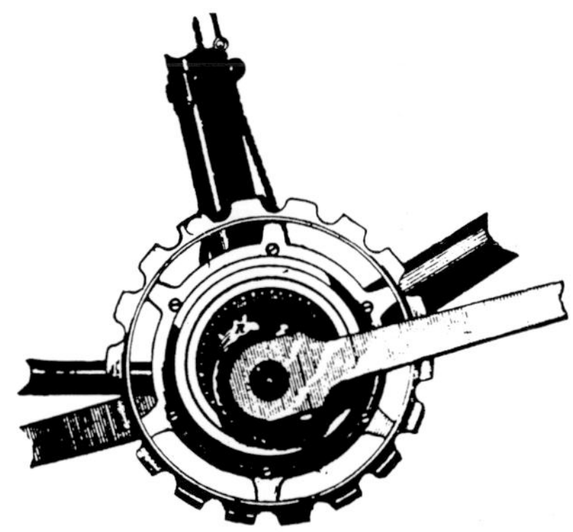

THE HY-LO CHANGEABLE GEAR

Above: Fig. 3.13. The 1894 Hy-Lo gear was probably the first variable gear for bicycles in the U.S. It was a 2-speed bottom-bracket gear with an eccentric stud-wheel, and appears to be similar to the British Collier gear.

Below: Fig. 3.14. The 1890 Transformateur Mercier was one of the first French bottom-bracket gears with three speeds.

machine. The *American Machinist* concurred:

> All things considered, the best results will be obtained by choosing a gear best adapted to the rider and the conditions under which he rides. He can then walk up hills which he cannot comfortably climb, and it will be better both for him and his machine that he should do so.

Below: Fig. 3.15. In 1896, Harry de Lyne Weed and Francis Webster Gridley, of New York, patented a two-sprocket gear. In the diagrams, a chain has been added to the original patent drawings to assist in explaining how the system worked.

Upper detail: In low gear, the chain drove a large outer sprocket.

Bottom detail: When the control rod was pulled, it operated cams that pushed the large sprocket downward. Now the chain also engaged the top tooth of the small inner sprocket, giving a high gear. Their design did not include a provision for chain slack.

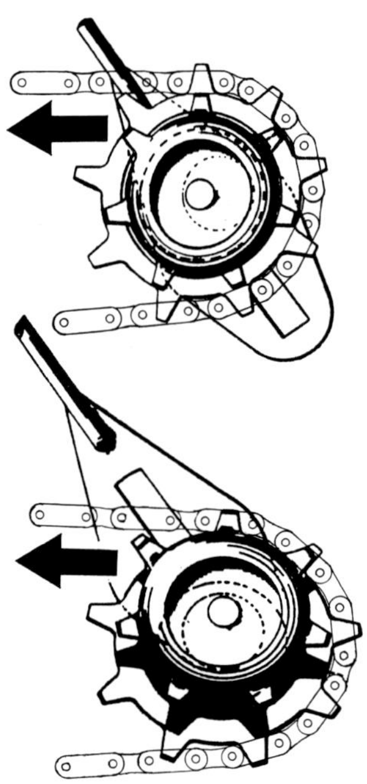

Nevertheless, by 1897, a number of 2-speed gear-trains were being sold in the U.S. Some of these were "chainless," with shaft- or rod-drive, avoiding the problems of bicycle chains clogged by mud. They included the following devices:

- A 2-speed bracket gear with shaft-drive, from the Pope Manufacturing Company.
- The St. Louis Chainless, driven by oscillating rods connected to the rear wheel.

In 1899, the Rambler bicycle from the U.S. was shown at the annual London Stanley Show, equipped with the Gormully & Jeffery (G&J) 2-speed gear. This gear had an eccentric driving wheel, just like the Collier bracket gear, but it fit inside the hub.

The same year, the Dieterich Gear Company of New York designed a completely automatic gearless transmission system that was patented in Britain. It was a variable-eccentricity lever-drive mechanism housed in an oversize bottom bracket.

### Lever-Drives: the Svea

Birger and Fredrik Ljundström produced the first Svea-velocipede. It operated on the same lever-drive principle as the American Star. Production began in Stockholm in 1893. The prototype Svea interested Alfred Nobel, the famous inventor of high explosives. Nobel established The New Cycle Company in Northfleet in Britain to manufacture the Svea. Production developed slowly, and when Nobel died in 1896, his will directed

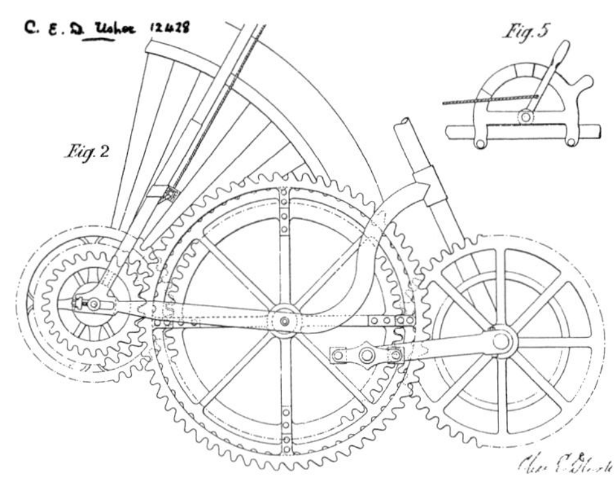


Above: Fig. 3.17. The 1896 patent drawing for Captain Charles Usher's 3-speed. The concept was heavier, more expensive, and less efficient than chain drives.

Right: Fig. 3.16. The Svea lever-drive bicycle was driven by two levers, each connected to a chain. A third chain raised one arm as the other was pushed down. By moving the driving chains to different notches up or down the driving arms, any one of five different gear ratios could be selected.

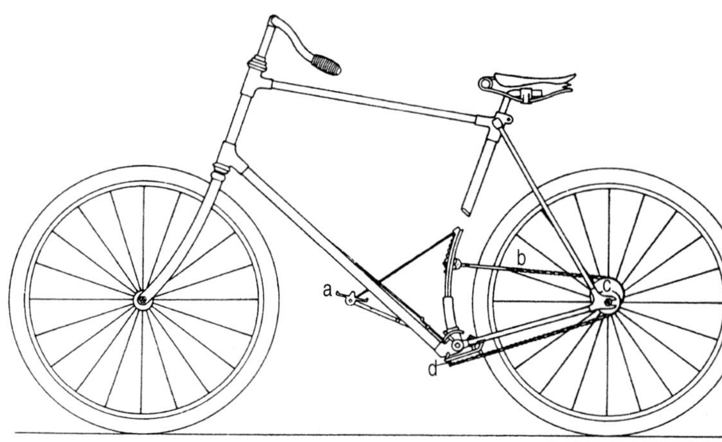

that his assets be sold to fund the Nobel Prize. By 1898, the company was bankrupt, but the Svea continued to be manufactured in Sweden. A Belgian correspondent reported to *Cycling* in 1904 that the Svea bicycle was in use by the Swedish and Norwegian armies.

### Australian Variable Gears

In 1896, two chainless variable gear mechanisms, the Triad and the Love, were patented in Melbourne. Both appeared to be practical designs, but they did not progress beyond prototypes. The public was invited to invest in companies to manufacture the bicycles, but the ventures failed in the stock market slump of the late 1890s.

### More British Multi-Speed Gears

In 1896, Wyatt's gear had a 3-speed driving wheel engaging a shaft-drive. Two other British variable gears—the Sharp and the Eite & Todd—operated through cogs on an axle just behind the bottom bracket.

The Kynoch 2-speed epicyclic bracket gear was based on A.W. Bevis' 1898 patent. It could be installed on any bicycle, and provided gears of 72 and 54 in. The Kynoch used elliptical pinions, resulting in a non-uniform pedal response. In 1900, the Planet two-speed hub was marketed in Britain.

### Interchangeable Chainwheels

In 1887, buyers of Raleigh bicycles could have them delivered with several extra interchangeable chainwheels. You selected a chainwheel according to the difficulty of the route you intended to follow, and then bolted it onto the crank spider. The drawback was

Above: Fig. 3.20. Eite & Todd's 1896 gear had a countershaft behind the bottom bracket. Either of two intermediate cogs could be swung into place to engage the high or the low gear.

Below: Fig. 3.21. In 1890, the Raleigh Cycle Company produced sets of interchangeable chainwheels. The chain had to be lengthened or shortened each time.

Above: Fig. 3.18. The 1887 Carroll Gear-to-Gear bicycle was made in Philadelphia. It came at the end of the U.S. bike boom and had little to recommend it.

Right: Fig. 3.19. The Triad bicycle used Usher's patent. The chainwheel drove three huge cogs mounted on the chainstay. Three more cogs were mounted on the rear axle and could slide sideways so that any one of them could be engaged. This provided gears of 50, 70, and 116 in.

that you also had to remove and reinstall the pedal, then adjust the position of the wheel in the rear fork-end as much as the frame allowed, or shorten or lengthen the chain.

Vélocio used such a bicycle for one season, but rather than bother changing the chainwheel, he mainly used one chainwheel that provided a 50-inch gear (4.00 m development). This option soon disappeared from the Raleigh catalog.

### Bolt-on Chainwheel and Sprocket Enlargers

At the 1890 Stanley Show in London, an inventor showed a set of two half-chainwheels that could be installed over the existing chainwheel to give a higher gear. Here again you had to lengthen or shorten the chain. In March 1890, Vélocio suggested that a movable chain rivet locked by a hook would enable the chain to be lengthened or shortened quickly. He put the idea into practice some years later.

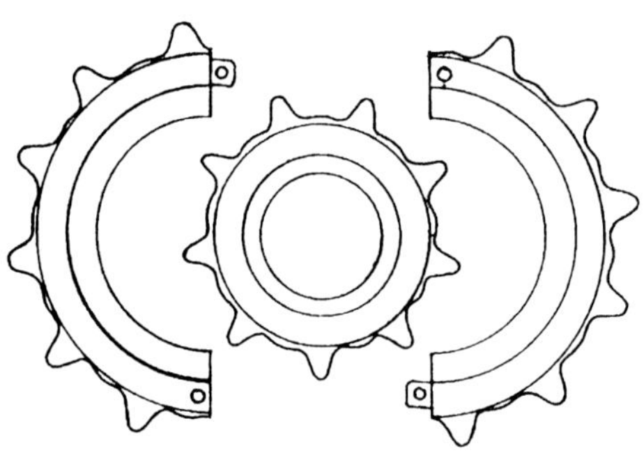

Above: Fig. 3.22. Monsieur Journaux, of Paris, produced a sprocket enlarger in 1890. A 9-tooth sprocket (teeth were then one-inch pitch, so this was equivalent to 18 teeth today) could be enlarged into a sprocket with 10, 11, or 12 teeth by bolting on two semi-circular pieces.

### The Jeanperrin System

Around 1895, Jeanperrin developed a mechanism that may have been the first front gear-change device. In four steps, the lever caused the following responses:

1. Released chain tension.
2. Shifted the chain to the other chainwheel.
3. Pushed the chainwheel so as to align it with the sprocket.
4. Re-tensioned the chain.

This mechanism was heavy, complicated, and unsuccessful.

Below: Fig. 3.23. Reversing the rear wheel. Monsieur Baujouan, an engineer and touring cyclist, assembled a bicycle for his own use and described it in *Le Cycliste* in 1890. The rear wheel had a sprocket of different size on either side, so he could have two different gears by reversing the wheel. Twenty years later, this idea became part of Tour de France lore.

### The Dijon Sprocket-changer

We know about this system only from a letter published in *Le Cycliste* in the May 1896, when someone signing "M. de St.-G." reported on a mechanism developed by a craftsman from Dijon:

> The entire bottom bracket is mounted eccentrically in a casing, and a half-turn of the cranks slackens the chain and places it on another rear wheel sprocket. In this double movement, the bottom bracket is displaced horizontally, so that the two toothed wheels work in the same plane.

The letter did not say whether the chain was placed manually or mechanically onto the second sprocket.

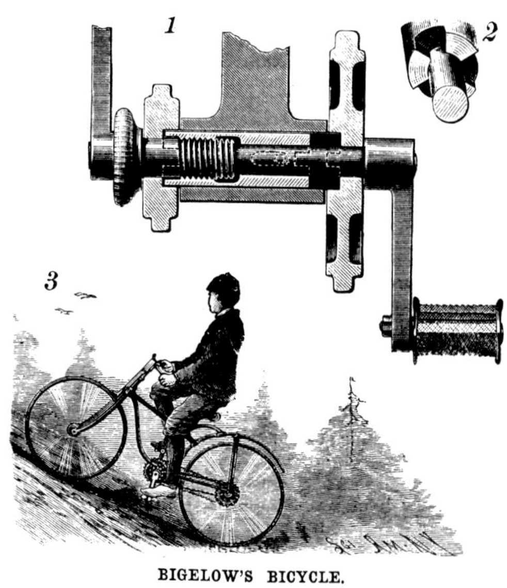


Above: Fig. 3.24. Frank R. Bigelow's 1891 bicycle was shown in *Scientific American.* A hand-screw engaged either the left-side or the right-side chainwheel.

## Double-Chain Drives

Another way to achieve two different gears is by having separate gear-trains on the left and the right side. Either chain could be made to drive, giving high gear or low gear. This double-chain, or *bichain*, method had already been used a few years earlier in Britain and Germany. In 1891, three bichain drives were patented in the U.S.:

- Horace B. Murdoch, of Michigan, proposed a lever on the seat tube for engaging the cranks with either the left or the right chainwheel.
- Gustave Egg, of Pennsylvania, proposed a bicycle with two chains, providing high or low gears. The high gear was to be brought into play in some way by the brake-actuating mechanism.
- Frank R. Bigelow, of New Jersey, proposed a crank axle with different sized chainwheels on either side.

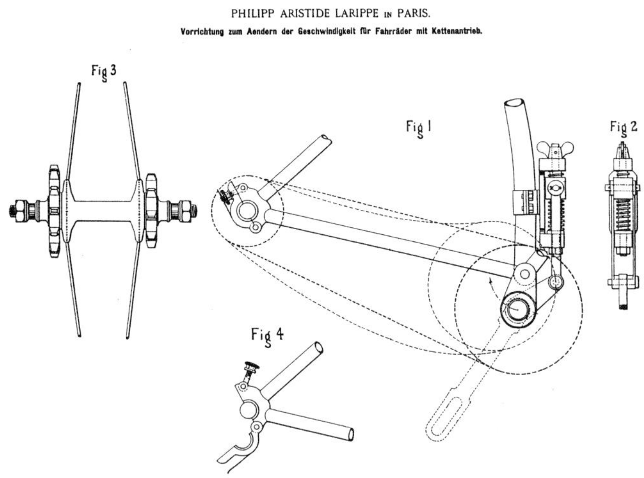


Right: Fig. 3.25. Philipp Aristide Larippe's 1891 patent for a reversible rear wheel.

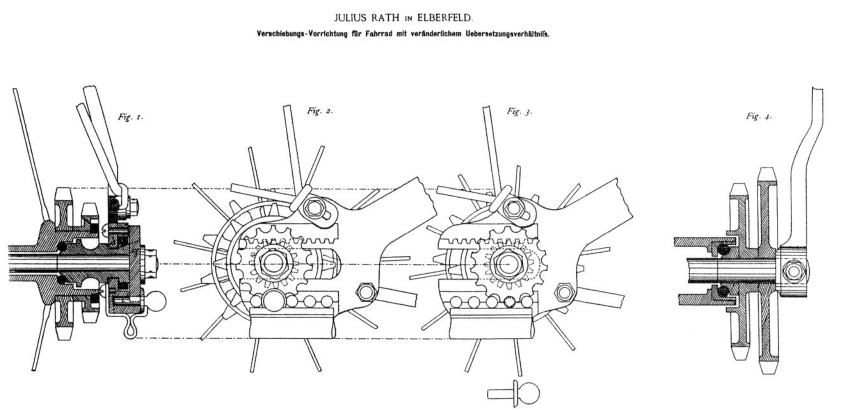


Above: Fig. 3.26. Julius Rath's 1891 patent for a double-chainwheel, 2-sprocket gear-train with four possible gears.

### The Slack Chain Problem

Today's derailleurs use a tension arm and a tension pulley to keep the chain taut. Before this simple and obvious solution was adopted, many inventors proposed multiple chainwheels and sprockets, but they had clumsy methods for moving the chain and dealing with the slack chain.

We found a dozen relevant patents in five countries for the period 1891–1896. None of these inventions were practical, because they did not allow changing gears while moving, but some of the ideas subsequently reappeared in different forms.

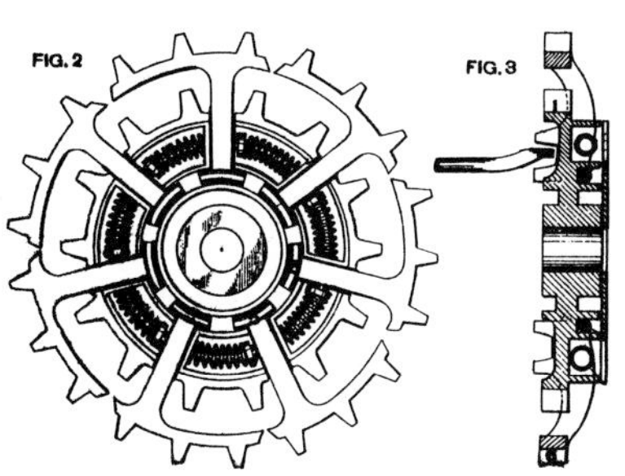


Left and Right: Figs. 3.27 and 3.28. John Wilson's 1892 patent for two chainwheels that could be swung aside, segment by segment.

In 1891, Philipp Aristide Larippe was granted a French patent to alter the gearing by having a different-sized sprocket on either side of the rear hub and simply reversing the rear wheel. He proposed quick-release rear fork-ends and a thumbscrew for moving the bottom bracket to take up chain slack. A simpler version of his technique later became legendary in the Tour de France.

In 1891, Julius Rath obtained a German patent for a double-chainwheel, two-sprocket gear-train with four possible gears. Chain slack could be taken up with a handwheel that wound the rear axle forward or back along a rack on the fork-end. Forty-five years later, in 1933, Tullio Campagnolo patented a similar fork-end rack to take up chain slack.

In 1892, John Wilson obtained a British patent for two chainwheels on the same side of the bottom bracket. The outer chainwheel could be swung aside, segment by segment, so that the chain would transfer to the inner chainwheel. As part of Wilson's ingenious proposal, the chain tension was taken up automatically by the rear wheel moving forward or back.

In 1893, Fritz Mortier, of Germany, proposed a novel way of

Right: Fig. 3.29. William Weatherill's 1894 patent had two chainwheels and two sprockets, with the chain running on any one of the four combinations.

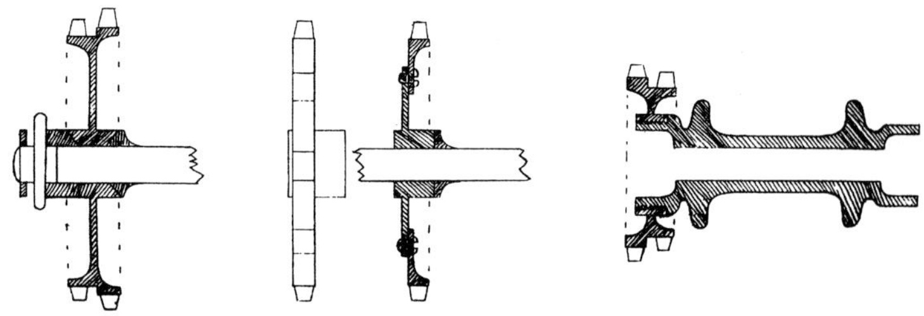

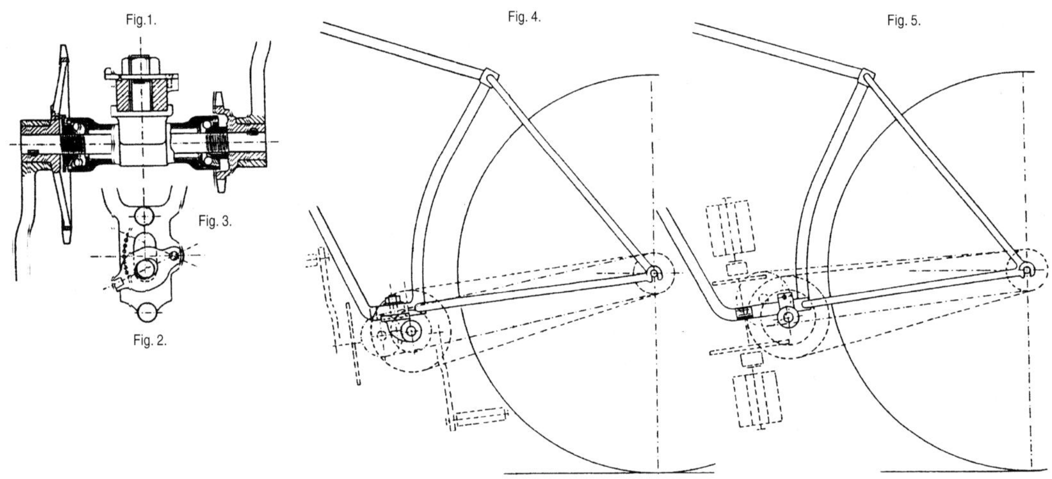


Left: Fig. 3.30. Fritz Mortier's 1893 patent described unbolting the entire bottom bracket and turning it upside-down.

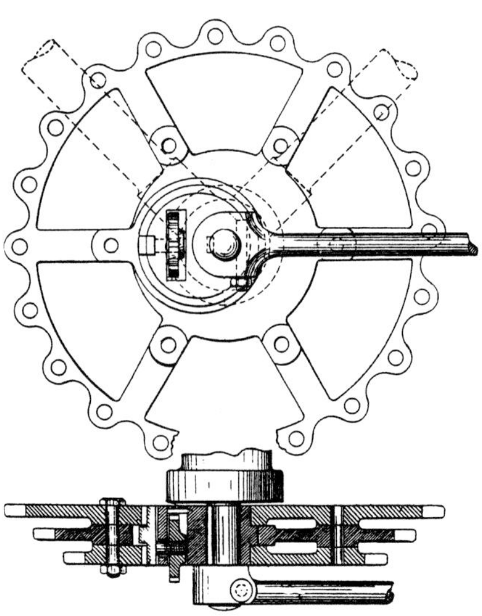

Left: Fig. 3.31. In Hugo Exner's 1895 triple-chainwheel patent, chain slack was taken up by an eccentric bottom bracket.

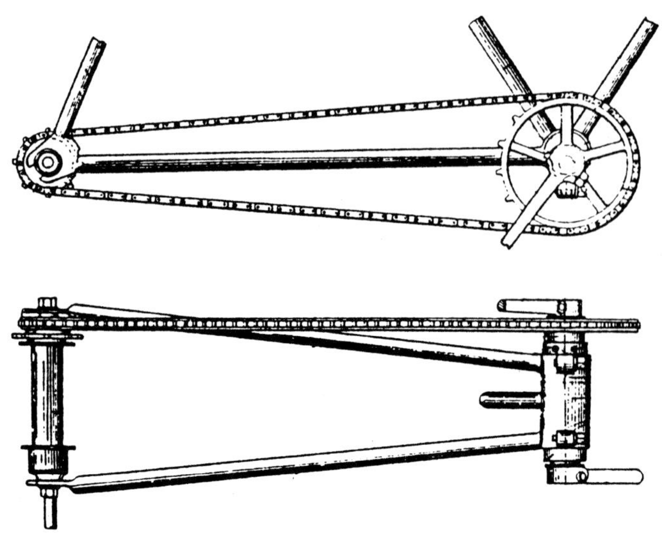

Right: Fig. 3.32. In 1894, Herbert Leonidas Bailey patented a 2-sprocket system with an eccentric bottom bracket to take up chain slack in the high gear.

changing from one chainwheel to another. The entire bottom bracket was unbolted and turned upside down to provide a different-sized chainwheel. The chain was placed onto the new chainwheel, the bottom bracket moved forward or backward to take up chain slack, and was then bolted into place again.

In 1894, William Weatherill of Manchester proposed two chainwheels and two sprockets, with the chain running on any one of the four combinations. The rider had to dismount to shift gears. Chain slack was taken up by sliding the wheel along the rear fork-end. The chainwheels and sprockets could be adjusted sideways, so that the chain would always run in line.

In 1894, Herbert Leonidas Bailey, of Chicago, proposed a two-sprocket system with a movable crank axle to take up chain slack. Two bolts under the bracket were loosened and an eccentric shell rotated inside the bottom bracket, similar to the arrangement used on modern tandems.

In 1895, W.M. Woodland, of Britain, proposed a system with the chain slack taken up by a long slotted rear fork-end. A quick-release device enabled the chain to be slackened, placed on a different chainwheel and sprocket combination, and then re-tightened.

In 1896, Carl Strüder and Emil Sohn, of Germany, patented a 3-sprocket cluster. The chain slack was taken up by sliding the bottom bracket forward or backward. This invention had a clever slant to it: As the bottom bracket slid forward along a diagonal chainstay, it also moved to the left. Thus the chainwheel stayed in line with the chosen sprocket and the chain ran straight in each of the three gears.

## Ed Berry's Bicycle

The American collector Ed Berry Jr. found a bicycle with solid rubber tires, a well-sprung saddle, and swept-back handlebars, suggesting that the bicycle was made between 1890 and 1895. Ed Berry purchased the bicycle in Santa Fe, New Mexico. Before that, it had been in a private museum in Houston, Texas.

Where did it come from originally? The threads on the Berry bicycle are metric sizes, and the chain links are stamped: SUPERIEURE TEMPRE J.G. No information about such a French bicycle has been found, but in 1938, Roland Sauvaget interviewed an elderly cyclist, Paul Ledieu, who recalled: "Around 1893, another builder also made a 2-speed with two sprockets and a single chain. The tension was adjusted at the bottom bracket, which swung for-

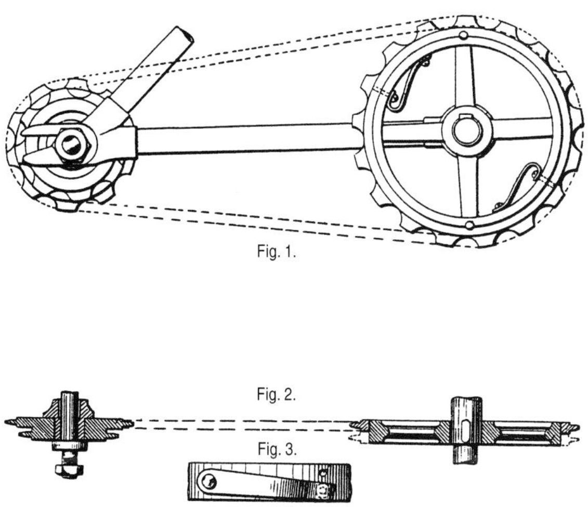


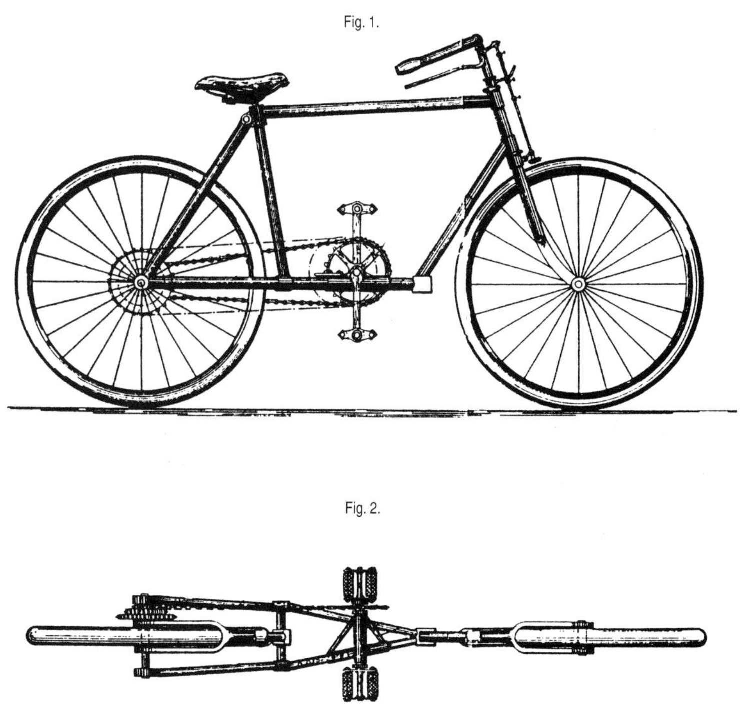


Above: Fig. 3.33. Wilhelm Meyer's 1895 patent had two sprockets. A slotted fork end allowed for chain slack. It maintained a straight chain line by sliding the chainwheel crown sideways.

Right: Fig. 3.34. Carl Strüder and Emil Sohn's 1896 patent had a 3-sprocket cluster with a sliding bottom bracket to take up chain slack.

ward or back." Was Ledieu recalling the Berry bicycle or something else? Perhaps this is the oldest derailleur bicycle, but we may never know for sure.

### Fichtel & Sachs

In 1895, Karl Fichtel and Ernst Sachs founded Schweinfurter Präzisions Kugellagerwerke Fichtel und Sachs (later to be called simply Fichtel & Sachs, or F&S), to produce ball bearings and bicycle hubs. By 1896, they had 70 employees and were making 50 to 70 hubs per day. They made their first freewheel hub in 1897.

Left: Fig. 3.35. Contributing author Ron Shepherd (left) with New York collector Ed Berry and his mysterious derailleur bicycle.

### Chainwheels and Sprockets with the Same Number of Teeth

In the 1890s, inventors often got around the slack chain problem by keeping the chain length in each gear constant. In the style of Schürer's variable tricycle gear (see Chapter 2), these designs shifted the chain across combinations of chainwheel and sprocket with the same total number of teeth.

In 1893, Bernard Cadot, of Paris, registered a patent in London for three chainwheels and three sprockets. To change gears the rider had to stop, unclip the chain, move it to another combination, and clip it together again. The chain always ran in a straight line. There was the same number of teeth for the chain to go around in each gear. In high gear

Above: Fig. 3.36. Ed Berry's bicycle is remarkable because it has an ingenious derailment system. When a lever on the top tube is operated, a guide moves the chain from one fixed sprocket to the other.

Right: Fig. 3.37. As the chain shifts sprockets, the bottom bracket on the Berry bicycle swings forward or backward to take up the chain slack.

there were 24 teeth on the chainwheel and 8 teeth on the sprocket, for a total of 32 teeth; in low gear, there were 16 teeth on the chainwheel and 16 teeth on the sprocket, again with a total of 32 teeth.

In 1897, Salomon Frank, a German engineer, patented a derailleur system in Britain. He proposed to run the chain around two guide pulleys on a frame in front of the chainwheels. By moving the frame, the chain could be swung to drive any one of the chainwheels.

In 1898, Richard Neumann was granted a German patent for a gear-train with three chainwheels. The outer and inner chainwheels could slide in line with the middle one to maintain a straight chain line.

### The First Tension Arm

A tension arm is a spring-tensioned device to take up chainslack caused by different-sized sprockets. In 1894, Linley, Biggs, and Archer introduced their Protean expanding chainwheel. The chainwheel consisted of two halves that could be spread apart. In the closed position, it corresponded to 18 teeth (1-inch pitch). Expanded it could correspond to 19, 20, and 21 teeth, giving four closely spaced gears. It was essentially an oval chainwheel, with one major axis for low gear and a major axis 90 degrees offset for high gear. It

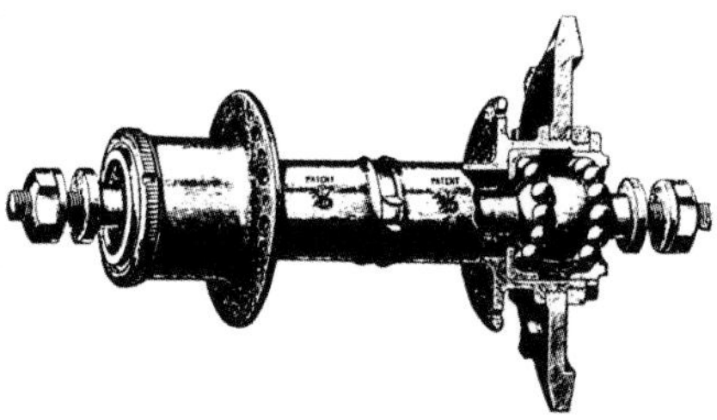

Above: Fig. 3.38. The first Fichtel & Sachs fixed-gear rear hub, 1896, clearly had adequate bearings.

Below: Fig. 3.39. Bernard Cadot's 1893 patent had three chainwheels and three sprockets. Each of the three gears had the same total number of teeth on the sprocket-and-chainwheel pair.

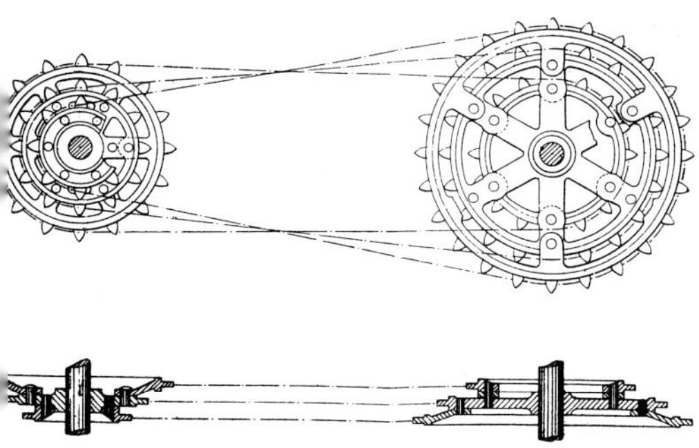

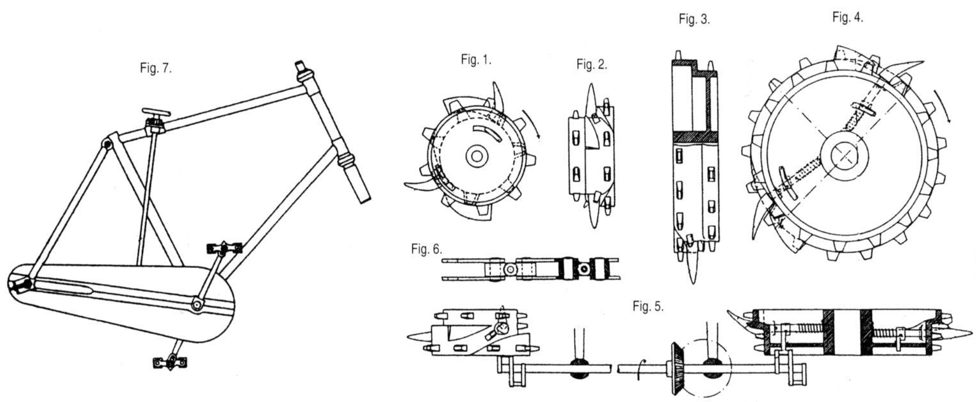


Above: Fig. 3.40. F. E. Hentschel's 1895 German patent for a double-chainwheel, double-sprocket arrangement inside a gearcase. When the rider turned a handwheel, the chain was guided onto either the inner or the outer combination. The special chain links swivelled sideways to ease shifting.

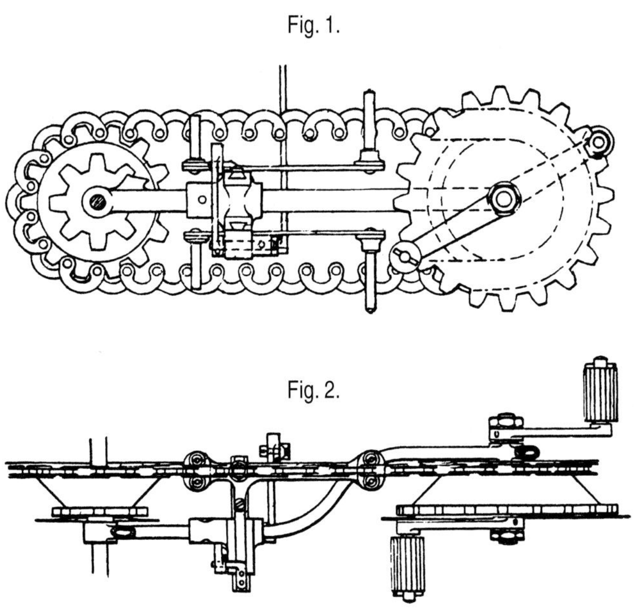


Right: Fig. 3.41. Eugen Föchtenberger's 1896 German patent for a paired chainwheel-sprocket arrangement. It used a special chain, consisting of arched links, so that the chain could slide sideways over the teeth. This was probably a plate or card chain. A set of guides simultaneously slid the chain from the inner chainwheel-sprocket pair to the outer pair or vice-versa.

was reported to be reliable. Linley and Biggs were prolific inventors, and they went on to produce the New Protean, which will be covered elsewhere in this chapter.

Just before the end of the 19th century, two British variable gear systems were patented and manufactured—the Gradient and the New Protean. They were genuine derailleurs, although not called such because the term *derailleur* had not yet been coined.

Left: Fig. 3.44. The Gradient 3-speed gear in the Science Museum, London. Hodgkinson patented the original version of this gear in 1896.

## The First Derailleurs

In 1895, Frenchman Jean Loubeyre patented a device for derailing the chain from one sprocket to another. He called his gear-changer the *Polyceler* (Multispeed). It had a rod-operated fork to push the chain sideways from one fixed sprocket to another. Two spring arms, one above the chainstay, the other below it, pushed on the chain. The arms took up slack when the chain was on the smaller sprocket. A bicycle equipped with a Polyceler was shown in the 1896 catalog of the Compagnie Générale des Cycles. Vélocio described this invention in his famous review of gear-change systems in *Le Cycliste* in 1919.

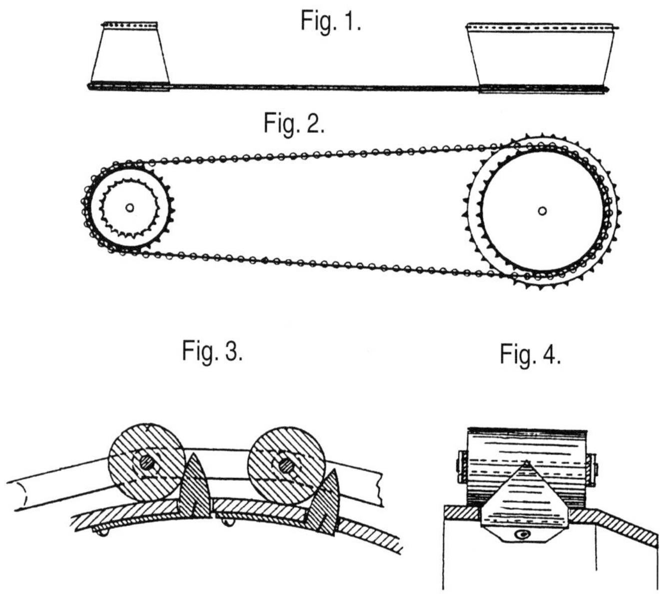


Right: Fig. 3.43. In 1898, Conrad Richter, of Germany, patented a paired chainwheel-and-sprocket system. His idea was that when the chain was pushed sideways against the teeth during a shift, the teeth would be pressed down, allowing room for the chain to slip into line.

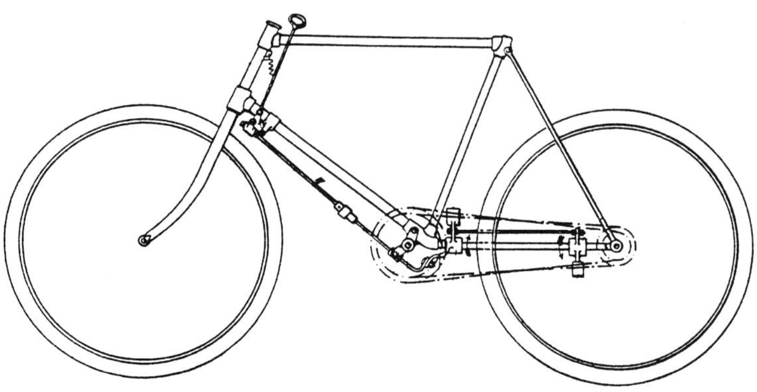

Above: Fig. 3.42. O. Schmidt's 1897 German patent for a paired chainwheel-sprocket changer mechanism. When a top-tube lever was turned, front and rear changers operated to derail the chain simultaneously across both the sprockets and the chainwheels.

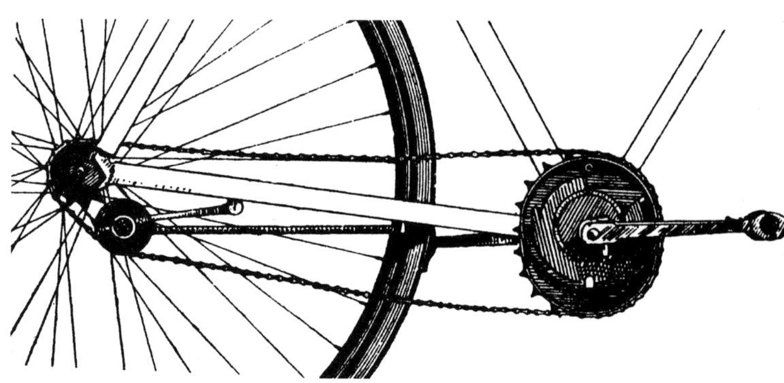

Above: Fig. 3.45. The 1884 Protean expanding chainwheel had the first practical tension arm for taking up chain slack.

They look conventional 100 years later. They were both reported to work very smoothly.

Edmund Hodgkinson's Gradient gear aroused great excitement at the 1899 Stanley Show in London. The *CTC Gazette* described it as "the most thaumaturgic feature of the Show." When the rider backpedaled, two T-shaped arms scooped the chain clear of the sprockets. Then a lever was moved to slide the sprockets sideways, so that the sprocket the rider wanted to use was under the chain. When the rider pedaled forward, the chain dropped onto the chosen sprocket. The chain was always perfectly in line.

Charles Linley patented the New Protean gear in London in 1899. It had two side-by-side sprockets and a special chain. Apparently, Linley did not realize that a bush-roller chain could jump from one sprocket to another. It was made and marketed by the Whippet Cycle Syndicate the same year. This was the most practical derailleur of the 19th century, yet it did not succeed in Britain. Whippet bicycles with New Protean gears were sold in France and attracted the attention of Vélocio and his followers.

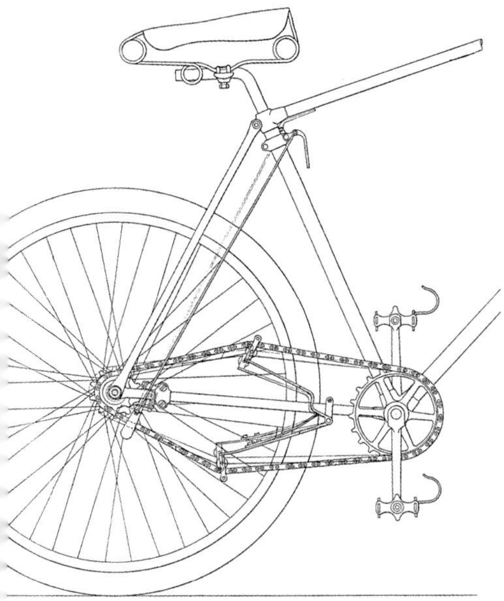

Above: Fig. 3.46. The 1895 Loubeyre 2-speed derailleur system.

Below: Fig. 3.47. The Gradient used a bush-roller chain. A drop-arm took up the chain slack. It probably shifted quite well, but it was expensive and it lost out to hub gears in the 20th century British market.

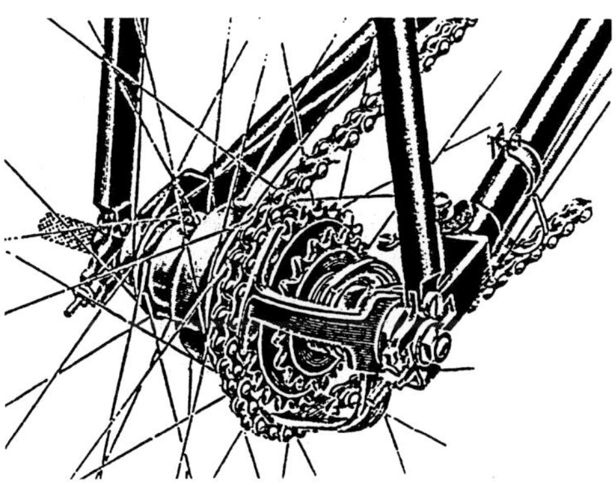

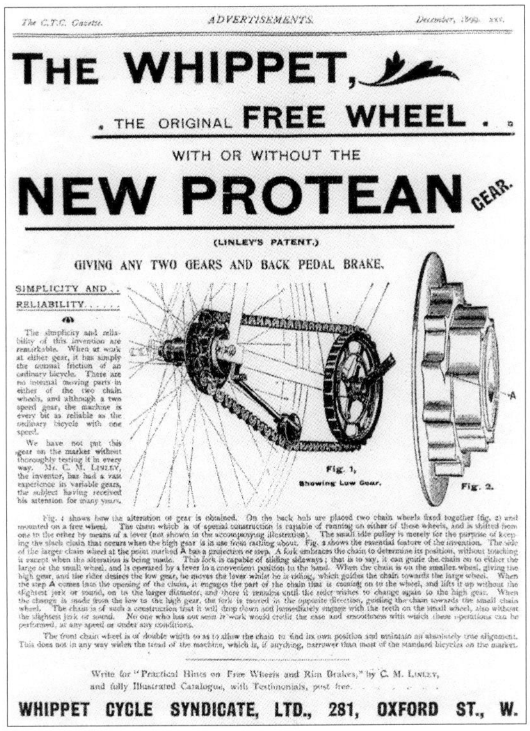

The C.T.C. Gazette. ADVERTISEMENTS. December, 1899. xxv.

THE WHIPPET,

. THE ORIGINAL FREE WHEEL . .

WITH OR WITHOUT THE

NEW PROTEAN GEAR.

(LINLEY'S PATENT.)

GIVING ANY TWO GEARS AND BACK PEDAL BRAKE.

SIMPLICITY AND . . RELIABILITY . . . . .

The simplicity and reliability of this invention are remarkable. When at work at either gear, it has simply the normal friction of an ordinary bicycle. There are no internal moving parts in either of the two chain wheels, and although a two speed gear, the machine is every bit as reliable as the ordinary bicycle with one speed.

We have not put this gear on the market without thoroughly testing it in every way. Mr. C. M. Linley, the inventor, has had a vast experience in variable gears, the subject having received his attention for many years.

Fig. 1, Showing Low Gear.

Fig. 2.

Fig. 1 shows how the alteration of gear is obtained. On the back hub are placed two chain wheels fixed together (fig. 2) and mounted on a free wheel. The chain which is of special construction is capable of running on either of these wheels, and is shifted from one to the other by means of a lever (not shown in the accompanying illustration). The small idle pulley is merely for the purpose of keeping the slack chain that occurs when the high gear is in use from rattling about. Fig. 2 shows the essential feature of the invention. The side of the larger chain wheel at the point marked A has a projection or step. A fork embraces the chain to determine its position, without touching it except when the alteration is being made. This fork is capable of sliding sideways; that is to say, it can guide the chain on to either the large or the small wheel, and is operated by a lever in a convenient position to the hand. When the chain is on the smaller wheel, giving the high gear, and the rider desires the low gear, he moves the lever whilst he is riding, which guides the chain towards the large wheel. When the step A comes into the opening of the chain, it engages the part of the chain that is coming on to the wheel, and lifts it up without the slightest jerk or sound, on to the larger diameter, and there it remains until the rider wishes to change again to the high gear. When the change is made from the low to the high gear, the fork is moved in the opposite direction, guiding the chain towards the small chain wheel. The chain is of such a construction that it will drop down and immediately engage with the teeth on the small wheel, also without the slightest jerk or sound. No one who has not seen it work would credit the ease and smoothness with which these operations can be performed, at any speed or under any conditions.

The front chain wheel is of double width so as to allow the chain to find its own position and maintain an absolutely true alignment. This does not in any way widen the tread of the machine, which is, if anything, narrower than most of the standard bicycles on the market.

Write for "Practical Hints on Free Wheels and Rim Brakes," by C. M. Linley, and fully Illustrated Catalogue, with Testimonials, post free.

WHIPPET CYCLE SYNDICATE, LTD., 281, OXFORD ST., W.

Above: Fig. 3.48. This 1899 Whippet ad explains the operation of the New Protean. The chainwheel and sprockets had flanges on each side to keep the chain from sliding off.

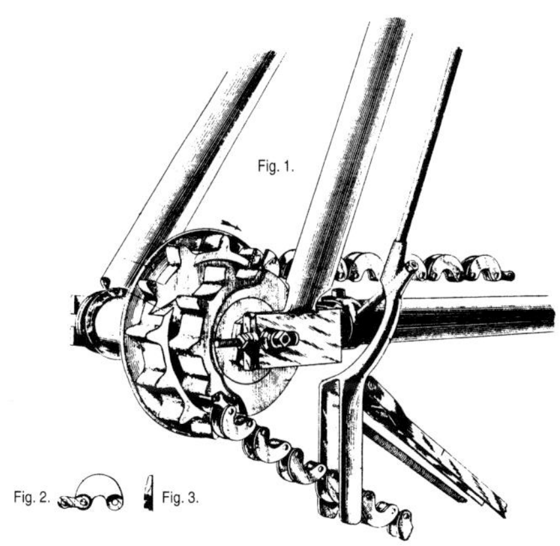


Left: Fig. 3.49. The 1899 New Protean used a beefed-up plate or card chain. The bottom of the plates allowed the chain to slide easily between sprockets. The chainwheel and the sprockets had flanges on either side to keep the chain from sliding off altogether. A swinging fork, controlled by means of a cable and a hand lever, derailed the chain. Chain slack was taken up by an arm attached to the derailing fork. A ramp cut into the larger sprocket eased shifting (a precursor of Shimano HyperGlide).

# 4

# Pre-Derailleurs, Epicyclics, and Exotics: 1900–1907

THE PERIOD from 1900 to the outbreak of World War I in 1914 was an exciting time for bicycle development. It defies a simple chronological listing of the innovations in bicycle gear-trains. However, around 1908, there was a fork in the road: English cyclists took the right fork and would pedal their 3-speed epicyclic hub gears for the next 40 years. French cyclists took the left fork and rode their derailleur gears right up to the present. Meanwhile, American cyclists fell down a hole in the road and continued to ride single-speed coaster-brake bicycles for the next 60 years.

Above: Fig. 4.1. William Reilly made the greatest contribution to the development of practical epicyclic hub gears.

This is a gross oversimplification, because all over the world most bicycles had just one speed. It took a long time for derailleur usage to grow beyond the small cadre of French *cyclotouristes*.

Most, but not all, of the significant developments of epicyclic hub gears took place before 1908, and interest in them was greatest in Britain. The first practical rear derailleurs appeared after 1908.

This chapter will cover the epicyclic gear-trains and the period from 1900 to 1907. Also covered here are the exotic gear-trains that were neither hub gears nor derailleurs, leaving rear derailleurs to be described in Chapter 5, which covers early derailleur developments and the period from 1908 to 1920.

Not all of the systems fit neatly into this chapter chronology, so there is some overlap in terms of national, chronological, and technical subject matter between these two chapters.

### Bicycles in the U.S.

In the early years of the twentieth century, bicycles were abandoned by the rich and famous. In the U.S., the end of the bicycle boom of the 1890s resulted in much-reduced sales and much lower prices. The American Bicycle Trust tried to cut production and thus maintain high prices, but they failed. With the collapse of the bicycle boom in 1903, bicycle innovation in gear-trains and everything else stopped for 30 years.

The innovative makers of quality bicycles, like Columbia and Overman, went out of business, and only low-priced bicycles survived.

### Bicycles in Europe

Cycling continued to be popular in Europe after the end of the 1890s bike boom. Standardized bicycle designs and mass-production methods lowered prices to the point where better-paid working people could afford a new bicycle, while many less well-paid workers could afford a second-hand bicycle. Millions cycled for transportation and recreation. Most bicycles had a single fixed gear. Although variable gearing devices were available in the early 1900s, they were rare.

In both Britain and France, development continued on practical and reliable variable gearing. Developments on either side of the Channel were noted by cyclists in both countries, particularly at the annual cycle shows—the Stanley Show in London and the Salon du Cycle in Paris. Distinct national preferences emerged: The British chose enclosed hub gears, while the French experimented with a wide range of devices.

## Variable Gears in Britain

At the start of the 20th century, there were five bicycle gear systems on the British market: the Collier bracket gear, the Hub and the Rambler two-speed hub gears, and the Gradient and Whippet derailleurs. These were the survivors of the myriad of patents that had been lodged in the 1890s, and the hundreds of prototypes that had been built, many of which were described in Chapter 3.

By 1901, The Hub 2-speed gear sold well and established an excellent reputation for smooth running and reliability. The critics who had described it as "clockwork" and distrusted its small size were proven wrong.

Also the Collier bracket gear had proven its reliability, and the users found it very satisfactory.

The G&J Rambler 2-speed gear was still available in 1901, but it was not mentioned in the cycling press thereafter.

### The Gradient Derailleur

Edmund Hodgkinson's Gradient seems to have worked well, although few were made. The following comments appeared in the *CTC Gazette* and *The Cyclist* in 1901 and 1902:

> In practice, the gear works very well, and changes beautifully. It has the great advantage that,

Left: Fig. 4.2. The 1908 Sears Roebuck Elgin King cost $14.95. With a coaster brake, it cost $3 more. Twenty-five years later, the standard American bike was almost unchanged from this.

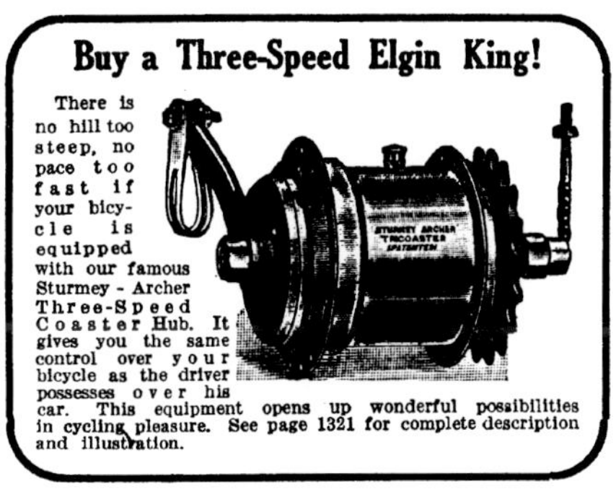


Above: Fig. 4.3. Six years later, in 1914, the Elgin King still cost $14.95, but for $6 more you could get it with this Sturmey-Archer 3-speed coaster-brake hub.

> save for the idle wheel, no more parts are at work with any one of the gearings than would be the case with the average (single-speed) machine.
>
> Hodgkinson's patent Gradient 3-speed gear has never been shown in detail before. Harry J. Swindley has been riding a bicycle fitted with a Gradient gear for some time, and he thinks highly of it.

On the Gradient gear, the smallest sprocket was on the inside, the largest sprocket in the middle, and a medium-sized sprocket on the outside. Despite the apparent complexity of the gear-change action, involving back-pedaling while coordinating movements of the slider that controlled the chain lifters and the lever that moved the sprockets sideways, the writer stated that "the movements are quickly and sweetly performed after a very little practice." Indeed, the writer had a considerable opinion of the Gradient.

> It will undoubtedly prove most useful on motor bicycles, as well as on pedal-propelled machines.
>
> We have had no practical experience with it, and it has been little pushed. Two riders of our acquaintance who have used the gear have spoken well of it.

Archibald Sharp, in his review of the 1901 Stanley Show, reported that the Gradient's shifting mechanism had been much improved. The chain lifters now engaged automatically, so the rider had only to move a single lever to shift gears.

The British penchant for neatness was revealed in the following letter:

> No one, not even the inventor, will claim that it improves the appearance of the bicycle. It does not. The chain-lifting and striking gear give a suggestion of rigging about the frame.

The Gradient was exhibited at the 1902 Stanley Show, but by October 1903 it had drawn no comment, although it was illustrated in a major feature on variable gears in *Cycling*. A 1903 review of variable gears in *The National Cyclists' Union Review and Official Record* did not mention the Gradient. It seems never to have been advertised to consumers, and the following year Hodgkinson sold the patent rights to Terrot et Compagnie in Dijon, France.

Left: Fig. 4.4. Edmund Hodgkinson's Gradient bicycle, as shown in the catalogue for the 1900 Stanley Show. This Gradient was fitted with a 44-tooth chainwheel and sprockets of 14, 18, and 24 teeth.

### The New Protean Derailleur

Charles Linley's New Protean was available only on Whippet bicycles. Apparently, it shifted well. A number of comments appeared in the *CTC Gazette*, *Cycling*, and *The Cyclist* from 1900 to 1902. Archibald Sharp reported that he had not used the New Protean. A reader said that it was a practical article. Riders praised the New Protean for its lack of friction. One reader, responding to another reader's comment, stated:

> Possibly, Mr. Walker has never had an opportunity of trying a 2-speed gear in which the friction of the bicycle is normal at both gears; that is to say, where no extra friction (such as that caused by internal cogwheels) exists. The extra friction in most of the wheels exists. The extra friction in most of the forms of 2-speed gears of the past was particularly undesirable. There are two points in 2-speed gears which I consider a sine qua non—first, no extra friction whatever at either gear; second, ability to change gear while pedaling without loss of time

> and consequent loss of momentum.
>
> The New Protean gear seems to me to leave nothing to be desired in a 2-speed gear. The friction is normal at both gears, there are no extra parts at work except the jockey (which is quite negligible), there is no limit to the range, the low gear is produced by throwing the chain into a larger sprocket on the hub (in the correct way it should be, and not by reducing the crank wheel), and, most important of all, the changes can be made while pedaling, and without the slightest loss of momentum.

Another reader, Chas. T. Cowley, reported covering "upward of 6,000 miles" (10,000 km) on his Whippet cycle with a New Protean Gear and had recently taken delivery of a new machine with the same gear. He had found "when riding with friends that the ability to change to the lower gear whilst ascending a hill gave me a great advantage, and enabled me to reach the top an easy winner." Cowley also praised the lack of friction compared with a Crypto epicyclic gear he had used with a tricycle, and added that,

> the ability to change to the higher gear whilst riding, so as to take advantage of favorable winds and level roads, is so great that after thirty years of cycling I find that I am now able to travel greater distances and at a faster rate than in my younger days.

### Derailleurs Fail in Britain

In late 1902, the Sturmey-Archer three-speed hub gear was introduced, and it soon became hugely popular. The cycling press became more critical of Britain's two derailleurs. *Cycling* and the *National Cyclists' Union Review* printed the following:

> The Protean is admittedly a bold device, but it has not taken the fancy of the public to a tithe of the extent which the Hub 2-speed gear has done, and, after all, an inventor should seek to please the public rather than to gratify the dictates of his own genius.

A writer with the pen name Spiral compared the gears on the market and ranked the Whippet gear (as the New Protean gear was now called) lowest for neatness and added: A dangling jockey chain has an insecure look about it. It looks as though it would easily slip off. Looks count. Riders never like to have anything uncommon about them, to attract the small boy's attention.

In his survey of the variable gears on the market, Walter Phillips commented on the "sweetness" of the shifting action of the Whippet and added:

> At first sight the gear looks clumsy and cumbersome; on further investigation it is only the shape of the chain which gives it this appearance, but it is necessary to adopt this form

Left: Fig. 4.5. The Whippet bicycle, equipped with the New Protean two-speed gear.

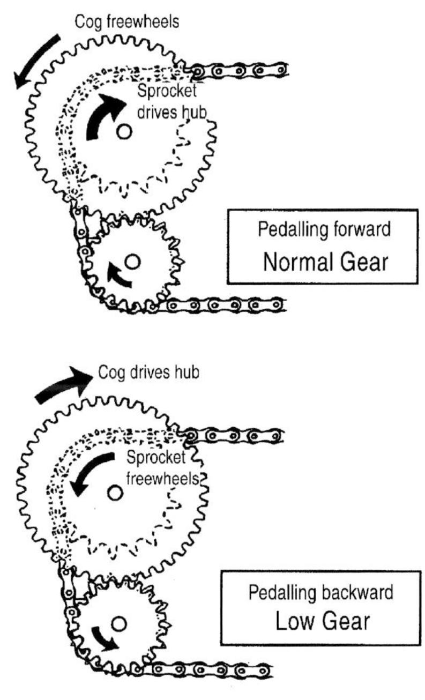


Right: Fig. 4.6 A and B. The 1900 Shepherd two-speed retro-direct gave a lower gear by pedaling backward. The chain went over an idler sprocket below the normal sprocket. A pair of meshing cogs was attached to the normal sprocket and the idler sprocket respectively.

> of chain to carry out what is required of it, viz. the shifting sideways on and off the teeth: it is in reality a toothed belt instead of an orthodox chain, and it should be considered as such.

Then, suddenly, there were no further reports of the Whippet or the Gradient. In reviews of variable gears published in *Cycling* in 1906, they did not rate a mention.

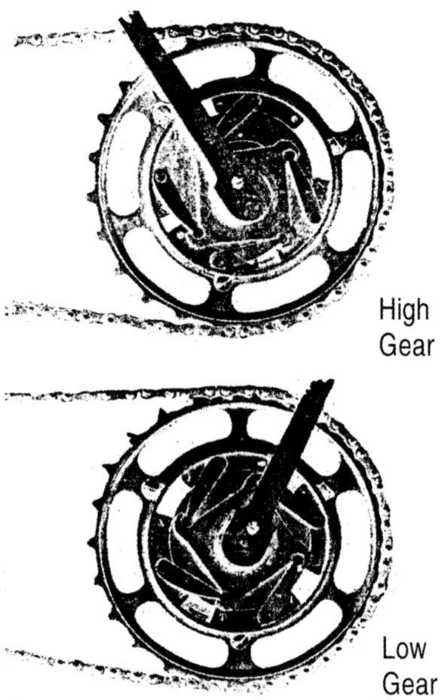


Above: Fig. 4.7 A and B. The 1903 Mitchell Any-Speed gear had a chainwheel that could be raised or lowered off-center. Four arms connected the crank axle to clutches on the chainwheel. As the chainwheel turned, the lower arms gripped their clutches and moved the chainwheel around, providing a higher gear.

So neither of the two pioneering British derailleurs survived in their home country. They both migrated to France: Terrot adopted the Gradient, while Vélocio obtained a Whippet gear and enthusiastically experimented with it. Indeed, he subsequently referred to all the early derailleur gears as "whippets." For the next quarter century, the British were happy with their hub gears and almost entirely forgot about derailleurs.

In 1907, Harry Hewitt Griffin described the Terrot copy of the Gradient as "a thoroughly French idea of altering the speed by chain changes." When F.T. Bidlake recalled the Whippet in an article in *Cycling* in 1913, he also got the details wrong.

### Unorthodox Variable Gears in Britain

Although derailleurs lost out, and hub gears took over, there was still much interest in other kinds of variable gearing in Britain, and a great number of workable gears were presented to the cycling public. Three alternatives to hub gears were presented at the 1900 and 1901 Stanley Shows:

- The Shepherd 2-speed gear was a retro-direct.
- The Gota cycle had a large cogwheel and an eccentric disc on the chainstay, with long rods connected to the cranks. By adjusting the rods the crank movement could be changed, thus altering the gearing.
- The Paradox expanding chainwheel was produced by W.D. Wansbrough, of Lincoln, and first marketed in 1903, after a giant model had been featured at the 1901 Stanley show. It was reported to be a considerable improvement over the 1894 Protean expanding chainwheel, both in terms of

Below: Fig. 4.9. The 1902 Garrard Two-Speed Gear, exhibited at the 1901 Stanley Show, was a neat epicyclic unit that could be screwed onto the rear hub of any bicycle. When engaged, it increased the gear by one-third.

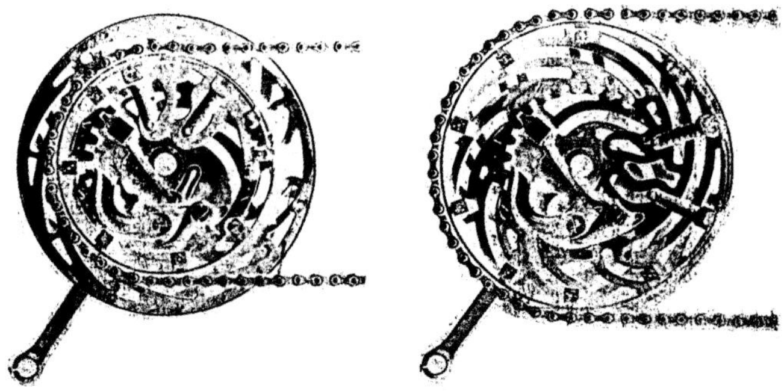

Left: Fig. 4.8 A and B. The 1903 Paradox chainwheel expanded in high gear (right) and contracted in low gear (left). It could be opened or closed in one-tooth increments from 20 to 26 teeth.

its range and its "perfect sweetness" of operation. However, it was short-lived, and production ceased in 1904.

### Archer's and Sturmey's Hub Gears

William Reilly, who had invented the highly successful Hub gear, continued to develop his ideas on variable gears. In 1902, he designed a three-speed hub. However, he was under contract with the Hub Two-Speed Gear Company, so he got a colleague, James Archer, to sign his patent applications.

In the same year, John James Henry Sturmey also patented a three-speed hub. Henry Sturmey was a science and mathematics teacher and a cycling journalist. Sturmey presented his hub during a famous talk on variable gearing, which he gave to the Cycle Engineers' Institute. Soon, Archer, Reilly, Sturmey, and others formed the Three-Speed Gear Syndicate.

Designing a three-speed hub was only half the story: someone had to manufacture it. Frank Bowden, chairman of the Raleigh Cycle Company, agreed to manufacture the hub for the new syndicate. The first Sturmey-Archer hubs were marketed in late 1902, and progressively modified to improve reliability. The remarkable enterprise called Sturmey-Archer continues to this day. However, the company is now owned by SunRace Roots, and the factory is now located in Taiwan. Tony Hadland published the story of Sturmey-Archer hubs in 1987.

## How Many Gears?

The appearance of the various multispeed gearing devices on the market prompted a great deal of discussion about "ideal" gearing. Should cyclists have one, two, three, or more gears? The common single-speed bicycle was cheaper, simpler, and lighter than a variable-speed model. Its advocates said it was just a matter of choosing the "ideal" gear size, which could be pedaled nimbly on the level, and pushed steadily uphill. On really steep hills, of course, you could get off and walk.

A popular rule-of-thumb held that the ideal gear size was "ten times the crank length." With the usual 6½-inch (165 mm) cranks of the period, this meant a gear size of 10 x 6½ = 65 in. Racing cyclists chose a slightly higher gear, typically 69 or 72 in.

To convince cyclists to use his 3-speed Gradient gear, Edmund Hodgkinson's brochure for the 1900 Stanley Show in London reasoned that life was divided into three categories of experience—the good, the bad, and the indifferent. A 3-speed gear gives a suitable gear for either good, bad, or indifferent circumstances, and no others exist.

On the other hand, manufacturers of 2-speed gears asserted that two gears were all you ever needed, for either "Hard Work" or "Easy Work." Moreover, the brochure for the Villiers two-speed hub gear of 1906 claimed that the bottom gear on a 3-speed was "so monotonous that it gets on one's nerves," and that such a low gear was "practically useless," so that many riders preferred to walk up hills.

Right: Fig. 4.10. This 1907 Golden Sunbeam with three-speed hub gear was the quintessential English roadster.

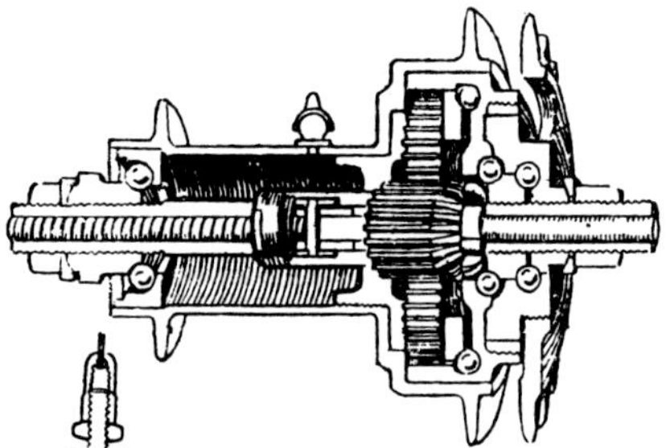
The Crabbe two-speed

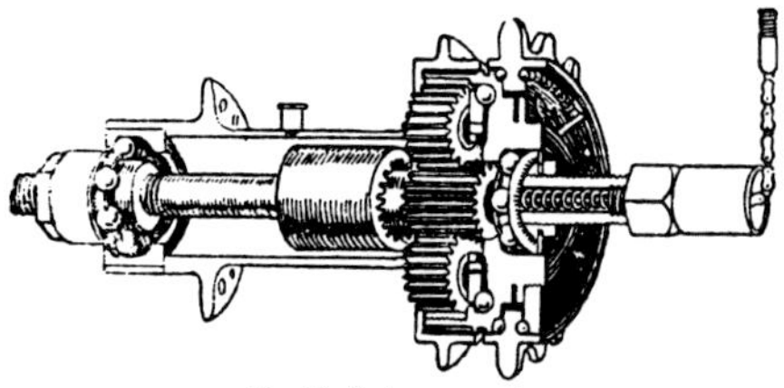
The Eadie two-speed

The Centaur two-speed

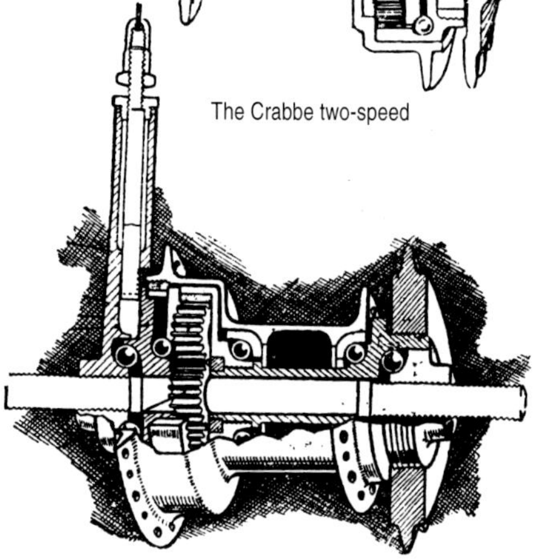
The James two-speed

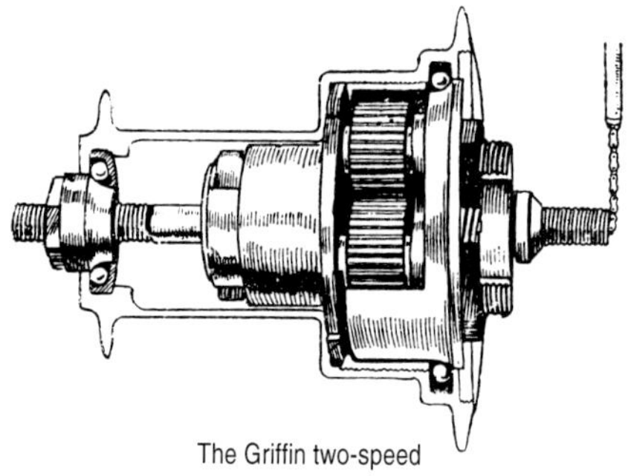
The Griffin two-speed

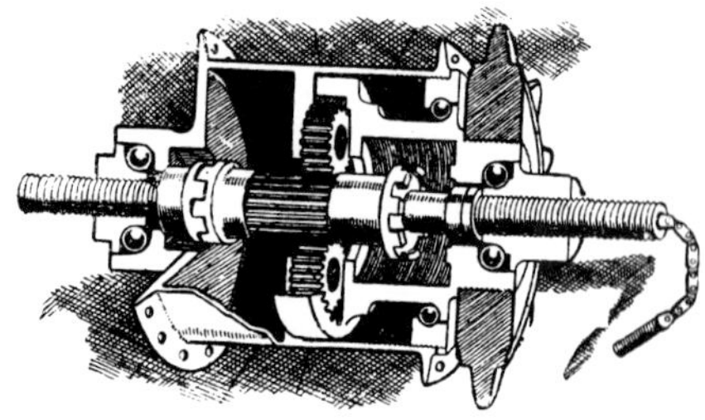
The Simplex two-speed

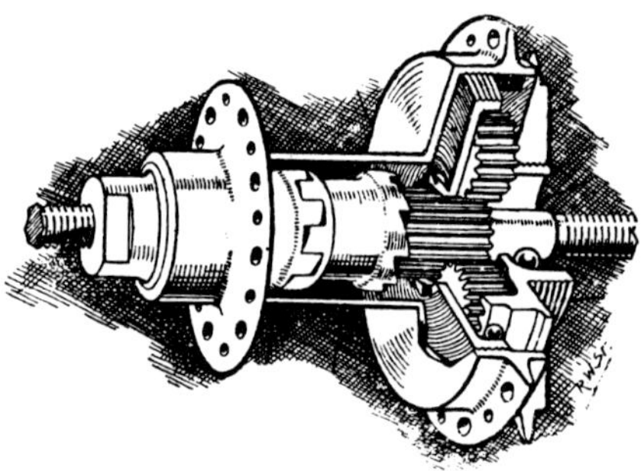
The Stanley two-speed

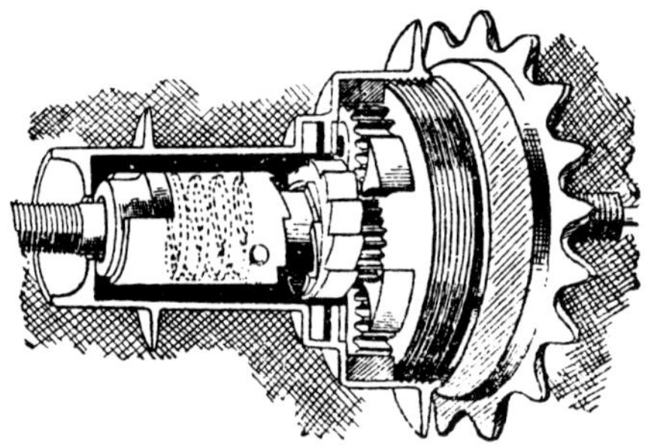
The Villiers two-speed

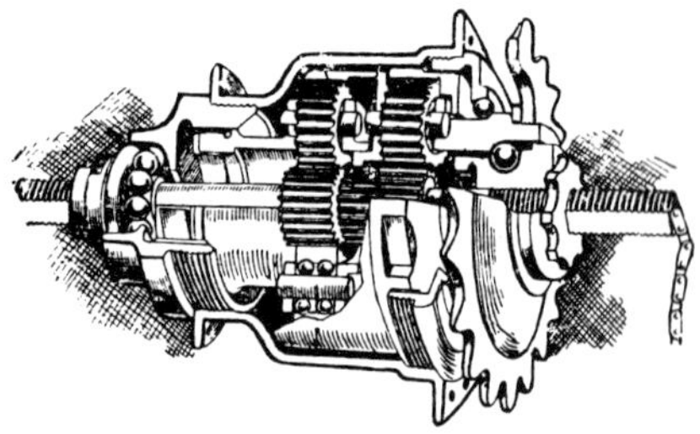
The Sturmey-Archer three-speed

The James two-speed bracket gear

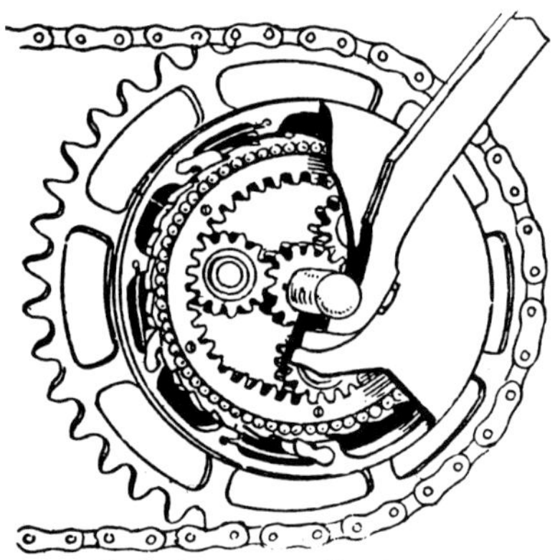
The Sunbeam two-speed bracket gear

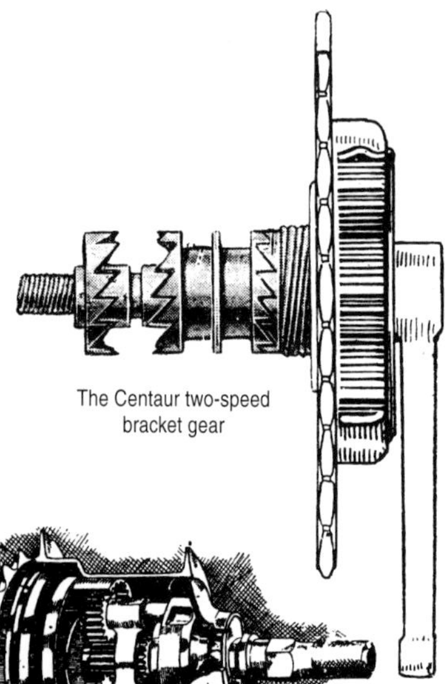
The Centaur two-speed bracket gear

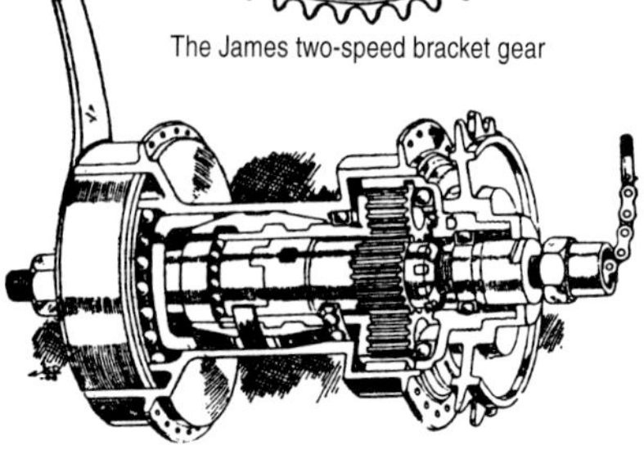
The Eadie two-speed coaster

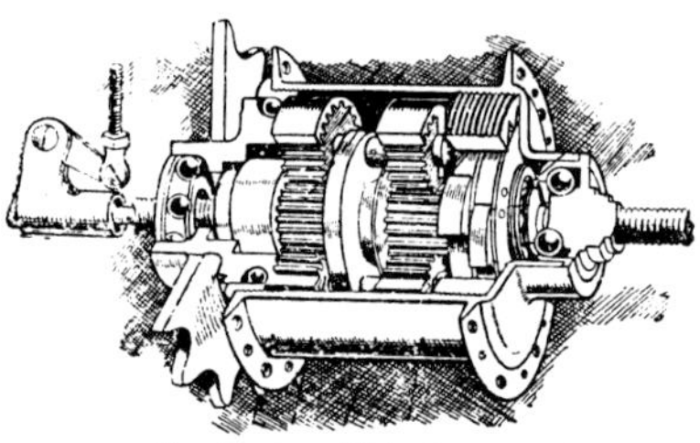
The Armstrong Triplex three-speed

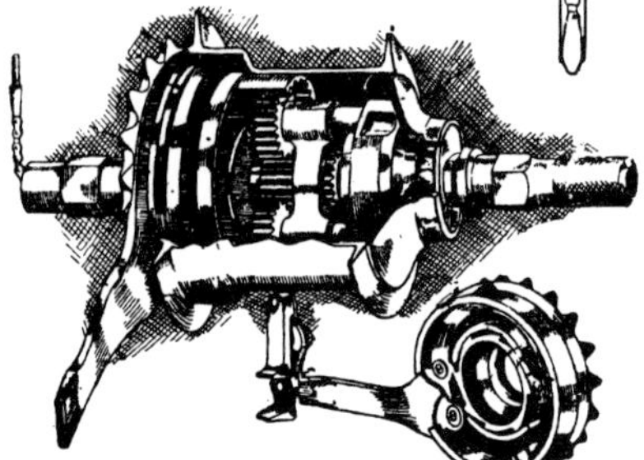
The Crabbe two-speed coaster

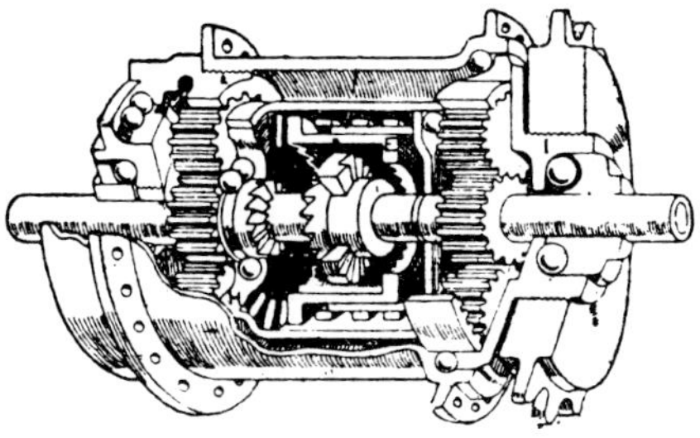
The Seabrook three-speed

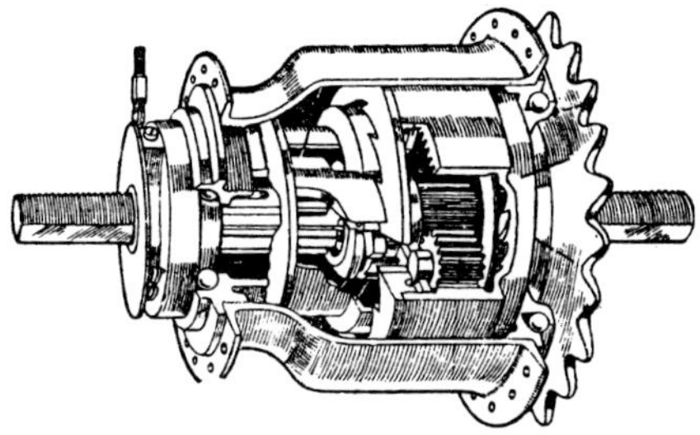
The Crabbe-Simplex three-speed

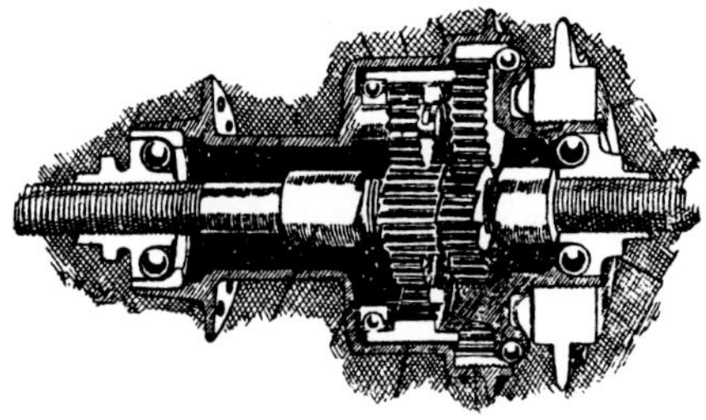
The Humber-Cordner three-speed

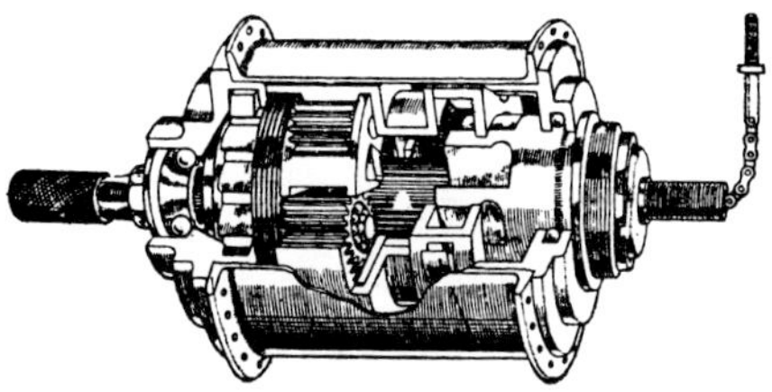
The B.S.A. three-speed

The Stanley three-speed

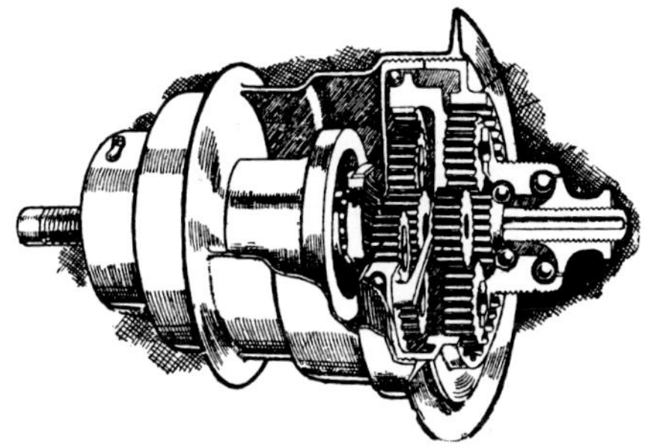
The Sunbeam three-speed

The Pedersen three-speed

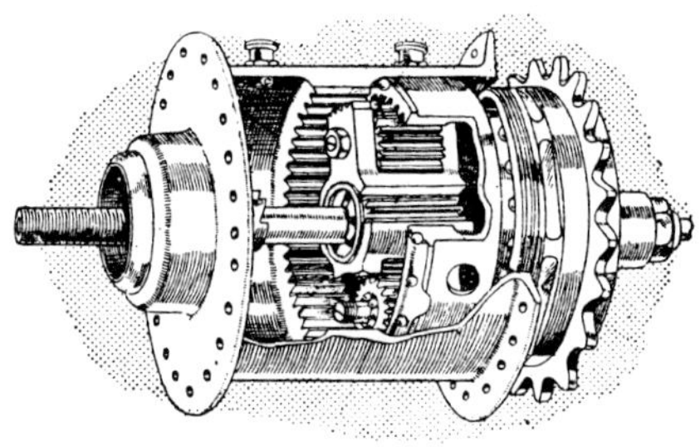
The Micrometer three-speed

The Optimus three-speed

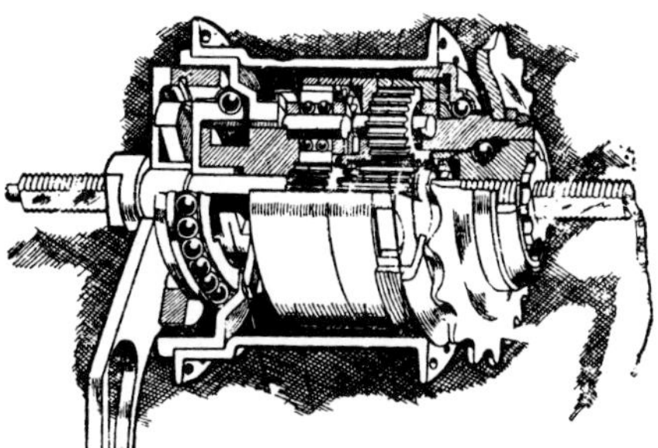
The Sturmey-Archer Tricoaster

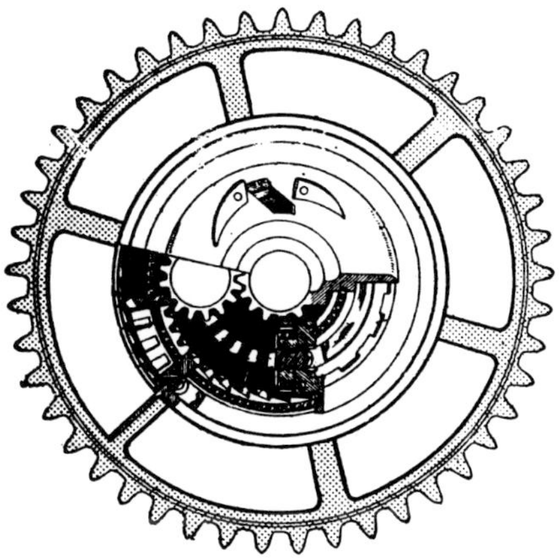
The James three-speed bracket gear

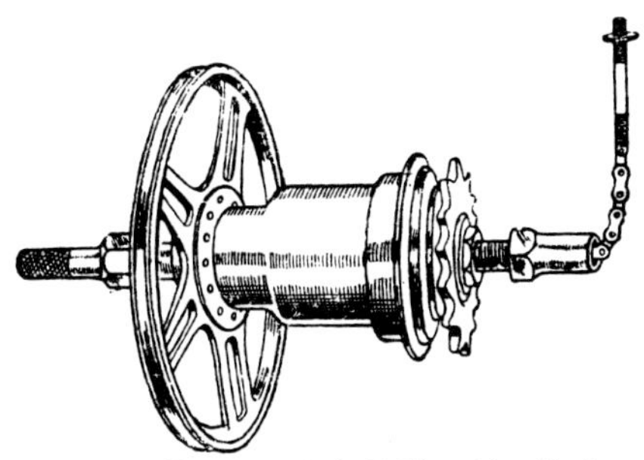
Triumph (Sturmey-Archer) with Triumph band brake

The All-Speed gear

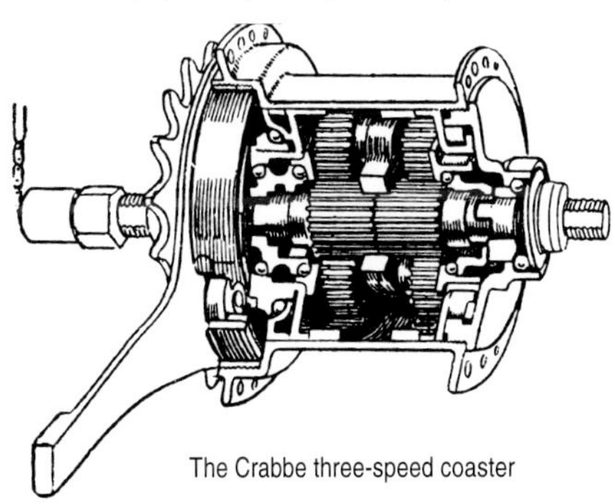
The Crabbe three-speed coaster

Above and facing page: Fig. 4.11. The range of epicyclic bicycle gears produced in Britain in the first decade of the 20th Century.

Hybrid gearing was soon proposed. For example, the 2-speed Garrard hub could be combined with the 7-speed Paradox chainwheel to give 14 speeds from 62 to 101 in. By adding a Sturmey-Archer hub, 42 speeds were possible! This was regarded as a laughable fancy at the time, but today, a century later, 30-speed gearing has become common.

Many people consider that the Marston Golden Sunbeam was the finest mass-produced British bicycle. In 1903, Sunbeam produced a 2-speed bottom-bracket gear, lubricated by oil in the chain case. In 1907, they made a three-speed hub gear. In 1909, they combined the two epicyclics into the Six-Speed, with gears of 49, 66, 72, 88, 96, and 129 in. Other sets of chainwheels and sprockets (hopefully with lower gears) could be ordered.

The brief appearance of the Dieterich gear had raised the possibility of a continuous range of gears, rather than just a finite number. Other infinitely-variable gears appeared—first the Mitchell Any-Speed bracket gear, and later the All-Speed hub.

The Dieterich was also automatic, and magazine articles thought that this was a desirable feature. Over the years, other automatic gears have been manufactured; the Deal Drive and the F&S Duomatic in the 1980s, the Browning electronic changer in the 1990s, and the Shimano Nexus today. Perhaps this attractive concept will finally become a practical reality.

Above: Fig. 4.12. The 1908 Marston Sunbeam 2-speed bottom-bracket gear with an oil-bath chain case

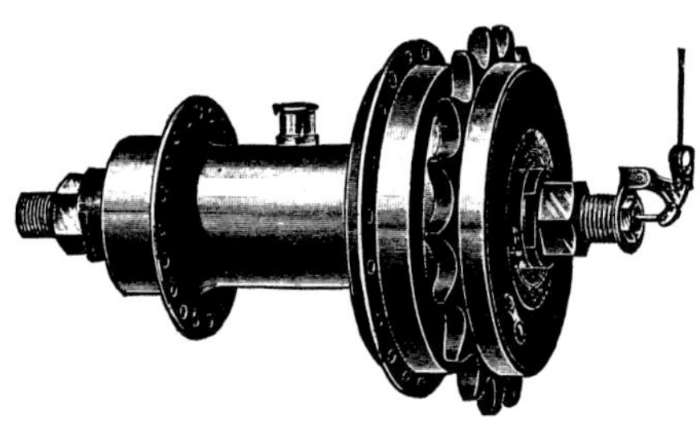

Right: Fig. 4.13. In 1906, John Fagan, of Dublin, invented this beautifully simple Eadie two-speed hub.

### Hub Gears Proliferate

Following the instant success of the Sturmey-Archer three-speed hub, many other British manufacturers marketed multi-gear hubs.

In 1904, the Dursley-Pedersen two- and three-speed hubs were produced. The same year, there were three new two-speed epicyclic bracket gears—the Sunbeam, the Hobart, and the Centaur. Variable Gears Ltd. made a chainwheel which could be continuously expanded over a range. The Peart three-speed gear was screwed onto an existing hub in a manner similar to the Garrard two-speed.

In 1906, the Eadie, Villiers, Micrometer, BSA, and New Departure Coaster two-speed hubs appeared.

By 1909, there were at least thirty variable gear systems, mostly epicyclics, on the British market.

However, by 1914, Sturmey-Archer had prevailed in the battle with its competitors. The 1910 Sturmey-Archer hub sold for less than one third of the 1903 price. In 1913, Sturmey-Archer 3-speed gear production reached 100,000 units. By the end of the First World War, few other hub gear makers were still in business. The demobbed British soldier in search of a new 3-speed was faced with a simple choice—Sturmey-Archer or BSA—and the BSA hub was made under a Sturmey-Archer license.

### The Amazing Vernon Blake

One of the most influential people on the British cycling scene was Vernon Blake. He was astonishingly talented, not only as a racing and touring cyclist, but also as a scientist, painter, sculptor, art expert, writer, engineer, inventor, linguist, traveler, and mountaineer.

Blake was born in 1875 and educated at King's College and University College, London, from which he graduated with degrees in chemistry and physics. He was a prominent member of the North Road Cycling Club while still a teenager. In 1894, he won the Anfield Cycling Club's annual 100-mile race. The same year, he paced F.T. Bidlake in a 24-hour tricycle race. Bidlake, who was 47 years old by this time, wanted to give up, but Blake helped him win second place, with 331 miles (533 km).

Around the same period, Blake rode from London to Constantinople (now Istanbul) and back, a round trip of about 3,000 miles (4,800 km). Blake was a skillful mountaineer and the first person to climb several peaks in Britain and in the French Alps. When only 21, he fell while climbing near Mont Blanc, and his ice axe pierced his lung. While convalescing in the south of France, Blake took up painting and was accepted into the prestigious Slade School of Art in London.

During the same period, he cycled across Europe and the Middle East. Blake had an amazing talent for picking up languages, and was reputed to be fluent in 14 different languages, ranging from French to Arabic. Disguised as an Arab, he traveled up the Nile to Khartoum before Lord Kitchener's British forces captured it in 1898. Blake studied painting with Matisse in Paris, where he met and married Marie Bonnin. They went to Florence, Italy, where Blake made a living as a painter and sculptor. In 1907, he became Director of the British Academy in Rome. In 1910 they returned to Paris.

Above: Fig. 4.14. Vernon Blake in 1919 on a bike with floating chain.

About this time, Blake met Vélocio (see below and Chapter 5) and accompanied him on several marathon rides, which he reported as "our French correspondent" for *Cycling*.

He preferred bicycles with a floating chain (described elsewhere in this chapter). He made several modifications to this design, and continued to use it for the rest of his life. Most other touring cyclists, including Vélocio, switched over to the improved derailleurs as they became available.

In the 1920s, Blake lived in the village of Les Baux in the south of France with his wife and children. There, he toured the rough unpaved mountain roads on a fat-tired bicycle remarkably similar to a modern mountain bike. His bike even had cantilever brakes of his own design.

Blake was commissioned to create several magnificent war memorials for villages throughout southern France. The work was exhausting. He lost the sight of one eye when a marble chip flew up. He suffered from recurrent fevers, contracted 30 years earlier, and from heart disease, and he died in 1930 at the age of 54. His old friend Bidlake wrote Blake's obituary for *Cycling*.

## Exotic Gear Systems

A number of different, and since-forgotten, methods of gearing were introduced during this period. The most interesting of these are covered in the following sections.

### The Delbruck System

Using his chainwheel and hub clutches, Vélocio (the pen name of the French writer and inventor Paul de Vivie, whose work will be covered in detail in Chapter 5) improved a gear-change system designed by a Belgian touring cy-

Above: Fig. 4.15. Vernon Blake on his 1930 "bad roads" bicycle, complete with floating chain, fat tires, and Blake's own cantilever brakes. Some consider it to have been the first mountain bike.

clist named Delbruck in 1900. It was a variant of the bichain principle and allowed the rider to select any of four gears without getting off the bike. Though complicated to explain, the Delbruck system was as easy to use as today's front and rear derailleurs.

The Delbruck system was probably the most ingenious combination of chainwheels, freewheels, and clutches ever produced. It was installed on the right-hand side of the bicycle. The two chainwheels were side-by-side but independent. The inner 20-tooth chainwheel was mounted on a freewheel working counterclockwise. The outer 26-tooth chainwheel was equipped with Vélocio's foot clutch and could be disconnected from the cranks and bottom-bracket axle, allowing it to revolve freely on a roller-bearing.

Just behind the chainwheels, there was an intermediate jackshaft carrying four fixed sprockets, inner to outer: 20-, 23-, 28-, and 13-toothed. The 20-tooth sprocket was permanently connected to the inner 20-tooth chainwheel by means of a chain. The 13-tooth sprocket was permanently connected to the outer 26-tooth chainwheel by means of another chain.

There were two freewheels, independent of each other, on the right-hand side of the hub. The inner 20-tooth freewheel was permanently connected by chain to the 23-tooth fixed sprocket of the intermediate spindle. The outer 16-tooth freewheel had a cable-operated clutch with a handlebar control to disengage it from the rear hub. It was permanently connected by chain to the 28-tooth fixed sprocket on the intermediate spindle.

With the normal wheel size of the time (about 28 in.), this system provided four gears of approximately 100, 65, 48, and 31 in. respectively—without having to get off the bike. The system was operated by means of handlebar-mounted or foot-operated controls. This represented significant progress compared to previous mechanisms. Here is the description of its operation:

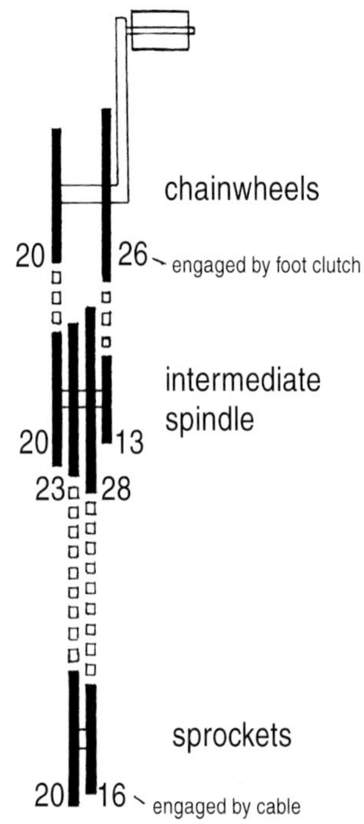


Above: Fig. 4.16. The arrangement of chainwheels, sprockets, and chains on the Delbruck System.

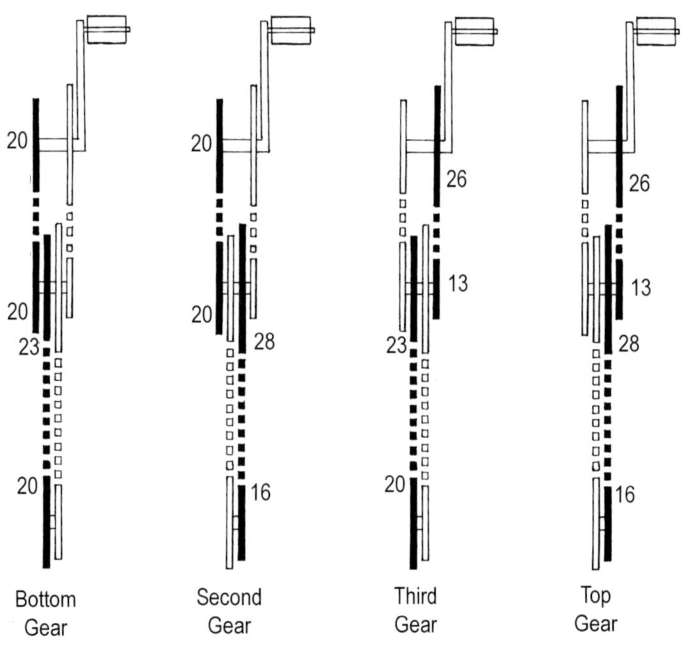


Left: Fig. 4.17. The Delbruck system going through the gears.

- Top gear. Outer chainwheel and outer freewheel connected. Drive from the 26-tooth chainwheel, through the 13- and 28-tooth intermediate sprockets to the outer 16-tooth freewheel. The inner chainwheel and inner freewheel idling.

- Third gear. Outer chainwheel connected, but outer rear freewheel disengaged. Drive from the 26-tooth chainwheel, through the 13- and 23-tooth intermediate sprockets to the inner 20-tooth freewheel. The inner chainwheel and outer freewheel idling.

- Second gear. Outer chainwheel disengaged and idling, outer 16-tooth freewheel connected once more. Drive from the inner 20-tooth chainwheel, through the intermediate 21- and 28-tooth sprockets, to the 16-tooth freewheel. Inner freewheel idling.

- Bottom gear. Outer chainwheel and outer freewheel both disengaged. Drive from the 20-tooth chainwheel through the intermediate 21- and 23-tooth sprockets to the 20-tooth freewheel. The 16-tooth freewheel disengaged.

Vélocio rode his Delbruck-equipped *polycyclette* bicycle long distances over muddy roads in the fall of 1903. He recommended it to his readers in 1904 and had great hopes for its commercial future. Unfortunately, the retro-direct system appeared just then, and it priced this polycyclette off the market. Vélocio was still using his personal polycyclette in 1926, and it is now in the St.-Étienne Museum of Art and Industry.

## Retro-Direct Gears

Retro-direct gears take us back to the 1869 Barberon & Meunier patent for their "double-speed velocipede." Thirty years later, when this patent had long been forgotten, an unknown bicycle tourist by the name of De Viviés redis-

Right: Fig. 4.20. The 1900 Delbruck System used two normal-length chains and two shorter ones, two chainwheels, two freewheels, and four fixed sprockets on an intermediate spindle. In this photo, the two long chains have been removed in order to more clearly show how it works.

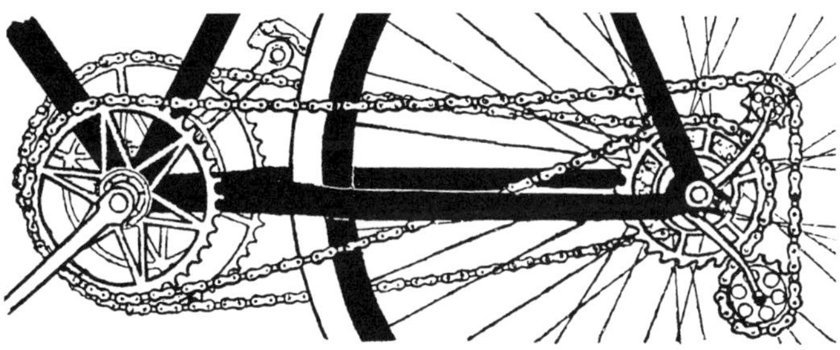


Right: Fig. 4.18. The 1900 Magnat & Debon retro-direct, showing how the left-side drive reversed the chain.

Above: Fig. 4.19. Following Magnat & Debon's lead, Hirondelle Cycles, of St-Étienne, made a retro-direct that was an exact copy of Barberon & Meunier's second (sprocket-and-chain) system of 1869, the patent for which had expired by then.

Below: Fig. 4.21. Magnat & Debon also produced this 3-speed, with one gear pedaling forward and two gears pedaling backward.

covered the principle. In 1899, he devised a mechanism with a chain on either side of the frame.

In this system, there is a second chainwheel-and-freewheel-drive on the left side of the bike (in addition to the usual drivetrain on the right side). When pedaled backward, the left side drive went from the bottom of the chainwheel to the top of the sprocket. An idling sprocket reversed the chain, which then ran back to the top of the chainwheel. When pedaled forward, the left side freewheeled. When pedaled backward, the retro-direct action drove the bicycle and the right side freewheeled. You passed from one drive to the other instantly and automatically.

Captain Perrache, an eminent member of the Touring Club de France and well informed on the technical aspects of cycling, was immediately won over. In a 1900 report, he tried to prove that pedaling backward was more efficient than pedaling forward. All his life, Perrache remained a fierce supporter of retro-direct. He advised the company of Magnat & Debon, in Grenoble, who produced the first commercial retro-direct of this period. It was a bichain with two reversing pulleys (instead of one as in De Viviés' system).

Other firms came up with their own mechanisms. Some reversed the movement by means of hub and bracket gears. Others used one or several chains. Still others used intermediate shafts. Had they survived, you could fill a museum with them!

The most popular arrangement had an idler sprocket below the chainstay. The single chain went around one drive sprocket, over the idler, then back over the other drive sprocket.

Vélocio had already devised this system in 1902, and some of his readers had homemade versions of it. In 1903, to Vélocio's great annoyance, Hirondelle patented the idea ahead of him. It won a gold medal at the 1905 Touring Club de France trials (more about these trials in Chapter 5). This system was probably the best known of all. It was reliable and straightforward in design, and was included in the Hirondelle catalog until 1939.

Vélocio tried out the retro-direct enthusiastically. But after 4,000 km (2,500 miles) of road testing, spread over four seasons, he concluded that the advantages were not as great as were claimed. In fact, simplicity was the retro-direct's only real benefit. There were courteous but unyielding exchanges in the reader's columns of *Le Cycliste* between advocates of the retro-direct and of the purely direct drive. Tired of Captain Perrache's incisive arguments, one fine day in 1903, Vélocio went off and climbed the rocky heights of Mont Ventoux—

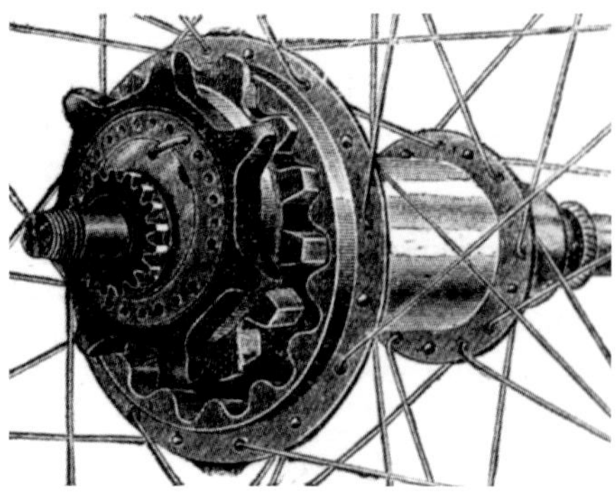

Above: Fig. 4.24. The 1901 Moyeu BiGear of 1901 had 2 or 3 sprockets on an eccentric cam. Rotating the cam brought one or the other sprocket in contact with the chain.

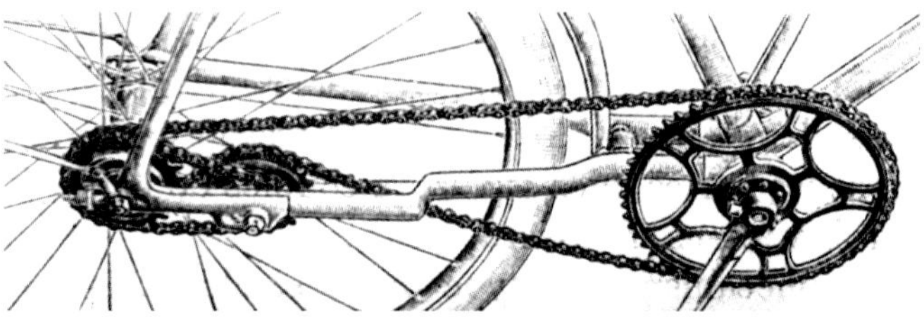

Above: Fig. 4.22. This Terrot retro-direct was a simplified version of their double-chain model.

Right: Fig. 4.23. The 1905 Pernod expandable chainwheel had four speeds over a two-to-one gear range.

6,300 ft. (1,909 m) without stopping—on a direct-drive bicycle!

### Expandable Chainwheels

A chainwheel that can be made larger or smaller while riding is another way of varying the gear ratio. In Britain, the Protean expandable chainwheel was produced by the Whippet Cycle Syndicate in 1894, and the Paradox expandable chainwheel was produced in 1902.

In 1905, Pernod, a bicycle manufacturer in St-Chamond, built a chainwheel which expanded from 14, to 18, to 22, and finally to 28 teeth (one-inch pitch, i.e. equivalent to 28, 36, 44, and 56 teeth on a modern bicycle with ½-in. pitch chain). Shifting required a fine touch that the cyclists of the time had apparently not yet acquired, and the system often broke down. Pernod improved it later, and the improved model was installed on an Automoto bicycle entered in the 1922 T.C.F. trials run on very rough roads (see Chapter 5). The bicycle completed the test without penalty, proving that the chainwheel had worked perfectly.

Expandable chainwheels returned half a century later with the Hagan Drive in the 1970s, and the Cambiogear and the automatic Deal Drive in the 1980s.

### French Hub and Bracket Gears

In the 1900s, French epicyclic gears were being developed that fit inside either the bottom bracket or the hub. At the time of Barberon & Meunier, Monsieur Hazard had already patented a 4-speed hub gear for boneshakers. However, the development of these "internal" gears occurred mainly in Britain. By 1909, there was an amazing variety of 2- and 3-speed models available. Combined with double chainwheels, they provided 4 or 6 speeds.

Many of these gears were imported into France. Fagan, Terrot, and Peugeot each produced two-speed hubs in France, while Hirondelle made a three-speed hub.

Retro-direct and hub gears were sometimes combined, such as on the Peugeot bicycle and the Hirondelle 6-speed presented at the Paris Cycle Show in 1904.

In 1899, Variand made a 2-speed bottom-bracket gear. There were also the 3-speed Brossard and the 2-speed Lancelot with a secondary shaft. The Lecarme Frères & Michael 2-speed mechanism was entirely enclosed in a case incorporated in the chainwheel, and could thus be installed on any frame. The Durieu company combined a bottom-bracket gear and a hub gear to give four speeds.

Most of these systems were made before 1905. They were short-lived because they usually required a special frame. Further, friction loss increased with wear and with the power applied. Precision machining and high-quality steel were essential. However,

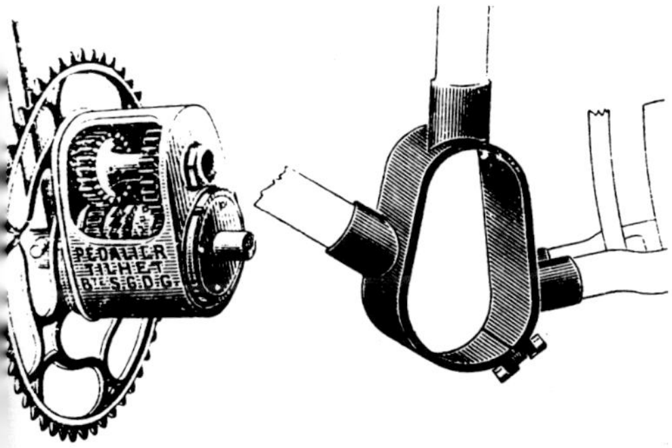

Above: Fig. 4.25. The Tilhet 2-speed bottom-bracket gear gave a low gear that was half the size of the high gear.

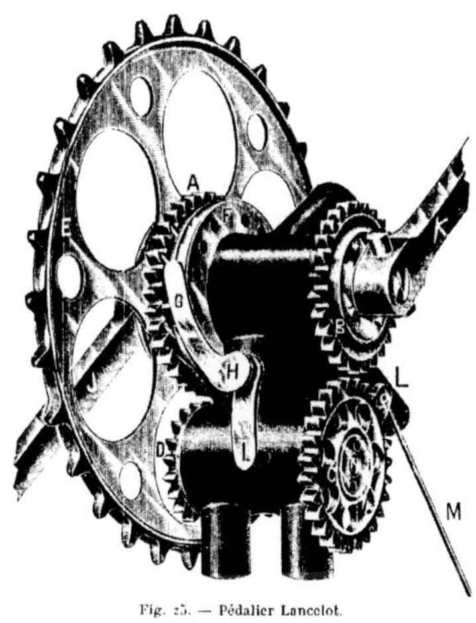

Above: Fig. 4.26. The 1899 Lancelot 2-speed bottom-bracket gear was cable-actuated.

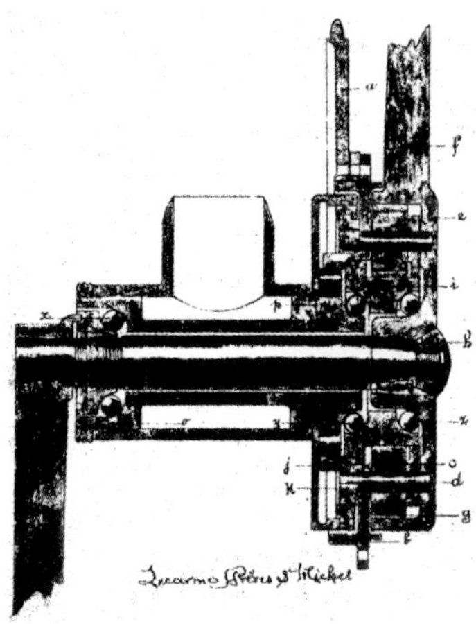

Right: Fig. 4.27. The 1899 Lecarme Frères & Michael was a 2-speed epicyclic bottom-bracket gear.

the Magnat-Debon company long enjoyed a good reputation for its 3-speed bottom-bracket gear.

### Chainless Bicycles (Acatènes)

The chainless bicycle, or *acatène,* was driven by a shaft that fit neatly inside the right-hand chainstay. In the 1890s, chainless drives were popular in France and the U.S. People liked the idea because it did away with the usually dirty, greasy chain. L'Acatène was the trademarked name of La Metropole, the company that contributed most to the growth of this system from 1894 onward. Durieu, a manufacturer near Paris, made a 2-speed acatène that won a silver medal at the TCF trials of 1902.

Vélocio bought a chainless Touricyclette in 1905 and was still using it in 1924, especially on days when he thought the roads would be too muddy for a chain-driven bicycle. Raymond Henry disassembled it to restore it. The enclosed mechanism was still in perfect condition, still lubricated, and practically without signs of wear! It had a large bevel gear cog instead of a chainwheel. The drive was through a shaft inside the chainstay (except that there was no chain) with a small bevel-gear at either end. The front cog remained engaged. A tiny change mechanism engaged the cog at the rear on one of two moveable cogs on the hub. To change gears, you back-pedaled a quarter of a turn. Precision workmanship was essential, which made

Above: Fig. 4.28. Vélocio and his 1905 chainless Touricyclette. This bicycle is now in the Museum of Art and Industry at St.-Étienne.

Above: Fig. 4.30. Gearing of the 1903 2-speed Columbia shaft-drive bicycle. High gear was 92 in., and low gear was 68 in.

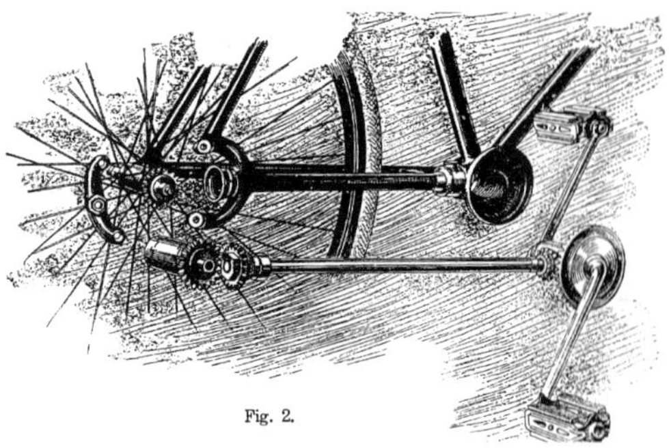


Right: Fig. 4.31. Exploded view of the 1903 Columbia shaft-drive, showing how the rear wheel was removed.

Above: Fig. 4.29. The 1895 Acatène shaft-drive. The illustration gives an idea of the precise machining required—and the difficulty of removing the rear wheel.

Below: Fig. 4.32. The elegant 1903 Columbia 2-speed shaft-drive bicycle had neither cables nor a chain to clutter its appearance. When the rider backpedaled a quarter turn, the other set of cogs clicked into place. The 2-speed model included front suspension and a coaster brake. It cost $100, versus $75 for the single-speed model.

the Touricyclette an excellent but expensive machine. The shaft-drive had advantages over the early chains, but it made it hard to mend or replace the rear tire, and it was more expensive.

In the U.S., Columbia made their first single-speed shaft-drive in 1898. The 2-speed model appeared in 1903.

### Lever-Driven Cycles

The drive system on lever-driven cycles, or, *lévocyclettes*, was quite different from that on other bikes. Instead of turning around in circles, the cranks rose and fell alternately, above and below the horizontal. A lever that moved forward and backward was fixed at right angles to each crank, more or less vertically. Each lever pulled a chain acting on a rear-wheel sprocket.

Probably the first lever cycle was the 1886 American Star. In the days of the safety bicycle, this concept was revived by Svea in Sweden, but first marketed in Britain under the name Alert. It had a gear range from 12 to 117 in. The first French *lévocyclettes* appeared in the last years of the century without much success. The Télécycle had gears ranging from practically nil up to about 88 in., but you had to get off the bike to shift.

From 1904 onward, Vélocio tried a Svea, covering some 2,500 to 3,000 km (1,500 to 1,800 miles) per season for some years. The Svea was improved in France by the Terrot company in Dijon, who were always interested in variable gears.

In 1905, Terrot produced their own lever cycle. The levers had 10 notches, each giving a different gear. Drive was through sliders linked to small lengths of chain acting on a freewheel sprocket. A second transverse chain raised one pedal when the other was pushed downward.

The lever cycle was comfortable to ride, but there were high and low "dead spots" when the movement of the levers reversed. Consequently, it was unsuitable for racing. It stayed in the Terrot catalog for many years, selling steadily, though in small numbers. Terrot even made lever-drive tandems.

Vélocio experimented unsuccessfully with several lever cycles of his own design. Some of them survive in the Museum of Art and Industry in St.-Étienne. In the 1980s, the Alenax lever cycle was marketed in the U.S. to an unappreciative market.

## Conclusions

We have looked at the multitude of gear-change mechanisms that were available in Britain and France in the first decade of the 20th century. The epicyclic hub gear survives to this day. Chapter Five will cover the first derailleurs.

None of the ingenious alternatives to hub gears or derailleurs found a significant market. They all disappeared because they were either impractical, unreliable, complicated, or expensive.

Left and right: Figs. 4.33 and 4.34. The 1905 Terrot Lévocyclette had 10 notches on the levers to give 10 different gears, selected by a rotating twistgrip on the handlebars. This may have been the first "10-speed," as well as the first twistgrip.

# 5 The First Derailleurs: 1908–1919

THE PRECEDING chapter covered hub gears and alternative bicycle gear-trains in the period from 1900 to 1907. This chapter covers the development of the first practical derailleurs. Most of these developments took place in France in the brief period from 1908 to the beginning of World War I. To avoid a lot of jumping back and forth, I have not followed a rigid chronological separation.

By 1914, the national preferences were firmly established, though the majority of cyclists the world over continued to ride single-speed machines. British cyclists would use single-speeds or three-speed hub gears. French cyclists would use single-speeds or derailleurs.

Left: Fig. 5.1. An undated photograph of Vélocio with one of his La Gauloise double-chain bicycles. The handlebars on this machine could be reversed to provide two different riding positions.

## Vélocio

Paul de Vivie, who became known by his pen name Vélocio, was the most significant force behind the development of derailleur gearing, both during this period and until his death in

1930. He and his followers established the superiority of derailleurs in the years immediately after 1908.

Vélocio's real name was Paul de Vivie. He was born in 1853 to impoverished members of a minor noble family. They ran a stagecoach post and inn at Pernes, a large village near Avignon. After his mother died, his father remarried and worked on his new wife's farm in Meyzieu, near Lyon.

Paul de Vivie's secondary education was in a private school at Lachassagne, near Lyon. In the 1870s, he set himself up as a commercial agent in the ribbon-making and weaving industry in St.-Étienne. He had a taste for outdoor sport and was a co-founder of a local cycling club. He began touring on a high bicycle around 1881, but switched over to the more secure tricycle. He adopted the safety bicycle soon after it appeared.

His trade took him to England, in particular to Coventry, where he noticed how the cycle industry was taking over the then failing weaving mills and was forging ahead at a rate far greater than in France. Back in St.-Étienne, he diversified into the cycle trade.

In 1886, he launched the Agence Générale Vélocipédique (General Bicycle Agency) and began importing the high-quality British cycles then so prized by French cyclists. The following year, he started publishing *Le Cycliste Forezien*, a magazine that catered to local cyclists and advertised his cycling goods. This publication was so successful beyond the region around St.-Étienne, that in 1888 its title was shortened to *Le Cycliste.*

*Le Cycliste* was essentially a one-man effort, with other articles and letters sent in by readers. He could write good French verse, and his readers appreciated the style of his articles. De Vivie signed his articles "Vélocio," a tag that stuck to him forever. His articles played a leading part in the rapid growth of the St.-Étienne cycle industry. In 1892, he launched his own La Gauloise bicycle brand. However, competition was already fierce in a small market. Vélocio was restricted to assembling custom bicycles for the discerning few in his very modest premises.

Not that he cared much. It provided him with a living. His real interests were testing and improving bicycle components, mainly gear-change mechanisms, long-distance cycling, vegetarianism, and writing about all these topics in his magazine. He coined the word *cyclotouriste* in 1889. When he was well over fifty, Vélocio completed 40-hour, 400 km (250-mile) rides and climbed high mountains. His yearly distance totals often topped 20,000 km (12,000 miles), and his exploits left people gasping. He soon became the figurehead of French bicycle touring, with many people considering themselves his disciples.

Vélocio was a picturesque figure, short and spare of stature, always dressed in home-made coarse woolen clothes: a jacket buttoned up to his chin, knickerbockers and thick woolen stockings, monks sandals no matter the season or the weather, and a wide-bottomed beret as headgear. Out on the road on summer days, he often doffed jacket and shirt. He was nearsighted and sported a pair of pince-nez with a safety ribbon. With a wide-flowing moustache and a balding head, he was easily recognized from afar. Though highly interested in bicycle technology, he was also a cultured man with a wide range of conversation. He read Latin, and was reading Seneca's *Letters to Lucilius* just before his death in 1930.

From 1888 onward, Vélocio tracked every improvement in bi-

Right: Fig. 5.2. Vélocio's bi-chain. By pressing a lever with his foot, he could switch the drive from one chain to the other. He could also shift the chain across chainwheels by hand, thus giving four useable gears. Note the quick-release link (bottom, just to the right of the tire) to adjust the chain length.

cycle technology. As soon as a new product appeared, he described it. He tested it in detail if he thought it was worth the trouble. He reported the tests in *Le Cycliste* and eagerly discussed the pros and cons of the bikes and their gearing systems with his readers.

He reviewed the catalogs and described the bikes exhibited at the Paris Cycle Show and the Stanley Show in London. He tried to improve numerous accessories, and then tested them himself, regarding on-the-road experience as more important than theoretical factors. His fundamental and all-pervading interest was in multispeed gearing.

Each spring and fall, Vélocio used to organize meetings in villages chosen for their peace and quiet. They were informal gatherings of bicycle tourists, quietly announced only in the columns of *Le Cycliste*. The cyclists compared new components and homemade gadgets mounted on their own and other's bicycles, tried them out, and discussed them at length. Albert Raimond's Cyclo gear first appeared at one of these meets. The riders usually had a snack together before setting off for home.

Of course, Vélocio was not the only innovator. He took up where others left off, and pushed their research further. Others did the same things with his ideas, but the difference was that his research was all-embracing. Although he revolutionized nothing, the extent and the continuity of his work in favor of multiple gears, and the fact that he reported all of his and others' innovations in *Le Cycliste* made him the key figure. He encouraged other inventors. Two of his younger friends, Joanny Panel and Albert Raimond, became dynamic leaders in the field of derailleur manufacture.

Vélocio would devise and select multispeed gear systems to suit his customers. Once, for the fun of it, he assembled a bicycle with 168 speeds. Being a lover of the simplest systems for his own riding, he used it only to demonstrate what was possible.

Jan Heine tested a 1909 La Gauloise bicycle with bichain, the mechanism depicted in Fig. 5.2. Shifting is easy: Stop pedaling, press the shift leer with the right foot to continue. Changing the chain to a different chainwheel-sprocket pair takes more time and would be done before or after a long grade.

## Steps in the Development of the Derailleur

Derailleurs had been entirely forgotten in Britain, but the concept was thriving across the Channel in France. By the start of the century, most of the important theoretical ideas for variable gearing had been thought of and patented, mainly in Britain. While the British opted for either single speed bicycles or three-speed hubs, in France all the other ideas continued to be vigorously and persistently experimented with.

Of the many suggestions for variable gearing, which really worked? Which devices were simple, easy to use, efficient, and reliable, even under adverse condi-

Left: Fig. 5.3. Vélocio and a group of his followers, known as "L'école Stéphanoise" (the St.-Etienne school), at one of their meetings at Chavanay, Rhône, in 1928.

tions? It was not on drawing boards or in the patent offices that the matter was decided, but in extended use on the muddy roads of rural France. In the following sections, contributing author Raymond Henry describes this period of extensive trials.

### Variable Gears in France

At the time, French bicycles were sold without accessories. Manufacturers were attracted by the publicity they obtained from racing, and they cared little about the touring market. The racing bikes or the everyday bikes that were available did not meet the needs of touring cyclists. The tourists wanted reliable and comfortable bicycles that allowed them to travel on any road without undue fatigue. It was left to small makers such as Vélocio to equip bicycles suitable for touring, and to experiment with variable gearing.

Racers stuck to single-speed fixed-gear bicycles, spinning rapidly on the flat and pushing strenuously up the hills. When the slope became too steep, they got off and ran with their bikes. When the first derailleurs appeared, the racers fiercely mocked the tourists who used them.

Henri Desgrange, the newspaper magnate who founded the Tour de France, admired pure athletic strength and scorned all innovation. He referred to Vélocio, an ardent promoter of variable gears, as "that fool with the little wheels." This was a reference to one of Vélocio's touring machines, equipped with 20-inch (500 mm) semi-balloon tires.

The UVF (Union Vélocipédique de France) was mainly interested in competitive cycling and they passed drastic regulations that systematically prevented technical innovation. Later, the Union Cycliste Internationale (UCI) also restricted bicycle design. So, bicycle tourists equipped with derailleurs who followed the Tour de France were able to reach the mountain passes ahead of the racers. Even the aging and terribly nearsighted Vélocio was able to follow them on the descents.

Little by little, the more open-minded racers appreciated that racing could not continue to ignore the derailleur. But the racers were not free to decide. They were caught up by rules and contracts. So the touring cyclists improved cycling equipment, and the racers adopted their innovations only much later.

### The TCF Technical Trials

In the early years of the 20th century, the new Touring Club de France (TCF) organized several technical trials to improve the standard of touring bicycles. First there was a brake test in 1901. Then in 1902 and 1905, the TCF organized more general trials to assess comfort, reliability, and

Right: Fig. 5.4. This 1902 poster shows the Terrot double-chain bicycle that was awarded the gold medal at the TCF trials of 1902. It is depicted at the top of the Tourmalet pass, where the trials were held. It is being ridden by Terrot's famous model, known as *l'Impertinente* (the impudent one).

above all, variable gears. The trials aroused a great deal of interest among bicycle tourists.

The 1902 trials were held in the Pyrenees. The course consisted of two circuits, both climbing the impressive Tourmalet mountain pass, elevation 7,262 ft. (2,214 m). Twenty-four different manufacturers participated, with a total of 46 bicycles. Before the actual trials, the bikes were subject to a two-day inspection period. During the event, there were fifty marshals around the course to note if any competitor went on foot, even for a few meters! On the hair-raising descent, local people were employed to put up "SLOW" notices at dangerous bends. Twenty people were appointed to clear the road of any cattle straying from the open pastures. The roadway surface was swept in places.

Some manufacturers employed professional racers to ride their bikes in the trials, but other manufacturers were represented by bicycle tourists. The racers, who were offered substantial speed bonuses, rode the course as fast as they could. The tourists

Left and above: Figs. 5.5 and 5.6. Hirondelle with retro-direct and two chainrings. It used one of the first front derailing devices.

Left: Fig. 5.7. A similar image was used in this 1912 poster for the Touricyclette shaft-drive 2-speed, which was awarded the silver medal in the 1902 TCF trials.

Above: Fig. 5.8. The 1902 Magnat & Debon retro-direct competed in the 1902 TCF trials.

Below: Fig. 5.9. The 1902 Hirondelle retro-direct with two chains, from the 1902 TCF trials.

took it easier, stopping at the checkpoints for snacks and a few minutes' rest. Nobody was able to ride all the way to the top of the Tourmalet—everyone walked the final section. Of the riders who did the complete double-circuit course, one tourist on a retro-direct Hirondelle walked only 2 km 1.25 miles) of the course. The weakest climber walked a total of 16 km (10 miles). Even among those who did only one circuit, nobody rode the whole course. The best climber walked 1 km (5/8 mile), the worst 20 km (12 miles). Muller, a racer on a Clément with a two-speed hub, came in first to the cheers of the crowd. Nevertheless, he had walked 11.7 km (about 7 miles).

There was a great variety of variable gears in these trials:

- Double chain (Terrot, Hurtu, Automoto, Hirondelle, and De Vivie);
- Triple chain (De Vivie);
- Chainless 2-speed (Peugeot Percutante and Durieu Touricyclette);
- Two-speed hubs (Clément, Gladiator, Simpson, La Française, Terrot, Peugeot, WFW);
- Bottom-bracket gears (Variand, Lancelot, Lecarme and Michel);
- Retro-directs (Hirondelle, Magnat & Debon and Floquart);
- Lever-action (Palmerantz-Svea, but not tested).

After the trials, 17 bikes were closely examined to check their condition. Finally, the gold medal was awarded to the Terrot company for a double-chain 4-speed bicycle. Many of these gear systems are described in detail later in this chapter.

Jan Heine tested the Hirondelle retro-direct depicted in Figs. 5.5 and 5.6. It had 40/50 chainwheels and 20- and 34-tooth sprockets, giving 4 speeds—two pedaling forward and another two pedaling backward. The transfer from forward to reverse pedaling turned out to be seamless, and front shifts were incredibly smooth.

The 1902 trials stimulated a great deal of innovation in vari-

Below: Fig. 5.10. The Magnat & Debon bicycle with 2-speed bottom-bracket gear was awarded a gold medal for its performance in the 1905 TCF trials.

Right: Fig. 5.11. Terrot won a gold medal for innovation in gear shifting in 1905. This poster shows the Terrot Model H, ridden by *L'Impertinente*, and the Terrot Lévocyclette, ridden by her companion. Terrot also entered a 4-speed retro-direct model.

able speed mechanisms. However, in a detailed report of the trials, the TCF noted that few firms had entered machines properly equipped for touring. Several inexperienced builders had installed gears too high for mountain roads. Some of them "had adapted for the event their usual machines built more for sport than for traveling." Other builders "had made commendable efforts but, unsure about what was expected of them, poorly helped by spare-part makers, and finally pressed by time, had entered only their current standard production models."

In 1905, the TCF organized new trials in the Alps. The object was to promote the development of the ideal touring bicycle. The entry invitation stated:

> Whatever his age and without any painful training, the cycle tourist must be able to tackle all roads, no matter what the gradients or the state of their surface. No breakdown must interrupt his journey. Consequently, every item of the bicycle must be reliable, easy of access, easy to adjust or to repair. [Frank Berto comment: A century later, these requirements still describe an ideal touring bicycle.]

The two-day competition over an arduous mountain course was hard for riders and machines alike. It included a total of 120 km (75 miles) of climbing. Walking was not permitted in these trials. Only 26 bicycles from 10 makers were at the start. They differed completely from the standard bicycles of the time, designed for sport on good, level roads. Firms that had made little or no technical effort in previous trials and relied only on employing well-known racers did not enter this event. Accessories were more adequate and better built. Following the event, all the bicycles were submitted to a detailed technical inspection, including dismantling important parts, notably variable gears. In contrast to the 1902 trials, no serious defects were found.

The Hirondelle also won a gold medal at the 1905 TCF trials for its simplified retro-direct

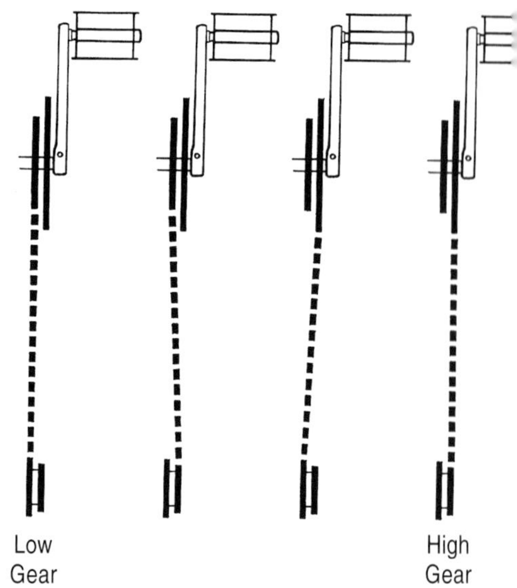


Above: Fig. 5.15. Vélocio's four speeds using out-of-line chainwheels and sprockets. Two gears were in line and two were out of line.

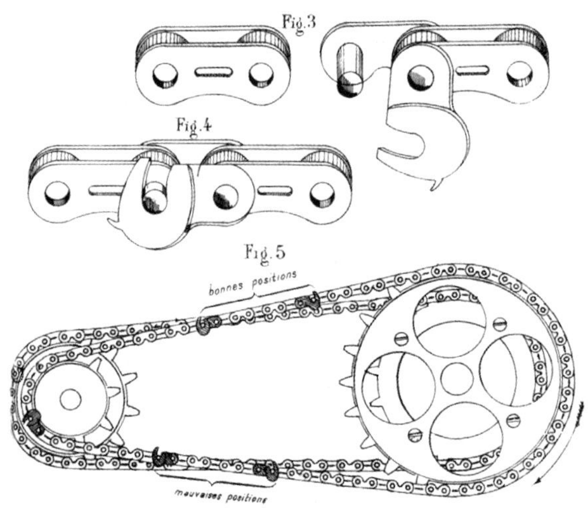


Right and below: Figs. 5.12 and 5.13. Vélocio's 1897 quick-release hook-and-rivet chain allowed him to move the chain from one chainwheel to the other in less than a minute.

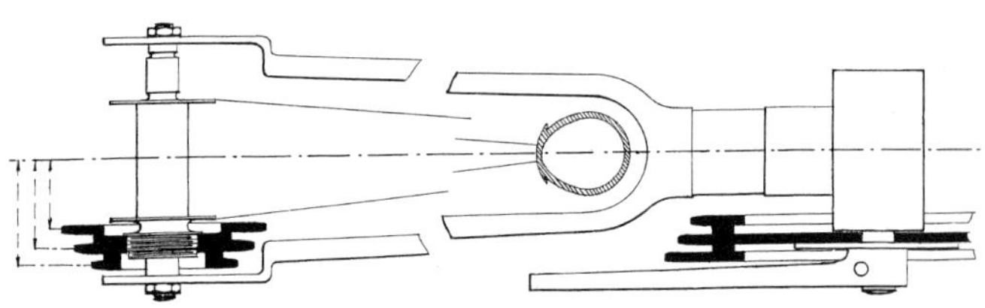

Right: Fig. 5.14. Vélocio's 1898 in-line 3-speed used a freewheel to allow a slack chain, which was shifted by hand.

(patent taken out in 1903). Durieux entered a double-chain retro-direct.

The Magnat & Debon bracket gear was now working properly, which had not been the case previously and there were other bracket gears. The British Raleigh machines were equipped with 3-speed Sturmey-Archer hubs, and the official report said no abnormal wear had been detected in them and that the control lever on the handlebars was easy to operate.

By organizing these trials, the TCF greatly stimulated bicycle design, and in particular the evolution of the classic French *randonneur* touring bicycle. Much later, Vélocio wrote that the 1902 and 1905 trials had created a great interest in multi-speed gearing. Technical trials by the TCF and other organizations continued to stimulate improvements in bicycle design right up to 1949.

### Vélocio's Early Gear Change Mechanisms

Around the turn of the century, there was an astonishing variety of gear mechanisms in France. Many inventors tried to come up with a simple and reliable way to alter the gearing on a bicycle while riding.

In *Le Cycliste*, Vélocio classified the existing gear-change mechanisms, many of which he had tested himself. The technical details needed for derailleur development were painstakingly evolved in these early devices.

### Displacing the Chain

Many readers in the columns of *Le Cycliste* stressed the need for an uncomplicated system of multispeed gearing. Vélocio took a personal interest in the problem, especially from 1896, when he began riding much greater distances himself. Vélocio was no drawing-board inventor. He relied on practical experiments and road experience. The readers of *Le Cycliste* helped him; though few in number, they formed a highly interested French and foreign elite of cycle tourists and professionals in the touring bicycle business.

In the July 1896 issue of *Le Cycliste*, Vélocio described a 2-speed version of his Gauloise bicycle. It was equipped with a chainwheel and sprocket on either side. The chainwheels and sprockets were of different sizes, but the total number of teeth was identical on both sides. For example, a 25-tooth chainwheel and 9-tooth sprocket on the right added up to 34. A 22-tooth chainwheel and 12-tooth sprocket on the left also added up to 34.

In theory, the same chain could be used on either side. To change gears, you unbolted the chain on one side and installed it on the other side. In practice, it worked less well than expected. The chain bolt soon wore loose and got lost.

Above: Fig. 5.16. In 1902, Terrot also produced this quick-release chain link for swapping the chain from one chainwheel-and-sprocket set to another.

Left: Fig. 5.17. Vélocio's 1901 Whippet bicycle with its New Protean derailleur survives in the St.-Étienne museum. It still works perfectly, but judging by the wear, Vélocio must have traveled many kilometers on it.

Right: Fig. 5.18. The "steel finger" marketed by Cyclo. Vélocio produced a number of "steel fingers" with ball bearings. Other people used homemade imitations, sometimes just a piece of thick steel wire.

Next, Vélocio tried a quick-release hook-and-rivet idea, of which he had first thought in 1890. Now he could move the chain from one chainwheel and sprocket set to another in less than a minute. About the same time, the same idea occurred to a TCF member named Bouny. The system seemed to work, so a few months later Vélocio installed double chainwheels and double sprockets on both sides of the bicycle, thus obtaining four speeds.

In an 1898 article in *Le Cycliste*, Vélocio shows a in-line 3-speed system. The caption reads: "A single-speed bicycle which we have adapted to three-speeds: 41, 66, and 94 in., changeable in a few seconds."

By 1900, freewheels were becoming common. Because a freewheel allows a slack chain, Vélocio used this possibility to fit a triple chainwheel and a triple freewheel, giving three sets of in-line sprocket and chainwheel combinations, each set totaling the same number of teeth (recall Bernard Cadot's design, described in Chapter 3). With thumb and index finger, Vélocio moved the chain from one set to another, thus obtaining three gears, each perfectly in line. Some readers of *Le Cycliste* also adopted this method.

Vélocio next tried out-of-line chainwheels and sprockets. Early chains were wide and stiff, and most people were convinced they had to run in a straight line. Disregarding this common belief, Vélocio tried a double chainwheel and a double sprocket on the same side of the bicycle. The chain was moved by hand onto every combination. This gave four different gears, with the chain pulling out of line on two of them. There was the same drawback as before. You had to adjust the wheel or lengthen or shorten the chain. You could also pinch a few links together with a bolt, an ugly solution because the chain went round with a loose loop hanging from it. With a fixed gear, the chain had to be taut all the way round.

### Vélocio's Whippet

In 1901, Vélocio had a Whippet sent to him from England. As

Above: Fig. 5.19. Chain unclogger on an Intégral bicycle. The chain unclogger, devised by Vernon Blake, consisted of small guides under the bottom bracket to stop a dirty loose chain from wrapping itself around the small chainwheel—referred to as "chain suck" by today's mountain bike riders.

Top right: Fig. 5.20. Detail of Vélocio's 1900–1905 Gauloise bichain bicycle with a foot clutch.

Bottom right: Fig. 5.21. Detail of Vélocio's 1909 Gauloise bichain bicycle with a foot clutch.

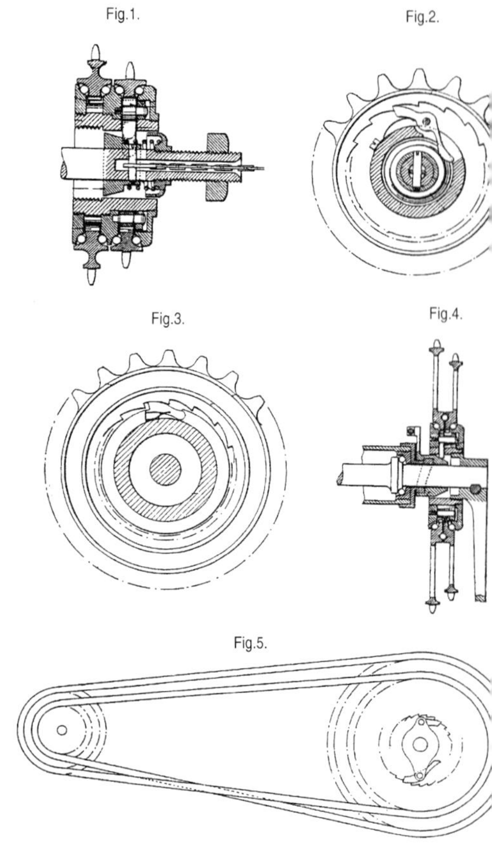


Above: Fig. 5.22. In 1907, Georges Py and Simon Botkowitz patented this complicated 4-speed system with two chainwheels and two sprockets.

usual, he tried to improve it, but the result was not to his satisfaction.

Despite all of his experimenting, Vélocio preferred the simplest solutions. He eventually removed the chainwheel changers from his personal machines, and just pushed the chain down from the large chainwheel to the small one with his heel, lifting it back up again by hand. He jokingly called this the "Whippet A." Whippet was the name that he gave to all derailleurs, that term not yet being in use. "A" stood for Avenir ("future"), because he thought that simplicity would prevail in the long run. In fact, up to the time of World War II, many bicycle tourists preferred to use their heel and fingers, or sometimes a hooked piece of steel wire, to shift the chain across the chainwheels.

### The Floating Chain

The floating chain was the ultimate in simplicity. Intent on reducing any form of drag, Vélocio put weaker and weaker springs on the tension arms he used with his variable gear devices. But under muddy conditions, this loose chain had a tendency to climb onto the tips of the sprocket teeth and slip.

One day in 1911, having suffered more than usual from these chain slippages, Vélocio thought of a way to stop the links from climbing on to the teeth. He installed a small steel roller, which he called the "steel finger," just below the freewheel. The thing worked! Of course, this required a single-sprocket freewheel. Finally, in January 1912, he removed the tension arm and pulley altogether, just letting the chain dangle. People warned him that he was losing his chain. Thus was born the "*chaine fottante,*" or "floating chain."

By combining a floating chain with a triple chainwheel and a rear wheel with a sprocket on either side of the wheel, the touring cyclist had the simplest possible six-speed gear set. Shifting across the chainwheels with the fingers gave three gears for flat country. Reversing the rear wheel gave three mountain-climbing gears. *Flottantistes* (riders who used the floating chain) generally used three chainwheels, but some had up to five.

### Bi-Chains with Foot Clutch

Vélocio used a British Sparkbrook tricycle with two chains. By operating a foot clutch, either one chain or the other could be engaged, giving a "power" gear for hills or a "speed" gear for level roads.

In 1900, he installed a bichain with a foot clutch to one of his Gauloise bicycles. He marketed two separate sets of chainwheels and freewheels. The larger chainwheel combined with a small freewheel sprocket, and the smaller chainwheel combined with a larger freewheel sprocket.

These two chainwheel and sprocket sets could be installed either on the same side of a bicycle or on opposite sides. The larger chainwheel had a clutch controlled by a foot-lever on the crank, engaging in a notch in the

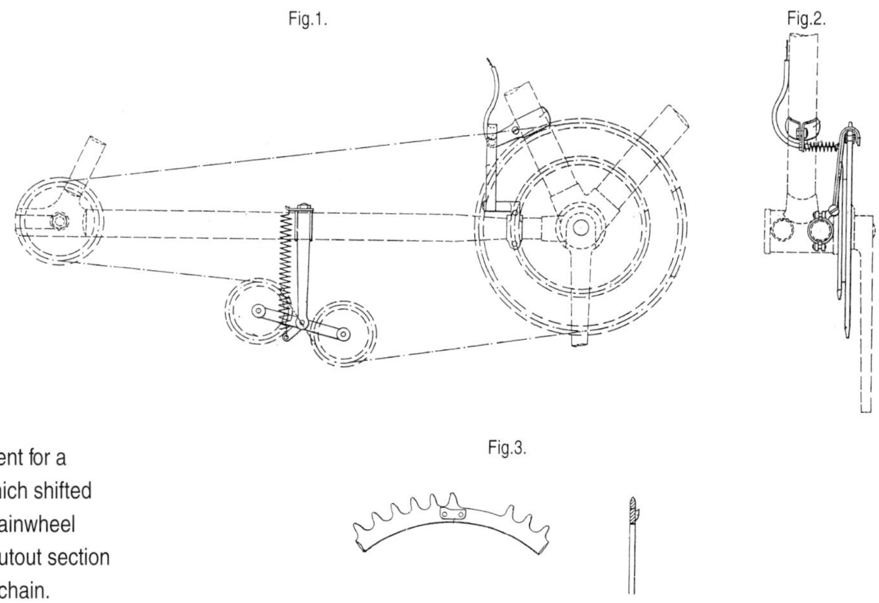


Right: Fig. 5.23. In 1908, Terrot & Co. obtained a patent for a derailing fork hinged on the right-hand chain stay, which shifted the chain across a double chainwheel. The larger chainwheel was fitted with an additional guide tooth and had a cutout section to assist with shifting. A double pulley tensioned the chain.

chainwheel. With the lever engaged, the large chainwheel gave the high gear. With the lever released, the small chainwheel was engaged, providing the low gear. Vélocio also used cable-controlled rear-hub clutches. Pins on a laterally sliding freewheel engaged slots on the hub.

As early as 1892, Lotz, of Nantes, and Gauthier, of Blois, had shown 2-speed bichain bicycles in their catalogs. With the Gauthier, you could disengage one of the chainwheels by gently sliding it to the right. There were other makers of bichain bicycles before 1900. All these mechanisms suffered from friction that reduced the advantage of changing gears. Few of these primitive bichain bicycles have survived.

Jan Heine road-tested a 1909 La Gauloise bichain bicycle. This bike is shown in Fig. 5.21. Shifting is simple: Stop pedaling, then press the shift lever with the right foot to continue. Changing the chain to a different chainwheel sprocket pair takes more time and would only be done before a long grade.

Hirondelle, in St.-Étienne, and Terrot, in Dijon, also produced bichain bicycles. A Terrot bichain bicycle won the TCF 1902 trials. It had four chainwheels: two on either side, two double rear sprockets, and chains with hooked rivets, so that they could be moved by hand from one set of gears to another. This provided two gears without dismounting and two more if you got off and battled with your chain.

Automoto and Hurtu cycles also entered bichain bicycles in the TCF trials. Vélocio combined his clutches with triple and even quadruple-chained bicycles, but the added weight made these solutions prohibitive. Jacquier and Bonnet-Guyonnet-Canonne also made simple, robust bichain mechanisms consisting of freely-revolving chainwheels, alternately clutched to the bottom-bracket axle.

In 1907, Georges Py and Simon Botkowitz, of Nancy, patented a clever 4-speed system. It used two chainwheels and two sprockets, all of which were mounted on ratchets. A single continuous chain went around both chainwheels and both sprockets. A small lever near the bottom bracket engaged either of the two chainwheels and a toggle chain inside the rear axle enabled the rider to engage either of the

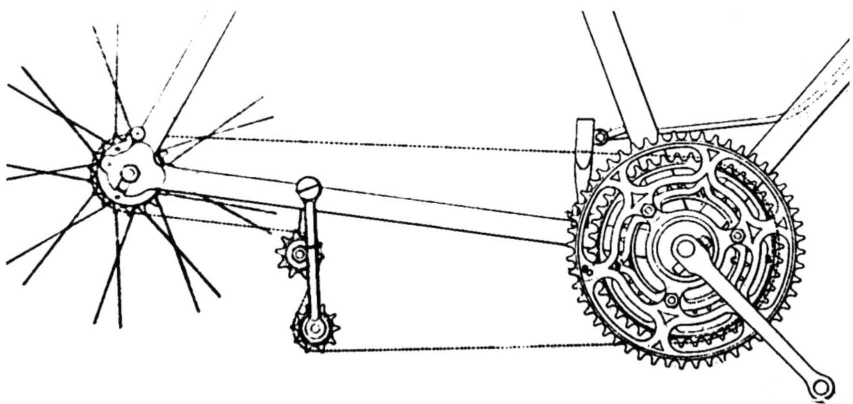

Above: Fig. 5.24. In 1910, Henri-Michel Matocq, of Paris, made this triple-chainwheel derailleur, similar to the Terrot design.

Left: Fig. 5.25. The 1905 Terrot Model H was an in-line derailleur which moved the freewheel sprockets sideways.

Above: Fig. 5.26. Derailleur detail of the 1905 Terrot Model H, showing how the chain lifter has raised the chain clear of the sprockets. Next, the sprockets moved sideways, and then the chain was lowered.

two sprockets. By selecting one of the chainwheels and one of the two sprockets to drive, the rider thus had a choice of four different gears. A bicycle with this 4-speed system appeared in the 1911 Alcyon catalog. However, there were friction losses in the intermediate gears, and the system never became popular.

### Vélocio's Early Derailleur Experiments

In the early 1900s, Vélocio and his followers had installed multiple chainwheels and freewheels with multiple sprockets to their bicycles. They had found that the chain did not have to be perfectly aligned in every gear. However, to change gears, it was necessary to stop, dismount, and move the chain by hand and to do something about the different chain lengths between gears.

Above: Fig. 5.27. The 1908 P.d'A. (Prével d'Arley) derailleur foreshadowed the Le Cyclo by nearly two decades.

### Tension Arm with Pulley

If the "in-line" rule could be disregarded, then three chainwheels and three sprockets could give nine theoretical gears and about seven usable gears. The slack chain problem remained.

Then Vélocio had the idea of taking up the slack chain by installing a fixed slotted arm with an adjustable pulley and wing nut. Soon after, he fitted an arm with a spring so that the pulley tensioned the chain automatically each time he shifted gears. Thus, Vélocio rediscovered the tension arm and pulley that the British Whippet Cycle Syndicate had used in 1894.

Above: Fig. 5.28. The 1908 Tri-Direct derailleur had a jockey pulley and a tension pulley, which were moved sideways by a horizontal arm pivoting around a bolt. A cable connected with a lever on the frame or the handlebars operated the arm. Chain tension was maintained by a coil spring. The derailleur required a special frame.

Below: Fig. 5.29. The Tri-Direct sprocket teeth were flat on one side to help the chain climb onto the larger sprockets. In those days, sprocket teeth were beveled on both sides. Boizot's 1909 patent showed beveled teeth with gaps between teeth and ramps to assist shifting, predating Shimano's HyperGlide by about 80 years.

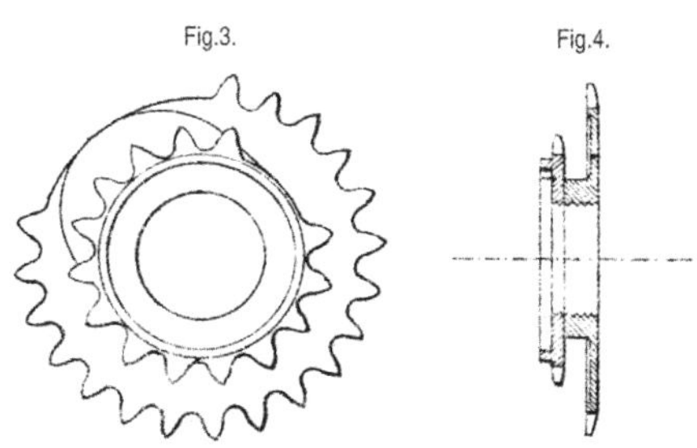


### The First Front Derailleurs

Vélocio described another experiment that was, essentially, the first French use of a front derailleur:

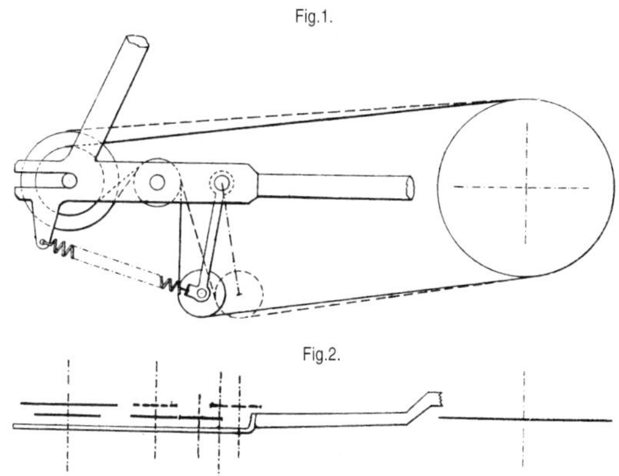


Above: Fig. 5.30. Boizot's 1908/1909 patent mentioned "*un organe dérailleur*" (a derailing device). This was probably the first use of the term "derailleur."

Below: Fig. 5.31. Boizot's 1910 Bi-Direct derailleur had only two speeds, but it could be installed on any frame.

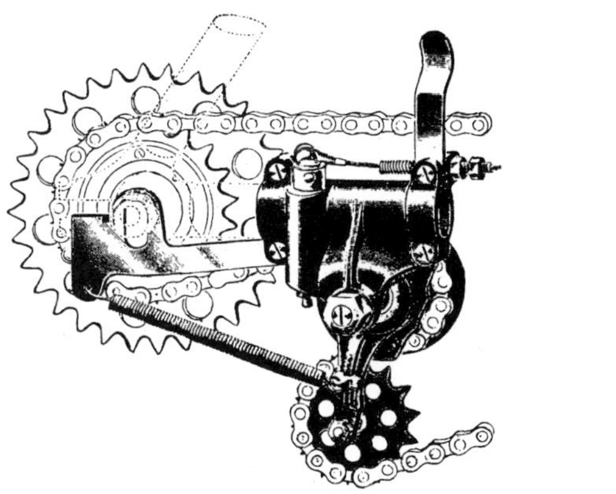

> I sometimes use a movable chain guide, which brings the upper run of the chain on to whichever of the three chainwheels I choose without touching the chain with my fingers.

Vélocio installed this device to a Gauloise, which had a brief moment of glory in 1910 with quadruple chainwheels and sprockets, thus providing 16 speeds. Other people were also interested in this idea, and individual bicycle tourists made homemade gadgets to serve as front derailleurs. In 1908, Marsot, a manufacturer in Lyon, invented a system that, like Hodgkinson's Gradient, lifted the chain, moved it sideways, and dropped it on one of three chainwheels. This complicated device was not successful.

### Early Rear Derailleurs

The Gradient and Whippet rear derailleurs were quickly forgotten in Britain once reliable three-speed hub gears became available. In France, however, the concept of derailleur gearing started to gain ground.

### The Terrot Model H

Terrot—always in the forefront of gearing developments—bought the patent for the Gradient from its inventor, Edmund Hodgkinson. In 1904, after making some improvements, Terrot installed the mechanism onto his Model H (Hodgkinson).

On the Terrot Model H, a double-cable control lever on the frame operated a mechanism that lifted the chain from the sprocket, then moved the freewheel sprockets sideways under the chain. After that, the chain fell back onto the chosen sprocket. The first freewheel had three sprockets. Later models had four.

The Terrot Model H did well in the TCF trials of 1905. However, it needed precision workmanship and precise cable adjustment. The Model H was an "in-line" derailleur. In this period, sprockets were usually $^3/_{16}$-inch (5 mm) thick and the matching wide chains did not like flexing sideways. Keeping the chain in line was seen as an advantage.

Terrot improved the control mechanism and called the new model the HE. A later model, the HT, no longer required chain lifters. Instead, the chain ran over two tension pulleys and was pushed onto the selected sprocket by sliding the whole freewheel sideways.

Left: Fig. 5.32. A 1912 Boizot bicycle with a 5-speed freewheel and an oval chainwheel.

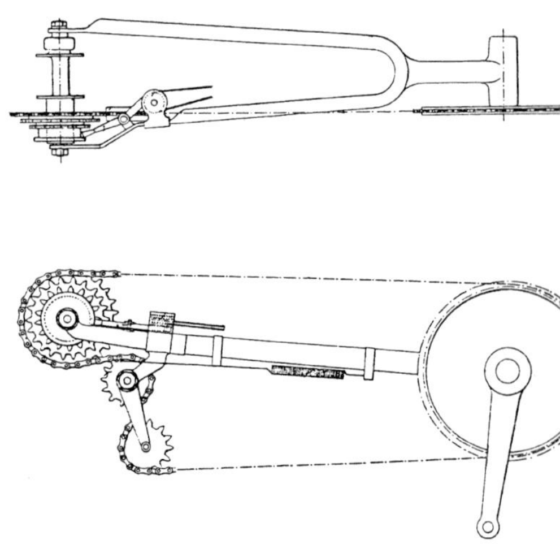

Above: Fig. 5.34. The 1909 patent drawing for Gabriel Hervier's in-line rear derailleur.

Right: Fig. 5.33. Gabriel Hervier's 1909 bicycle and rear derailleur did not differ much from today's designs.

### Vélocio's "Fire Pokers"

Around 1905–1906, Vélocio tried a gear-change method that his friends mockingly called his "fire pokers." He used two thin rods to shift the chain. One rod moved the chain across the two chainwheels when pedaling forward, and the other shifted the chain across triple sprockets when pedaling backward. The machine had six speeds. In fact, the rods were primitive front and rear derailleurs. The chain was kept taut by a spring-loaded tension pulley on the lower run of chain.

### 1908, the Pivotal Year

In 1908, four new rear derailleurs appeared from Perret, Prével d'Arley (P.d'A.), Tri-Direct, and Boizot. Commenting on the almost simultaneous appearance of these diverse mechanisms, Vélocio wrote: "When an idea is in the air, it germinates in many brains!"

Perret, in Boèn-sur-Lignon, a small town north of St.-Étienne, offered a 2-speed mechanism with a swinging fork to push the chain from one sprocket to the other. A single tension pulley absorbed the chain slack from the small difference between the two sprockets.

The P.d'A. derailleur was first marketed in Lyon by its inventor, then by Proust & Duvoy, an engineering company that made car wheels. The P.d'A. derailleur had a top-tube lever that rotated a rod with a bent lower end. It acted like a finger pushing the changer mechanism. The chain tension was maintained by a coil spring in a circular slot in the jockey-pulley housing. It was simple and robust. The freewheel had 18, 24, and 36-tooth sprockets. After taking out his French patent, Prével d'Arley took out another patent in Belgium for an improved cable-operated derailleur. He also invented a suspension fork. He was shy, modest and, like many inventors, had little business sense. His derailleur was manufactured until World War I, and then it disappeared. Disappointed by its lack of success, Prével d'Arley turned to archaeology. He died in 1926.

The Tri-Direct derailleur was made by Charles Boizot & Co., bicycle manufacturers in Puteaux, a Paris suburb. Tri-Direct buyers could specify sprockets between 14 and 32 teeth. The Tri-Direct was robust. One of Boizot's customers, Monsieur Boucher, took a long trouble-free journey in the Carpathian Mountains, Galicia, and Hungary, returning through the Dolomites, southern Tyrol, and along the Mediterranean coast.

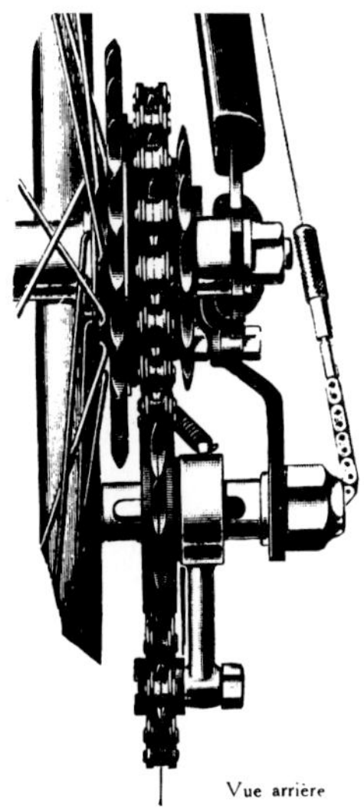


Above: Fig. 5.35. The first Le Chemineau derailleur of 1912 was mounted on a separate clamp. Later models required a bracket welded to the frame, and were only available on Panel's Le Chemineau bicycles. Panel also made a budget derailleur, L'Izoard (named after a pass in the Alps), which could be retrofitted to any bicycle.

Left: Fig. 5.36. The *École Stephanoise* at the Col du Grand Bois in 1909. Joanny Panel is on the far right.

Boizot was a versatile inventor. He also marketed a rim which provided better braking, an improved chain, a three-speed hub, a rear hub with built-in freewheel, a front hub, a bottom bracket with annular bearings, and an oval chainwheel.

Boizot's 1912 derailleur had a 5-speed freewheel with sprockets from 14 to 36 teeth. The derailleur maintained a constant chain gap over this wide range, with the jockey pulley a constant distance from whichever sprocket the chain was on. The chainwheel moved sideways on the bottom bracket axle, so that the chain was always in line. This derailleur was marketed until the late 1930s as Le Montagnard (The Mountaineer).

### The Gauloise

Vélocio wrote, "In the course of tests that I pursued on the Gauloise after the 1905 TCF trials, I also adopted the transverse-action shift-sprocket." Around 1908, he added four sprockets, a jockey pulley, a tension arm, and a tension pulley. A cable-operated the mechanism and a return spring reversed the movement when the cable was eased. Although he could not have known it at the time, Vélocio had made the first prototype of today's rear derailleurs.

### Gabriel Hervier

In 1909, Gabriel Hervier, a bicycle maker in Rive-de-Gier (a small town near St.-Étienne), produced an in-line derailleur. He borrowed the idea of a three-sprocket freewheel sliding sideways on the hub from Terrot's Model H. His freewheel was pushed sideways by a metal finger moved by two cables, so there was no need for a return spring.

### Joanny Panel

In September 1910, the Amicale Touriste Forezienne, Vélocio's club of keen bicycle tourists, organized a contest between "*polys*" (multi-speed gear enthusiasts) and "*monos*" (those who preferred single-speed bicycles). The course, in the mountains near St.-Étienne, was 232 km (144 miles) long and included 3,000 m (10,000 ft.) of climbing.

There were 26 competitors. The three riders on single-speed machines were dropped on the first hill. Alfred Faure, a good Tour de France racer, used two single sprockets and reversed his wheel by hand. There were plenty of bichains, a tri-chain, and two bicycles *transport de chain* (with chain shift). One was a 16-speed Gauloise (with 12 or 14 useable gears), and the other a 6-speed Hervier.

Most of the competitors dropped out as the race went on. A friend of Vélocio's, Joanny Panel, a young clerk from St.-Étienne, finally won on the Gauloise in 10 hours 20 minutes. Alfred Faure finished fourth, an hour later. Joanny Panel would soon move on to become known as the inventor of the Chemineau derailleur and as a bicycle builder.

### Le Chemineau

By 1908, Vélocio and the other makers had succeeded in establishing practical rear derailleurs, but they still had to find acceptance with the general public. Joanny Panel contributed the most to making the derailleur

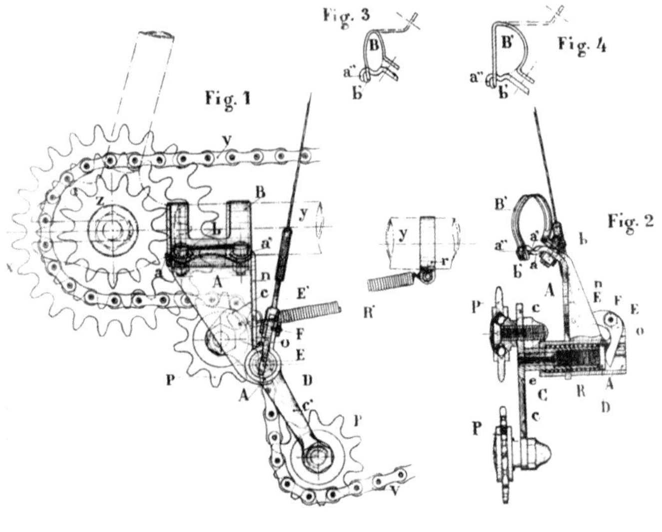


Left: Fig. 5.37. The 1919 patent drawing of the Le Chemineau derailleur. The cable pulls the bell crank, which pushes the jockey pulley inward onto the larger sprockets. The spring pushes the jockey pulley outward.

known outside the small world of French bicycle tourists.

Panel was born in 1887, the year Vélocio launched his magazine *Le Cycliste*, in St.-Paul-en-Jarez, a small village near St.-Étienne, the large industrial town where he spent all his active life. He was not yet 20 years old when he joined Vélocio's informal group of bicycle tourists, which came to be known as the *École Stéphanoise* (St.-Étienne School). Most of them were keen and experienced cyclists. As the name suggested, they considered themselves to be members of a philosophical school after the manner of the ancient Greeks. Young Panel was small and wiry, and soon became a talented cyclist. He admired Vélocio, this small fifty-year old man, who was cultured, temperate, persuasive, and capable of riding forty hours at a stretch. Vélocio appreciated Panel as an intelligent and passionate young man who rode extremely well and understood the advantages of multiple gearing. They remained life-long friends.

In 1909, Panel was a clerk with the arms manufacturing company of Rivolier Père et Fils (RPF). The company had just started making bicycles, and Albert Raimond directed the bicycle department. (More will be said about Raimond later.) Panel's victory in the 1910 touring bicycle trial, riding the derailleur-equipped Gauloise, fired Raimond's imagination.

As Vélocio said later of Panel, "This success opened the way for him." Panel made the Gauloise derailleur lighter and more adaptable to all bicycles. He shortened the long tension arm and improved its action. With Vélocio's blessing, Panel took out a patent for his new derailleur with the help of the draftsman Claudius Bouillier. Panel left RPF, and set himself up as a bicycle manufacturer in 1911. The "road and touring bicycles" which he offered his customers, though of a distinct style, were similar to those made by RPF. Panel's bicycles were not custom-made but good quality production models. He chose Le Chemineau (The Tramp) as the brand name for both his bicycles and his derailleurs.

The first Le Chemineau derailleurs used a hollow cylinder containing a short sliding shaft. I call this the "plunger type." The shaft carried the jockey pulley, the tension arm, and the tension pulley. The shaft was pushed inward by a coil spring and pulled back outward by a small chain at the end of a control cable. This pull-chain system was later adopted by Simplex, Huret, and nearly every other maker, and remained in use for fifty years. The cable was connected to a hand lever installed on the top tube. Notches in the lever housing held the lever against the pull of the spring. A long coil spring, fixed at the other end on the chainstay, provided tension for the chain.

In 1912, Panel took part in the Tour de France in the Independent category. After all, at the 1910 St.-Étienne trials, he had done much better than Alfred Faure, who had been 13th in the 1908 Tour. Panel's aim was not to compete with the champions, but

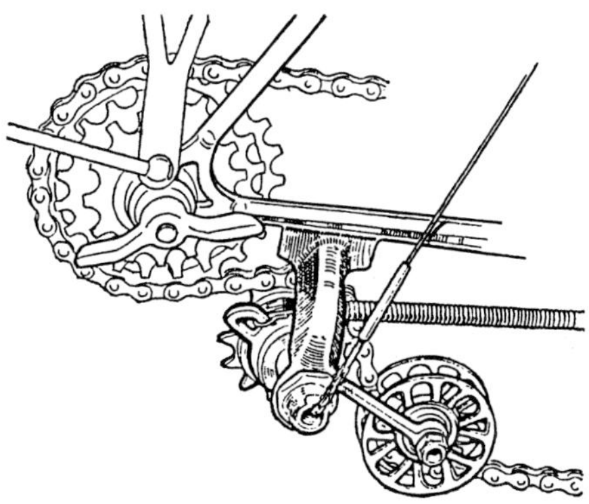

Above: Fig. 5.38. A 1946 Daniel Rebour drawing of Le Chemineau. Its design was little changed from the 1912 model. The last advertisement in *Le Cycle* appeared in 1949, which is surely the longevity record for a derailleur design.

Right: Fig. 5.39. The 1912 Le Chemineau touring bicycle. The Chemineau catalogue included 12 different models of bicycles, as well as spare parts and technical information.

to complete the Tour de France course on a fully equipped touring bicycle with wired-on tires, luggage rack, fenders, acetylene lighting, a double chainwheel, and a derailleur. He carried luggage, because Independents had to be completely self-sufficient. His bicycle weighed 19 kg (42 lbs.). His ambition was to complete "*La Grande Boucle*" (the Big Loop, as the Tour de France was popularly called) without undue fatigue—and above all to show the superiority of his Chemineau derailleur.

In the early Tour de France races, daily distances were as much as 500 km (300 miles). Because of the great length of the stages, the racers rode only every other day. This gave Panel time to write up his "Impressions of the Road," which were published in *La Loire Républicaine*, a St.-Étienne daily newspaper. His articles, of course signed "Le Chemineau," were very popular. The readers were very proud of their local boy, and waxed enthusiastic about the Tour. Unfortunately, near the start of Belfort–Chamonix stage, another cyclist crashed into Panel, bending two of his frame tubes. Following a makeshift repair, he arrived after the time limit. He had been in 26th place in the Independent classification.

Panel had another try in the 1913 Tour de France. This time, he got lost at night during the 470 km (290 mile) stage from Brest to La Rochelle and was again eliminated. He continued the Tour unofficially and arrived in Paris 12 stages later, at the same time as the other racers.

On his third attempt, in 1914, a serious fall in the first stage forced him to abandon the Tour once again. This injury left Panel with a permanent limp, but at least it prevented him from having to fight in World War I, from which many racers did not return.

Despite his misfortunes, Panel's exploits attracted public attention to himself and to his Chemineau bicycles and derailleurs. He continued cycling and organizing cycling events, mainly tours. He made many friends among the bicycle tourists he met on the road. He formed his own cycle-touring club, the Chemineau-Touriste Stéphanois. He organized Chemineau Day, a timed hill-climb followed by a meal at an inn. He participated in technical trials and sponsored regional and professional racers. He also organized long-distance non-stop rides, such as from St.-Étienne to the Lautaret Pass and back

Left Fig. 5.40. Joanny Panel and Benoit Faure at the St.-Étienne velodrome after Faure's 1929 Tour de France victory.

Benoit Faure was employed at Panel's factory. He was nicknamed "The Mouse" because of his short stature and liveliness. Faure was a very good climber in the mountains. Derailleurs were allowed in the Tour de France in 1929 and 1930, and Faure twice won the Tour de France in the Independent category wearing the Le Chemineau colors. This was fame! Faure and Panel were carried in triumph through St.-Étienne. However, the boss did not pamper him: Work was work, and after a race, Faure was sent back to his workbench at the factory like anybody else.

Panel also built an aircraft with pedals driving a propeller, and tried it out on the steep slopes around St.-Étienne. When the machine and pilot finished up against a tree, Panel spoke no more of flying. He tried to adapt cycling to water transport and mounted tandems on aluminum floats, but the salt water corroded the chains, and this experiment was not a success either.

Panel was an interesting character. He was colorful, picturesque, lovable, and generous, though sometimes niggardly. He could be warm and engaging, and was always straightforward. In 1926, Vélocio said that Joanny

Panel had done more for the popularity of the derailleur than any other manufacturer. He was short-tempered and moody at times, and he could be obstinate, as can be seen from the incident of the Vélocio monument incident which will be covered later in this chapter. For a long time, Panel refused to sell his derailleur except on bicycles of his own make. This commercial error left the field open to his formidable rival, Albert Raimond and his Le Cyclo derailleur. Panel continued to market his Le Chemineau derailleur almost unchanged until 1949.

### Vélocio's New Derailleur

Vélocio willingly allowed Panel to adapt his 1908 derailleur, because by 1910–1911, Vélocio was at work on an improved model. Some of his machines in the St.-Étienne Museum of Art and Industry are equipped with this derailleur.

The weak point of the Chemineau design was that the spring pushed the chain inward onto the large sprocket. If the cable or the small chain broke, the chain flew into the spokes. In Vélocio's new derailleur, the control cable pulled the chain onto the larger sprockets and the spring shifted the chain outward onto the small sprockets. Instead of pulling on the jockey pulley, the control cable pulled on a small lever. The lever had a central pivot, and its other end pushed the jockey pulley toward the large sprocket. The lever looked like a clothes peg, and Vélocio jokingly called this the "clothes peg principle." The return spring needed less tension to drop the chain down to the smaller sprockets and the system was mechanically more satisfactory.

### The Monument to Vélocio

Vélocio was still very fit when he was killed in a traffic accident in 1930, at the age of 77. The following year, a monument was raised in his memory at the 1,127 m (3,700 ft.) high Col du Grand Bois, which Vélocio had regularly climbed.

There is a story about this monument. Vélocio had become known as the "apostle of *cyclotourisme*." The inscription on the monument was to use those very words, but local trade rivalries intervened. Cyclo was the name of a very successful derailleur made by Albert Raimond, who was also an influential member of the monument committee.

Joanny Panel, manufacturer of the older, partly superceded, Chemineau derailleur, always believed that Raimond had stolen his ideas to produce the Cyclo. Panel hated Raimond. Secretly, Panel bought the land at the top of the pass where the monument was to be erected. He then offered it free to the committee—provided the word "Cyclo" was not included in the inscription. Panel was adamant, and the committee had to give in. That's why the monument is inscribed to the memory of *"Paul de Vivie, apôtre de la polymultipliée"* (apostle of multi-speed gearing). This won't mean much to the visitors who stop under the fir trees at the top of the pass to admire the monument.

The Museum of Art and Industry at St.-Étienne has a unique collection of Vélocio's experimental bicycles. Many have unusual variable gears. In the days when he kept them in an untidy back room, his friends used to call them his "heap of scrap iron." There is a story about these bicycles. Not one of them has a saddle. When Vélocio died, his friend Albert Raimond stored his bicy-

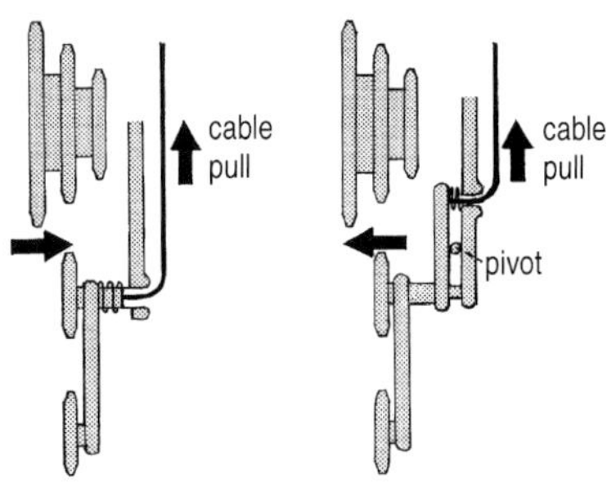


Above: Fig. 5.41. Vélocio's 1910 derailleur used a pivot to reverse the cable and spring action.

Right: Fig. 5.42. Vélocio's 1910 rear derailleur was installed on the wheel axle and was held in place by the axle nut. Chain tension was maintained by a long coil spring below the chainstay, as on the Chemineau.

cles. When the German troops retreated north at the end of World War II, they commandeered all available vehicles, including bicycles. However, Raimond had removed the saddles from Vélocio's machines, thus making them unridable, and of no interest to the Germans.

Raimond donated the bicycles to the museum shortly before his death in 1953. People thought there had been a collection of Vélocio's old bikes, but nobody knew what had become of them. Successive museum curators just knew that there was a heap of nondescript old bikes of unknown origin rusting away in a corner.

By the 1980s, nobody knew that Vélocio's bikes still existed, except Raimond's son and Raymond Henry, who went to the museum twice to try and find these bikes without success. After several years, he wrote to the curator, Nadine Besse, and explained Vélocio's importance. She said that there was a pile of rusty old bicycles in the basement, so he went to the museum the next week and found that they really were Vélocio's old bicycles. They have been under restoration ever since, and several are depicted in this book.

Vélocio remains the key figure of the most interesting period in the history of multispeed gearing. Many cycling experts rejected the first gear-changers, but they caught the attention of a few pioneers who were on the lookout for technical breakthroughs. Vélocio lived through the period of endless investigation, trial, and error. The TCF trials came and went. In his tiny workshop in St.-Étienne and out on the road, Vélocio continued his own experiments, feeding the eager interest of connoisseurs through his unique magazine *Le Cycliste*.

The derailleur was not born in a day, and these obscure beginnings are part of its story. Vélocio was the first historian of bicycle gears. In 1919, he wrote a long article, serialized in *Le Cycliste*, which inevitably features at the top of any bibliography on the subject. At the time, Vélocio had ridden 26 different multispeed gearing systems. Vélocio was still experimenting, cycling, and writing when the derailleur became accepted as a valuable bicycle component. Jan Heine provided the following summary of these articles:

1. Vélocio was not concerned about chain line, and he mocked the British obsession. He had experimented with a floating sprocket on the rear hub and found that it would not maintain its optimal position.

2. Vélocio was concerned about tension in the lower chain. He felt that this added directly to the pedaling force. [Frank Berto comment: Vélocio was wrong. Chain tension works in both directions, so it balances out.]

3. Vélocio felt that a good spread of gears was needed, which could be achieved with 3 speeds or, better yet, 6 speeds.

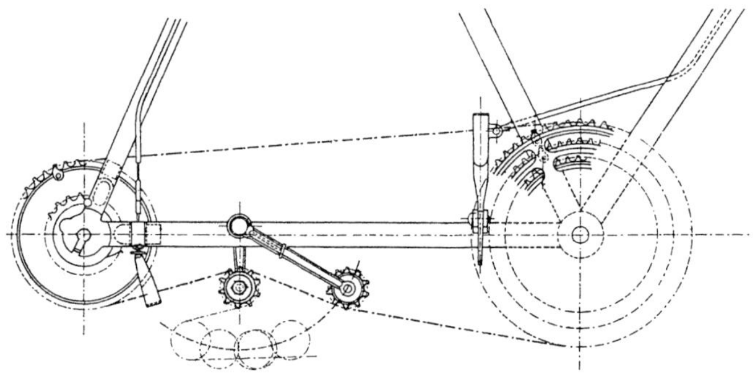

Above: Fig. 5.44. Matocq's 1912 patent provided both front and rear derailleurs, but it did not lead to a commercial product.

Left: Fig. 5.43. Inauguration of the monument to Vélocio at the Col du Grand Bois, 1931.

## Other Gear Changers

Rear derailleurs were being rapidly improved by a small number of inventors and craftsmen. However, the big firms did not adopt them, because they preferred to use well-defined patented gear systems.

Peugeot, for example, adopted the 2-speed system invented by Badois. With this novel system, the bike started off in low gear, and at a certain speed—around 13 km/h (8 mph)—centrifugal force and the inertia of two sliding weights automatically shifted it into high gear.

### Matocq's Front and Rear Combination

After patenting his front changer, Henri-Michel Matocq, of Paris, added a rear derailleur with another fork acting on the lower run of the chain to shift the chain from one sprocket to another.

Matocq presented his gearing system at the Concours des Petits Fabricants et Inventeurs Français (Forum for Small French Manufacturers and Inventors) in 1911. He patented the system in 1912. It had additional "shark teeth" to assist the chain to shift across the wide chainwheel range.

### Audouard's In-line Derailleur

The 4-speed Audouard rear derailleur was directly copied from Terrot's sliding freewheel Model H, but it had a rigid control rod. In later versions, two cables replaced the control rod. One cable pulled the freewheel outward. The other cable passed through the hollow axle and pulled the freewheel inward. With two cables, no spring was needed.

### And More

In about 1912, Albert Raimond created his Le Routier (The Roadster) rear derailleur. Little is known about its design, except that it used a single cable and a return spring and it could shift across 15 teeth on the freewheel. An amateur rider from St.-Étienne used one in the 1912 and 1913 Tour de France.

According to Vélocio, numerous other derailleurs appeared after 1912, but they caused little stir at the time and left little or no trace since. *L'Alpin* was launched in 1914 with a circular tension spring in a housing, as in the Prevel d'Arley derailleur. All further development ceased during World War I.

In 1918, Romain Gaston Chauveau, of Paris, introduced an interesting 3-speed derailleur. The body of the derailleur moved on a helical ramp, operated by a small chain and a return spring. However, the outbreak of World War I prevented it from being marketed.

## Variable Gears and Racing

For a long time, racing cyclists used single-speed fixed-gear bicycles. In 1910, the great mountain stages were first introduced in the Tour de France, and a few racers tried various systems, either on their own initiative or because they were paid to do so by their sponsors. The organizers of the Tour took no notice, and the sponsoring publication, "*L'Auto*, ignored such innovations in its accounts of the race. Knowledge of multi-speed gearing was still in its infancy.

Though hub and bracket gears were not designed for racing, Petit-Breton won several stages of the Tour of Italy using a

Right: Fig. 5.45. The 8-speed Audouard bicycle had two chainwheels and four sprockets. The derailleur made the rear shifts, and the chain was shifted between the chainwheels by hand —or with the foot.

Sturmey-Archer three-speed hub. He also used one in the 1911 and 1913 Tour de France. In the 1913 Tour de France, Brocco used a 2-speed Cornet hub.

Henri Alavoine, an independent, used a Tilhet chainwheel in the 1910 Tour, and he was paid by the inventor to use a P.d'A. derailleur in the 1911 Tour.

It is interesting to speculate as to why the riders using three-speed hub gears did not destroy the competition in the mountain stages. Racing cyclists were unfamiliar with gearing, and they rarely knew how to match gear size to the steepness of the grade. However, Petit Breton stated that three-speed hubs were "friction boxes" (even though he had successfully used them himself).

Emile Georget, a champion road racer, and Emile Friol, a champion track racer, both rode La Française Diamant bicycles. For the first time in any race in the 1910 Tour de France, Georget had a rear wheel with different-sized sprockets on either side. Cycle tourists had already used this technique ten years earlier. Georget stopped at the foot of the Ballon d'Alsace mountain, turned his wheel around, and then easily overtook the other racers on the climb. He crossed the summit well clear of the others and won at Belfort with a 4-minute lead. The following year, all the racers adopted Georget's idea.

### Speed and Distance Records

In 1907, Marcel Berthet raised the hour record to 41.1 km (25.55 miles). In 1914, Oscar Egg raised it to 44.25 km (27.5 miles). Oscar Egg's record stood for 19 years. These records, and all of the official one-hour records until the 1970s, were set on classic single-speed, fixed-gear track bicycles with no streamlining.

In 1913, Marcel Berthet rode the streamlined Vélo Torpille at 52.2 km/h (32.4 mph) for 5 km (3.1 miles). A match race between two streamlined bicycles was held in Berlin in 1914. Since streamlined bicycles were clearly faster, the Union Cycliste Internationale (UCI) banned streamlining of any kind after 1913.

When Joanny Panel turned up at the start of the 1913 Tour with his fully-equipped 6-speed Chemineau bicycle, he was ridiculed. The laughs stopped when he passed several professional racers on the hills.

Henri Desgrange made the fixed wheel compulsory in a stage of the 1913 Tour, thus eliminating derailleurs that year. Later, he banned gear changing other than by reversing the rear wheel in the Professional section of the Tour.

Although the Tour de France riders could not switch gears on the road (derailleurs), they could switch bicycles. A rider would switch to a bicycle with a low gear at the bottom of a long hill and then trade it off at the top of the hill for the descent. At the bottom of the grade, he would trade for his level-road bicycle. This was in addition to turning the wheel. The larger teams pro-

Left: Fig. 5.46. Poster showing Émile Georget and Émile Friol. Georget was the first racer to use a reversible wheel with sprockets on both sides (though not shown here).

Above: Fig. 5.47. Lucien Petit-Breton rode several stages of the 1913 Tour de France using a Sturmey-Archer three-speed hub. Apparently, the prohibition of multispeed gearing devices was not strictly enforced.

vided their stars with more bicycles.

Usually, gear-change devices were permitted in local races. But many were technically unsound and the objects of scorn. The racer's brute strength required simple and reliable gear changers. Reversing the rear wheel certainly met these criteria. Though stopping to turn the wheel around lost several seconds, it created no friction loss and it had the status of being used by the Tour de France aces (even though they had no choice).

## More Touring Trials

There had been no touring bicycle trials since 1905. Louis Roudaire, who was editor of the Parisian magazine *Le Cyclotouriste* and a member of the well-known Audax Club Parisien, recognized the benefits of further trials. Roudaire and some of his friends organized trials, the *Championat de la Polymultipliée* (multispeed gearing championships), at Chanteloup, near Paris in 1913 and 1914.

As the name implied, this trial was limited to bicycles equipped with multispeed gears. Its aim was to promote the best mechanisms. The course was 100 km (62 miles) long, with total elevation gain of 2,000 m (6,000 ft.). The route consisted of 10 laps, each with a climb up the steep hill at Chanteloup les Vignes. There was a time limit of 5 hours. Both years, some companies employed professional racers for this arduous test. These Polymultipliée trials were resumed after World War I.

Left: Fig. 5.48. The race between streamlined bicycles in Berlin in 1914. Note that the bike on the left even has disk wheels.

*Table 5.1. Tour de France winners, 1903–1919*

| Year | Rider (Country) | Bicycle | Derailleur |
|---|---|---|---|
| 1903 | Maurice Garin (France) | La Française | forbidden |
| 1904 | Henri Cornet (France) | Cycles J.C. | forbidden |
| 1905 | Louis Trousselier (France) | Peugeot | forbidden |
| 1906 | René Pottier (France) | Peugeot | forbidden |
| 1907 | Lucien Petite-Breton (France) | Peugeot | forbidden |
| 1908 | Lucien Petite-Breton (France) | Peugeot | forbidden |
| 1909 | François Faber (Luxembourg) | Alcyon | forbidden |
| 1910 | Octave Lapize (France) | Alcyon | forbidden |
| 1911 | Gustave Garrigou (France) | Alcyon | forbidden |
| 1912 | Odile Defraye (Belgium) | Alcyon | forbidden |
| 1913 | Philippe Thys (Belgium) | Peugeot | forbidden |
| 1914 | Philippe Thys (Belgium) | Peugeot | forbidden |
| 1915–18 | (No races held) | | |
| 1919 | Firmin Lambot (Belgium) | La Sportive | forbidden |

### Other Countries

The 1903 F&S (Fichtel & Sachs) Torpedo was a freewheel rear hub with a backpedal coaster brake. It was named after the new weapon of the German Navy. For the next 90 years, virtually all F&S hubs

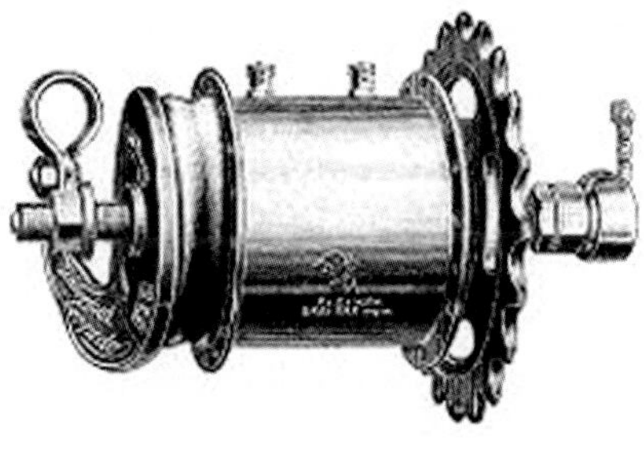

Above: Fig. 5.49. The 1912 Fichtel & Sachs 4-speed Universal Torpedo included a coaster brake (back-pedaling brake). It preceded the Sturmey-Archer 4-speed hub by 27 years.

included a backpedal coaster brake. Although there were freewheel and drum brake versions, German cyclists preferred back pedal braking.

The Doppel-Torpedo two-speed hub appeared in 1904. It was a huge success, and by 1906, F&S had 1,640 employees, and annual production reached 480,000 hubs.

In 1912, F&S made the 4-speed Universal Torpedo with three sets of planetaries. It did not reappear after World War I.

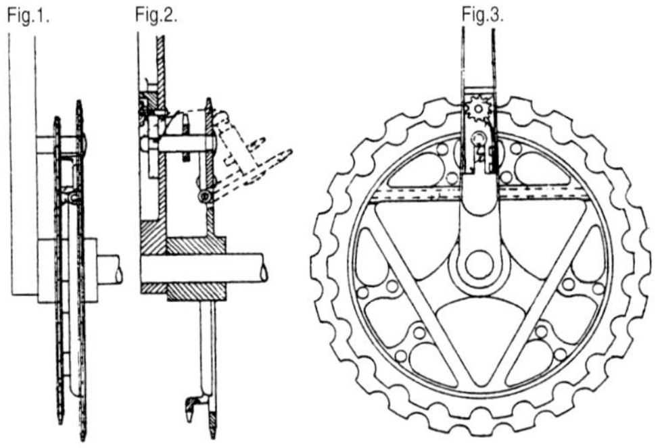


## Derailleurs Versus Hub Gears

There are a number of different explanations for the question "Why did the British choose hub gears while the French chose derailleurs?" Some of these are listed below in the words of contributing authors Ron Shepherd and Rayond Henry, as well as Frank Berto's own comments.

### 1. Raymond Henry's Explanation

Raymond Henry was a keynote speaker at the 2003 Taiwan Bicycle Industry Research and Development Conference. There, he explained the different European customs as follows:

Left: Fig. 5.50. In 1901, Fritz Barth, of Emden, Germany, patented a hinged segment on the outer ring of a double chainwheel. By swinging this segment outward, the chain could be removed without loosening the rear wheel axle.

In England, with its rather rainy climate and muddy roads, bicycles had chain–guards and variable gears housed in the rear hub. Apart from this adaptation to climate, bicycle touring remained much the same as in France. However, bicycles evolved little in England.

In Holland, a flat country where cycling was easy, everyday utilitarian cycling became general and sturdy, uncomplicated 1-speed bicycles built for long usage became customary. It was much the same in Germany. Dutch and German cyclists touring in France could easily be recognized by their heavy bicycles, ill adapted to touring or to mountain roads.

In Italy, at the beginning of the 20th century, the middle class was not very numerous, and the warmer climate was less favorable to the strenuous activity of cycling. With the advent of the automobile, anyone who could afford a car bought

Below: Fig. 5.51. In 1916, Fichtel & Sachs patented this adjustable front chainwheel. A stud wheel was attached to the cranks in place of the usual chainwheel. The chain ran over two front sprockets, driven by the stud wheel. Apparently, this arrangement enabled slack to be taken up when the chain was shifted from one rear sprocket to the other.

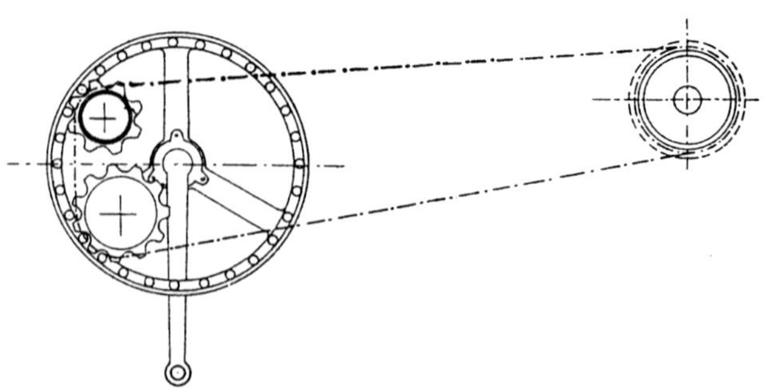

Above: Fig. 5.52. Alois Polivka, a Viennese bicycle maker and locksmith, advertised this double-chain 2-speed bicycle in a 1903–1904 handbook.

one as soon as he could. In the poorer South, ordinary people just went on foot. After World War I, bicycles became cheaper, and they became the usual means of transport of the lower classes. The standard unsophisticated machines of that period are still in use in the country and on the farms.

On the other hand, cycle racing became popular in many European countries and people flocked to see races. Racing cyclists found a path to social promotion. Former humble operatives became well-paid champions. In this somewhat simplified pattern of use—racing and utilitarian— there was little commercial or recreational space left for pleasure or touring bicycles.

In Spain, it was much the same. Renowned racing cyclists, many of whom gathered laurels in the mountain stages of the Tour de France, were cheered by crowds in a country where cycling was mainly practiced around the major towns.

But in France, the picture was different. There was an important social middle-class of civil servants, shopkeepers, and factory foremen. There were also good roads, a temperate climate, a great variety of landscapes, and other inducements to touring. Under such favorable conditions, recreational and touring cycling became popular. They formed associations, which were dissatisfied with the common run of bicycles offered by the industry, and they organized technical trials to select bicycles better adapted to their touring needs.

They wanted machines halfway between the racing and the utilitarian, with the ease and efficiency of racing bicycles, and more comfort than utilitarian bicycles. They also wanted better brakes and they especially wanted reliable variable gearing. From 1901 to 1949, the technical trials gave rise to touring bicycles of such high-quality that some of the improvements were fed back to racing machines.

Left: Fig. 5.53. Fichtel & Sachs poster advertising the Torpedo coaster-brake hub, 1911.

Right: Fig. 5.54. Bianchi made a series of bicycles for the Italian army. This 1914 Bianchi Model 14 folding bike was designed to carry a machine gun tripod. Some models had two sprockets and a slot in the rear axle. It was a simple matter to loosen the axle nuts, slide the axle along the slot, and slip the chain onto the other sprocket.

### 2. Ron Shepherd's Explanation

Shepherd believed that, in this period, the roads in both countries were unpaved and often muddy. Enclosed gearing would be of great practical advantage in France as it was in England. Perhaps it was the British penchant for tidiness that made them prefer the enclosed hub gear, while the French "let it all hang out." British reviewers criticized both the Gradient and Whippet gears for their appearance. They also disliked the untidiness of the French mechanisms. Describing the bicycle gears at the Paris Motor and Cycle Show of 1907 in the January 1908 *CTC Gazette,* H.H. Griffin wrote:

> The native applications do not call for favourable comment when viewed by British eyes, so used are we to neat and compact gears self contained in the hub. Evidently French riders are neither particular as regards appearance or weight.

In December 1911, *Cycling* commented:

> To the Britisher, accustomed to the epicyclic principle of gearing, with all the mechanism hidden within the hub shell, the crude methods of obtaining increase and reduction adopted by the French appear ludicrous.

In France, however, part of the attraction of the bicycle was that you could see how it worked. J. Bertaut wrote:

> The beauty of the bicycle resides in its sincerity. It conceals nothing. All its workings are open and visible.

H. Sauvalle wrote to *Cycling* magazine from France:

> (...) such was the demand for a reliable variable gear that devices allowing the reduction of gearing or the contrary have been put on the market with good results. They may appear strange and intricate to the English cyclist—the great lover of neatness—but we judge them by their practical value. We wanted low gears, so low that they would be useless in your country (I mean gears of 30 in. and even lower), with which we might climb, without descending from our often heavily laden mount, slopes rising from 8 to 12 percent or more for miles. Our French variable gear devices give us the low gears we require, with wide ranges of speeds. Now we combine them with your two- or three-speed hubs, with excellent results.

Racing cyclists remained suspicious of hub gears and dubbed them "friction boxes." Champions Lucien Petit-Breton, who won the Tour de France in 1907 and 1908, and Octave Lapize, who won in 1910, regarded the Sturmey-Archer hub as a toy:

> A beautiful toy, certainly, but we would each need two pairs of legs to compensate for the loss of power caused by the friction of the cogs!

Left: Fig. 5.55. With the 1915 Military Sunbeam, the cyclist was prepared for anything. It did not come with gears, but it did boast a rear hub with a "Patent Divided Axle."

### 3. Frank Berto's Comments

Raymond Henry gave his talk in Taiwan five years after Ron Shepherd wrote his explanation. In the interval, Henry had written his book on the French bicycle industry. I think that Raymond Henry is correct in saying that the roads, the weather, and the scenery were better in France. The larger number of middle-class buyers was also a significant factor.

However, I think that the Touring Club de France and the magazine *Le Cycliste* were more significant factors. They did a much better job of encouraging new developments than their British counterparts. There was no British Vélocio. When superior multispeed gear-train products became available in France, they thrived, because there was a market of well-informed and interested buyers, whereas in England, new developments remained obscure even to those who might be interested.

# 6 Practical Derailleurs: 1920–1929

FOLLOWING World War I, the European countries struggled to re-establish their shattered industries. Bicycles were used extensively for practical transportation, and as the economies improved, for sport and recreation. Most bicycles remained simple single-speeds, but in France and Italy, small manufacturers created reliable and effective derailleurs that could be retrofitted, i.e., installed on existing bicycles. In Britain, three-speed hubs were used increasingly on touring—and even racing—bicycles.

In this chapter, Raymond Henry describes the pioneers in France; Ron Shepherd, Tony Hadland, and Gordon Selby describe development in Italy and Britain; and Frank Berto fills in the rest.

## Cycling in France in the 1920s

During this decade, the characteristics of the French cyclist changed a great deal. The mass popularity of French cycling that had begun in the closing years of the previous century was now an accepted fact of daily life and had been so for a number of years. Bicycle tourists were no longer members of the middle-class, because most of those now traveled by car. Instead, the cycle tourists of this period were mainly shopkeepers, craftsmen, minor

Right: Fig. 6.1. André de Boubers, the president of the FFSC, photographed in Avignon with his floating-chain bicycle in 1929. That was also the last year in which a floating-chain machine was entered in the technical trials.

civil servants, white-collar workers, and manual workers. In general, they were lower-income employees. For them, the bicycle provided affordable Sunday outings in the country and a useful and inexpensive means of travel.

After the nightmare of World War I, there was an interest in new recreational activities. People who had been confined to the cities by the war were eager to travel to the countryside. Outdoor activities became popular. Naturists appeared on deserted beaches.

In 1923, the Fédération Française des Sociétés de Cyclotourisme (FFSC) was formed, and bicycle camping developed. In 1928, in St.-Étienne on Vélocio Day (the great annual gathering of bicycle tourists still held to today), there was even a demonstration of tent pitching in the glade at the Col du Grand Bois. Cyclists became more and more numerous on the roads on Sundays. Tandems also became popular during this period.

However, times were hard, and cyclists used whatever they could afford. Often this was only a single-speed machine ill-suited for the hills, or with a double-sprocket rear wheel that had to be reversed at the beginning of mountain climbs. A homemade double chainwheel with a floating chain was another cheap solution. If you had a little more to spend, you used a 2- or 3-sprocket freewheel and shifted the chain by hand. Dedicated tourists, of course, had a derailleur or a floating chain.

## Touring Trials Continue

After France recovered from World War I, the Championat de la Polymultipliée began again in 1921. This time, professional racers were banned, because it was thought they would win through their athletic ability, even with inferior gearing. The Poly de Chanteloup, as these trial would later be called, were particularly interesting, because they recorded the changes that took place in the use of the different gear-change systems.

*Table 6.1. Results of Poly de Chanteloup trials, 1913, 1914, and 1921*

| Year | 1913 | 1914 | 1921 |
|---|---|---|---|
| Starters | 62 | 44 | 32 |
| Finishers in 5 hours | 16 | 14 | 5 |
| Finishers with extra hour | N/A | N/A | + 11* |
| Hub gears at the start | 51 | 26 | 8 |
| Hub gears at the finish | 11 | 7 | 1 + 2* |
| Derailleurs at the start | 6 | 13 | 24 |
| Derailleurs at the finish | 2 | 5 | 4 + 9* |
| Bichains at the start | 3 | 2 | 0 |

* The figures preceded by a + sign in the 1921 column are the numbers of riders who did not finish within the specified time, but did finish after being granted an additional hour. There were 2 hub-gear riders and 9 derailleur-gear riders, for a total of 11 riders.

The number of participants dropped in 1914 because the circuit was so difficult. It dropped again in 1921 because professionals were banned. This could also explain the smaller number of hub gears entered in 1921, as their makers could no longer rely on superior athletic performance to produce a victory. By 1921, only British three-speed hub gears and French derailleurs remained in the competition. Generally speaking, derailleurs won the battle, and bichains disappeared.

Vélocio organized trials between cycle tourists on bicycles with derailleurs and professional racers on bicycles with fixed gears. The derailleurs usually won. After one such contest, Henri Desgrange, the organizer of the Tour de France said:

> I applaud this test, but I still feel that variable gears are only for people over 45. Is it not better to triumph by the strength of your muscles than by the artifice of the derailleur? Come on, fellows. Let's say that the test was a fine demonstration—for our grandparents! As for me, give me a fixed gear!

## The Floating Chain

Floating chains still persisted, although there were few at the Poly de Chanteloup. The small number of *flottantistes* increased, especially among readers of *Le Cycliste*. They reached a peak just after World War I. Perhaps the war restrictions encouraged the use of this simple system. The users were enthusiastic, although thinly spread throughout France.

There was a meet of *flottantistes* early 1920 in the Paris area, mainly to discuss improvements to remedy the drawbacks of the floating chain. How could they keep the chain from jumping off the sprockets or chainwheels, from getting wedged between freewheel and spokes, or from getting stuck on the smaller chainwheel? The gear was usually shifted at the chainwheel. Many used the Matocq front shifter without the tension pulley. Others used do-it-yourself gadgets or Chelloise rods, a simple gadget devised by two brothers of that name.

The floating chain had supporters right into the early 1930s.

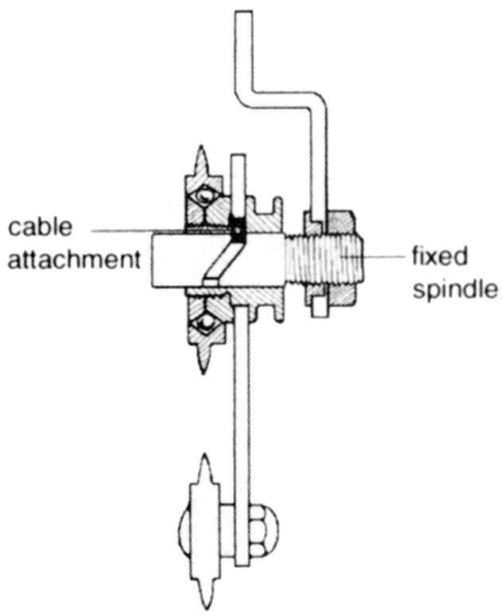


Above: Fig. 6.2. Cross-section of the 1924 Le Cyclo derailleur mechanism, showing how the pin on the sleeve engaged the spiral slot cut into the shaft. Rotating the sleeve moved the jockey pulley in or out.

André de Boubers, president of the Fédération Française des Sociétés de Cyclotourisme, used one on an Intégral bicycle for several years. The last use of a floating chain in a trial was in 1929, when Bimbinet rode an Intégral bicycle built by Bachelier. The cycle tourists knew all about the floating chain, but few used one.

Just after the war, the Touring Club de France, or TCF, organized three 1-week trials for touring bicycles. The first was in the Vosges mountains in 1920, the next one was in 1922 in the Massif Central, and the third one was in 1924 in the Dauphiné Alps. These trials demonstrated the reliability of derailleurs under extreme conditions on very rough mountain roads. In the 1924 trials, there was a trend toward an increased number of gears. After 1924, the TCF lost interest in bicycle touring and became essentially a motoring association. Even during the time of the TCF trials, the Poly de Chanteloup, which was held near Paris, had more press coverage, so it overshadowed the TCF trials, which may have been one reason why they were discontinued.

Right: Fig. 6.3. The 1924 prototype of the 2-speed Le Cyclo derailleur used two cables to rotate the sleeve and move it along the fixed shaft, aligning the jockey pulley with the freewheel sprocket. The tension arm and the tension pulley were connected to the sleeve. A coil spring attached to the chainstay maintained chain tension. The two-cable design did away with the return spring that was a weak point of the Chemineau and similar derailleurs.

Rear derailleur designs improved, and many models appeared, mainly from small manufacturers. However, they were used mainly by cycle tourists. Joanny Panel, Vélocio, and others promoted derailleurs, but their heyday came later.

## Albert Raimond and Le Cyclo

Albert Raimond was born in 1885 to a modest family in a village in the Massif Central (a highland region in the center of France). He showed mechanical aptitude after primary school and was apprenticed in St.-Étienne, a large industrial town that offered jobs to people from the poor villages in the surrounding mountains. He worked his way up to a responsible position in the gun-making company of Rivolier Père et Fils (RPF), where Joanny Panel had also worked. Raimond was thin, sensitive, discreet, and generous, with a serious and thoughtful demeanor. It was said that he was a dedicated worker and was good

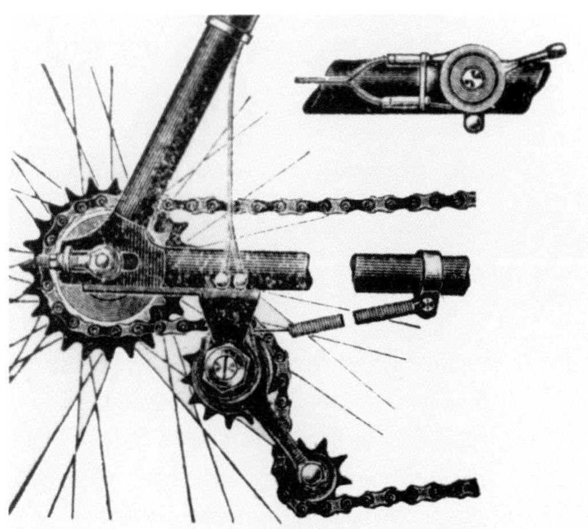

company in public, though sometimes not easy to live with.

Like Vélocio and Panel, Raimond was a practicing touring cyclist. He was not as strong a rider as his friends and he discovered that multispeed gearing helped him to keep up. In 1909, Raimond persuaded RPF to enter into bicycle production, and he was put in charge of that section of the company.

Gun-making was seasonal, with sales confined to fall and winter. Bicycles sold mainly in spring and summer. Adding bicycle manufacture kept the work force occupied year round. The involvement of gun manufacturers in the production of bicycles or components was common in the St.-Étienne region and also occurred in other countries, examples being BSA (Birmingham Small Arms) and Royal Enfield in Britain, Zeus in Spain, and Miyata in Japan.

As already mentioned in Chapter 5, Panel left RPF in 1911 and set himself up on his own as a bicycle manufacturer. Panel agreed to supply Le Chemineau derailleurs to RPF. Raimond was two years older than Panel. He was another strong character, and the two men did not get along. Panel was sometimes slow in supplying derailleurs, and Raimond was exasperated by these delays. Raimond was a skilled metal worker, and he considered making a derailleur to replace the Le Chemineau derailleurs on RPF bicycles. Before the war, he had designed a rear derailleur, dubbed Le Routier (the Roadster), but very little is known of this derailleur, even by his family.

Raimond enlisted in the army, and after the war he returned to his old job. A few years later, he devised a new, improved rear derailleur. It was inspired by a mechanism he had seen on machine guns. This unnamed derailleur, 2-speed at first, was installed on a few RPF bicycles in 1923. Raimond and Claudius Bouillier obtained a patent and, with encouragement of RPF's management, Raimond marketed it officially under the name Le Cyclo in April 1924.

Left: Fig. 6.6. This 1929 photo shows, left to right in the front row: Hyacinthe Roumanille, Vélocio, Thérèse Roumanille, and Albert Raimond with his young son. Roumanille and his niece were well-known bicycle tourists of the period, and close friends of Vélocio's.

Above: Fig. 6.4. The definitive 1925 Le Cyclo Standard rear derailleur was a 3-speed. The spiral slot was now cut in the outer sleeve. There was a choice between a toothed and a flanged alloy tension pulley. This model remained virtually unchanged for the next 25 years.

Right: Fig. 6.5. A 1924 RPF bicycle with a Le Cyclo rear derailleur. Two RPF machines with Cyclo-RPF geartrains were entered in the 1924 Poly de Chanteloup. Change the drop handlebars and remove the fenders, and you have the precursor of the mountain bike.

While working at RPF, Raimond built Le Cyclo derailleurs in his spare time. The parts were made by local craftsmen working on demand, a common practice in St.-Étienne at the time. Assembly was done by RPF workers after working hours, or by Raimond himself and his children in the evening or on Sundays. The freewheels were made by other local manufacturers. The cutting of the chainwheels was done by one craftsman and the riveting by another.

The Le Cyclo was inexpensive, and it came with a two-sprocket freewheel. It targeted the bicycle touring market, where Le Chemineau held a strong position. Joanny Panel never forgave Albert Raimond for this deliberate competition.

In January 1925, Raimond brought out a 3-speed version of the Le Cyclo. On this model, the spiral slot was in the sliding shell and the pin on the fixed shaft. You could choose between a toothed or a deep-flanged tension pulley. Later, the two separate cables were replaced by a longer single cable equipped with a fixed peg that engaged a hole in the sliding shell. The two ends of the cable were tightened and fixed on the control lever by means of a screw.

From 1924 onward, Le Cyclo derailleurs were entered in all of the technical trials and tests for touring bicycles and tandems. Le Cyclo took first place at the Poly de Chanteloup in 1925, 1926, 1927, and 1928 and at other touring bicycle trials in France, and in Italy at Turin in 1928 and 1929. In the 1931 Tour de France, Le Cyclo took second and fourth place in the Touriste-Routier (Tourist Roadster) category, where variable gears were permitted.

Joanny Panel continued to fit Le Chemineau derailleurs only onto the bicycles he manufactured. Responding to the competition from Cyclo and Simplex, he produced a budget version, L'Izoard, which could be retrofitted to any bicycle, but Albert Raimond's Le Cyclo nearly swept Joanny Panel's Le Chemineau off the market.

In 1925, Raimond produced a second chainwheel that was easy to install, followed in 1926 by the Rosa double chainwheel, named after his wife, and in 1931 by the Rosa triple chainwheel. In 1927, Raimond's flanged pulley was the

Above: Fig. 6.7. This 1928 advertisement for Panel's L'Izoard derailleur is from *Le Cycliste.*

Right: Fig. 6.8. The L'Izoard derailleur was named after this mountain pass, the 2,360 m (7,740 ft.) high Col d'Izoard in the French Alps.

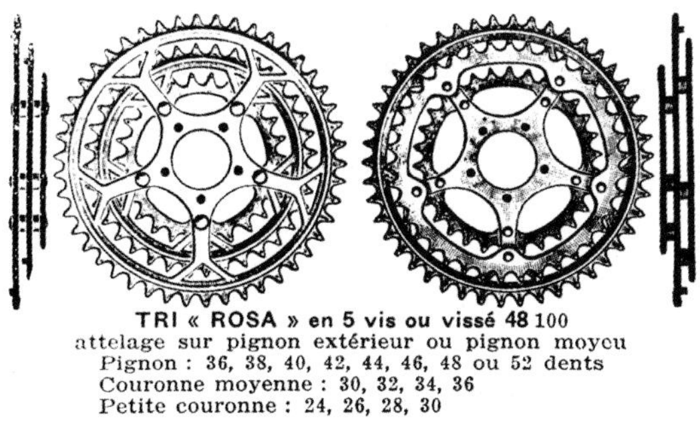


Above: Fig. 6.9. The 1925 Rosa chainwheels were named after Albert Raimond's wife. Today, 85 years later, you can still buy TA chainwheels with the same attachment bolt pattern.

Below: Fig. 6.10. The 1931 Simplex 3-speed derailleur was very similar to the 1928 2-speed model.

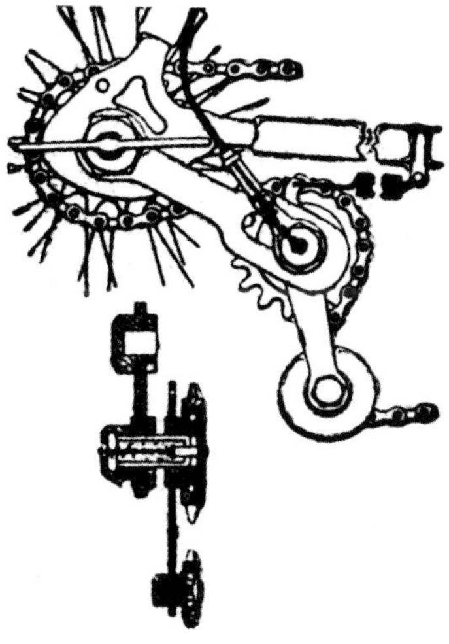

first use of aluminum in a derailleur. At this time, rear frame fork-ends were horizontal slots, so he marketed small stops that screwed into the fork-ends to eliminate the need to adjust the wheel each time it was placed in the frame. He also made the first vertical-dropout fork-ends for frames to be used with Le Cyclo derailleurs. This improvement soon became standard on touring bicycles, and a few years later it was also adopted on racing bikes.

Finally, Raimond left RPF and set himself up in modest premises with three workers. In 1928, he began selling his derailleurs in Britain with the help of Louis Camillis, who became his British agent.

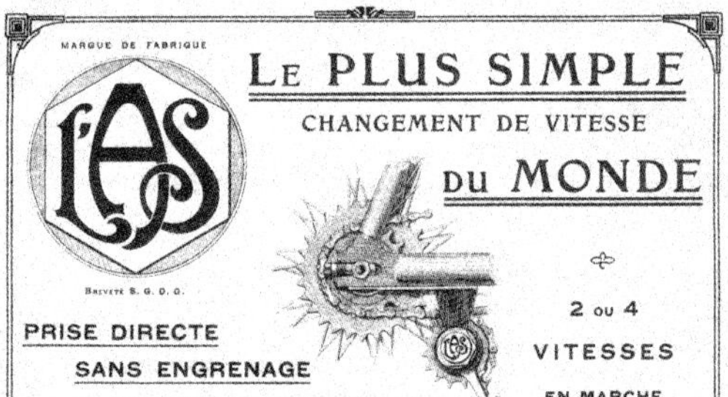


Above: Fig. 6.11. The 1920 L'As (the Ace) was a 2-speed derailleur that was shifted by back-pedaling.

## Lucien Juy and Le Simplex

Le Cyclo's main competition came from Le Simplex. As its name indicated, this new derailleur was intended to be simple. The name was also a sly dig at the Le Cyclo derailleur with its helical ramp and its intricate cable movements, which was often a problem for less-experienced mechanics.

Le Simplex was produced by Lucien Juy, who ran a bicycle shop in Dijon, the main city of the Burgundy region. Juy was neither a bicycle tourist nor a racer himself, but he was a passionate fan of bicycle racing. He understood that sooner or later, bicycle racers and officials would overcome their scorn for the derailleur. He also knew that racers were wary of the apparent intricacy of the Le Chemineau and Le Cyclo devices, and decided that something simpler and more rugged was needed.

Juy produced his first 2-speed derailleur in 1928. It worked on the same principle as the Chemineau: There was the same jockey pulley with a pull cable and return spring, a tension arm with a tension pulley, and a long coil spring attached to the chainstay. The major difference was that the mechanism was installed on the rear axle and was secured by the axle nut. When you took the wheel out, the derailleur came with it. The advantage was that the distance between the freewheel and the derailleur was adjusted once and for all by means of a slot in the bracket. This model did not make much of a name for itself, perhaps because it offered little more than the better-known Le Chemineau.

Nevertheless, Juy persisted in promoting his derailleur to the racers and the managers of the

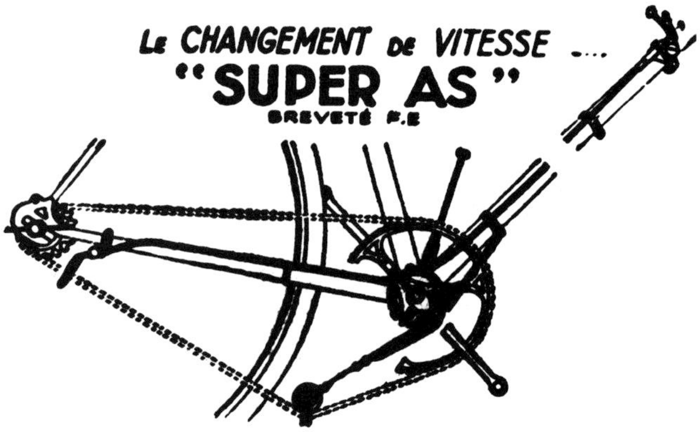


Right: Fig. 6.12. The 1923 Super As from the Automoto catalog. The large lever was used to control chain tension.

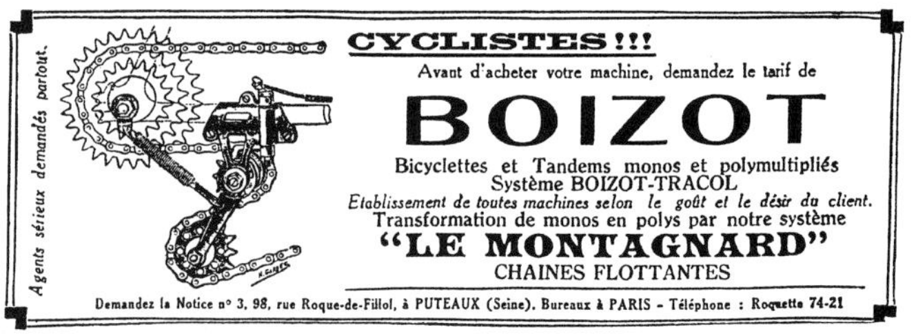


Left: Fig. 6.13. Advertisement for Boizot's Le Montagnard, from a 1925 issue of *La Pédale*.

top teams. The day before the 1928 Paris–Roubaix race, he finally convinced the Alcyon team manager to adopt the Simplex derailleur, and spent the whole night installing his derailleurs on the team bikes. The next morning, the racers refused to face the muddy cobbled roads of the "Hell of the North," as the Paris–Roubaix race was (and still is to this day) popularly called, with this fragile and untried gadget that came off with the wheel and got in your way when you had to change a tire during the race. A very disappointed Juy had to remove all the derailleurs. He returned with them to Dijon with rage in his heart. He eventually did enjoy great success with his Simplex derailleurs, but that was not until the 1930s.

## Other French Derailleurs

At least five other companies made rear derailleurs in France in the 1920s. These included L'As (the Ace) Super-As (Super Ace), Boizot's Le Montagnard (the Mountaineer), Le Crack, and Tracol's L'Ecureuil (the Squirrel).

In 1920, Couzon, a manual worker, and Bouillet, a racing cyclist, invented an "automatic" 2-speed derailleur called L'As (the Ace). It was manufactured by Brunet and Company at Grand Croix, near St.-Étienne. This derailleur looked conventional with two pulleys, a tension arm and a coil-spring, but it had no control cable or rod. The gearshift was made by backpedaling a quarter-

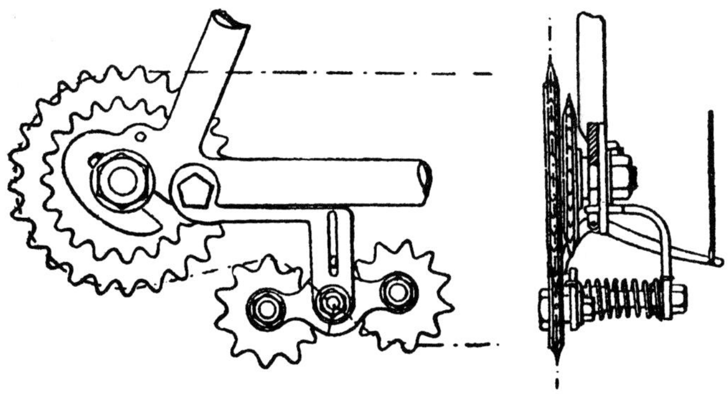

Above Fig. 6.16. The 1924 Le Crack 2-speed derailleur had a jockey pulley and a tension pulley. They were pulled outward by a cable for high gear. A return-spring engaged low gear.

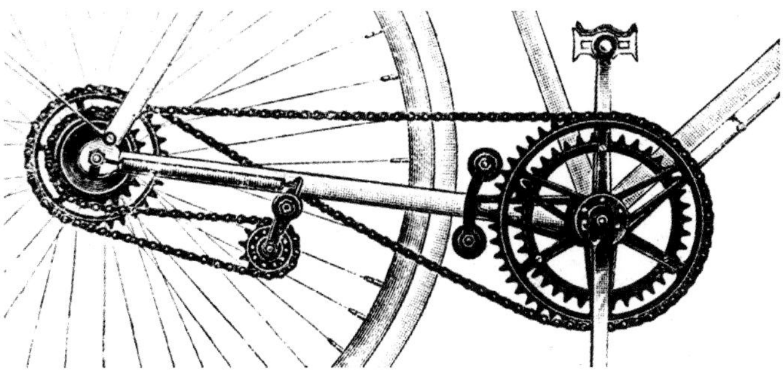

Above: Fig. 6.14. Close-up of the Hirondelle retro-direct with a front changer. A lever on the frame operated a rod that moved one pulley or the other to contact the chain and shift it from one chainwheel to the other.

Below: Fig. 6.15. From 1924 up until World War II, Hirondelle equipped its retro-direct bicycles with two chainwheels and a front changer.

Above: Fig. 6.17. This typical 1922 Tour de France bicycle had a fixed gear and a reversible rear wheel, and only a front brake. It weighed 11 kg (24 lbs.). These were for men of iron, who had to deal with stages as long as 400 km (250 miles) and steep, unpaved mountain passes—on bicycles without gearing.

turn, like the Touricyclette. The main virtues of L'As were simplicity of operation and robustness, but it was quite heavy.

In 1925, Charles Boizot, of Pars, refined his 1910 Bi-Direct rear derailleur to handle 3 sprockets. It still had the tension spring attached to the rear axle. In 1929, Boizot revised his Le Montagnard derailleur to maintain chain tension with a long coil spring attached to the chainstay.

### Front Derailleurs

In this period, most bicycles had a single chainwheel. The only alternatives available were the Hirondelle front changer of 1924, and Raimond's 1925 double chainwheels, with the chain shifted by hand. There was research into multiple chainwheels and front changers for touring-bicycle trials, and occasional special orders from cycle-touring connoisseurs.

Above: Fig. 6.18. The 1928 Intégral front derailleur. Note how the two chainwheels run between the chainstays and the split tubes of the main frame.

Around 1928, the Intégral front derailleur excited interest amongst bicycle tourists. It was usually equipped with three chainwheels, but it could have as many as five. It was cable-operated and consisted of a lever with a central pivot and a fork at its lower end. The chain was pulled onto the larger chainwheels, while a spring pushed it back onto the smaller ones. The bicycle itself was of an unusual experimental design, with the chainwheels inside a divided frame, and with a floating chain. It was a failure in the Chanteloup trials, but André de Boubers, president of the FFSC, used one for many years.

From 1924 up until World War II, Hirondelle equipped its retrodirect bicycles with two chainwheels and a front changer. The 1924 Hirondelle catalog gave this cryptic description of how the changer worked: "Two conical pulleys, on a two-way arm moving around the bottom bracket, make the chain pass from one chainwheel to the other."

Jan Heine road-tested a 1920s Hirondelle with a 4-speed retrodirect and a front derailleur. Changing to back-pedaling was seamless but Jan could not back-pedal faster than 40 RPM. The front derailleur shifts were incredibly smooth in both directions.

## Variable Gears in Italy

As early as 1914, Bianchi had devised a way to move the position of the rear wheel, enabling the chain to run on different-sized sprockets. Other Italian companies and individuals were also working on gear-change devices during this period.

### Vittoria, the First Italian Derailleur

In the 1920s, Italian racers, like their French counterparts, changed gears by stopping, get-

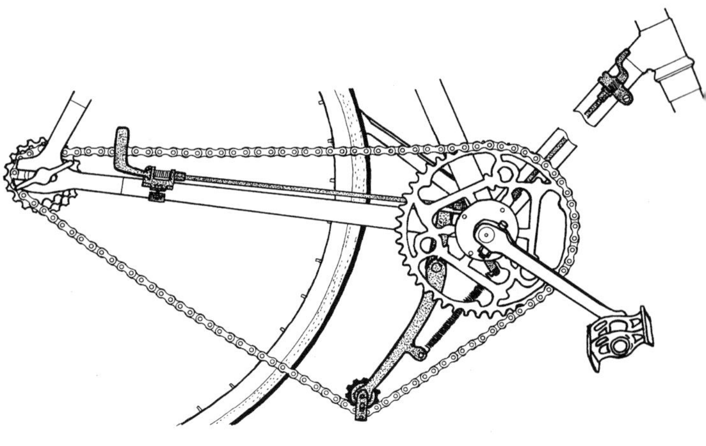

Right: Fig. 6.19. Tommaso Nieddu's 1928 Vittoria derailleur installed on an Ancora ("anchor") bicycle. Note the anchor-motif cutouts in the chainwheel.

ting off their bikes, reversing the rear wheel and shifting the chain onto another sprocket. Italian Alfredo Binda still used this technique to win the 1927 world road cycling championship. The Nieddu brothers, Tommaso and Amedeo, developed the first Italian derailleur system in about 1927.

Tommaso was a postal clerk in Turin, and an avid cyclist, who loved to tinker with bicycles, a practice he continued until his death in 1984, at the age of 92.

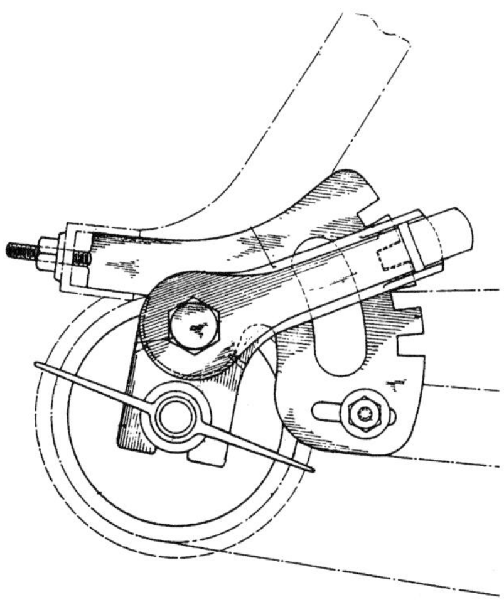

He was especially interested in bicycle gears—having developed 3-speed freewheels as early as 1912.The Nieddu family began to supply Vittoria derailleurs to Italian amateur racers from 1928 to 1930. The Vittoria consisted of a derailing fork mounted on top of the chainstay. To shift gears, the rider pedaled backward, and swung a cranked lever on the downtube. This lever rotated a long rod that twisted the derailing fork from side to side, moving the chain to the next sprocket. A spring-loaded arm, mounted just behind the bottom bracket, maintained the chain tension.

When Vittoria derailleurs began to get good results in races, the advantage of being able to shift gears while moving became apparent, even to the professional racers.

Left: Fig. 6.20. Dr. Bruno Carattoli, of Perugia, patented this small lever that would move the rear dropout forward or backward to allow the chain to be put on either of two different sprockets. A later version allowed for up to four speeds.

## Bicycle Gearing in Britain

The British economy recovered slowly from the devastation of World War I. Bicycles were used extensively for everyday transport and increasingly for recreation and sport. During the 1920s, some five million bicycles were produced annually in Britain. Many had three-speed hubs. The

Above: Fig. 6.21. A famed tricycle record-breaker of the 19th Century, F.T. Bidlake was now a respected elder spokesman, with a regular column in the weekly *Cycling*. Bidlake wrote several articles stressing the importance of the presumed "correct" chain line.

Left: Fig. 6.22. Despite the limited use of variable gearing, cyclists toured all over Britain and continental Europe. There was little motor traffic, and touring by bicycle to unknown places was a grand adventure. Throughout the first half of the century, Frank Patterson drew much-loved rural scenes for *Cycling*. Nearly all of his sketches include a lone male cyclist on a single-speed bicycle.

least expensive utility models were single-speeds, with either a fixed sprocket or a freewheel. Half-inch (12.7 mm) pitch chains were used and the chain width fit sprockets and chainwheels with a thickness of 1/8 inch (3.2 mm). The British Gradient and Whippet derailleurs, which had been marketed at the turn of the century, had long faded from memory.

In a review of variable gears in *Cycling* in 1924, Henry Sturmey, a journalist and an early partner in Sturmey-Archer, did not mention derailleur gears at all, except to note:

> In France, especially in the Alpine districts, quite a number of somewhat quaint and crude devices are in use for the purpose of giving the rider more gears—in one case as many as seven—and an annual reliability trial is held over a mountainous route, mainly for the purpose of testing these gears.

News about French derailleurs, and about the French variable-gearing trials was frequently reported in *Cycling*. Nevertheless, British manufacturers continued to advertise mainly 3-speed or single-speed machines.

Touring cyclists used a "lightweight" design with 26x1-inch wheels, a smaller frame, and drop-handlebars. Typical gearing consisted of either a 44-, 46-, or 48-tooth chainwheel and a single 18-tooth sprocket. With a three-speed hub, a slightly larger sprocket would be used. The emphasis was on smooth steady pedaling uphill and "nimble" pedaling downhill. When hills became too steep, cyclists got off and walked. In Britain, many roads dating back to Roman times go straight up the hillsides. Riding a bicycle up such gradients is impossible without very low gears.

Cycling reported on tours in Britain and abroad. In 1928, two London cyclists climbed the 9,055-ft. (2,760 m) Stelvio Pass. They used hub gears with bottom gears of 32 and 34-in. respectively, enabling them to pedal all the way, whereas turn-of-the-century cyclists trudged up the Alpine passes pushing their bicycles.

Racing in Britain consisted largely of time trialing, where cyclists raced individually over a distance of 25, 50, or 100 miles, or for times of 12 or 24 hours. Flat courses were preferred, and riders tried to maintain an even speed. Under such conditions, variable gears were of little advantage.

Many touring and racing bicycles were equipped with two-sided hubs with, for example, a 16-tooth sprocket on one side and an 18-tooth sprocket on the other. The rear axle was held with wingnuts, so that the wheel could be removed and reversed without tools. However, this reversing was seldom done out on the road. At the start of the racing season, "medium-gear" events were held, with cyclists restricted to gears of 72 in. or

Left: Fig. 6.23. *Cycling* photo of a bicycle on the Grimsel Pass in the Alps. Although it may have had a hub gear, it clearly did not have derailleur gearing.

Above: Fig. 6.24. Louis Camillis was a Britain-based importer of French products, who represented Le Cyclo from 1928 onward.

even to 63 in. This was supposed to encourage the development of fast pedaling style and avoid knee strain in the cool spring weather.

By the late 1920s, double-sprocket freewheels were available. By installing a double sprocket on either side of the rear hub, a total of four different sprockets was possible. Many British cyclists shunned double sprockets because they involved running the chain out of line.

As early as 1913, F.T. Bidlake concluded an article on "The Importance of Chain Alignment" as follows:

> I cannot rid myself of a deeply-rooted preference for rigid adherence to absolutely true alignment for obtaining the finest results.

In 1929, Bidlake was still writing:

> I must abstain from enthusiasm about a chain that runs crookedly and has to be of inferior character to undertake the job.

In another article, he wrote:

> With derailleurs we have to start off with an admission that precision is impossible, and the allegation that it is superfluous if we resort to chains designed and urged to run off the rails.

Le Cyclo derailleurs became popular, particularly for tandems. Solo touring was almost entirely a male activity. Tandems enabled women to ride behind their male companions. Hub gears often failed in tandem service.

Derailleur use in Britain in the 1920s was restricted to a small number of cycling enthusiasts. The first Le Cyclo derailleurs, which had been purchased from Vélocio, were imported into England in 1926. Louis Camillis, an import agent for cycle and motorcycle components, represented Cyclo in Britain. In 1928, several British bicycle enthusiasts wishing to try out French derailleurs, especially the Le Cyclo, approached him. His friend Victor Bastide, a Parisian craftsman, advised Camillis that the Le Cyclo was not a practical device, so Camillis discouraged orders. In 1929, after numerous requests, he finally accepted a firm pre-paid order for 25 Le Cyclo derailleurs. He received further orders, and had sold 150 by the end of 1929.

When he first made them available, many British bicycle mechanics had trouble attaching them, as the standard rear hubs and rear axles were too short to accommodate three sprockets and the frames were too narrow. Albert Raimond, Le Cyclo's manufacturer, encouraged Camillis to persist, and sent over special hubs. Eventually a few small British bicycle makers adapted their designs to accommodate French Cyclo derailleurs.

## Cycling in theUnited States

The automobile, particularly the Model T Ford, triumphed over the bicycle in the U.S. The first 1908 Model T sold for $850. By 1915, the price was down to

Left: Fig. 6.25. In 1921, a new Model T Ford touring car cost $355.

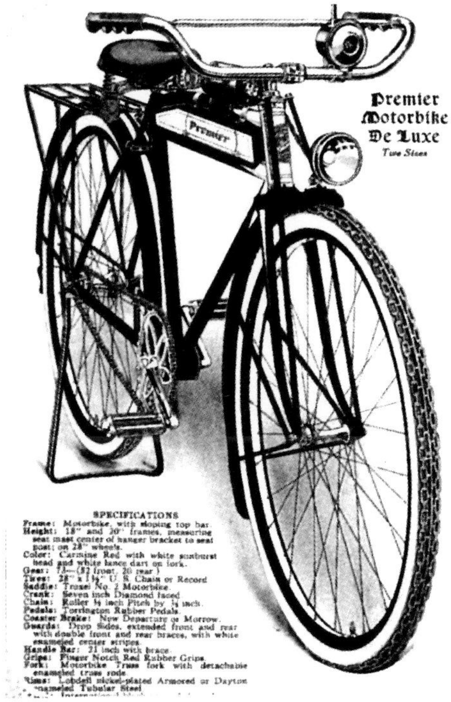


Right: Fig. 6.26. The 1920 Ranger Motor Bike (just a name, because it did not have a motor) was styled like a motorcycle to appeal to the American teenager of the time. Shipping weight was 70 lbs. (32 kg).

$440 and 370,000 were sold. In 1920, more than one million Model T cars were sold. In the mid-1920s, the price went as low as $290, and 2 million Model T's were sold annually. The American bicycle market decided to concentrate on the juvenile market, making bicycles with dummy gas tanks, cross-braced handlebars, fork braces, fat tires, heavy fenders, motorcycle saddles, rear carriers, headlamps, and battery cases.

## Germany

There was no significant interest in derailleurs in Germany during this period. However, there was significant development in hub gearing.

Fichtel & Sachs (F&S) were the largest hub gear makers in Germany. They first made the legendary F&S Torpedo three-speed hub gear in 1925. At this time, their hub gears were more advanced than what Sturmey-Archer was making in Great Britain.

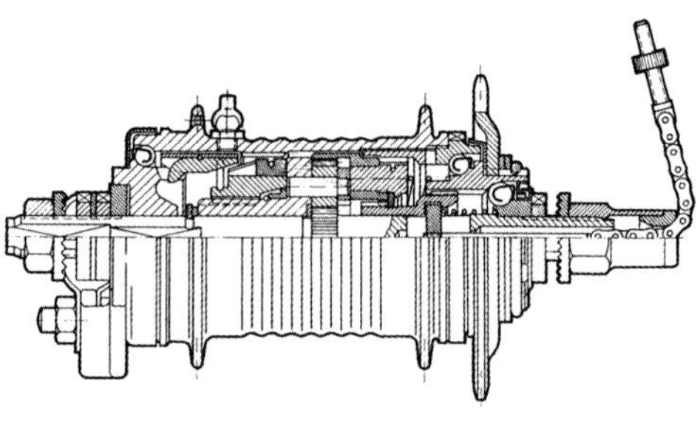

Above: Fig. 6.27. The 1925 Fichtel & Sachs Torpedo three-speed hub gear with coaster brake.

## Japan

Japan had a significant domestic bicycle industry in the 1920s, producing a million bicycles per year in 1925 and 1.5 million in 1929. By the mid-1920s, Japan had become a major exporter to the Asian market.

In 1912, Shikanosuke Maeda founded the Maeda Iron Works in Kawachii-Nagano. Shortly afterward they moved to Sakai. near Osaka. They made sprockets and parts for agricultural machinery. They also made bicycle sprockets and freewheels, calling their bicycle products 8.8.8., which (just like Shimano's 3.3.3.) allegedly was meant to look like "B.S.A.," the hallmark of quality in the eyes of discerning Japanese cyclists of the time.

In 1921, Shozaburo Shimano established the Shimano Iron Works in Sakai City, Osaka. Their first product was a bicycle free-

Right: Fig. 6.28. Shikanosuke Maeda, founder of Maeda Iron Works, manufacturer of SunTour products, in 1912.

Below: Fig. 6.29. Shimano's 3.3.3., SunTour's 8.8.8., and two other Japanese companies' product designations as derived from the BSA trademark. This explanation appeared in Maeda Industries' 60-year anniversary publication.

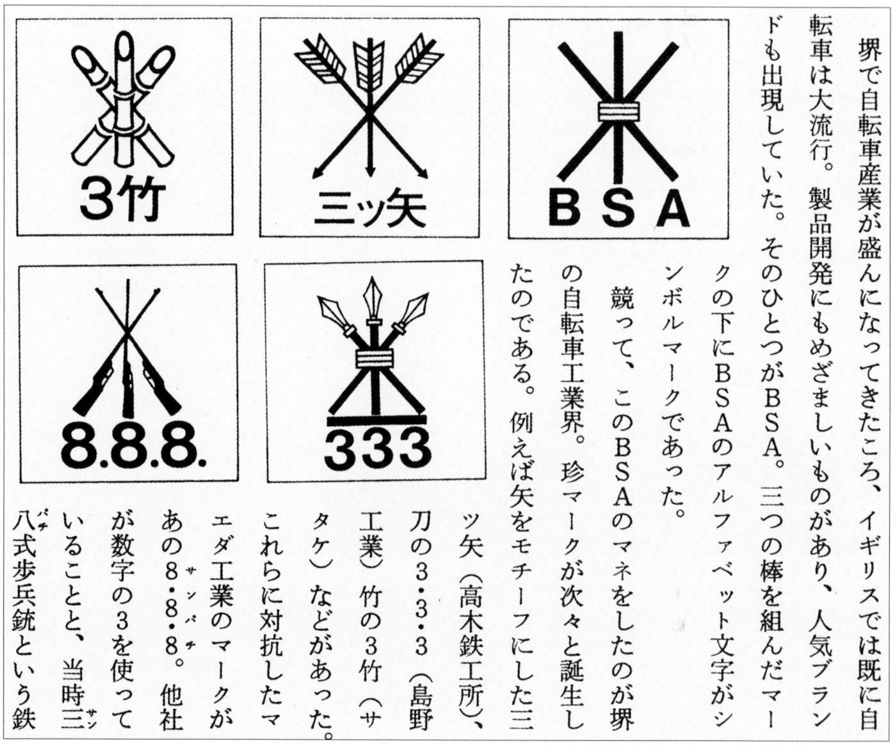


堺で自転車産業が盛んになってきたころ、イギリスでは既に自転車は大流行。製品開発にもめざましいものがあり、人気ブランドも出現していた。そのひとつがBSA。三つの棒を組んだマークの下にBSAのアルファベット文字がシンボルマークであった。

競って、このBSAのマネをしたのが堺の自転車工業界。珍マークが次々と誕生したのである。例えば矢をモチーフにした三ツ矢（高木鉄工所）、刀の3・3・3（島野工業）竹の3竹（サタケ）などがあった。

これらに対抗したマエダ工業のマークがあの8・8・8。他社が数字の3を使っていることと、当時三八式歩兵銃という鉄

Above: Fig. 6.30. Shozaburo Shimano (left) founded Shimano Iron Works in 1921.

Below: Fig. 6.31. The first Shimano product was this 3.3.3. freewheel, introduced in 1921.

wheel. By 1922, production of Shimano 3.3.3. freewheels had reached 3,000 per month. The 1927 depression caused Shimano to close its doors. They reopened after the crisis and built a new 25,000 square meter (290,000 sq. ft.) factory.

## Bicycle Racing

Throughout this period, the use of derailleurs was still not common in bicycle racing. In fact, their use had been prohibited in the Tour de France since 1919 and would remain so until 1937. The racer's only options were still reversing the rear wheel, with different size sprockets on both sides, or switching out bikes for steep climbs.

### Speed Records

In 1924 the motor-paced speed record, held by Mile-a-Minute Murphy since 1899 was improved to 133 km/h (83 mph) by Gustave Ganay of France. The new record was set at Daytona Beach.

Right: Fig. 6.33. Typical for racers of the period, Australian Hubert Opperman rode fixed-wheel bicycles, i.e. without derailleurs, or even a freewheel.

Above: Fig. 6.32. In 1924, Gustave Ganay set a new paced speed record of 133 km/h (83 mph) at Daytona Beach.

*Table 6.2. Tour de France winners, 1920–1929*

| Year | Winner | Bicycle | Derailleur |
|---|---|---|---|
| 1920 | Philip Thys (Belgium) | La Sportive | forbidden |
| 1921 | Léon Scieur (Belgium) | La Sportive | forbidden |
| 1922 | Firmin Lambot (Belgium) | Peugeot | forbidden |
| 1923 | Henri Péllisier (France) | Automoto | forbidden |
| 1924 | Ottavio Bottechia (Italy) | Automoto | forbidden |
| 1925 | Ottavio Bottechia (Italy) | Automoto | forbidden |
| 1926 | Lucien Buysse (Belgium) | Automoto | forbidden |
| 1927 | Nicolas Frantz (Luxembourg) | Alcyon | forbidden |
| 1928 | Nicolas Frantz (Luxembourg) | Alcyon | forbidden |
| 1929 | Maurice Dewaële (Belgium) | Alcyon | forbidden |

# 7

# The Golden Age: 1930–1939

THE 1930s was the golden age for bicycle touring and for derailleurs in Europe. Increasing affluence, and legislation that allowed more leisure time, enabled people in Europe to explore the countryside by bicycle. Dozens of different brands of derailleurs appeared on the market.

These new devices were gradually accepted by touring cyclists, and eventually by racers. In this chapter, we describe the development and use of derailleur gears as it spread from the incubator of mainly French bicycle touring clubs to other countries around the world.

Left: Fig. 7.1. French *cyclo-touristes* in the 1930s often used the most advanced derailleurs that were being developed at the time. This photo shows Régina Gambier (center) and others on a cross-country randonnée stopped in Abbeville in 1931

## Cycling in France in the 1930s

Cycling increased greatly in popularity in France during the 1930s. The membership of the Fédération Française des Sociétés de Cyclotourisme (FFSC) quadrupled between 1928 and 1936.

The year 1936 was important in the social history of France, and also in the history of cycletouring. The great depression of the early thirties was still being felt. Workers were dissatisfied with working conditions, and they wanted better pay and shorter working hours. Shortly after the general elections, there was general strike throughout France. In response, the Popular Front government passed laws that mandated a 40-hour workweek and 15 days of paid vacation annually.

This released a flood of new vacationers onto the French roads. Although they had more time, their income had not increased significantly, so they had to be frugal. The bicycle was the obvious means of escape, with a trailer for their luggage, and a tent for overnight accommodation. One thing they could afford was the luxury of a rear derailleur. Numerous new bicycle brands appeared, usually offering racing and touring models, as well as tandems.

The growing number of serious riders gave added support for the technical trials for touring bicycle of the time. The small makers, who were usually also practicing cyclists, experimented to improve their bicycles. Aluminum alloys were widely used from 1935 onward, and touring equipment became considerably lighter. The increasing numbers of bicycle tourists provided a market for improved equipment.

Front derailleurs did not sell well, as most cyclists considered the three to five gears provided by a rear derailleur sufficient.

## Bicycle Racing

Bicycle racers stayed faithful to steel components, but the derailleur was no longer a discredited accessory and was now allowed in most races. Henri Desgrange had long opposed variable gears, but he retired as the organizer of the Tour de France in 1936, and the new organizers permitted derailleurs starting with the 1937 Tour. Given the enormous popularity of the Tour, this was hailed as the ultimate victory for the rear derailleur.

The generic Tour de France bicycles were made by Alcyon, and the frame decals said *L'Auto*. The 1937 model had a Super Champion derailleur, a 47-tooth chainwheel, and a 4-speed freewheel with 15-, 16-, 17-, and 18-tooth sprockets. This would have given it a high gear of 85 in. and a low gear of 70 in.

### Speed and Distance Records

In 1933, Maurice Richard raised the unpaced one-hour record to 44.78 km (27.83 miles), beating Oscar Egg's 1914 record; and in 1937, Maurice Archambaud raised it to 45.84 km (28.49 miles).

In 1933, Marcel Berthet went 50 km (31 miles) in one hour in a Vélodyne streamlined bicycle. In 1939, François Faure went 50.5 km

*Table 7.1. Tour de France winners, 1930–1939*

| Year | Rider (Country) | Bicycle | Derailleur |
|---|---|---|---|
| 1930 | André Leducq (France) | generic* | forbidden |
| 1931 | Antonin Magne (France) | generic* | forbidden |
| 1932 | André Leducq (France) | generic* | forbidden |
| 1933 | Georges Speicher (France) | generic* | forbidden |
| 1934 | Antonin Magne (France) | generic* | forbidden |
| 1935 | Romain Maes (Belgium) | generic* | forbidden |
| 1936 | Sylvère Maes (Belgium) | generic* | forbidden |
| 1937 | Roger Lapébie (France) | generic* | Super Champion |
| 1938 | Gino Bartali (Italy) | generic* | Vittoria Margherita |
| 1939 | Sylvère Maes (Belgium) | generic* | Super Champion |

* Henri Desgrange tried to limit the influence of the manufacturers of bicycles and components. For that reason, in this decade of national teams, all riders were supplied with a standard bicycle, but they could use their own saddle and handlebars.

Below: Fig. 7.2. In 1941, Alfred Letourner went 108.9 mph (175 km/h). His Schwinn bicycle with 1-inch-pitch chain had a 57-tooth chainwheel and a 6-tooth sprocket, resulting in a 252-in. gear.

(31.4 miles) in one hour in a Vélocar streamlined recumbent bicycle.

In 1941, Alfred Letourner covered a mile in 33.05 seconds behind a hood on a race car — that's 108.9 mph (175 km/h).

## French Derailleurs

In the following sections, we'll take a look at the major players in the development of the derailleur in France during these years.

Above: Fig. 7.3. Albert Raimond at a display of Vélocio's bicycles at a *Semaine du Cycle* (a regional cycle show) in St.-Étienne in 1931.

Below: Fig. 7.5. The 1931 Cyclo Tank rear hub with a 4-speed freewheel allowed the rider to replace a spoke by separating the freewheel body from the pawls by hand.

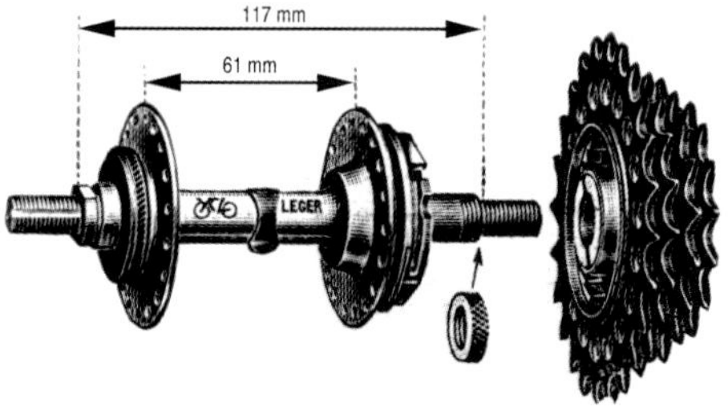


**MOYEU CYCLO-TANK,** 4 vitesses, léger, vélo, poids complet 0 k. 700

### Le Cyclo

In 1930, Albert Raimond produced the 4-sprocket Le Cyclo Dural rear derailleur, made entirely of aluminum alloy, and the Rosa front derailleur.

When his friend Vélocio died in 1930, Raimond bought his company and equipment, including *Le Cycliste*. He moved into Vélocio's building, although it was hardly bigger than the one he left.

Raimond expanded his activities overseas. When Britain levied heavy customs duties on his products, he formed a joint partnership with his agent Louis Camillis to open The Cyclo Gear Company in Birmingham, England. Le Cyclo derailleurs for the British market (as well as some other exports) were made in Britain from 1932 onward.

Feeling the competition from Lucien Juy's 1931 racing derailleur, Raimond bought the Witmy patent and marketed it. The Witmy's main advantage was that it could be installed on any bicycle. In French, Witmy is pronounced "veetmee", the same as the words "*vite mis*," meaning quickly fitted. However, it did not shift well, and soon the joke went, "Quickly fitted—quickly removed!"

Raimond's generous PR policy went beyond good business sense. He was unfailingly generous to local cycle-touring clubs and to the FFSC, the national cycle touring association of the

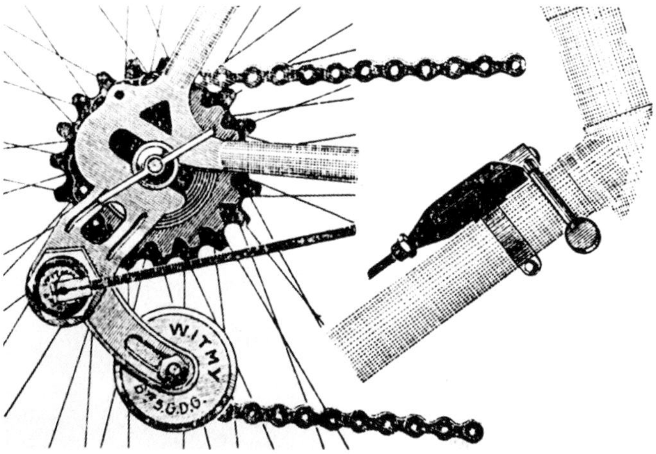


Left: Fig. 7.4. The 1933 Cyclo Witmy 2-speed derailleur, like all other single-pulley derailleurs, had limited capacity and shifted poorly.

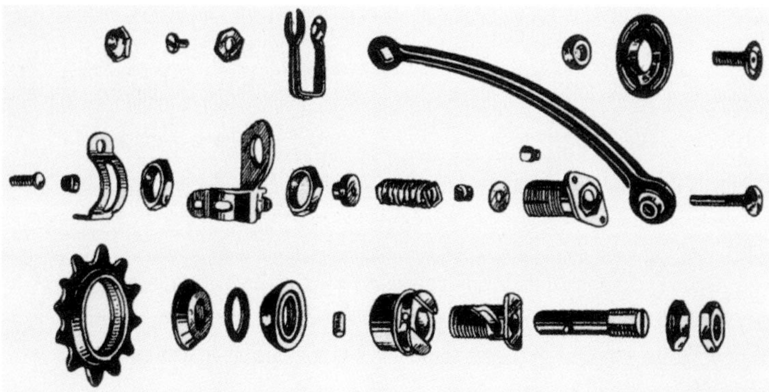

Right: Fig. 7.6. Exploded view of the 1935 Super Cyclo derailleur.

time (the name has changed since). He never forgot that he had been an ardent bicycle tourist, and to the last he was a tourist at heart.

At the end of 1935, Raimond produced two derailleurs—the Super Cyclo and the Poly—designed specially for racing cyclists. By this time, racers were using derailleurs in many events, rebelling against Desgrange's prohibition of multispeed gearing in the Tour de France.

The Super Cyclo had a fixed jockey pulley, moved by a helicoid actuator, while the chain was tensioned by an arm swinging under the bottom bracket. The Poly used the same type of tension arm, but it used a derailing fork. It was difficult to adjust the derailing fork so that it did not scrape after a gear change. The

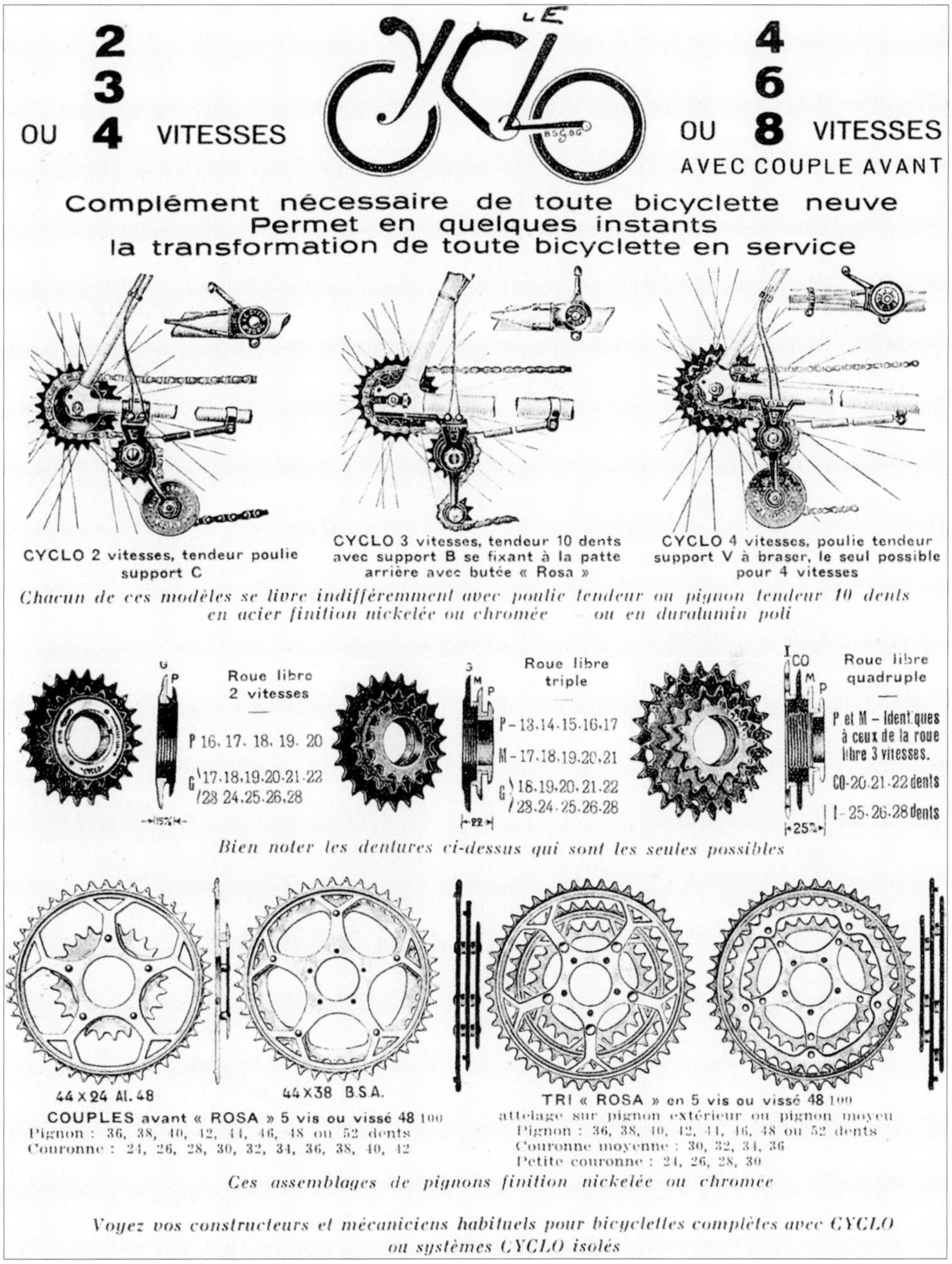


Above: Fig. 7.7. Cyclo's 1933 range of touring derailleurs, freewheels, and chainwheels. Note that the tension pulley was available both as a toothed sprocket and as a flanged wheel.

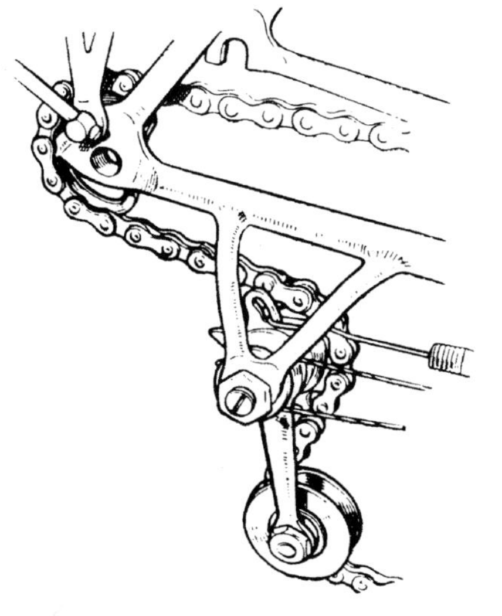

Above: Fig. 7.9. The Cyclo Touriste was a lighter version of the Cyclo Standard, with a smaller jockey pulley and cable drum, a flanged tension pulley, and a Dural (aluminum alloy) tension arm. It accommodated up to 5 speeds. First shown in 1937, it was eventually produced in the new factory starting in 1939.

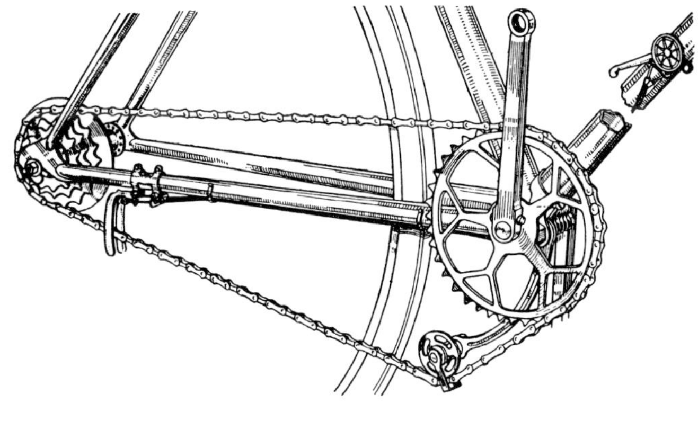

Right: Fig. 7.8. The 1935 French Cyclo Poly was similar to the 1936 Super Champion, but it used a twin-cable operator to control the derailing fork.

British Cyclo Ace was a modified version of the French Cyclo Poly.

The names of Cyclo derailleurs after 1935 can be confusing. When the Cyclo Company made only one derailleur, it did not matter whether it was called Cyclo or Le Cyclo. By 1935, both the British Cyclo and the French Cyclo factories were making different derailleurs and sometimes they used the same name for a different derailleur. It is easy to confuse the British and French names. Over the years, French Cyclo's basic two-cable, helical-operated touring derailleur was called Le Cyclo, Cyclotouriste, or Standard. From this chapter onward, it will be called the Cyclo Standard.

The year 1938 was a turning point for the Cyclo company. They moved into a new, large and well-equipped factory in the small town of St.-Priest-en-Jarez, near St.-Étienne. As the company developed, Raimond increased the work force. From 1938 on, most of Cyclo's components were made in-house.

Everything seemed to be going well for Cyclo, but it did not last. The economic slump of the thirties was being felt. The cost of labor and the cost of living went up, which raised the price of derailleurs. Everybody knew that a war with Germany was coming, so people were saving rather than spending. Bicycle touring increased, but so did competition: Some very good new derailleurs appeared on the market to compete with Cyclo.

Above: Fig. 7.10. This 1938 Cyclo advertisement shows a young man on a loaded touring bicycle with a Cyclo gear train teasing his father with a fixed gear about the difficulty of the climb.

Above: Fig. 7.11. Cyclo Rosa front derailleur on a 1936 Longoni tandem.

## Simplex

In 1931, the basic Simplex rear derailleur was upgraded to three speeds. The 2-speed Sports model was a typical Simplex derailleur, with a single pulley to shift and tension the chain. Although it shifted poorly, it weighed only 150 g (5½ oz.), and

Above: Fig. 7. 12. This 1936 Uldry touring bike with Cyclo Standard rear derailleur had two chainrings, but no front derailleur, so a chainguard could be fitted.

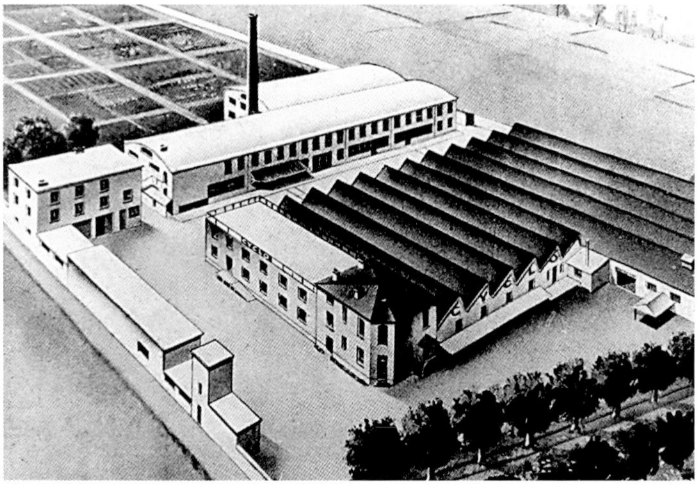

Left: Fig. 7.13. The Cyclo factory at Saint-Priest-en-Jarez, near St.-Étienne, where most of the components were made in-house.

it was easy to remove the rear wheel to change a flat tire. From 1931 to 1934, Simplex won more than one hundred victories each year. Juy claimed a production of 40,000 units in 1933, and 60,000 in 1934. Simplex did much to overcome the anti-derailleur prejudice in racing circles. Well-known riders, such as Maurice Archambaud, Speicher, Le Greves, René Vietto, and Antonin Magne, used Simplex derailleurs.

The meeting between Lucien Juy and René Vietto was a turning point for Simplex. Vietto was a promising young professional racer. He had just won four mountain stages in the 1934 Tour de France (without a derailleur, because variable gears were still forbidden) and had become very popular. He would later be nicknamed "King René." His thoughtful approach to racing equipment differed from the attitude of most other racers. He made or adopted many innovations. Juy and Vietto were destined to work well together. Vietto brought fame to Simplex. All through 1935, Vietto raced with a Simplex derailleur.

The 1934 Simplex Tour de France used a bell-crank to reverse the cable action. Pulling on the control cable did the hard work of pushing the chain up onto the larger sprockets. The spring pulled the chain back down. The mounting did not clutter up the rear dropout because it was fixed on a lug that was brazed on the chainstay. It could be unhitched and swung back for easy rear wheel removal. There was hardly any increase in chain length and

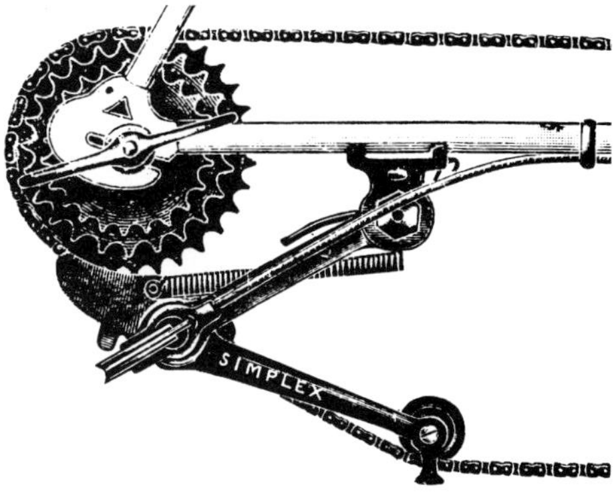


Above: Fig. 7.16. The 1934 Simplex Tour de France was a better touring derailleur than the 1934 Champion de France, because it had three speeds, a longer tension arm, and a maximum range of ten teeth.

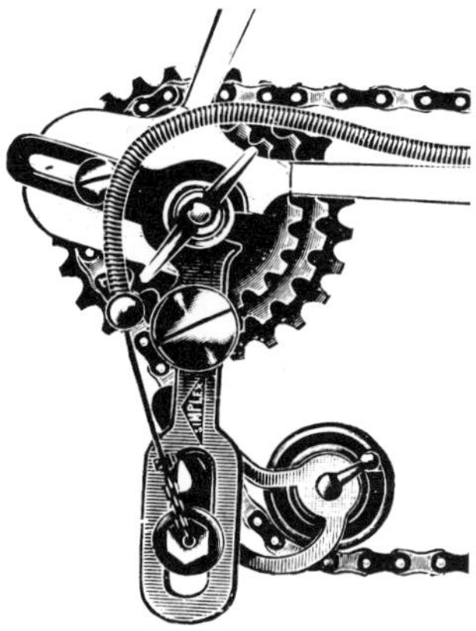

Above: Fig. 7.15. The 1935 Simplex Champion de France was the 1934 single-pulley racing derailleur reworked to give three speeds with a maximum range of 5 teeth. Instead of being bolted onto a lug, it was installed on a bracket that also acted as a stop in the dropout, thus automatically adjusting the wheel position.

Below: Fig. 7.17. The Simplex Champion 35 was a refinement of the 1934 Simplex Champion de France. It still had a single pulley, but the two side plates guiding the chain were larger. The same spring moved the jockey pulley and the cage backward, tensioning the chain.

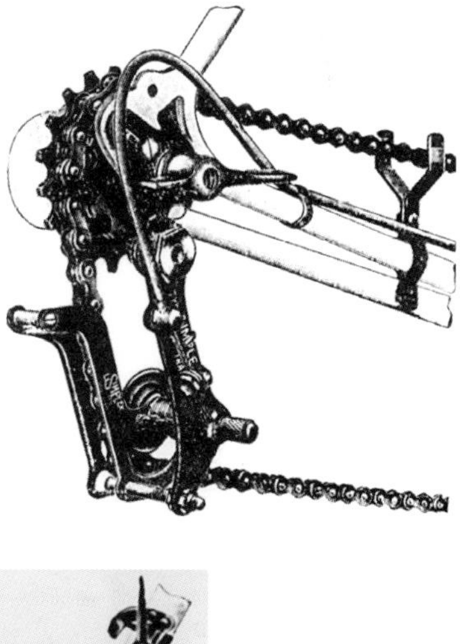

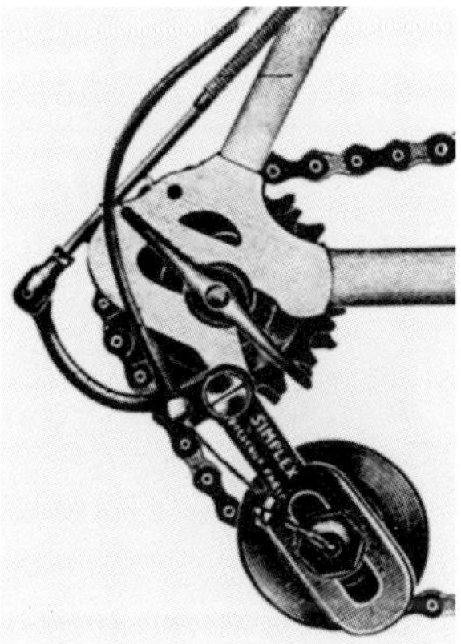

Above: Fig. 7.14. The 1934 two-speed Simplex Champion de France was a better derailleur than the earlier Tour de France. It was installed on a lug on the rear dropout, allowing easy wheel removal. It used a single flanged pulley. It suited the needs of the racing cyclists of the time.

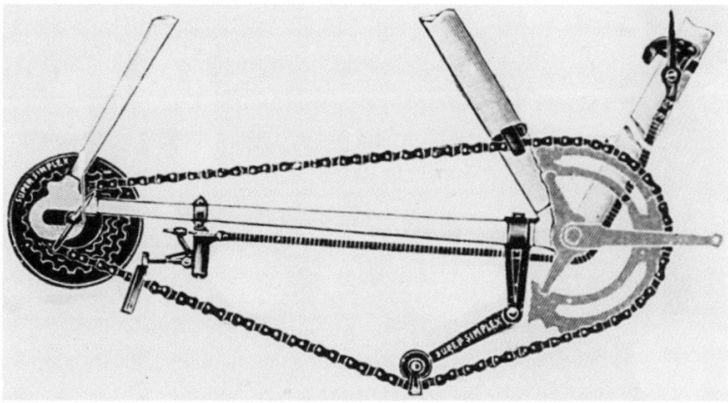

Left: Fig. 7.18. The 1935 Super Simplex had a tension arm and a derailing fork, which was clearly copied from Oscar Egg's Super Champion.

the tension spring was neatly tucked away. Though promoted as a racing derailleur, it had enough capacity for touring, and it was used mainly by tourists. Mademoiselle Lily Sergueiew, aged 22, went on a long tour through Europe using this derailleur.

A year later, in 1935, Simplex introduced two new racing derailleurs and two new touring derailleurs. The 1935 Champion de France was an upgrade of the 1934 Champion de France.

The Champion 35 was Simplex' classic single-pulley racing derailleur. It first appeared in 1935 and it remained in production with minor cosmetic changes until 1948. Over the years, the name changed to Champion 36, and finally to Champion du Monde. By today's standards, it was a fragile, poorly shifting derailleur with limited capacity. However, it was inexpensive and profitable. Lucien Juy used some of his profits to support major teams.

In 1935, the 1932 two-pulley Simplex touring derailleur was renamed the Route. Various versions gave 2, 3, or 4 speeds. It could be installed on the rear axle or the chainstay. A latch allowed the chain to be loosened for wheel removal.

The 1935 Super Simplex was an entirely new model that was available as either a 3-speed or a 4-speed. It had a capacity of 8 teeth. The Super Simplex was copied from Oscar Egg's Super Champion (see next section). The derailing fork of the Super Simplex was mounted on a parallelogram that kept the fork in line with the chain and gave smoother shifting. This was probably the first use of the parallelogram principle on a derailleur. However, some riders thought that the parallelogram linkages were a weakness. Later versions without the parallelogram were called The Selection.

By 1936, Simplex' range of derailleurs could handle 3, 4, or even 5 speeds, using the narrow $3/32$-inch (2 mm) chain. Their freewheel capacity was up to 10 teeth for the racing models and up to 14 teeth for the Route models. By 1936, Lucien Juy was producing a thousand derailleurs a day.

Below: Fig. 7.19. Antonin Magne is leading the 1934 Tour de France, with René Vietto on his left. Derailleurs were not yet allowed in the Tour at the time, but in other races, where derailleurs were permitted, both used Simplex derailleurs.

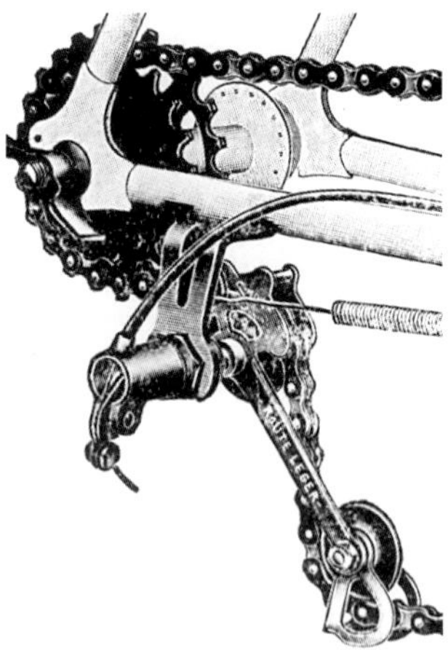

Above: Fig. 7.20. The 1936 Simplex Route was a touring derailleur with a long tension arm and narrow pulleys. It had a bell-crank to reverse the cable action, giving positive shifting to the larger sprockets. At last, Simplex diverged from the Chemineau ancestry of the 1928 Simplex. This model changed little over the next twenty years.

Right: Fig. 7.21. In 1939, Simplex produced hard-rubber pulleys, precursors of the plastic pulleys introduced in the late 1950s.

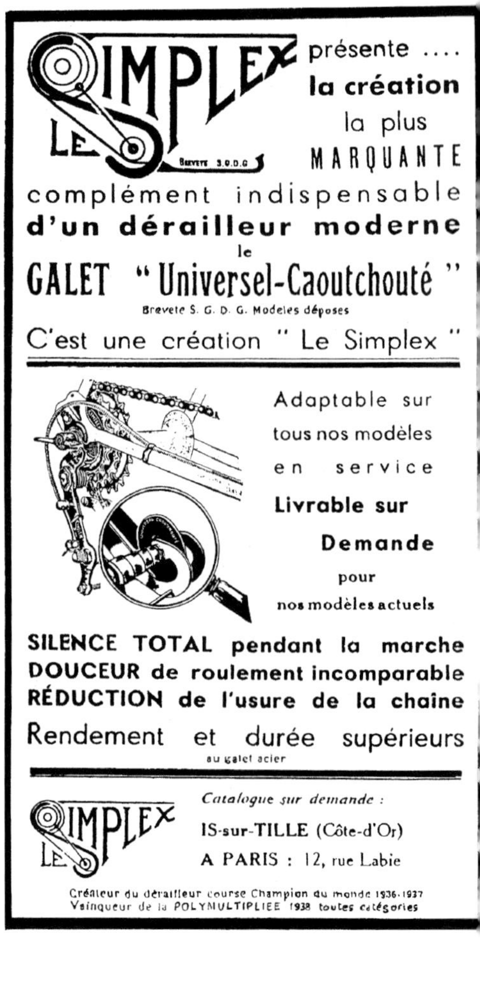

Simplex made minor cosmetic changes to their various models most years, but they changed model names almost every year. Other manufacturers changed model names, but no one changed as often or as confusingly as Simplex. These name changes plague today's derailleur collectors. When Simplex derailleurs were exported to Britain, the British names were not always identical. to those of the same derailleur in France.

Left: Fig. 7.22. In 1936, Simplex introduced 5-speed freewheels with narrow $^3/_{32}$-in. (2 mm) sprockets.

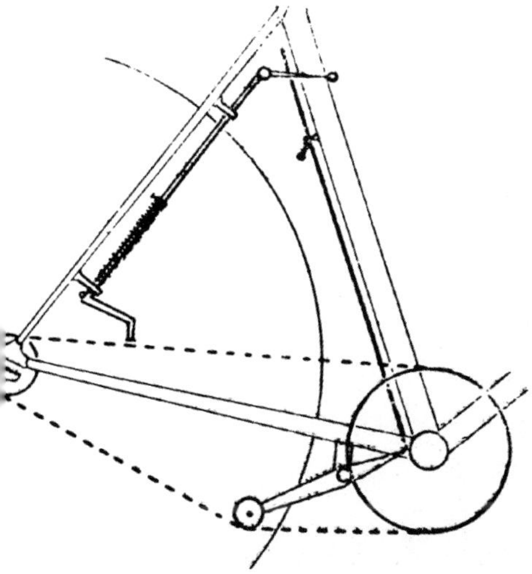

Above: Fig. 7.23. Oscar Egg's 1932 Champion derailleur provided two or three gears and allowed a range of 8 teeth. The following season, leading racers used this model to win several major races.

## Oscar Egg and the Super Champion

Oscar Egg was born in Switzerland in 1890. He became famous as a track racer, but he also won a number of road races. He set the world hour record three times—in 1912, 1913, and 1914—at the Buffalo track in Paris. His last record of 44.47 km (27.5 miles) had stood for 19 years.

Egg presented his Champion derailleur at the 1932 Paris Salon du Cycle (Bicycle Show). It had two separate parts. The derailing fork that shifted the chain sideways was at the back and was operated by a control rod fixed along the seat-stay. The sprung tension arm was fixed below the chainstay near the bottom bracket and was adjusted by a

Above: Fig. 7.24. This 1935 advertisement for the Super Champion said that it was available in a 3-speed racing model and a 4-speed touring model.

Left: Fig. 7.25. This souvenir photo of Oscar Egg dates from 1919. Egg's shop was in the Avenue de la Grande Armée, the fashionable street in Paris for the top bicycle brands. The racing machines he sold gained prestige from his fame as a record-breaker.

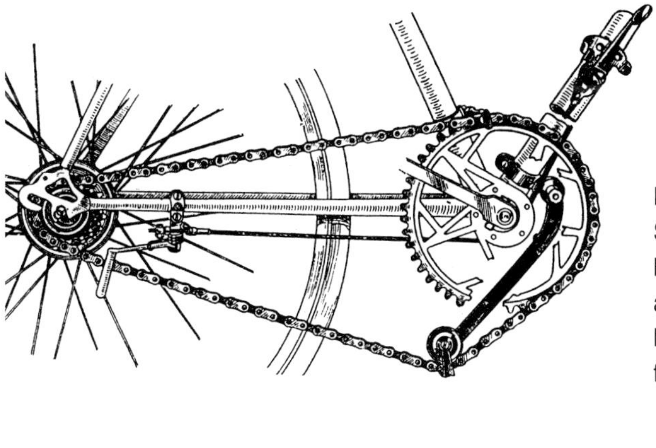

Left: Fig. 7.26. The 1936 Super Champion derailleur had an overall capacity of about 15 teeth and could handle a wide-range freewheel.

rod running down the seat-tube. The chain ran over a pulley at the end of the tension arm.

To change gears, you unlocked the tension arm and backpedaled while operating the derailing fork. After the shift, you relocked the tension arm. Oscar Egg had unwittingly revived Vélocio's 30-year old experiments, or he may have copied the 1928 Vittoria.

In 1933, Maurice Richard increased the unpaced one-hour record to 44.777 km (27.8 miles). His feat revived interest in Oscar Egg's previous one-hour record and in his derailleur. The 1933 model was the first to be called Super Champion. Improvements were soon made, and the 1934/1935 Super Champion was an impressive derailleur.

TOUR DE FRANCE 1937
Le vainqueur, Roger LAPÈBIE entourè de ses camarades GAMARD et CLOAREC

Above: Fig. 7.27. Roger Lapébie (center), winner of the 1937 Tour de France, with team members Gamard and Cloarec. Note their Super Champion derailleurs.

Below: Fig. 7.28. The 1937 Super Champion Professional Tour de France model with chain detensioner. A motor-pace racer, François Vallée, worked as factory foreman at Egg's factory. He devised a chain detensioner for the Super Champion with a wing nut on the shift lever to operate a cable attached to the tension arm.

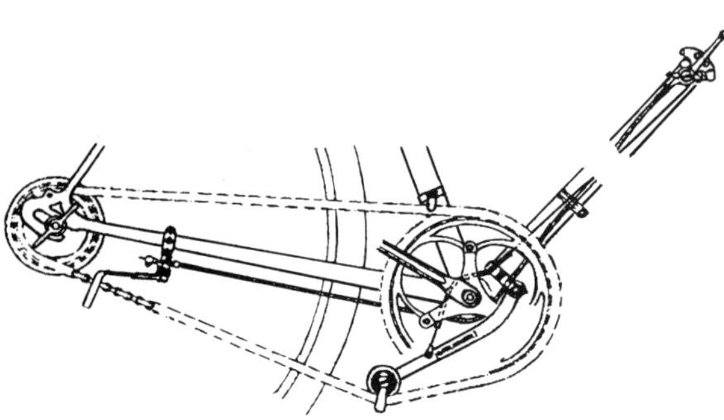

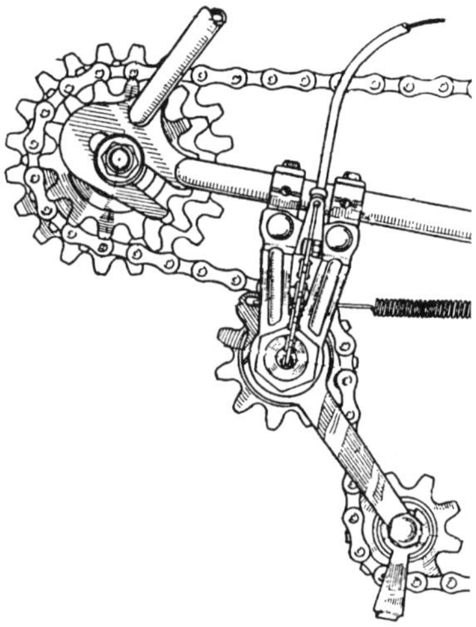

Above: Fig. 7.29. The 1938 Super Champion Grand Sport had a capacity of 30 teeth and was suitable for touring. It was a copy of the age-old Chemineau, with a plunger, a pull-chain operator, and a return spring.

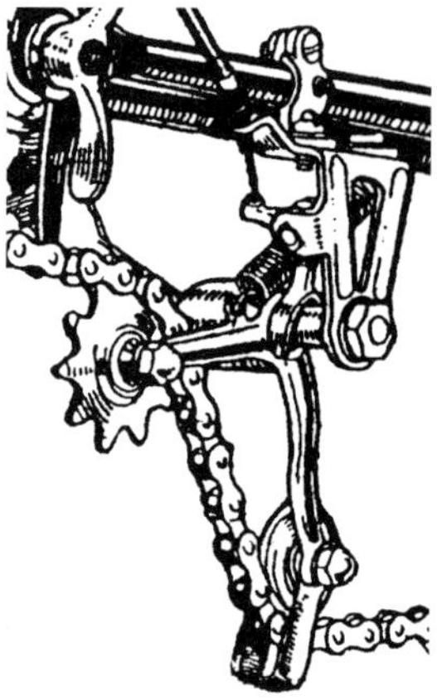

Above: Fig. 7.31. The 1937 Super Champion Route was a two-pulley three-speed touring derailleur with a capacity of 16 teeth.

Below: Fig. 7.32. The 1939 Super Champion du Monde 39 was a single-pulley racing derailleur similar to the Simplex Champion du Monde. It had a capacity of 8–10 teeth.

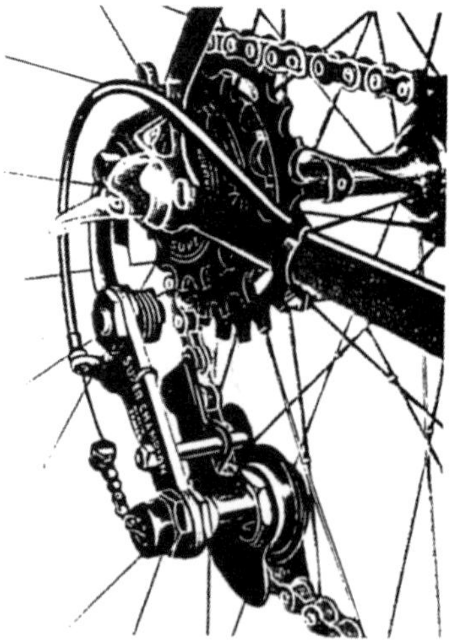

Right: Fig. 7.30. The Belgian racer Sylvère Maes, leader of the *Escadron Noir* (Black Team), won the 1939 Tour de France. Note the characteristic chain profile of the Super Champion tension arm. Maes had also won the Tour in 1936, when derailleurs were still banned.

On the first models, the derailing fork and the tension arm were clamped to the chainstay. On later models, the tension arm was attached to the downtube near the bottom bracket.

Jane Heine tested a 1934 Super Champion derailleur on a 1939 Oscar Egg bicycle with an 18-19-20 freewheel. The detented quadrant for the shift had three holes and two stops. You shift by moving forward two holes and back one (overshifting). The chain moves swiftly and silently over the one-tooth steps on the freewheel. Shifting to the middle sprocket is a bit harder. If you don't overshift enough, the chain simply rides on the sprocket.

The Super Champion pleased the racing fraternity because it was reliable and easy to use. It had no jockey pulley to bend and drag the chain. And this the de-

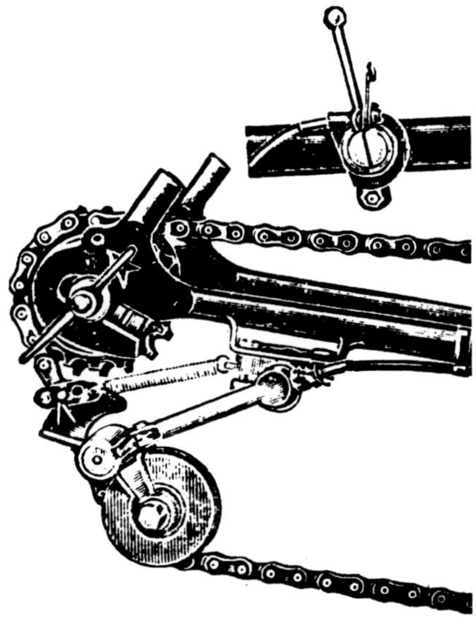

Above: Fig. 7.33. This 1932 Huret model was a typical single-pulley racing derailleur of the era. Georges Speicher won the World Championship with a Huret derailleur.

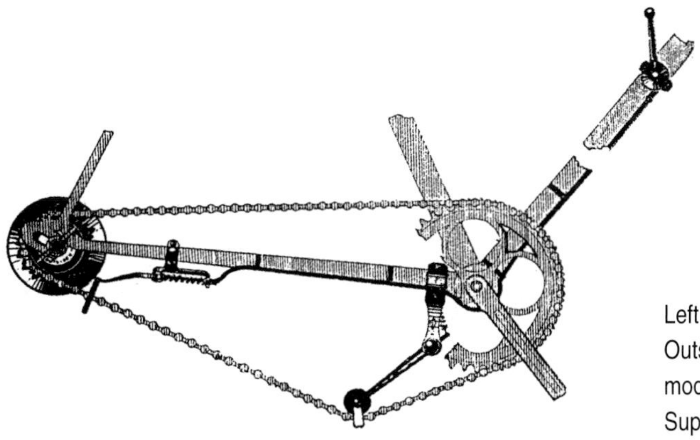

Left: Fig. 7.36. This 3-speed Outsider Course (racing model) was one of many Super Champion copies.

Right: Fig. 7.34. André Huret started making derailleurs in 1930.

Right: Fig. 7.37. The Lewis bicycle and derailleur.

Above: Fig. 7.35. A 1932 Huret advertisement. Translation: "SPEED CHANGER (for Racing) HURET 2-3-4 SPEEDS (by reversing the wheel). Permits a range of 5 teeth and intermediate sizes. PULLEY equipped with a lining that reduces chain wear. Extra-Fast Removal and Installation. IS USED BY THE BEST-KNOWN RACERS. Request the descriptive brochure."

Below: Fig. 7.38 Close-up of the Lewis derailleur. If the derailing fork got damaged, it could be folded out of the way under the chainstay.

railleur did not get in the way when you removed the rear wheel—a frequent occurrence in those days of fragile racing tires and poor roads.

The success of the Super Champion added more pressure to allow derailleurs in the Tour de France. When variable gearing was finally permitted in 1937, it was stipulated that all the racers must have the same type of gear-changer. The French chose the Super Champion and were followed by the Belgians, Spaniards, and Germans. Only the Italians obtained approval to use their own Vittoria Margherita derailleurs. The average overall speed suddenly increased in the Tour, with the last rider averaging about the same speed as the winner of the year before. Roger Lapébie won the 1937 Tour de France, but the real winner was the derailleur.

There were three models: the basic steel Super Champion Standard, the aluminum alloy Super Champion Course, and the Super Champion Professional, with the chain detensioner. After Lapébie's 1937 victory, this model was called the Tour de France.

The Super Champion derailleur had an overall capacity of about 15 teeth. It was also suitable for touring. It could handle a wide-range freewheel (13–28

Below: Fig. 7.39. On the 1936 Gri-Gri Course racing derailleur, the tension arm and the derailing fork were clamped to the right chainstay.

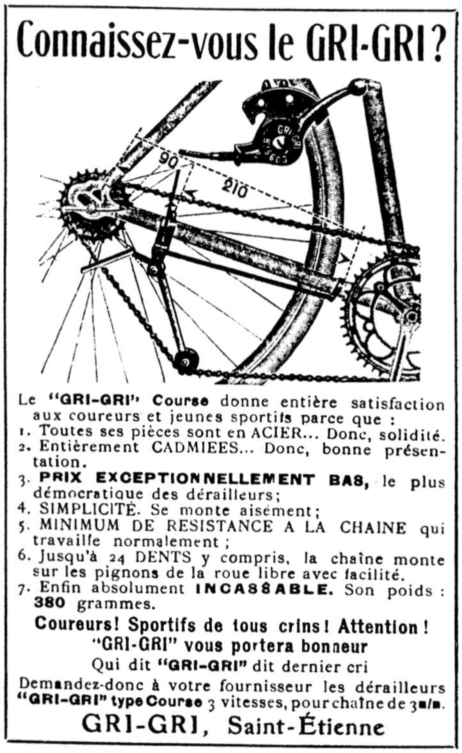


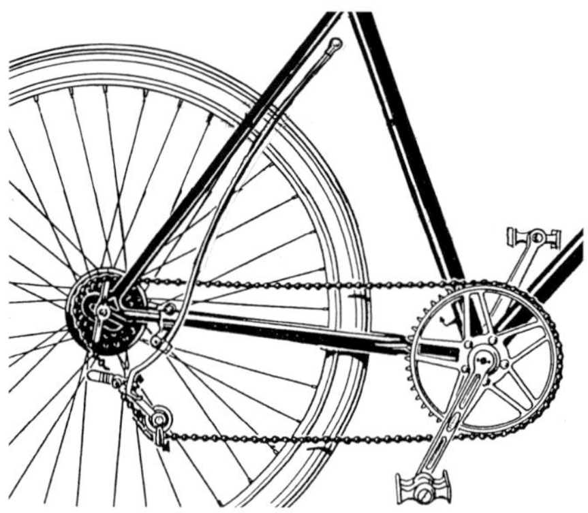

Above: Fig. 7.40. In 1933, Mercier designed a rod-operated rear derailleur for Francis Pélissier, a well-known champion racer. A rod along the seatstay carried a pulley that both shifted and tensioned the chain. Stronglight bought the design in 1935.

Above: Fig. 7.41. Joseph Chavanne and his Pelletier bicycle with Super-Meilhor derailleur. On the 1936 Super-Meilhor, the spiral spring was replaced by a coil spring under the chainstay.

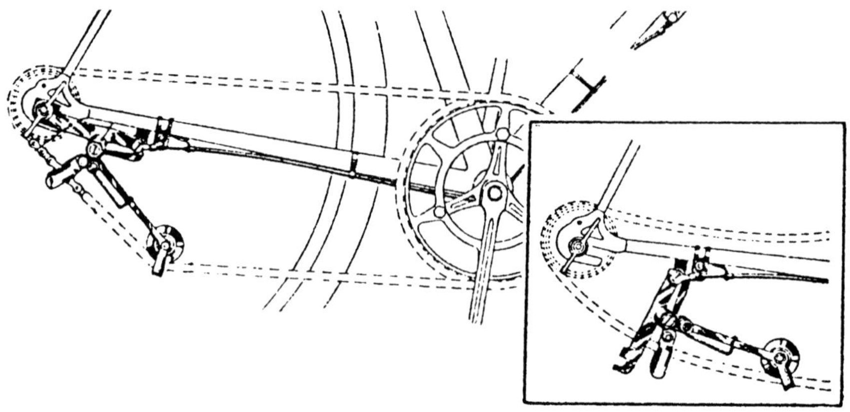

Fig. 7.42. The 1935 Le Vainqueur (The Winner) was a cable-controlled derailleur with a short tension arm articulated near the chainstay. The inset shows how the derailleur swung away when the wing nuts were released.

teeth), and accommodate a close-range ("half-step") double chainwheel. It spawned numerous copies.

You needed a deft flick of the finger to shift gears with the Super Champion. When you missed a shift, the sound of the chain on the fork put your teeth on edge. Nevertheless, the Super Champion was a huge success in the years just before World War II. It was used by many champion racers and accumulated resounding victories, earning enormous publicity for Egg. It was assembled in a large modern factory in Levallois-Perret, a suburb of Paris. The components were made by different outside specialist workshops. In 1934, 45,000 derailleurs were made. In 1936, 343,209 derailleurs were made. The factory could make 4,000 derailleurs a day. By 1939, more than a million Super Champions had been produced.

Recognizing the limitations of derailleurs with a tension arm and a derailing fork, Oscar Egg expanded his range in 1937 to include a conventional two-pulley touring derailleur, the Super Champion Route. This was replaced in 1938 by the Grand Sport. The 1939 Champion du Monde 39 was a single-pulley racing derailleur. The latter two were made under license from Cyclo.

Above: Fig. 7.43. The 1931 Le Lautaret was made by the Charvin company, of St.-Étienne. It was a cable-operated plunger-type derailleur, with the jockey pulley and the tension pulley operated via a bell-crank. The offset of the jockey pulley adapted to the sprocket diameter. The Le Gri-Gri Route was a later version of the Le Lautaret with a flanged jockey pulley. In 1938, the name was changed back to Le Lautaret.

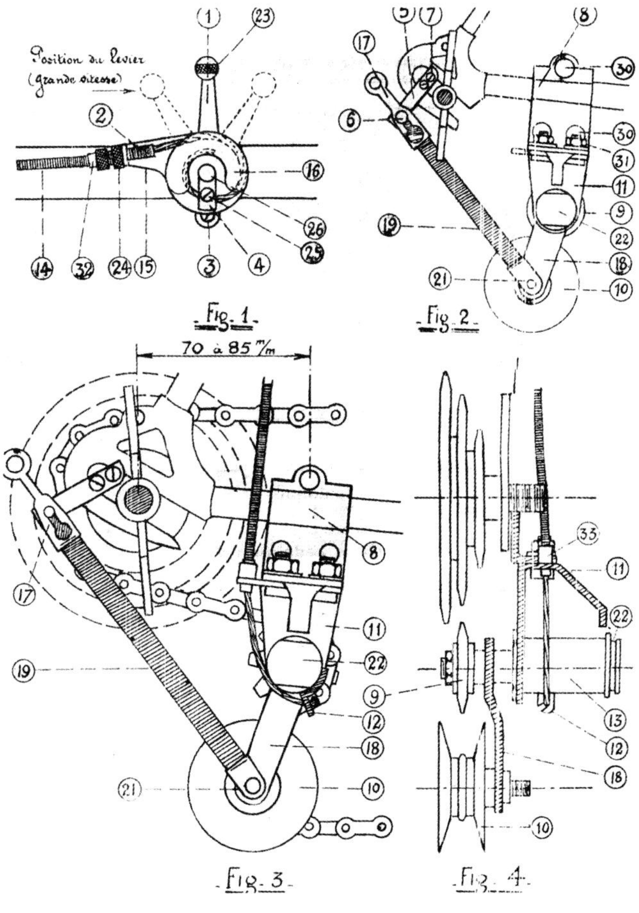


Right: Fig. 7.44. The 1934 Mercier Valor was a single-cable copy of the Cyclo Standard with a bit less capacity. Fig. 2 shows the single-pulley racing model, while Figs. 3 and 4 show the two-pulley touring model.

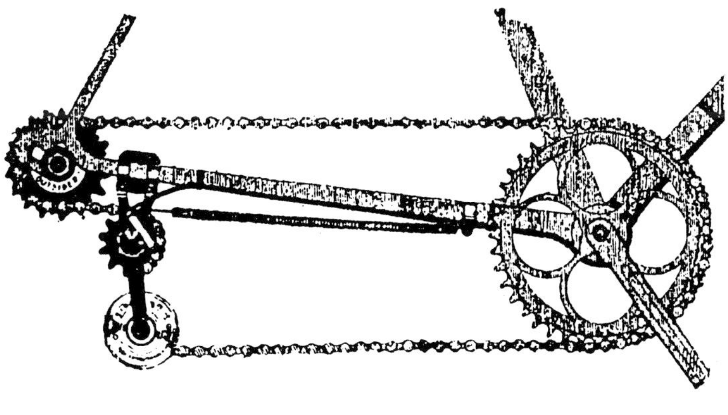

Left: Fig. 7.45. The 1935 Outsider Route (Road) used the control cable to rotate a gear wheel that engaged a rack on a sliding rod that moved the jockey mechanism back and forth.

## Other French Derailleurs

Joanny Panel's Chemineau, Albert Raymond's Cyclo, Lucien Juy's Simplex, and Oscar Egg's Super Champion were the "big four" derailleur makers in the inter-war period, but they had plenty of competition. Other firms marketed interesting models, which can be grouped into the following categories:

- Racing derailleurs with a single pulley near the freewheel

Derailleurs of this type included the Super As, the 1939 Super-Rapid Record Léger (light), the 1931 Simplex Sport, and its ensuing variations: the Le Cyclo Witmy, the Simplex Champion 36, the Huret single-pulley model, and the 1939 Super Champion Champion du Monde.

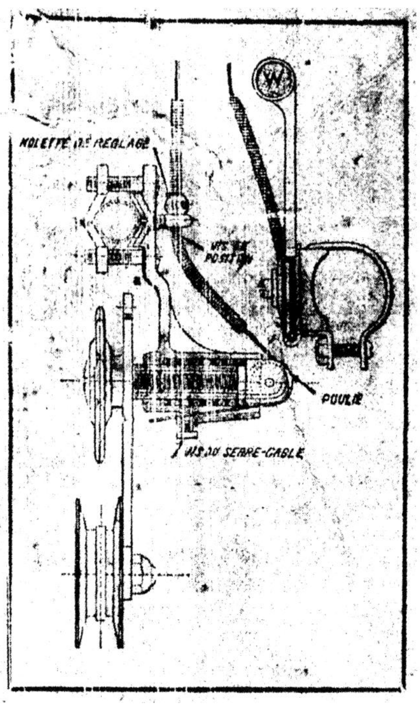


- Racing derailleurs with a derailing fork and separate tension arm and tension pulley under the chainstay or bottom bracket

This was the design used by Oscar Egg for the Super Champion. Several manufacturers imitated this popular system, including Cyclo Poly, Super Prior, Super-Rapid, Super Simplex, Lewis, Outsider, Super As, (nothing to do with the one mentioned previously).

- Double-cable touring derailleurs

The successful Cyclo design spawned many variants. The 1935 Meilhor was made by Pelletier, a small bicycle manufacturer in Lyon. It could hardly have been simpler to build or easier to install. The derailleur was fixed on a bracket brazed on the chainstay. The bracket carried a tube with a rod sliding inside. The two cables, guided by a single pulley fixed at either end of the rod, pulled it one way or the other. The rod carried the double-pulley changer mechanism at its inner end. A spiral spring maintained chain tension.

Left Fig. 7.46. The 1937 Westminster Route (Road) clamped onto the chainstay. The jockey pulley was coaxial with the top pivot.

Right: Fig. 7.47. The 1937 Rota was made in Lyon by Victor Simon. It used a plunger and a return spring. The tension arm was limited in its movement by stops to reduce chain tension and resistance. The Rota was exported to the U.S.

The Sirocco derailleur ran a single cable to the shift lever after going around a pulley in the changer mechanism. The cable was clamped to this pulley. The cable turned the pulley that acted on a cam, and pulled or pushed a rod carrying the combined jockey and tension pulleys. Another Sirocco derailleur exactly copied the rack of the Outsider Route, but without its return spring. A long rod, held by a collar attached to the frame, operated the cylinder.

The 1939 Hirondelle derailleur of the Manufacture d'Armes et Cycles de St.-Étienne was a low-quality copy of the traditional Cyclo Standard. The only apparent difference was the steel cap stamped "MF" on its outer end.

- Touring derailleurs with a single cable and return spring

Touring derailleurs showed more innovation because they needed a wider range than racing derailleurs. All of the 1930s touring derailleurs provided at least three speeds. Le Cyclo offered four

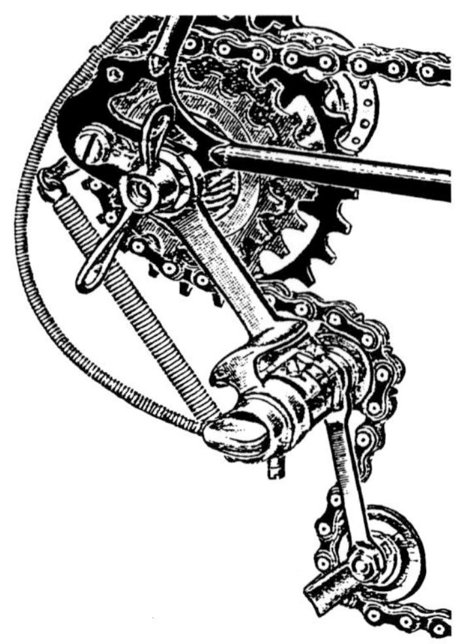

speeds after 1930, then five speeds after the introduction of narrow 3/32-inch (2 mm) chains.

Two different derailleurs were named after famous Tour de France mountain passes: The Col du Lautaret is a 2,057 m (6,700 ft.) high Alpine pass near the Italian border. The Col du Tourmalet is a 2,115 m (6,940 ft.) pass in southern France. Pelletier in Paris made the 3-speed Le Tourmalet derailleur. It could handle a 13–28-tooth freewheel.

In 1934, Mercier produced 3- and 4-speed Valor derailleurs with combined spring- and helical-ramp-action.

In 1937, Jean Léonard of St.-Étienne brought out his Westminster Route rear derailleur. The sliding rod was pulled toward the spokes by the cable, and the spring pushed it the other way. In 1938, Jacquemard, a St.-Étienne builder of hubs and drum brakes, produced the Ideal derailleur. Its special feature was that the sliding rod was at an angle of 30 degrees. This angle made the jockey pulley move close to the smaller sprockets, which improved shifting. There was a spiral tension spring attached to the lower side of the rear dropout by a bolt. Adjustment was by a screw. This derailleur was ahead of its time. Jacquemard also made a racing model. Throughout this decade, Joanny Panel continued to manufacture his Le Chemineau and L'Izoard derailleurs.

## Unique Designs

A number of different designs that fell outside the categories just described were introduced during this period, some of which are depicted on these pages.

The Vélectrik by Blanchard-Grange (BGA), a St.-Étienne cycle maker, was ugly but effective. A lever on the downtube was linked under the bottom bracket to a rod that moved the derailleur mechanism along an oblique slot under the chainstay. This placed the jockey pulley in line and close to the chosen sprocket. The long tension spring was coiled around the operating rod.

The Funiculo was made by Jacques Schulz in La Garenne-Colombes, near Paris, in 1933. Schulz was a prolific inventor. In addition to the Funiculo derailleur, he made monotube bicycle frames, alloy rims, and very powerful brakes. The Funiculo

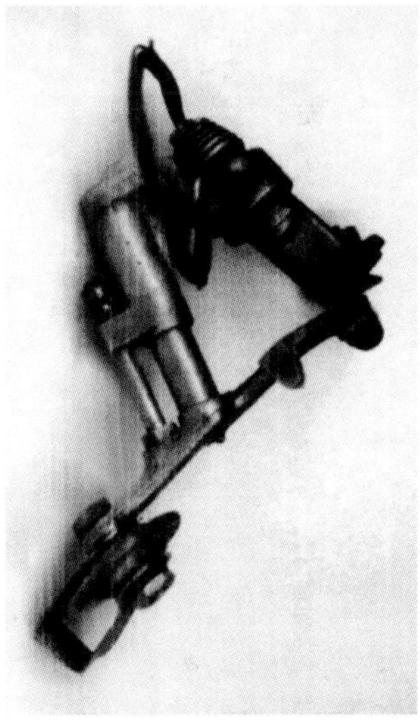

Above: Fig. 7.49. The 1937/1938 Jacquemard Ideal rear derailleur had a slanted plunger to track the freewheel sprockets. This single-pulley model was presumably meant for racing.

Below: Fig. 7.48. The 1933 Rectiligne-Caminade was an in-line derailleur, a direct descendant of Terrot's Model H and the Audouard.

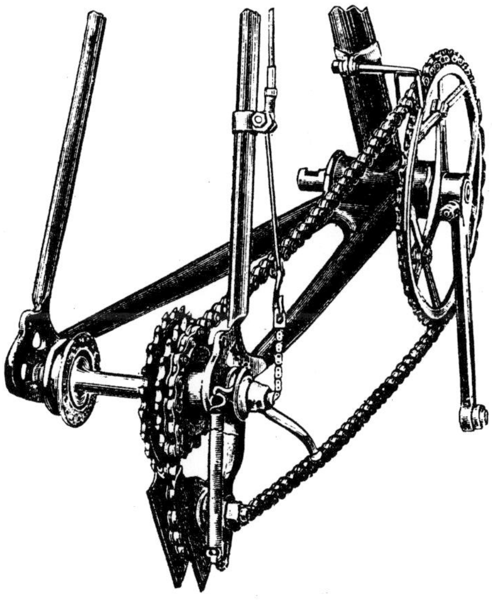

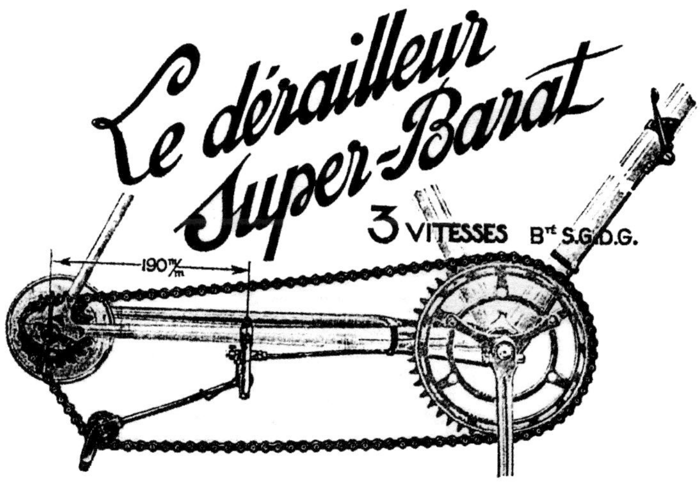


Right: Fig. 7.50. The Manufrance Super-Barat derailleur was a hybrid design that mounted under the chainstay. It consisted of a long tension arm that was swung from side to side by a cable, shifting the chain across three sprockets.

was an indexed derailleur that had a "floating parallelogram" similar to the 1975 Huret Duopar. This kept the jockey pulley close to the sprockets.

The first 3-speed version had 16-, 24-, and 40-tooth sprockets. A later 4-speed version had 16-, 20-, 28-, and 40-tooth sprockets.

Above: Fig. 7.51. The BGA Vélectrik derailleur wasn't pretty, but it was effective. (See also Fig. 7.52.)

The 40-tooth sprocket was probably the largest ever used with a rear derailleur. The 2½ to 1 range between sprockets allowed touring with a single chainwheel. It also reduced the strain on the chain caused by small chainwheels. Despite the large steps between sprockets, the chain shifted precisely and promptly. A plate pushing up on the chain assisted the shift up to the larger sprockets.

An interesting feature was the notched arm on the rear axle. By shifting into top gear and undoing the wing nut, the derailleur could be swung back and locked into the notched arm. This prevented the gear from being operated by curious onlookers when the bike was unattended. The Funiculo could be installed on any bicycle without modification to the frame or the rear wheel. However, Schultz installed Funiculo derailleurs only on his own handbuilt bicycles and tandems. Total production of this technical tour de force was very small, and it ceased before WW-II. Only two Schulz bicycles are known to have survived.

Jan Heine road-tested a Funiculo derailleur on a 1935 Schulz. He found that it shifted extremely well—even the huge shift from the 28- to the 40-tooth sprocket. The detented

Above: Fig. 7.55. In 1935, Caminade introduced a touring model with a double chainwheel called the Six-Speed Caminade. It was widely advertised, but the response was poor. Riders were suspicious of sliding freewheels.

Right: Fig. 7.52. The BGA bicycle, made by Blanchard-Grange (BGA), a St.-Étienne cycle maker, with its Vélectrik derailleur. Note the long control lever.

Left: Fig. 7.53. The 1935 Schulz aluminum bicycle contained numerous handmade components. It was expensive and few were sold.

Right: 7.54. Schultz' 1933 Funiculo derailleur could handle a 40-tooth sprocket, probably the record for a production bicycle.

shift lever provided indexed shifting in 1935.

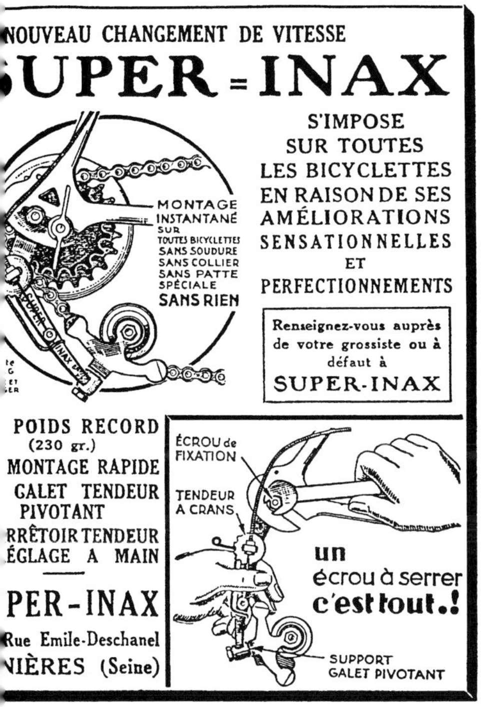


The Rectiligne-Caminade was an in-line derailleur, a direct descendant of Terrot's Model H and the Audouard with their sliding freewheels. Caminade was established at Bois-Colombes, near Paris. His single-pulley derailleur had three speeds. A cable, pull-chain, and spring slid the freewheel along the axle, maintaining chain alignment.

Left: Fig. 7.56. The 1938 Super Inax was a single-pulley derailleur with an elegant cam action. When the control lever was moved, the cable pulled the derailleur arm sideways. The single pulley was mounted on a swivel, and small chain guides kept the pulley aligned. This French advertisement for the Super Inax claimed it could be installed on any bicycle, with only one nut to tighten.

## The First Parallelogram Rear Derailleur

The 5-speed Nivex derailleur deserves special mention. This brilliant design was first made in 1938 in Villemomble, near Paris. A double cable actuated a horizontal parallelogram carrying a cage with two pulleys. Once installed on the chainstay, with the cables connected, the derailleur needed no further adjustment. The chain was tensioned by a coil spring connected by a small link to the tension arm in such a

Below: Fig. 7.57. The 1938 five-speed Nivex was the first parallelogram derailleur. It was the best of the pre-World-War-II touring derailleurs and shifted splendidly.

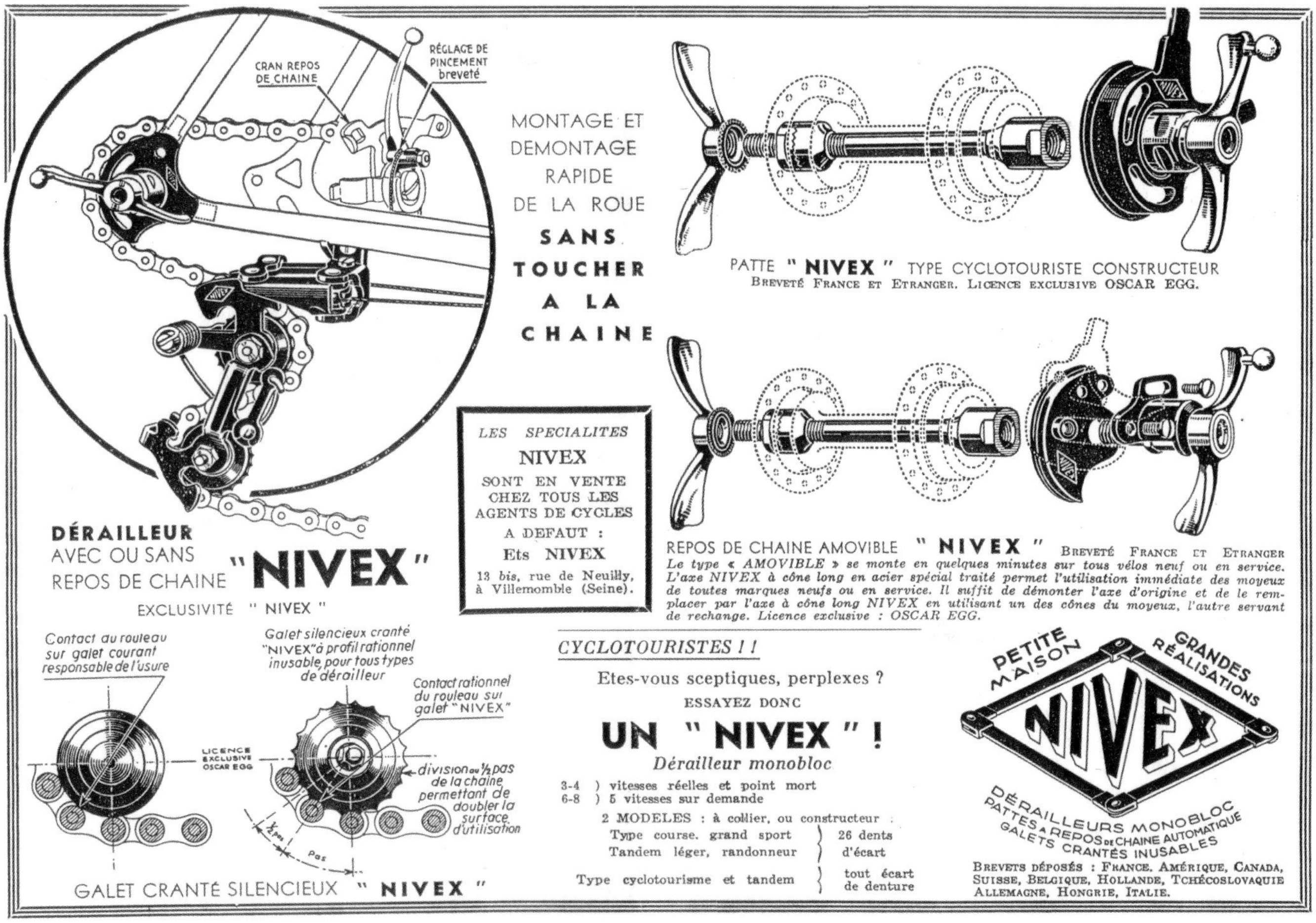

way that tension was light and uniform, regardless of the position of the cage. It had a capacity of 28 teeth. The fortunate few who used this derailleur praised it loudly. It appeared during the economic recession just before World War II and was made by a small manufacturer who did not market it properly. Singer subsequently bought Nivex.

The Nivex quick-release rear dropout was a splendid timesaver at a time when punctures were common. It allowed the rear wheel to be removed without touching the dirty chain. The usual rear dropout was replaced by a semi-circular chain rest attached to the frame. The rear axle was shortened, exposing the internal thread of the cone. A special wing nut with a stub axle screwed into the internal thread of the cone. To take the wheel out, you shifted the chain onto the chain rest. Then you simply loosened the two wing nuts and the wheel dropped like a ripe apple. The chain was left behind, draped around the chain rest. The only disadvantage was that you had to have a dropout spacing equivalent to an extra sprocket to make room for the chain rest. In 1939, Oscar Egg bought the licence for this dropout design.

Ernest Csuka, the well-known builder of Singer cycles, told Raymond Henry that at the 1947 or 1948 Paris Cycle Show, Tullio Campagnolo took a searching look at the Nivex derailleur and bought two of them. Some years later, Campagnolo brought out the Gran Sport parallelogram derailleur.

Manufacturers were constantly copying one another. Oscar Egg and Lucien Juy copied and were copied. Progress lay in the successive improvements of others' ideas. Campagnolo showed genius when he revised the Nivex to make a racing derailleur. The Gran Sport was ahead of its time, but the Nivex was probably the first parallelogram rear derailleur.

### Front Derailleurs in France in the 1930s

Before World War II, few cyclists knew that double or triple chainwheels existed. They were mainly used by cycle tourists, but even these enlightened few rarely used them with a front derailleur. They used their smaller chainwheels only when they came to a long hill. At that point, they were happy to move the chain down on to the small chainwheel with their heel, and when they reached the summit, to hoist it back up by hand. This was the method used by Vélocio with the good old Whippet.

Only after the Second World War, most of the tourists began to realize that front shifting was useful

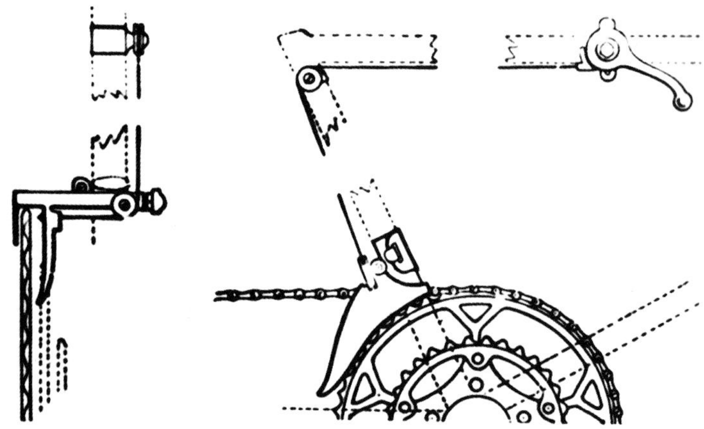

Above: Fig. 7.58. The 1931 Le Chat front derailleur had a rod-guided steel cage.

Left: Fig. 7.59. This 1938 top-of-the-line Reyhand touring bicycle had no front derailleur. Most bicycle tourists of the period did not see the need for one. They used their heels to push the chain to the smaller chainwheel while riding, and their fingers to shift to the bigger one when stopped.

even in non-mountainous terrain. As usual, the racers made the same discovery several years later.

The Rosa front derailleur, made by Albert Raimond in 1930, was the first really effective front derailleur that could handle a wide range of chainwheel sizes. It worked well and was not too expensive.

The brochure for the 1934 Concours de la Meilleure Machine (Competition for the Best Machine) organized in the Auvergne mentioned Simplex and Huret front derailleurs, which indicates that these two firms were at least experimenting with front derailleurs at this time. Simplex first listed a front derailleur in their 1935 catalog.

The Super Champion front derailleur appeared in 1937. It used a cable to pull a bell-crank. The other end of the crank pushed the cage and chain onto the large chainwheel. The cage was rod-guided and the guide sleeve held the return spring.

The Lewis front derailleur was produced just before the war and was also made after the war. It was designed for wide-range double chainwheels on touring bicycles and tandems.

The Le Chat (The Cat) front derailleur was marketed in 1931 by R. Léger, of Vitry-sur-Seine. It had a rod-guided steel cage. The

Above: Fig. 7.60. As of 1938, the Simplex front derailleur used a plunger and a return spring. The cable acted on a pull-chain.

Above: Fig. 7.62. Reyhand front derailleur on a 1938 Reyhand touring bicycle.

Right: Fig. 7.61. The 1939 Lewis front derailleur was designed for wide-range double chainwheels on touring bicycles and tandems.

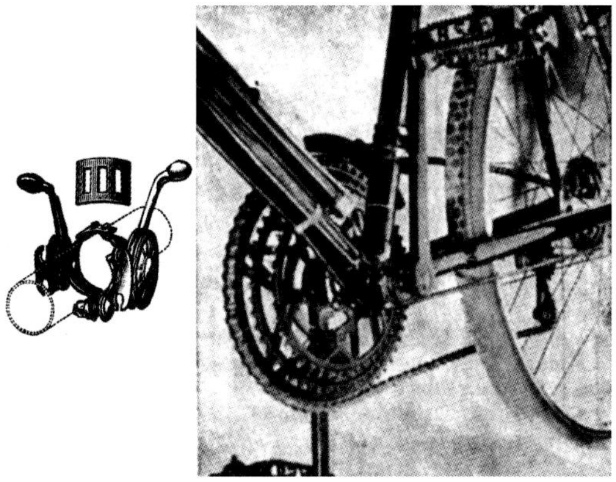

Above and below: Figs. 7.64 A & B and 7.65. The 1939 Rosa front derailleur could handle a triple crankset. It used the same double-pull cable as the Cyclo rear derailleur. The cable rotated a cog engaged in a rack on a sliding rod. The rod moved the derailing cage in and out.

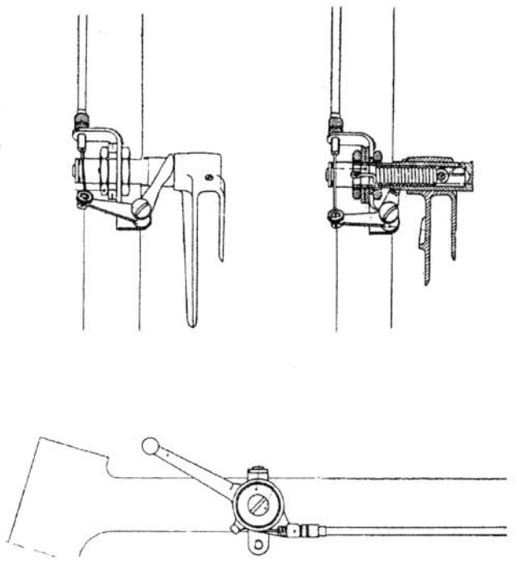

Left: Fig. 7.63. The 1937 Super Champion front derailleur used a cable to pull a bell-crank.

return spring was in the guide tube. A cable from the control lever passed over a series of pulleys and pulled the cage from the big chainwheel to the small one It survived for a long time on modest sales.

### Other Gear-Chance Devices in France in the 1930s

Many other interesting gear-change mechanisms besides derailleurs were used in France during this decade.

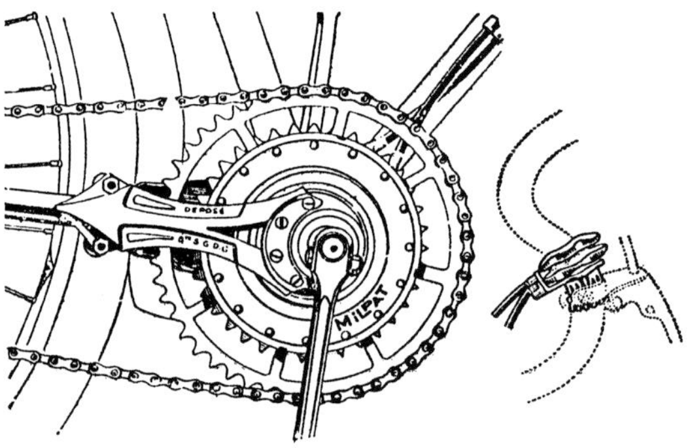

Above: Fig. 7.66. The 1935 Milpat expanding chainwheel used an eccentric to double or halve the effective size of the chainwheel.

Right: Fig. 7.67. The Relais Préférence was a cable-actuated 2-speed epicyclic freewheel.

From 1932 to 1938, Lailé and Robert sold the Varlax expanding chainwheel, which was claimed to provide 30-speed gearing.

The Milpat expanding chainwheel was made in 1935 by Claude & Manet, in Malakoff, a suburb of Paris. Its inventors claimed that it could not be explained in words and that they were so confident that the mechanism was sealed. The curiosity factor certainly helped sales. The Milpat was a more sophisticated version of the eccentric chainwheel used in the Mitchell All-Speed Gear. It weighed about a kilogram (more than two pounds) which worked against it.

Weight, complication, friction loss, and cost prevented expanding chainwheel systems from achieving the success hoped for them by their clever inventors.

The British Villiers two-speed hub gear was marketed by Brampton chains in France from 1934. Hub gears were commercial failures in France.

The floating chain disappeared in the early 1930s. One tenacious cyclist, Paul Treille, from Roanne, a town in central France, continued to use a floating-chain Intégral bicycle after 1945. His bike was stolen several times during the war, when bicycles were priceless—but never for long, because everybody in town knew who was the rightful owner that eccentric machine.

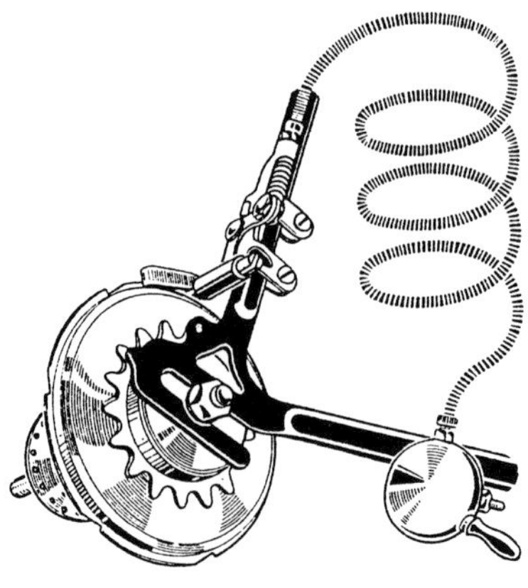

Right: Fig. 7.68. The Le Sphynx was another 2-speed epicyclic freewheel.

## Britain Rediscovers Derailleurs

In the 1930s, there was a modest economic upturn in Britain, and the work week was reduced to 48 hours. Many working people had time to explore the countryside. Club cycling and camping was a popular and affordable way to do this. The Cyclists' Touring Club grew to 25,000 members, and the Youth Hostels Association was formed, paralleling similar outdoor movements in France and Germany. At the start of the 1930s, British manufacturers were producing one million bicycles each year. By 1936, annual production had increased to over two million.

Above: Fig. 7.69. The Bilis, remembered only for its victory in the 1925 Poly de Chanteloup, was another expanding chainwheel.

Prior to 1936, the British cycling magazines and most British cyclists considered that derailleurs were something that no "real" rider would use. Thanks to the efforts of Sir Hubert Opperman, British cyclists rediscovered derailleurs around 1937. It helped that Britain was coming out of the depression. The change of the Tour de France regulations also helped. By the end of this decade, more than twenty different derailleur models were either imported or manufactured domestically.

## British Cyclo

Louis Camillis came to Britain in the 1920s, and his company, CAM, represented several French cycle and motorcycle firms, including Le Cyclo.

In 1930, Camillis published his first British Cyclo catalog, offering the Le Cyclo double-cable 2-speed and 3-speed rear derailleurs, freewheels from 13 to 28 teeth, and double chainwheels from 24 to 52 teeth. Presumably

un Cycle and Fittings Co., Ltd. 4

low fitted with the special strutted racket for greater rigidity, the Sun Manxman" should appeal even more trongly to the enthusiastic clubman. Vith an upright frame and attractive nd neat cutaway lugs it conforms to nodern standards and the specification ncludes Dunlop Silver Sprite tyres, rooks saddle, Monitor Super Cam rakes and a choice of Sturmey-Archer A.M.," Cyclo or Simplex three-speed gear. The price is £8 8s.

Raleigh Cycle Co., Ltd. . 45

An ever popular model, the Raleigh Record Ace, price, £9 15s. This is the model used by S. H. Ferris when he broke the 24-hour record recently. Another attractive machine on this stand is the "Charles Holland Continental." A number of other sports and tourist machines are also shown.

Enfield Cycle Co., Ltd. . 6

A typical Royal Enfield club model fitted with the Easy-change three-speed gear. Notable on this stand is the new racing machine named the "Continental" as used by Harry Hill in breaking the 50 miles record. For the every-day rider there is the "Fitness" model for both men and women.

Elswick Hopper Cycle Co., Ltd. . . . . . . . . 59

The Elswick Sports Light Roadster, price £4 19s. 6d. Amongst the sports machines there are many to interest club riders. Attractive colour schemes and considerable improvements in frame design make this stand one that should not be missed.

Left: Fig. 7.72. This page from the November 1938 issue of *Cycling* magazine shows four bicycles from the 1939 London Show. The Sun and Enfield bicycles were available with derailleurs. Raleigh supplied only bikes with fixed gears or Sturmey Archer hub gears.

Left: Fig. 7.70. Louis Camillis, Manager of British Cyclo, shown here in a 1938 cartoon by GAY in *The Bicycle*.

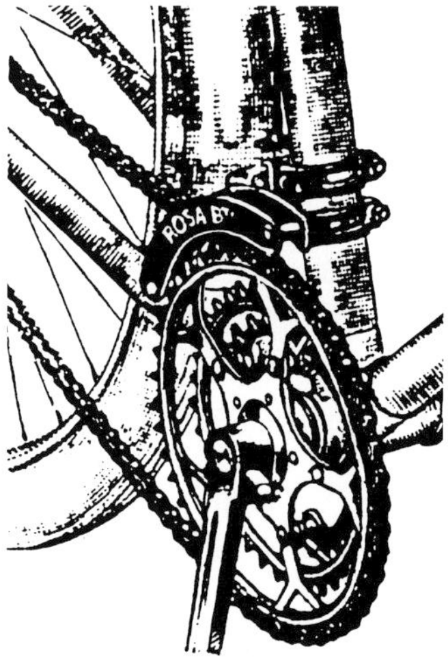

Right: Fig. 7.71. The Rosa triple chainwheels and Rosa front derailleur were first shown in the British Cyclo catalogue installed on solo bicycles in 1935.

the chain was shifted across these chainwheels by hand. The next British Cyclo catalog (issued in 1931 for the 1932 season) added a 4-speed rear derailleur and the Rosa front derailleur.

When Britain went off the gold standard in 1931, British Cyclo's success was threatened by severe import restrictions. For that reason, Albert Raimond encouraged Camillis to manufacture derailleurs locally. By the start of 1932, British Cyclo had rented space in a small factory on Bath Street in Birmingham, and was manufacturing a full range of gears. In 1933, Camillis, in partnership with Raimond, established the Cyclo Gear Co. Ltd. Cyclo was now promoted as a British product rather than as a "foreign" device. The following announcement appeared in *Sport and Play* in April 1932:

> Users of the Cyclo Variable Gears will be interested to know that these are now made in England, in fact in Birmingham, where all good things are made. Our friend, Louis Camillis, has established a Factory in Bath Street, and all these good people who cavil at anything except things made in England will have no cause to feel disloyal for not buying British. The "CYCLO" Gear, which has proved itself such a useful

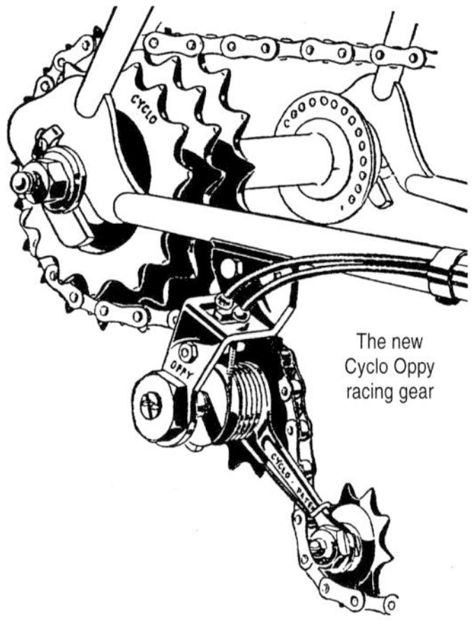


Above: Fig. 7.73. The 1938 British Cyclo Oppy, named after Hubert Oppenheimer, was a medium-ratio racing version of the Standard and was available until 1949.

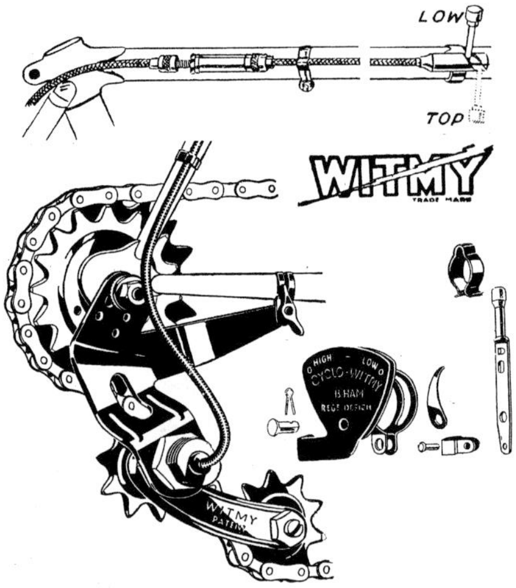


Right: Fig. 7.74. The British Cyclo 2-speed Witmy rear derailleur was added to the range of Cyclo products in 1934 and was available until 1947.

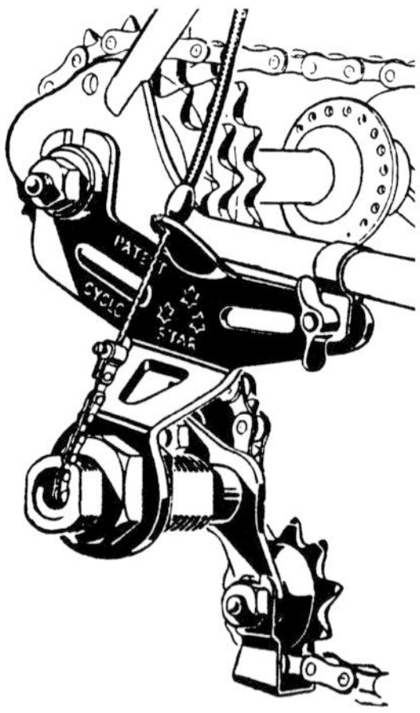

Above: Fig. 7.75. The 1938 British Cyclo Star, available until 1947, had two pulleys offset from the pivot and a single cable. It had 3 speeds, with a range from 13 to 23 teeth. It could be used with a 12–23 tooth Cyclo unit hub.

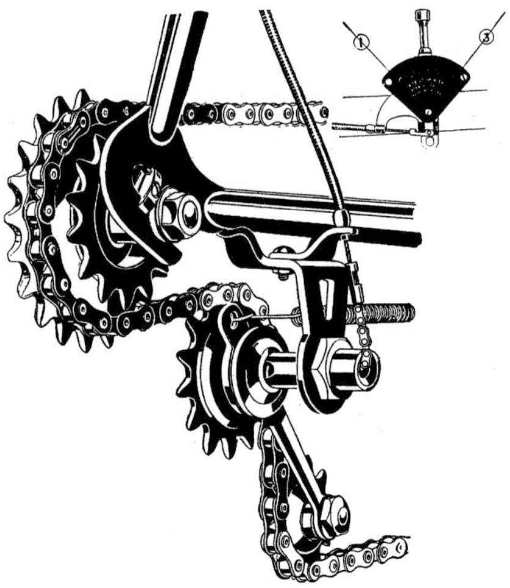

Left: Fig. 7.76. The 1938 British Cyclo Sinclo was a cheaper version of the Standard, produced for riders who preferred single-cable operation, but Cyclo still claimed that the original double-cable system was superior.

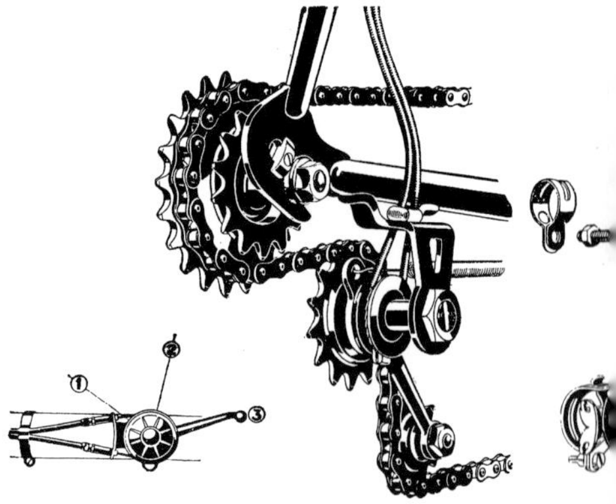

Above: Fig. 7.77. The 1932 British Cyclo Standard rear derailleur was available until 1958 in 2-, 3-, and 4-speed versions, using standard ½-inch pitch x ⅛-inch wide chains. These robust derailleurs were particularly suitable for tandem use.

Below: Fig. 7.78. Two different tension pulleys were available—a toothed and a flanged pulley. The flanged pulley was noisier.

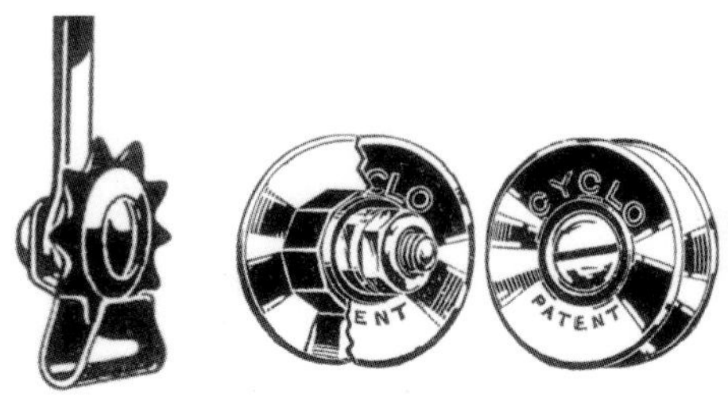

fitment, is now made in its entirety at the Bath Street establishment and all spares can be had from there also.

Raimond applied for British patents for his front derailleur in 1931. The 1933 patent for a single-cable, return-spring rear derailleur became the Sinclo.

By 1934, British Cyclo had 50 employees, and the catalog had many testimonials from satisfied and enthusiastic users. For example, C.A.P. wrote the following:

My riding was done on a 3-speed with gears 76, 63 and 48 (except that I substituted a 41 as the low while touring Cornwall and Devon), and these gears I found very satisfactory. With the low, hills up to 1 in 4, I found I could climb as a matter of course. On club runs it is nothing to see the "Derailleur Boys" ascend previously unclimbable hills in a body. [Frank Berto comment: This gearing would have been obtained by using a 44-tooth chainwheel with 14-, 18-, and 24- (or 28-) tooth sprockets.]

C.A.P.'s claim of riding up a 1 in 4 (25%) slope was a little optimistic. With a cyclist's entire weight on the pedal, i.e. in the "honking" mode, the maximum rideable gear is seven times the

Right and below: Figs. 7.81 and 7.82. The L'Izoard derailleur was marketed in Britain as the Pelissier.

Left: Fig. 7.80. Between 1934 and 1939, British Cyclo made their own retro-direct prototype, similar to that made by Manufacture Française d'Armes et Cycles, of St.-Étienne. Apparently, a few gears were made to test the market, which failed to materialize.

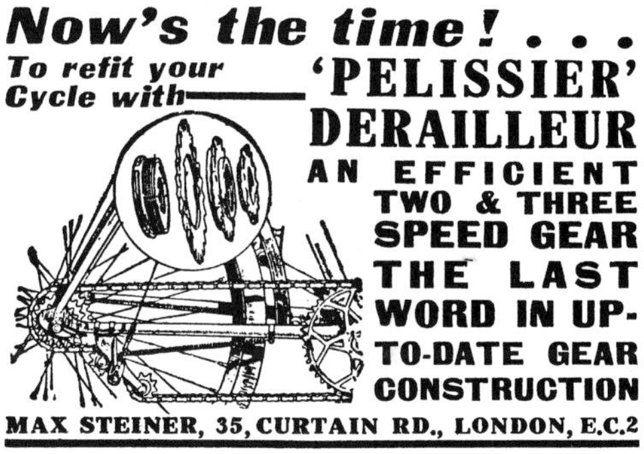


Below: Fig. 7.79. Early in 1930, *Cycling* published several articles on derailleurs, the first showing a Saxon tandem fitted with Joanny Panel's L'Izoard derailleur.

Above: Fig. 7.83. This Grubb tandem, advertised in *Cycling* in 1930, appears to be equipped with a Pelissier (L'Izoard) derailleur.

slope denominator. With his low gear of 41, C.A.P. could, with maximum effort, have ridden up a slope of 1 in 6. The ideal gear for touring is only twice the denominator. Thus a gear of 41 is comfortable to pedal up a slope of only about 1 in 20 (5%).

In 1933, the F.W. Evans bicycle shop in London installed British Cyclo Standard rear derailleurs and Rosa front derailleurs. A tandem built for Medwin H. Clutterbuck had front chainwheels of 40 and 60 teeth. In 1989, Clutterbuck recalled:

> The Cyclo Rosa chainwheel derailleur was an excellent device; operated by rack and pinion and twin-wire cable it was very positive in action. I still have two in good working order.

Disaster struck in early 1934, when a fire burned out the Bath Street factory. More than 3,000 derailleurs were damaged or destroyed. A new factory was found a mile north on Potters Hill, and manufacture restarted by mid-1934.

The 2-speed Witmy rear derailleur was added to the range of British Cyclo products in 1934. It had two pulleys, but it still had a capacity of only 4 teeth difference between high gear and low gear (e.g. 21 and 25 teeth sprockets). It remained in production until 1947.

Combining a triple chainwheel and a 4-speed freewheel, it was possible for the British cycle tourist of the early 1930s to have a wide gear range with ten useful gears, such as the "half-step-

Above: Fig. 7.84. Out-of-the-saddle riding, called "honking" in Britain, allows cyclists to pedal up otherwise unridable hills without very low gears.

Below: Fig. 7.85. With a cyclist's entire weight on the pedal, the maximum rideable gear in inches is seven times the slope denominator. Thus, a 1-in-10 incline requires a 70-inch gear.

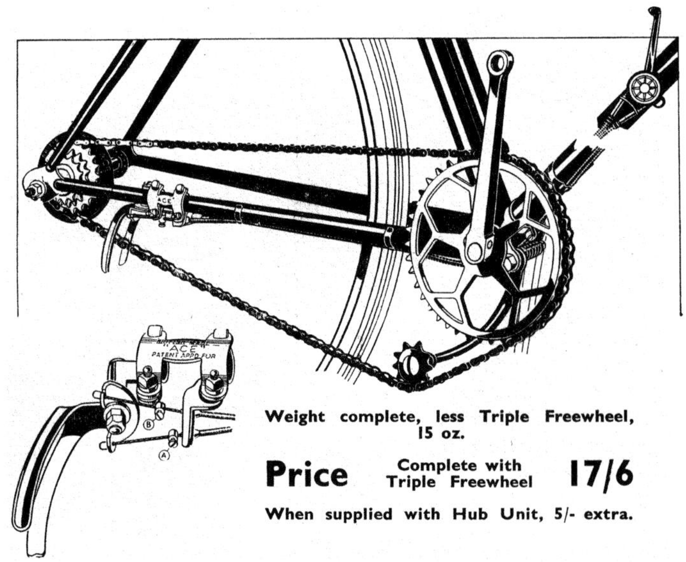


Right: Fig. 7.86. The 1936 British Cyclo Ace, which was available until 1954, was based on the French Cyclo Poly, with the tension roller replaced by a sprocket.

Left: Fig. 7.87. British Cyclo rear derailleur and Rosa front derailleur, in the collection of the Science Museum, London.

plus-granny" arrangement described in Table 7.2.

The term "half-step plus granny," though not used at the time, today refers to a triple-chainwheel setup on which intermediate gears are achieved by shifting between the two larger chainwheels, while a range of much lower gears is reached by shifting to the smallest chainwheel.

However, contemporary accounts suggest that triple, and even double, chainwheels were uncommon. Most derailleur-equipped tourists of this era simply had a 3-speed Cyclo Standard on the back. It would be thirty years before "gear freaks" would calculate and install optimum half-step combinations of sprockets and chainwheels, on which usable intermediate gears between shifts with the rear derailleur can be made by shifting the front derailleur between chainwheels.

From 1936 to 1939, the Cyclo Gear Company Ltd. (referred to as "British Cyclo" in this book) introduced five new derailleurs. Some were copies of French Cyclo designs and some were designed and made in Britain. In 1936, Brit-

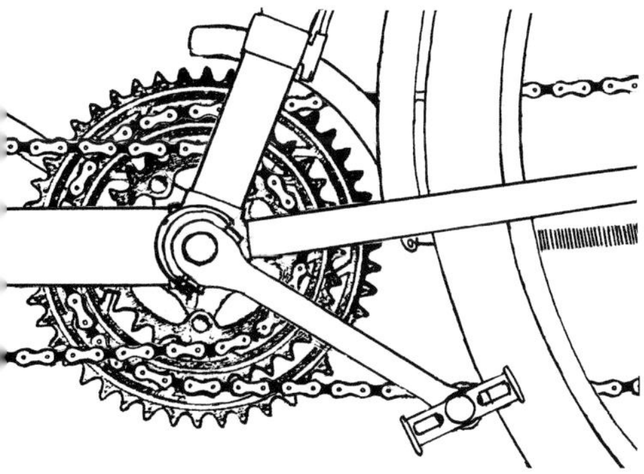

Above: Fig. 7.88. British Cyclo introduced the Rosa triple chainwheels in 1934. They were originally designed for mounting on the rear bottom bracket of a tandem. The two outer chainwheels drove the rear wheel. The inside chainwheel carried the separate "timing" chain from the front chainwheel.

Below: Fig. 7.89. The 1938 British Cyclo Poly was a single-pulley 3-speed derailleur suitable only for close ratios, with a maximum difference of 4 sprocket teeth. It was operated by a single cable.

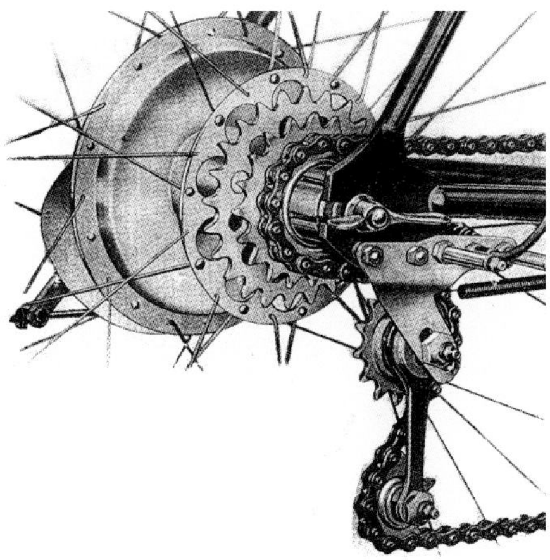

Above: Fig. 7.90. The 1932 TriVelox was a sliding-sprocket derailleur with three sprockets, available from 14 to 26 teeth.

Below: Fig. 7.91. The TriVelox B was a conventional moving-chain derailleur. A cable pulled a bell-crank to push the jockey pulley over onto the larger sprockets.

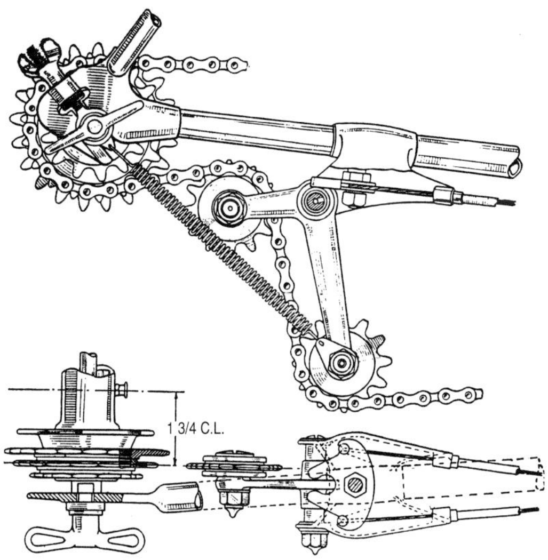


Above: Fig. 7.92. The Enfield Company, of Redditch, introduced a chainstay-mounted double-cable, double-pulley derailleur with a tension spring attached to the rear dropout in 1936. It was supplied only on Royal Enfield bicycles.

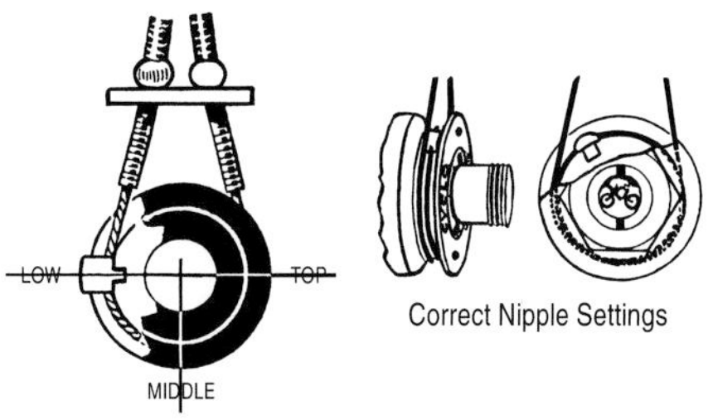


Above: Fig. 7.93 A and B. Many cyclists broke cables on the 3-speed Cyclo Standard, probably caused by inexpert installation. The 3-speed model just looped the cable around the groove on the sliding shell, and if it was installed at the wrong angle, the cable could break at the nipple. The 4-speed model had a longer cable, which went in a complete turn around the groove on the sliding shell and had few breakage problems.

ish Cyclo launched the Ace, which was the French Cyclo Poly with a derailing fork and a tension arm under the bottom bracket. In 1937, British Cyclo produced their own version of the Poly, a single-pulley 3-speed derailleur suitable only for close ratios.

In 1938, they produced the Sinclo, and in 1939 the Oppy (named after Hubert Opperman). Both were variants of the original Cyclo Standard. The Sinclo used a single cable. It did not sell well and was replaced by the Star in 1938. The Oppy retained the double cable, but revised the mounting and improved the cosmetics with a chrome-plated bracket and silver cables.

To demonstrate the reliability of the Star, Miss Evelyn Hamilton rode a century (100 miles, or 160 km) every day for 100 days, compiling a total of 10,704 miles (17,223 km).

The Poly, introduced in 1938, was a single-pulley 3-speed derailleur suitable only for close ratios, with a maximum difference of 4 sprocket teeth. It operated by a single cable, which pulled a plunger sliding the single pulley sideways. Veteran cyclist Derek Roberts recalls that the Poly was fiendishly difficult to adjust so that the chain did not ride up and miss a tooth or two.

The Oppy was a medium-ratio racing version of the Standard. It retained the double cable, but replaced the long tension spring with a coil spring around the piston. The bracket was tilted forward so that the twin cables could be routed along the chainstay. It was available with either a steel toothed tension pulley or a flanged aluminum pulley.

Albert Raimond persuaded Louis Camillis to approach Hubert Opperman about using Cyclo derailleurs. As soon as variable gears were permitted in the Tour de France, Opperman be-

Above: Fig. 7.94. The 1937 A.B.C. Gear was a plunger-type rear derailleur with two pulleys—a toothed jockey pulley and a flanged tension pulley. A bell-crank reversed the action, so that the cable pulled the chain onto the low gear. Like most derailleurs of the era, it used 1/8-in. wide sprockets.

*Table 7.2. Gears achieved with 1930s "half-step-plus-granny" gearing**

| No. of teeth on chainring | No. of teeth on sprocket | | | |
|---|---|---|---|---|
| | 13 | 17 | 22 | 28 |
| 40 teeth | 80 in. | 62 in. | 47 in. | N/A |
| 36 teeth | 72 in. | 55 in. | 43 in. | 33 in. |
| 24 teeth | N/A | 37 in. | 28 in. | 22 in. |

* See the text on page 133 for an explanation of the term "half-step-plus-granny."

Below: Fig. 7.95. This 1930 Mernick touring bike also appears to be equipped with a Pelissier (L'Izoard) derailleur. Note the slack steering geometry and the large fork rake, which provided stability on poor roads.

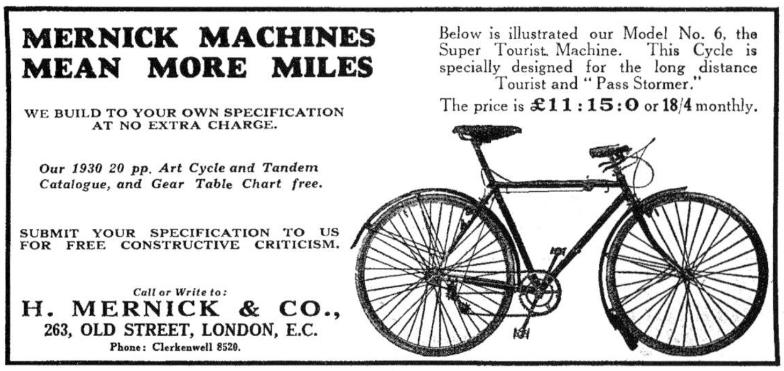


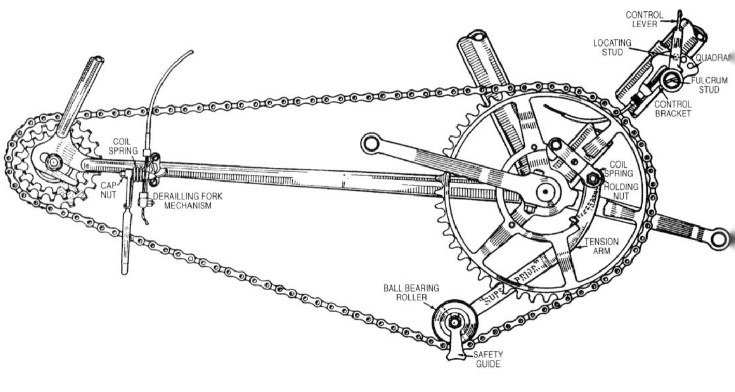


Above: Fig. 7.96. The 1938 Super Prior was a Super Champion copy imported from France by Hicking, of Hayes.

came an enthusiastic convert. In his autobiography he recalled:

> Around this period, cycling had received the impetus of an eternal tail wind from a new and most important development. An improved, robust and light Cyclo gear had become an accepted accessory. Despite early frantic efforts to debar the "changement de vitesse," in 1932, the Continent had capitulated and permitted their use in the classics. It, and others who copied, revolutionized speed, and to my mind, it became the most dramatic contribution to easier and faster pedaling since the creation of the pneumatic tire. The readily accessible top bar lever, moving the chain smoothly on to each gear ratio without the time-wasting stoppage to loosen wing nuts or turn a wheel, dramatically altered the pattern of road racing. Prior to its introduction it was understood that subject to location a good rider in an extended race could puncture twice and, by chasing on a higher gear than was used by the peloton, still overtake and win.
>
> But as even the magnificent Antonin Magne and the equally endowed Georges Speicher wryly admitted, following personal disillusionment in 1933, "Since the changement vitesse, one puncture and Pouf! we never see them again!" The Cyclo was invented and perfected by a Monsieur Albert Raimond, of St.-Étienne, with its English representative and manufacturer, one Louis Camillis in Birmingham. The keen Continental commercial instinct of both was impregnated with a peculiarly French sentiment for cycling. It was a joy to do business with them and to feel the benefit of the

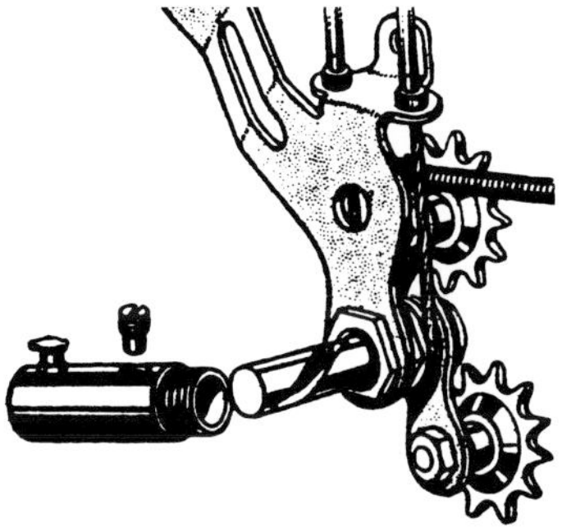

Above: Fig. 7.97. The 1938 Victor helical derailleur was produced by the Taybrook Manufacturing Co. Ltd. of Birmingham. It was similar to the Morgan Helical and was available in two models—one operated by a rod, the other by a double cable (shown here).

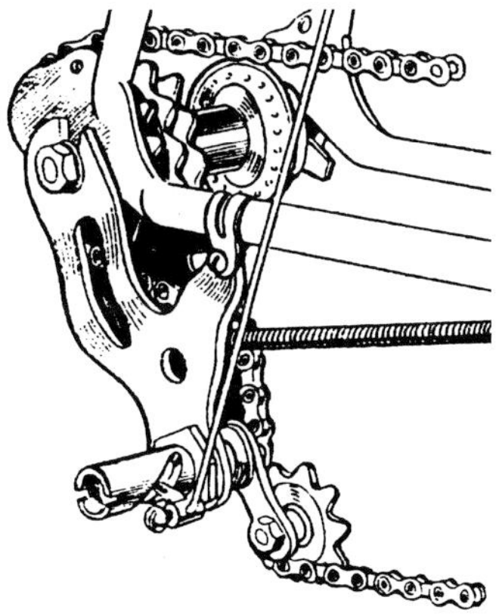

Right: Fig. 7.98. The 1937 Morgan Helical derailleur, manufactured by J.T. Morgan of Cardiff, was a two-pulley derailleur that mounted on a bracket bolted to the rear dropout and the chainstay. A rod turned the piston in a helical slot that moved the jockey pulley. It was clearly copied from the Cyclo Standard.

Left: Fig. 7.99. English club cyclists enthusiastically adopted the Cyclo Standard rear derailleur. It was reliable and durable. Harry Suddaby installed a Cyclo to a tandem that he and his wife used for fifty years.

Below: Fig. 7.100. Cyclo advertisement,1943. The onset of the Second World War in 1939 put an end to bicycle manufacture in Britain. Many factories, including Cyclo's, converted to wartime production. Young men were called up for war service. Many club cyclists, rated as being very fit, were assigned to the Royal Air Force (RAF). They continued time trialing in cycling clubs established at RAF bases around Britain and overseas.

new development they merchandised.

Hubert Opperman, "Oppy," broke many point-to-point records in Britain and Australia on Malvern Star bicycles equipped with Cyclo derailleurs, while another Australian, Ossie Nicholson, set yearly distance records of 43,997 miles (70,806 km) in 1933, and 62,856 miles (101,135 km) in 1937 on a Cyclo-equipped bicycle. These well-publicized records helped establish the reputation of Cyclo and of derailleurs in general in English-speaking countries.

Opperman is credited with making out-of-the-saddle riding acceptable in Britain. He had competed in the Tour de France in 1928 and again in 1931, an era when variable gears were still banned in this race, and to climb hills on a single-speed machine, the Continental racers would stand up on the pedals, pedaling forcefully but smoothly in a style called "*en danseuse*" (dancing) by the French, and referred to as "honking" by British cyclists.

### Other Derailleurs in Britain

Joanny Panel was one of the earliest exporters of derailleurs to Britain. His derailleurs were renamed Pélissier after the Pélissier brothers, who were famous racing cyclists of the 1920s—and probably never used derailleur gears.

By 1936, the full range of Simplex derailleurs was being exported to Britain. It included the single-roller Champion du Monde Professional, the derailing fork Super Simplex, and the Light Tourist.

The Constrictor Tyre and Accessories Company, of London, marketed Osgear derailleurs in Britain. These were actually Super Champions produced in France by Oscar Egg. Osgear sometimes used different names for these derailleurs. Thus, the Super Champion Standard became the Osgear Course. The Super Champion Professional with the chain detensioner was called Osgear Professional in Britain after Super Champion changed the name to Tour de France. The Grand Sport. and the Route used the same names in both Britain and France. The Osgear derailleurs are covered under Oscar Egg. Osgears became popular with both racing and touring cyclists in Britain before and after World War II, and they were used with double chainwheels and front derailleurs.

In 1931, Alec Shuttleworth and William Hill patented an inline derailleur, with sliding sprockets, to satisfy the British preference for straight chain lines. This harked back to the sliding sprockets of Edmund Hodgkinson's 1900 Gradient gear. The Triumph Company, of Coventry. manufactured this new design as the TriVelox in 1932 and used it on Triumph bicycles.

There were numerous TriVelox versions. The first version had a fork that shifted the sprockets sideways. Later versions had a

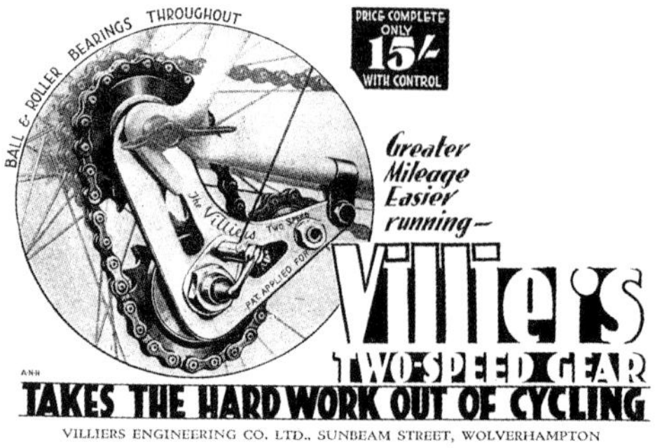


Above: Fig. 7.102. The Villiers 2-speed gear was produced in Wolverhampton. When the cable was pulled, a small driving chain was engaged to override the rear sprocket and produce a higher gear. The Villiers was heavy but cheap, and it could easily be installed on any bike.

Left: Fig. 7.101. This 1930 Perry derailleur appears to be a copy of the Cyclo Standard.

Facing page: Fig. 7.103. This page from the 1940 Raleigh export catalog shows the different Sturmey-Archer hub gears and drum brakes that were available with Raleigh bicycles.

hollow axle with a rod inside that pulled the sprockets sideways. Unfortunately, the derailleur/hub combination was heavy and expensive. To install an in-line TriVelox, the rear wheel had to be rebuilt and the fork-ends spread. The TriVelox was often used on tandems, where the weight was less important.

By 1937, the Triumph Company had introduced a range of simpler TriVelox B models which bent the chain sideways like conventional derailleurs. The TriVelox B still used a special hub with an integral freewheel and three sprockets with a range from 12 to 28 teeth. The last Model C used a conventional freewheel that was screwed onto a standard hub.

### *Other Variable Gears in Britain*

Derailleurs had to compete for popularity with a wide variety of other variable gear systems. Of these, hub gears, especially those made by Sturmey-Archer, were the overwhelming favorites.

The Sturmey-Archer three-speed hub was the popular choice for the British cyclist who wanted more than one speed. In 1939, Sturmey-Archer offered a 4-speed hub. The Cyclists' Touring Club acclaimed the 4-speed hub as "the greatest improvement in cycle design or equipment during the year."

Raleigh, the largest bicycle maker in Britain, owned Sturmey-Archer. Raleigh did not offer a derailleur-equipped bicycle before World War II. By 1941, when production ceased, Sturmey-Archer offered 2-, 3-, and 4-speed hub gears in various combinations of ultra-close, close, medium, and wide ratios.

Sturmey-Archer produced three-speed hubs with either wide-ratio (33% increase, 25% decrease) for utility use, medium-

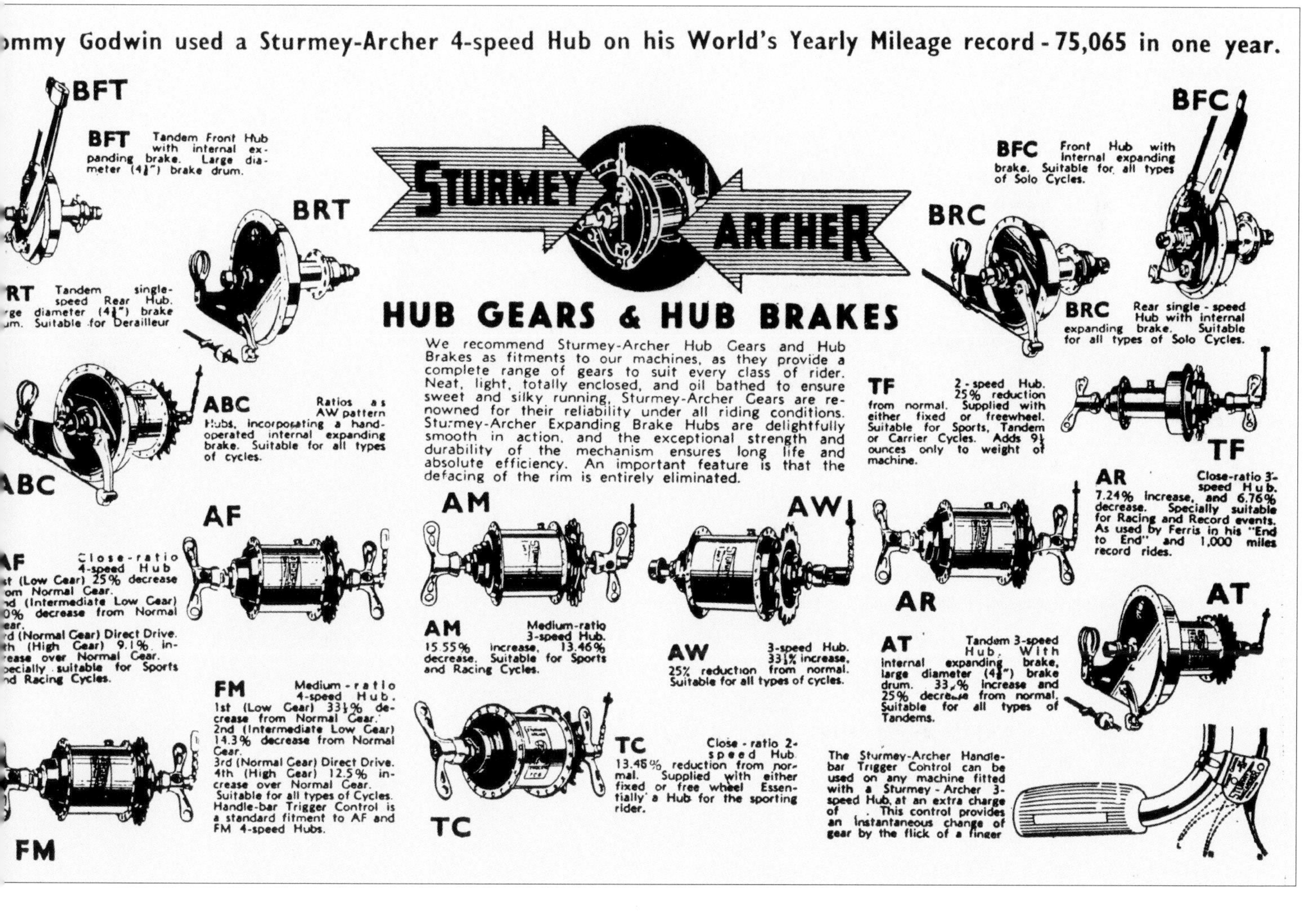

ratio (16.6% increase, 14.3% decrease), or close-ratio (12.5% increase, 11% decrease) for racing. The wide-ratio hubs were also available with either a hub brake or a back-pedal (coaster) brake. By 1939 Sturmey-Archer also marketed various four-speed hubs, both for racing and for touring.

The James 2-speed bracket gear and the German Adler 3-speed bracket gear were also available. W. Neale, of Bletchley, manufactured an expanding chainwheel called the Nealeson.

Despite the variety of variable gears, their acceptance by British cyclists was slow. In 1939, Ralph Dougherty became the first cyclist to ride 25 miles (40 km) "inside the hour" on British roads—still using a single fixed gear. However, in his 1939 handbook, Kuklos (William Fitzwater Wray) wrote, "A touring bicycle with only one speed is now a rarity; it might almost as well have only one pedal!" This was a remarkable change from the start of the decade, when the multi-speed bicycle was a rarity.

Above: Fig. 7.104. The 1939 Sturmey-Archer KB wide-ratio three-speed hub with drum brake.

Below: Fig. 7.105. BSA's DP (Dual Purpose) gear could be switched from freewheel to fixed-wheel operation.

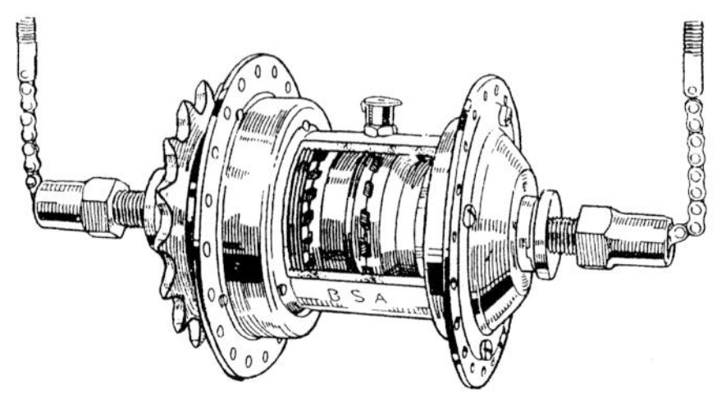

## Italian Derailleurs

In this section, we'll take a look at derailleur developments in Italy during the last decade preceding the Second World War. Campagnolo and Vittoria were the two most significant Italian derailleur makers of this period.

### Vittoria

The Nieddu brothers persisted with their Vittoria derailleur in the 1930s and developed an improved model. In 1932, Nieddu broke into the ranks of professional cyclists when Alfredo Binda used a Vittoria derailleur to win the world championship in Rome over the heavily-favored Learco Guerra.

The mechanism worked like a car stick shift. A long rigid lever moved forward and back between two serrated guides in the main frame triangle. The lever swivelled just behind the bottom bracket. The lower end of the lever held a tension pulley, which pressed on the chain. Binda shifted gears by pushing the lever forward to slacken the chain tension, then he pedaled backward and pushed the chain across with his fingers, and finally he pulled the lever back to re-tension the chain. After Binda's win, Vittoria derailleurs were widely adopted by Italian professionals and other racers around the world.

Above: Fig. 7.106. Detail of a 1935 BSA bicycle equipped with the new Vittoria Margherita derailleur on display in the Italian Galbiati museum. It never appeared in a BSA catalog.

Right: Fig. 7.107. Gino Bartali riding in the 1938 Tour de France. Note the Vittoria Margherita derailleur.

A 1934 Vittoria advertisement claimed, "At 230 grams, our derailleur is the lightest, simplest, most economical—the best in the world."

In 1935, the Nieddus launched a new model, the Vittoria Margherita derailleur, named in honor of Princess Margherita. The publicity for the new Vittoria Margherita invoked "Imperial Italy, leader of the world." Vittoria claimed to have invented the tension arm and pulley, and to have won "all the championships and great bicycle races of the world."

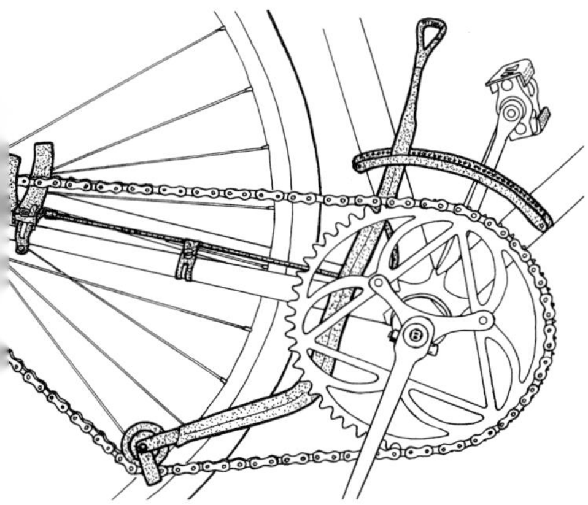

Above: Fig. 7.108. The 1935 Vittoria Margherita derailleur used a knob on top of the lever to operate two curved plates above the chainstay. These plates shifted the chain between sprockets.

The Margherita was simple, robust, and inexpensive. It allowed the rider to adjust the chain tension, and it worked when clogged with mud. It could "skim" over cobbles and sidewalks without breaking the tension arm, which happened with the low-mounted Super Champion. The rear wheel could easily be removed without the jockey pulley getting in the way. On the other hand, you had to backpedal to shift gears, and considerable skill was needed to operate the knob that controlled the shifting plates.

Gino Bartali's decisive victory in the 1938 Tour de France stunned the readers of the French magazine *Miroir des Sports*. They protested that his crushing victory resulted from the Italian team's use of Vittoria derailleurs, while the other teams were required to use Super Champions.

In 1939, Vittoria produced a 5-speed Giro de Francia (Tour de France) model. Prior to World War II, many races were won using the Margherita derailleur, including the following:

- Giro d'Italia: 1936, 1937 (Gino Bartali), 1938, 1939 (Giovanni Valetti), 1940 (Fausto Coppi)
- Milan–San Remo: 1938 (Giuseppe Olmo), 1939, 1940 (Gino Bartali)

### Ghiggini, Forgotten Genius

Francesco Ghiggini (1896–1989) was a mechanic and naval engineer employed in the northern Italian city of La Spezia. In his spare time, Ghiggini shut himself up in his tiny workshop and invented mechanical devices. By 1937, he had built a derailleur that was clearly based on the Vittoria Margherita, but with the great advantage of shifting gears while pedaling forward. Ghiggini patented this design in 1937.

He modified the design slightly in a further patent in 1938. As on all single pulley de-

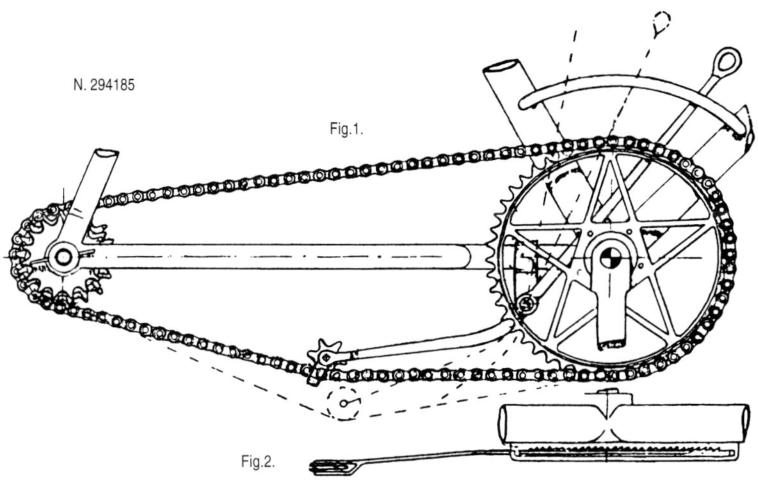


Left: Fig. 7.109. The patent drawing for the 1932 Vittoria derailleur.

Above: Fig. 7.110. Vittoria made sprockets with teeth cut away at the back to ease shifting while backpedaling. Vittoria's 3- and 4-speed freewheels were made entirely of aluminum alloy, and weighed as little as 100 grams (3½ oz.).

railleurs, the chain would "skip." Ghiggini improved his design by adding a second pulley. With a double pulley and an articulated parallelogram, it was one of the forerunners of modern rear derailleurs. However, Ghiggini did not get any feedback from real time users. Not satisfied with his original design, he made a derailleur with a swinging arm attached to the rear axle. The arm had a single pulley that both shifted the chain and took up the chain tension. With hindsight, this was a step backward.

Ghiggini did not get around to marketing his invention, and his prototype derailleur was left in a box in his workshop. In the 1950s, Campagnolo bought his brilliant design.

You can choose between the French (Nivex) version and the Italian (Ghiggini) version of the invention of the modern parallelogram derailleur, but that story is for Chapter 8.

### Campagnolo

Meanwhile another cyclist-mechanic, Tullio Campagnolo, developed an entirely different design. Campagnolo was a successful amateur bicycle racer in his twenties. In 1924, he was racing over the Croce d'Aune Pass near Padova. A snowstorm swirled as the leading bunch quickly dismounted to turn their wheels around in preparation for the steep climb ahead. His fingers were so numb with cold that he could not undo the wing nuts on his rear axle and had to climb the mountain in high gear. He kept thinking "*Bisogna cambiá qualcossa de drio*" (something needs to be changed in the rear). At least he did not have to stop at the top of the mountain to shift his chain back into high gear for the descent, and eventually finished in fourth place. Young Tullio Campagnolo had the inspiration that led to the invention of the quick-release. The Campagnolo company was founded in 1933 to manufacture quick-release hubs and the first rod-operated changers.

At the time of his fateful ride over the Croce d'Aune pass, Campagnolo was helping out at his father's hardware shop in Vicenza. In a storeroom in the back, he loved to work away with file and vise, making bicycle parts. He designed a cam-operated quick-release for the rear hub to replace the wing nuts used on racing bicycles. He patented his quick-release in 1930 and began promoting it. Today's quick-release hubs are based on Campagnolo's original concept.

His thinking went further. Why not revise the design so that when the rear wheel is loosened, it stays aligned in the frame? He filed small teeth around the rear axle and matching teeth on the top surface of the fork-end. Now when the wheel was released, it could roll forward or back and remain in alignment. Campagnolo kept the details of his device secret until he had applied for a patent in 1933. His patent stated:

> To permit a change of gear, it is clear that to shift for example from a sprocket with 18 teeth

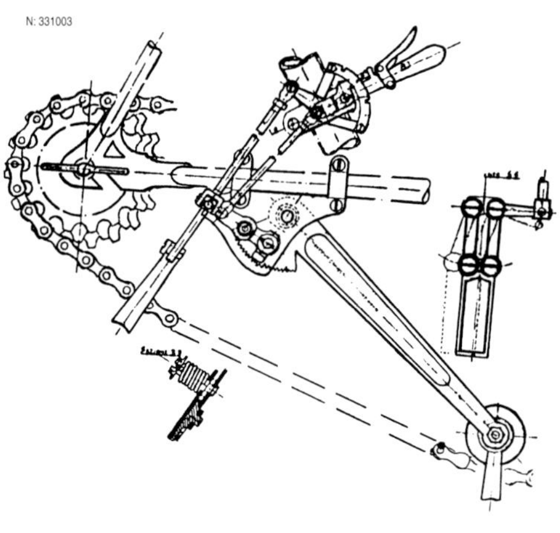


Left: Fig. 7.111. Ghiggini's 1937 patent showed a lever mounted on the upper part of the seat tube and connected by a long rod to a derailing fork under the chainstay. The derailing fork was mounted on an articulated parallelogram, so that the fork stayed vertical as it was moved sideways. A long tension arm took up the chain slack.

Right: Fig. 7.112. Ghiggini's third patent of 1941. Note the articulated parallelogram that maintained the pulley in a vertical plane.

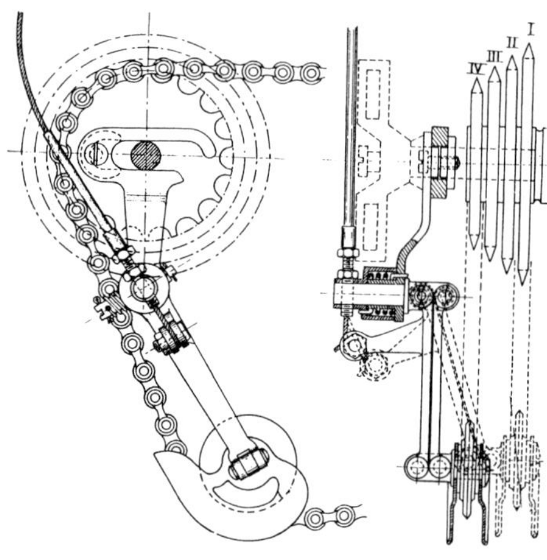

to one with 20 teeth, it is necessary to bring the rear wheel closer to the pedals.

Apparently, Campagnolo did not appreciate that the chain slack could be taken up with a spring-tension arm, as had been developed in Britain nearly forty years earlier. It was not until the end of the 1930s that he borrowed money from his sister Amelia and went to Paris to look at the rival French derailleurs. Or perhaps he may have tried to avoid pulleys and their friction.

Once his patent was secure, Campagnolo began the slow task of making personal approaches to notable Italian racers to win their respect and assurance that they would use his derailleur. In 1935, he set off with heavy suitcases full of his metal masterpieces.

His derailleurs were awkward to operate, but lightweight and virtually crashproof, and there was no pulley friction. The first riders to fit his derailleurs were Rafaelle Di Paco and Aldo Bini. Later on, he was able to win the support of the aging champion Alfredo Binda. On the spot, Campagnolo pulled out his file and converted Binda's machine to accept his derailleur. Binda straddled the bike and rode it around the garden, shifting gears perfectly. Soon Campagnolo was selling all the quick-release hubs, derailleurs, and pedals he could make. In 1940, he hired a mechanic, Enrico Piccolo, as his first full-time employee.

The definitive history of Campagnolo has not been written. Everything before World War II is folklore. The story of Tullio converting Binda's bicycle on the spot with a file is hard to accept. The early hand-made Campagnolo "cambio a bacchetta" (rod changer) prototypes resembled the postwar Campagnolo Corsa 1001.There are no diagrams or pictures of these early hand-made Campagnolo "*cambio a bacchetta*" (rod changer) prototypes. They probably resembled the post-war versions of the Campagnolo Corsa.

## Germany and Austria

In 1932, two German road racers, Fritz Scheller and Walter Lohmann competed in the World Amateur Cycling Championships in Rome. Their bikes used the famous Fichtel & Sachs Torpedo coaster hub. Four world champions had successfully used this hub since 1927. All of the other competitors in the road race used the new Italian Vittoria 3-speed derailleur that allowed them to change gears without getting off their bikes. The race became a farce. Lohmann, the faster of the two Germans, finished five minutes behind the victorious Giuseppe Martano.

Above: Fig. 7.113. A photo allegedly showing Tullio Campagnolo in the snowy race over the Croce d'Aune Pass in 1924. Although it may indeed have been taken at the Croce d'Aune, this is clearly a staged photograph, not taken during any race.

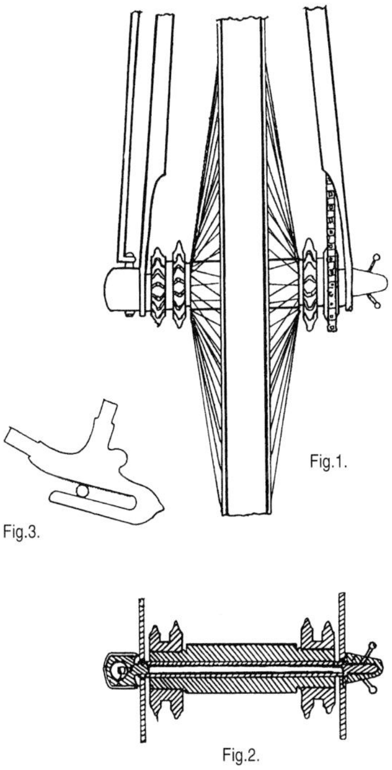


Left: Fig. 7.114. Diagrams from Tullio Campagnolo's 1933 patent for his quick-release mechanism. In this patent, Campagnolo pointed out that with a long lever, the quick-release could be operated while riding. He also mentioned adding a derailing fork.

Also in 1932, two new derailleurs were patented in Germany. The Durex was designed by Berlin cycle racer Erich Lucas and manufactured by Präzisions-Werke Bielefeld (PWB), a bicycle component manufacturer established in 1903. Shifting gears with the first Durex was difficult and messy: You pushed a lever with your right heel to loosen the chain. Then you bent down and used your forefinger and thumb to switch the chain to one of the three sprockets while backpedaling. Lastly, you tightened the chain again with the right heel.

The 1933 Durex was slightly improved. It was listed in German bicycle catalogs until about 1935. It seems to have been designed and used only for racing.

The Wanderer derailleur was patented in 1932. It was made by the well-known bicycle manufacturer Wanderer-Werke in Schönau, Chemnitz, Germany. The derailleur was supplied with the Wanderer Model 7 racing bike between 1935 and 1938. It is not known if the derailleur was sold separately.

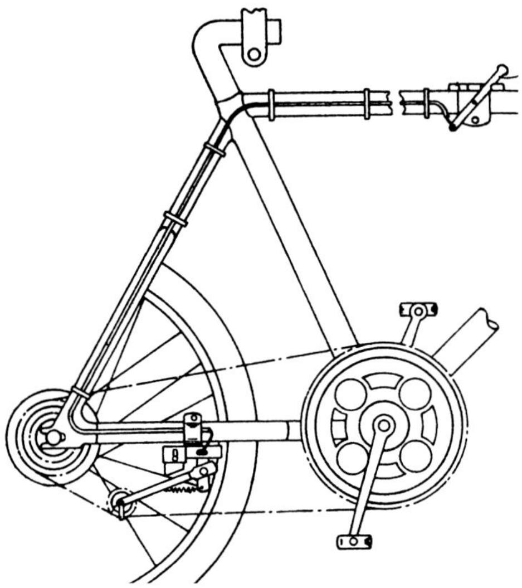

Above: Fig. 7.115. The 1933 Durex patent drawing shows a single-pulley derailleur that was operated by a lever and a cable.

Below: Fig. 7.116. The 1933 Austrian patent drawing for the Err-Zett derailleur looks like the first Vittoria Margherita. A lever above the front chainwheel moved through a special gate to shift gears and take up chain slack. The improved 1934 Err-Zett was in Viennese bicycle dealer catalogs until 1935. No Err-Zett derailleurs seem to have survived.

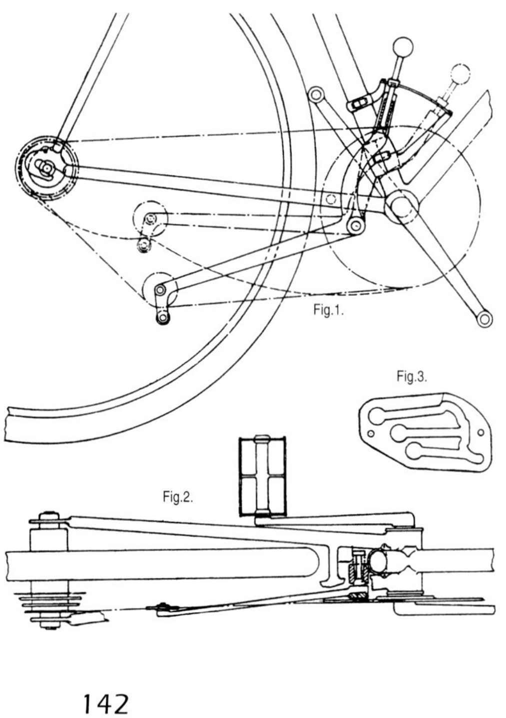

Above: Fig. 7.117. The German Durex derailleur was little more than a tension arm and a pulley. It was shifted by hand. However, it was cheap, light (only 200 grams, or 7 oz.), and easy to install.

Right: Fig. 7.118. The J.Wi.S expanding chain varied its length by using elliptical chain pins. A narrow coil spring went around the entire chain, held by hooks onto each link.

Above: Fig. 7.119. The Wanderer derailleur used a single pulley. Turning the control lever shifted the chain onto one of three sprockets. A helical spring on the control shaft maintained the chain tension automatically. Note the disc to keep the chain and the derailleur out of the spokes.

Above: Fig. 7.120. Detail of a 1938 German officer's bicycle equipped with a J.Wi.S derailleur and expanding chain.

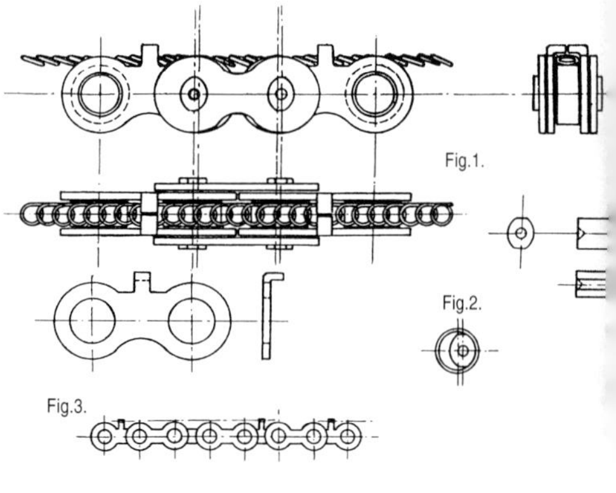

Fichtel & Sachs (F&S), of Schweinfurt, Germany, manufactured light aluminum racing hubs, including a freewheel hub. They made Germany's most successful derailleurs from 1935 onward. Their derailleurs could be used without the Torpedo coaster-brake hub, but it was company policy to combine them. This was unfortunate, since with a coaster brake must all the chain slack must first be taken up before the brake becomes effective. This also limited the capacity of the derailleur. F&S and most of the other German bicycle and component makers supported and financed professional cycle racing in Germany. Through sponsorship, the companies decided which bikes and equipment would be used. F&S made sure that their Torpedo hub and derailleur was used by most German bicycle racers from the 1930s to the 1950s. Their derailleur was also widely used on

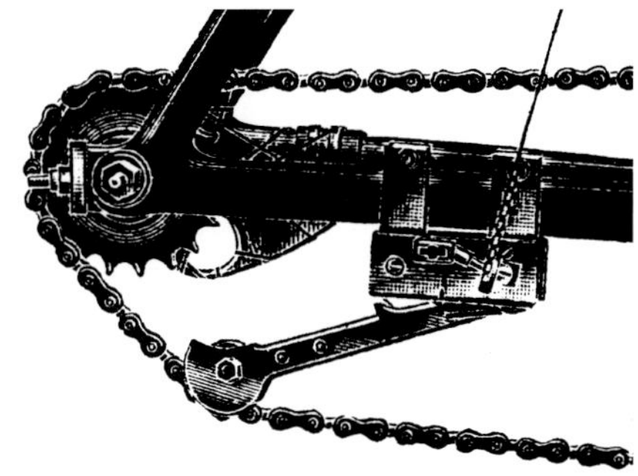

Above: Fig. 7.124. The 1937 Alpion derailleur was patented in Austria. It used a 16- and a 20-tooth sprocket. It was advertised as cheap and reliable. It could be used with a coaster brake. No surviving specimen is known.

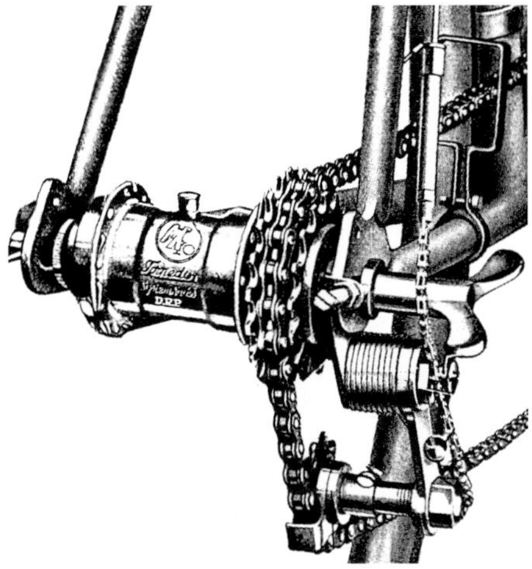

Above: Fig. 7.121. The 1935 Fichtel & Sachs single-pulley derailleur was installed under the right dropout. Note the cage above the chainstay to hold the slack chain when back-pedal braking.

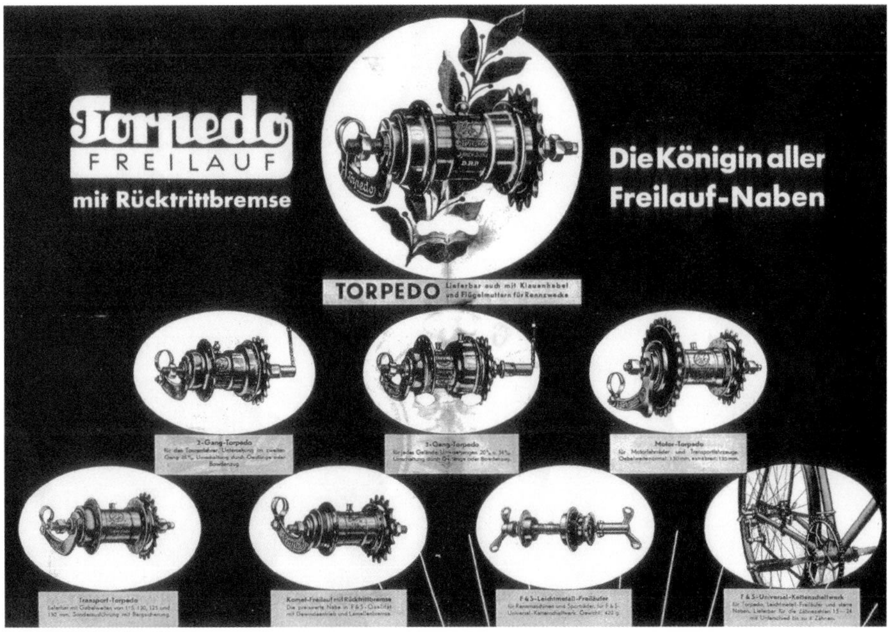


Above: Fig. 7.125. The 1939 Fichtel & Sachs catalog showed Torpedo hubs with single-speed, 2-speed, 3-speed, and moped models. All had coaster brakes.

Left: Fig. 7.122. The 1940 Fichtel & Sachs Aigle (Eagle) single-pulley derailleur moved in and out by a rotating helicoid.

Right: Fig. 7.123. The 1934 SU single-pulley derailleur used a double cable, operated by a shift lever on the top tube. The lever is stamped "SUWE SPORT-WERKE."

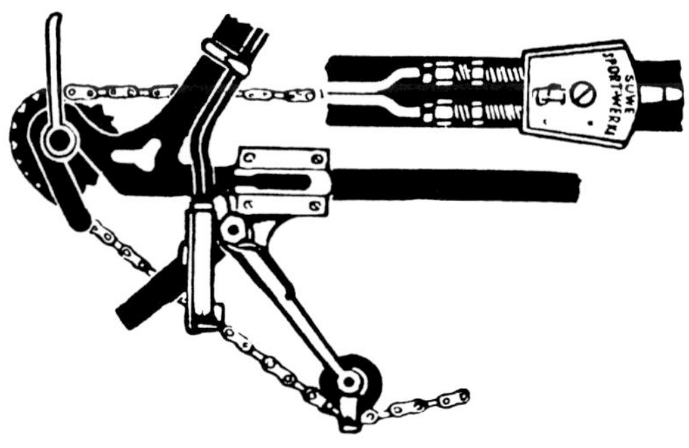

touring bikes. Around 1940, it was also promoted for women's bikes.

The Austrian company Suwe-werk-Siegmund Werner's SU 3-speed derailleur was sold in Austria from 1934 to 1937. It was available either nickel- or chrome-plated. After World War II, Suwewerk-Siegmund Werner in Vienna became Austria's most important derailleur maker.

The remarkable J.Wi.S. 3-speed derailleur was produced by Johann Winklhofer & Sohn in Munich from the 1930s until the early 1950s. It looked like a Super Champion-style selector fork, with the shift lever on the downtube, the top tube, or the handlebars, but there was no mechanism to take-up the slack chain. The secret was that the chain stretched.

Robert Pollak of Vienna was the inventor of the extendable chain. The chain allowed a total variation of 4 teeth between sprockets. At that time, touring bikes normally used 17-19-21-tooth sprockets and racing bikes used 16-18-20-tooth sprockets. The system worked only with the J.Wi.S 3-speed freewheels with a shark-tooth profile. This enabled the shortened links on the lower run to expand smoothly to the normal ½-inch (12.7 mm) pitch. The patents for the special chain and sprocket teeth date from 1937/1938.

Otto Haupt of Leipzig patented a 2-speed gearbox in 1931, which could be bolted onto the chainstay of any bicycle. The normal chain drove from the chainwheel to the top sprocket on the gearbox. Another short chain drove from the bottom sprocket of the gearbox to the hub sprocket. Two pairs of sprockets, joined by a third chain, were inside the gearbox. Either pair could be engaged by pulling or pushing a knob.

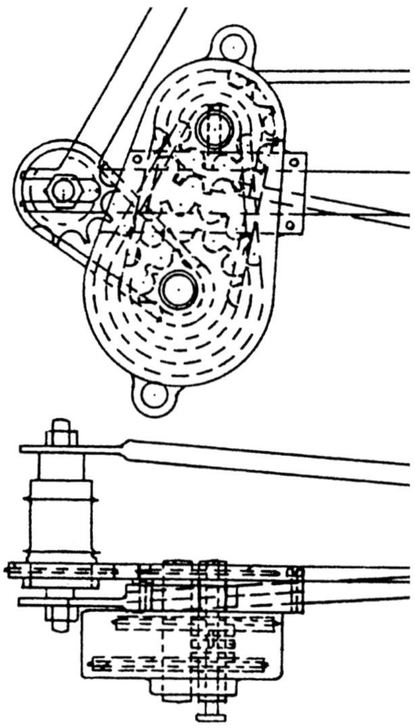

Above: Fig. 7.126. Otto Haupt's 1931 two-speed gearbox. It was conceived as an after-market accessory that could be bolted onto the chainstay of any bicycle.

Above: Fig. 7.128. This mid-1930s single-pulley Wissner 3-speed derailleur was made by Metallwarenfabrik H. Wissner, Zella-Mehlis, in Thüringen, Germany. It mounted directly on the rear wheel axle with a special wing nut, stamped with a trefoil leaf and "WISSNER."

Below: Fig. 7.129. This two-pulley Wissner derailleur has no markings. It was advertised in a 1939 catalog for both touring and sport.

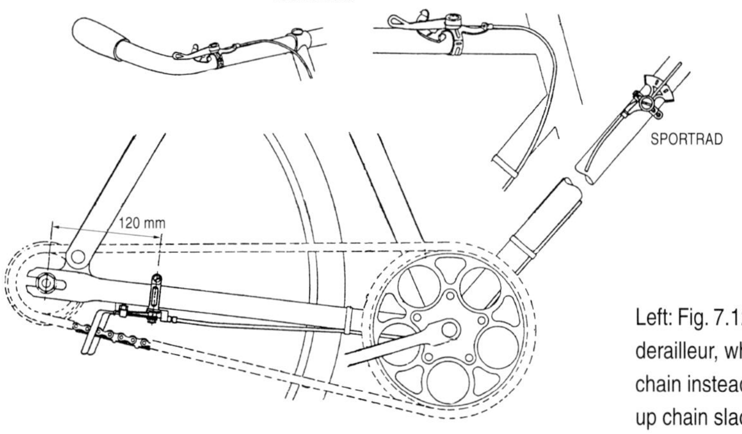


Left: Fig. 7.127. The J.Wi. S 3-speed derailleur, which used an expanding chain instead of a tension arm to take up chain slack.

## Other European Countries

The Spanish bicycle component company Zeus Industrial recounted its early history in the 1973 catalog. Zeus' founder, Don Nicolas de Arregui, began making bicycle components in the town of Eibar in 1926. According to the catalog, he undertook "the creation and development of the first derailleur based on the mechanics of the articulated parallelogram."

A prototype of this derailleur, which had presumably survived the turmoil of the Spanish Civil War of 1936–39, was displayed nearly forty years later at a Brussels show in 1971. This dates the Zeus prototype to the early 1930s, long before the use of the articulated parallelogram by Nivex. Zeus was still attending U.S. bicycle shows in 1984.

I tried to track down this prototype derailleur, or a picture of it, or any proof that it ever existed. I concluded that it was probably part of Zeus mythology. The Zeus company ceased to make components a few years later.

In 1932, Novitas Fabrik Elektrischer Apparate A.G., a Zurich company, patented and manufactured a rear derailleur. It was advertised in 1934/1935 catalogs, including a special version for use with the Torpedo coaster hub. Several specimens have survived.

Also in 1932, another Swiss derailleur was patented. This Pedor 3-speed rear derailleur was an early system with a detented (i.e. indexed) shift lever. The rear derailleur operated by a helicoid. It had two pulleys. The tension arm locked into one of three notches on the clamp that fixed the derailleur to the chainstay.

Jhaga [Editor's note: it's actually Ihage, but to all the world, that "I" looks like a "J" in the trademark] made a 3-speed derailleur in Belgium. It clamped onto the chainstay with the tension spring secured by the rear axle nut. A cable ran over a pulley next to the jockey pulley sprocket and pulled the jockey pulley onto the larger sprockets. An enclosed return spring pulled the jockey pulley to the smaller sprocket.

## United States

Bicycle use was low in the U.S. in the 1930s. Annual sales were only one bicycle per 500 people. Cars outsold bicycles ten to one. Bicycles were made for children and teenagers too young to drive. They were styled to look like motorcycles and weighed up to 70 lbs. (32 kg), fully equipped.

Virtually all bicycles sold in the U.S. during this period were

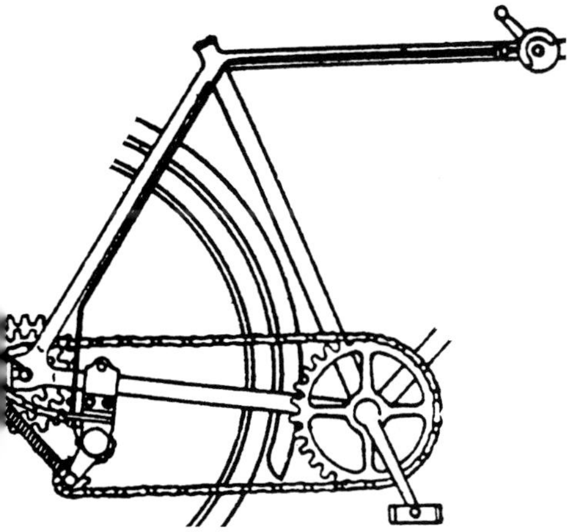

Above: Fig. 7.130. The indexed control lever for the Pedor rear derailleur.

Left: top and bottom: Figs. 7.131 and 7.132. The 1932 Pedor 3-speed rear derailleur system with a detented (i.e. indexed) shift lever. The derailleur was operated by a helicoid and had two pulleys.

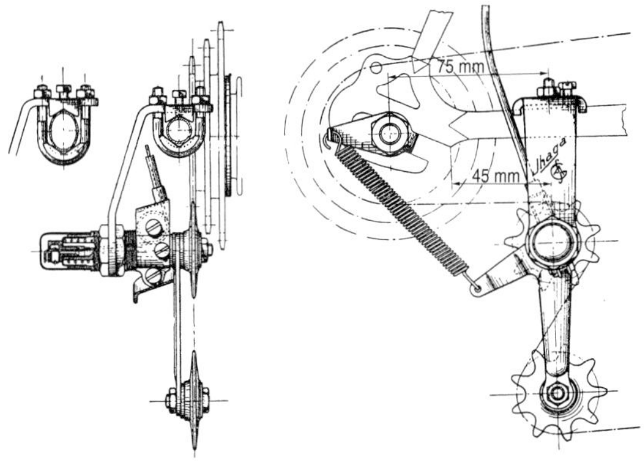


Right: Fig. 7.133. The Belgian Jhage (actually "Ihage") 3-speed derailleur.

domestically made. They had a single speed, with either a ½-in. pitch x ⅛-in. width roller-chain or 1-in. pitch x 3/16-in. width block-chain, usually with back-pedal coaster hub with a freewheel. Bicycle racers used a block chain with a fixed gear.

The only concessions to multiple-speed gearing were the 2-speed coaster-brake hubs made by New Departure, Morrow, and Musselman. All three used planetary gears and were available as conversion kits. The New Departure was cable-operated and gave a 40% reduction. The Musselman was lever-operated and gave a 28% reduction.

Schwinn was the technological leader in the U.S. In 1934, they introduced clincher balloon tires, a huge improvement over the old single tube tires that had been used in the U.S. for 40 years. President Frank W. Schwinn was convinced that he could sell adult bicycles to Americans. In 1935, he went to Britain and returned with several Raleigh bicycles. At his urging, Schwinn started building lightweight adult bicycles. They were not big sellers, but they opened the door.

You could not buy a U.S. factory-made derailleur bicycle; but at least by the late 1930s, imported derailleur gears could be ordered from a New York mail order company. The 1938 catalog of the California Cycle Company Inc. of New York listed the full range of British Cyclo gears. They included the standard Cyclo 2-, 3-, and 4-speed rear derailleurs, plus the Ace, Witmy, Poly, and Star rear derailleurs, as well as the Rosa front derailleur with double and triple chainwheels. The catalog also listed the Villiers 2-speed hub, the Sturmey-Archer and Torpedo three-speed hubs, and the Adler 3-speed bottom-bracket gear. A hub gear was shown combined with a 2-speed Cyclo to give a six-speed combination.

The 1940 catalog of the Hudson Cycle Equipment Corporation of New York showed two models of the French Rota derailleurs.

In his book *Hearts of Lions*, Peter Nye wrote:

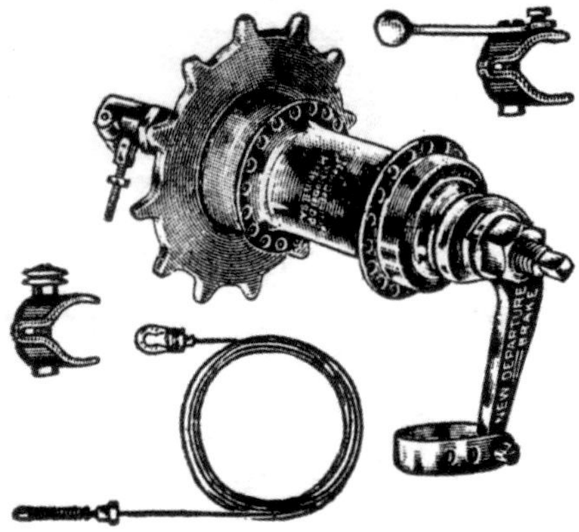

Above: Fig. 7.134. The New Departure two-speed hub was operated by a cable and a handlebar lever.

Left: Fig. 7.135. The California Cycle Company listing of British Cyclo Poly and Rosa derailleurs in its 1938 catalog.

Right: Fig. 7.136. The 1940 Rota 333 derailleur was mounted on the rear axle. It was a single-pulley 3-speed. Most U.S. bikes had coaster brakes, which operate by pedaling backward about a quarter of a turn. Narrow-range single-pulley derailleurs had very little chain slack, so they could be used with a coaster brake.

Model No. 333

American racers resisted derailleurs partly because they were difficult to come by, but mostly because the prevailing wisdom said that fixed gear riders were smoother pedal- ers.

The first racer to use a derailleur in the U.S. may have been James Armando in 1929 during a 62-mile (100 km) race in New York. Jack Simes II was among the spectators lining the road when they discovered that Armando was riding a bicycle with three speeds. Simes recalled the moment with enthusiasm:

Armando rode a Cyclo derailleur, made in England. It was a crude contraption that literally had a complete figure eight in the back to keep the chain tension, and it sounded like a coffee grinder. But the crowd just loved it. They kept yelling at him, "Throw it in high gear! Let's see what you've got! Throw it in high!"

## Conclusions

The derailleur came of age and became completely practical in this decade. In the process, a great many alternatives were tried and found wanting. The best derailleurs were the Nivex and the Cyclo Standard. They shifted almost as smoothly as the pre-index-shifting derailleurs of the 1980s.

### Touring Derailleurs

Three outstanding touring derailleurs were invented in this period—Funiculo, Spirax and Nivex. They shifted better than the pre-index-shifting derailleurs of the 1980s. The Funiculo and the Spirax did not reappear after WW-II. The Nivex lasted for a few years on a small number of expensive touring bikes.

Only the Cyclo Standard lasted into the 1960s but it was a dated design. Today's touring bicycles have indexed shifting and much improved chains, sprockets and chainwheels. However, no current bicycle allows the easy rear wheel removal that Nivex provided.

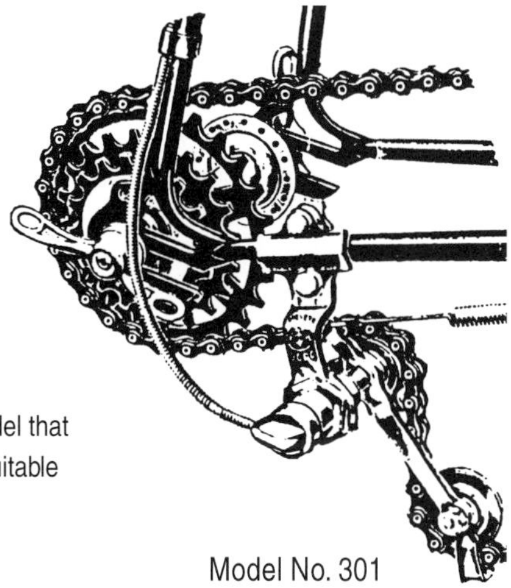
Model No. 301

Right: Fig. 7.137. The 1940 Rota 301 derailleur was mounted on the chainstay. It was a double-pulley model that could handle three or four sprockets, but it was not suitable for use with a coaster brake.

Below: Fig. 7.138. The Musselman two-speed hub was operated by a lever behind the saddle.

Right: Fig. 7.139. In 1938, Jim and Elizabeth Young rode their English tandem equipped with a Cyclo Standard 3-speed derailleur coast to coast from San Francisco to Baltimore and back. The tandem had a 52-tooth rear chainwheel and a 15-17-26-tooth freewheel.

### Racing Derailleurs

The 1930s racing derailleurs suffered from the false assumptions of the period:

- Chains have to be perfectly aligned. [Although the wide, stiff chains of the period gave some support to this theory, it is not true in general.]
- Derailleur pulleys add friction. [The amount of friction is insignificant.]
- Looping the slack chain around two pulleys adds friction. [It does not.]
- The tension of the return chain adds inefficiency. [It's offset by the tension at the top.]

The racing derailleurs were developed to meet these assumptions. The drop arm, Super Champion-type models shifted poorly because the derailing fork was a poor substitute for a jockey pulley. The single-pulley models had a limited range. They were better than a fixed gear, but their capacity was so small that most riders still walked up steep hills. The best-selling derailleurs were made to a price, and they were quite flimsy by today's standards.

Left: Fig. 7.140. The 1938 Nivex allowed easy wheel removal with a split axle. It was probably the best derailleur of the pre-WW-II era.

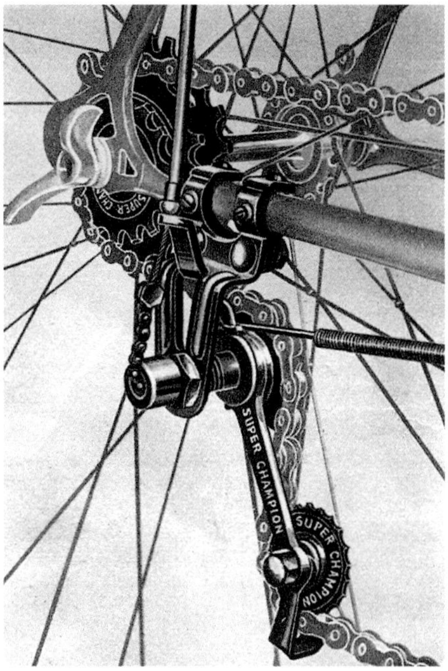


Right: Fig. 7.141. The 1938 Super Champion Grand Sport Léger was a 2-pulley touring derailleur with a 30-tooth capacity.

# 8 The Post-War Years: 1945–1954

IMMEDIATELY after the war, bicycle production around the world quickly increased to fill the gap left when the bicycle factories were converted to manufacture munitions. The first need was for replacement items like tires and tubes to keep the existing bicycles running. Then there was a huge pent-up demand for new bicycles. Many of the old bicycle-manufacturing machines at the factories were outdated. Since no new machinery was available, bicycles were made with whatever equipment had survived.

In 1946, *Le Cycle* estimated that there were 17 million bicycles in Germany, 12 million in Britain, 10 million in France, 10 million in the U.S., 7 million in Japan, 5 million in Italy, and 4 million in Holland.

In the early post-war years in Europe, the vast majority of bicycles were used to provide transportation for people who could not afford motor vehicles or the gasoline to fuel them. Not many of these early bicycles had gearing, but a few top-of-the-line geared bicycles were made for the various specialized national markets.

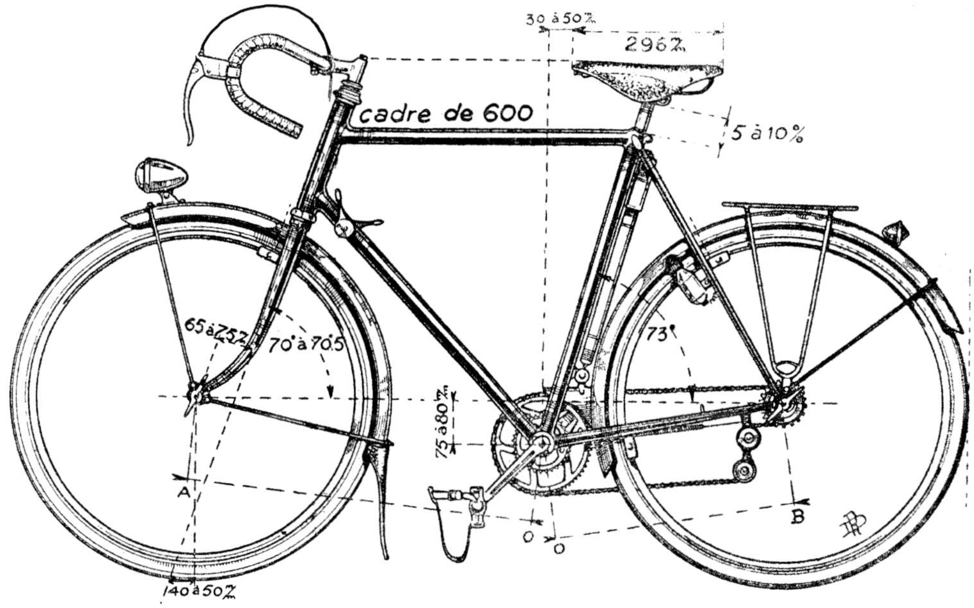


Above: Fig. 8.1. In 1946, the French Cycle Touring Federation specified a standard touring bicycle. It had 46/26 chainwheels, a 16-19-24 freewheel, and a Cyclo Standard rear derailleur.

Although France was the home of the derailleur bicycle, there were also derailleur users in Britain, Italy, and Belgium. Buyers of quality bicycles in Britain generally preferred three-speed hubs, but there was a good market for derailleurs. Most of the innovations in bicycle design came from the small custom frame-builders in France, Britain, and Italy.

As the decade progressed, competition developed between people-powered two-wheelers and motor-powered two-wheelers. It was not at all clear which would win. By the end of the decade, bicycle production and bicycle use declined as the post-war prosperity spread around the world, allowing millions of cyclists to replace their utility bicycles with mopeds, scooters, motorcycles, or small cars. Some car owners commuted by bicycle and saved the car for weekends. However, in Britain and in most other European countries, the majority of the people who were riding bicycles could not afford anything else. Bicycling picked up a "working-class" stigma, that lasted well into the 1960s.

The pre-war derailleur companies in France, Italy, and Britain resumed manufacture. Dozens of new small derailleur companies sprang up in the late 1940s. By 1950, there were more than fifty derailleur companies. As demand for bicycles declined, a shakeout started, and by the end of the decade, Simplex, Huret, Campagnolo, French Cyclo, and British Cyclo were the only significant survivors.

Derailleur performance was limited by the stiff 1/8-inch (3.2 mm) chains used at the time. Over the decade, more flexible 3/32-inch (2.38 mm) wide derailleur chains gradually took over. The number of freewheel sprockets went from three to four, and then to five. The dead hand of Vélocio and his aversion to pulleys continued to influence derailleur design, especially in Italy. There were good touring derailleurs with jockey and tension pulleys, but the racers felt (wrongly) that these had too much friction. This resulted in the survival of numerous single-pulley or fork-operated racing rear derailleurs. By and large, they all had limited capacity and shifted poorly.

Bicycle racing quickly resumed after the war. The derailleur innovations developed before the war by touring cyclists gradually worked their way over to the racing community, and by 1955, racers could choose from a much better class of derailleurs.

The 1951 Campagnolo Gran Sport was the breakthrough derailleur. It set such a high standard for quality and reliability, that few of the older makers or the old designs could compete.

The Simplex Tour de France, which was mass-produced from 1947 to 1958, set a different standard: It was light and flimsy and limited in capacity, but it was cheap and heavily marketed. Millions of them were sold.

Left: Fig. 8.2. This illustration of a Singer B.R.A. (Brevet Randonneur des Alpes) model is from a 1946 Alex Singer advertisement. It had 46/26 chainwheels, a 4-sprocket 14–26 freewheel, a Cyclo Standard rear derailleur, and a Simplex front derailleur operated by means of a cable from a shift lever under the seat. With 650B tires (about 26-in. diameter, 42 mm wide), it weighed 12 kg (26 lbs.). This is one of the earliest detailed Daniel Rebour line drawings. Note that he still showed the spokes (which he soon after stopped doing).

## Bicycle Racing

In France, the strong tradition of bicycle racing, especially the multi-day massed-start stage races revived quickly after the war.

Racing, in addition to cycle-touring, was now becoming a crucible for bicycle and component developments. Reading the ads, you might believe that races are won not by riders but by bicycles or by components like tires or derailleurs.

Of course, races are won by athletes; however, the derailleur makers paid the racers considerable sums to showcase their products. Advertisements of winning components were an effective sales tool. On the other hand, if enough top racers complained loudly enough about the inadequacies of a particular design, it might lead to a new racing derailleur. This was certainly the case with the Simplex 543 and the Campagnolo Gran Sport.

### The Tour de France

Amongst the stage races, the three-week Tour de France is the most popular sporting event. An abbreviated five-day race was held in 1946, but the first real post-war Tour de France was held in 1947.

The 1947 Tour de France was a bit of a fiasco. To ensure that a Frenchman would win, the Italian team was allowed to enter only second- and third-category racers. Coppi and Bartali, the best racers of the day, were told to stay home. Just the same, by staying inconspicuously in the pack and avoiding stage sprints, two Italians, Brambilla and Ronconi, had a comfortable lead going into the last day.

It would have been ruinous to French cycling prestige, and to *L'Équipe,* the sponsor, if two Italian domestiques would have arrived first and second in Paris. The Tour organizers decided that two Frenchmen, Robic and Fachleitner, should be first and second. The Belgian team was paid to pace the two Frenchmen to the finish. Brambilla lived in France, and supposedly his Italian teammates would not work for him. Tullio Campagnolo was delegated to tell the two Italians of the decision, and to suitably reward them for their disappointment. Under the circumstances, the Simplex derailleurs on Robic's winning Genial Lucifer bicycle were not a contributing factor in his victory.

### Speed and Distance Records

In 1942, Fausto Coppi raised the unpaced one-hour record to 45.85 km (28.4 miles). His record lasted for 13 years.

In 1951, José Meiffret raised the paced speed record to 109.2 mph (175.7 km/h) behind a Talbot race car with a rear wind screen.

*Table 8.1. Tour de France winners, 1946–1954*

| Year | Rider (Country) | Bicycle | Derailleur |
|---|---|---|---|
| 1946 | Lazarides (France) | France Sport | Campagnolo |
| 1947 | Jean Robic (France) | Génial-Lucifer | Simplex |
| 1948 | Gino Bartali (Italy) | Legnano | Campagnolo |
| 1949 | Fausto Coppi (Italy) | Bianchi | Simplex |
| 1950 | Ferdi Kübler (Switzerland) | Fréjus | Simplex |
| 1951 | Hugo Koblet (Switzerland) | la Perle | Campagnolo |
| 1952 | Fausto Coppi (Italy) | Bianchi | Campagnolo |
| 1953 | Louison Bobet (France) | Stella | Huret |
| 1954 | Louison Bobet (France) | Stella | Huret |

Right: Fig. 8.3. Tullio Campagnolo (left), Gino Bartali (center), and Fausto Coppi (right). Bartali was in the Italian army during his prime. Coppi, "*il Campionissimo,*" was the outstanding rider of this decade.

## France

France produced more than a million bicycles a year during the German occupation, and about ten million French bicycles were still in service at the end of the war. The French Cyclo factory, which had had trouble filling its order books in 1939, was in full production in 1942. However, Cyclo produced far more freewheels than derailleurs during the war.

The industry shut down from D-Day until France was liberated. France's bicycle industry revived quickly after the war, although there were problems, mainly due to shortages of high-quality steel, tires, and ball bearings.

France's bicycle industry was blessed with competition. Numerous companies made components such as cranksets, brakes, hubs, freewheels, handlebars, and stems. The challenge of derailleurs attracted entrepreneurs, and there were many other makers besides Simplex, Huret, French Cyclo, and Super Champion, who were the major derailleur manufacturers. For much of this decade, French custom bicycle builders made front derailleurs by hand rather than buying them.

The war was over, and the French were liberated, but they were not well off. Many had lost everything. Times were hard. So, like it or not, the great majority of people continued to ride bicycles until the 1950s.

France was well served by its bicycling organizations. The French Cycle Touring Federation proposed a "standard" touring bicycle in 1946. It was much more suitable for loaded touring than many of today's "credit card" tourers.

Almost everyone went to work by bicycle. On summer Sundays, young men went to swimming spots by bike. For a holiday, they loaded a tent onto the luggage rack. A 3-speed derailleur bicycle was sufficient.

French bicycle production grew steadily from 350,000 in 1944 to a peak of 1,300,000 in 1949. Then mopeds took over, and bicycle production declined steadily to 750,000 in 1954.

Little by little, under the influence of the technical touring competitions, roadster bicycles appeared in shop windows and catalogs. The 1950s French roadster had drop-handlebars, wide tires, lights, fenders, two

Right: Fig. 8.4. José Meiffret's record-setting bicycle was made by M. Yvars. It used a special 130-tooth chainwheel made by TA.

Left: Fig. 8.5. This René Herse Dilettante is an entry-level racing bicycle from the 1950s. It had a Herse double crankset and front derailleur and a Simplex rear derailleur. Note that Rebour no longer bothers showing spokes.

luggage racks, and eight speeds. The front derailleur was lever-operated; the rear derailleur was an inexpensive road model (usually Simplex). The gearing invariably used a 48/36 crankset and a 14–22 freewheel. These roadsters may have been inspired by the technical contests, but they were not lightweights.

Lighter "semi-racer" bikes were displayed beside the roadsters in the bike shop window. They had 700 x 28C wheels (not 650 x 35B, as used on roadsters), but they did come with fenders, lights, a luggage rack, and bottom-of-the-line racing derailleurs. Their gearing used a 50/46 crankset and a 14–20 freewheel. The semi-racer was a compromise between the real racing bikes that most young men dreamt of and the utilitarian bicycles that their parents considered useful. Finally, the bike shops also carried "real" racing bikes. The quality varied with the price. In the 1950s, racers used a five-sprocket freewheel with 52/46 chainwheels.

In 1947, a French bicycle industry task force standardized the key bicycle dimensions. Standardization was a major advantage for the French bicycle and component makers. These industry standards were the basis of the "French" screw threads, which were found on imported French bicycles in the 1970s, to the consternation of American bicycle mechanics.

By 1954, derailleurs were installed on the whole range of French bicycles—utility roadsters, semi-racers, racers, tourers, and "randonneur" bicycles. *Randonneur* is a French term that has no English equivalent, but "sports-tourist" and "club racer" are approximate translations.

The numerous French custom builders soon resumed production of quality touring bicycles. In the 1940s and 1950s, if a Frenchman wanted a good racing, touring, or randonneur bicycle, he usually ordered it from his local bicycle shop, whose owner often built frames and assembled finished bicycles.

Top-quality touring bicycles were not found in corner bike shops. They had to be ordered from Paris, Grenoble, or Lyon, where they were made to measure by a specialist builder. The builder was often a touring cyclist himself, and an artist with the brazing torch. You could order a bicycle to your exact specifications, equipped with all the accessories of your choice.

Quality touring bicycles usually had 15 speeds, a brazed-on Cyclo Standard rear derailleur, and a handmade front derailleur. Occasionally, Nivex or Spirax rear derailleurs were installed. You could order 18 speeds, but this abundance of gears was considered superfluous by most riders and builders.

Bicycle usage and sales grew steadily until 1953, when there was a serious slump and bicycle touring became less popular. France had compulsory bicycle registration, which provided a reliable census. The number of bi-

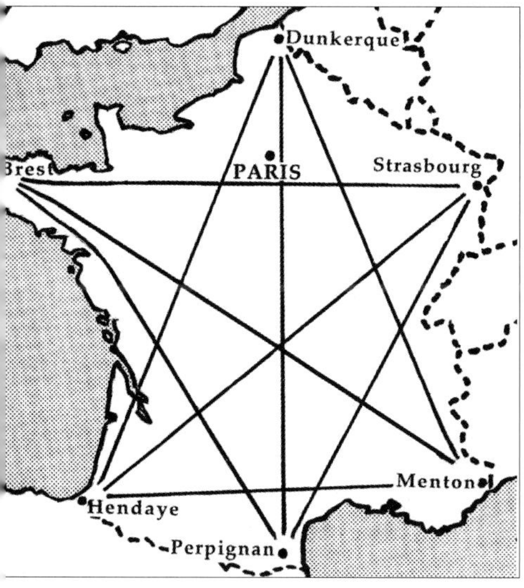


Left: Fig. 8.6. The nine diagonal rides across France. The country is shaped like a hexagon, and serious randonneurs ride diagonals, from one corner to opposite corners. Contributing author Raymond Henry has ridden all nine diagonals.

Right: Fig. 8.7. Illustrator Daniel Rebour and his new bride, Simone. It's 1948, and they are about to set off on their honeymoon: Paris–Brest–Paris, setting the new tandem record at 61 hrs. 56 min. along the way.

cycles in use fell from 11.3 million in 1953 to 10.6 million in 1954. Simplex, Huret, and French Cyclo were the main derailleur suppliers to the surviving racing and utility markets.

In 1948, *cyclomoteurs* (50 cc mopeds with pedals), *vélomoteurs* (100 cc lightweight motorcycles), or *motocyclettes* (175 cc or 250 cc motorcycles), made their noisy entrance onto the French roads. They were followed several years later by small cars. The French had yearned for comfort all through the difficult years. Once a luxury, the bicycle became ordinary. Worse, it became the vehicle of the poor—those who could not afford a moped or a car.

### The Poly Races and Touring Bicycle Trials

In 1946, *Le Cycle* revived the *Criterium de la Bicyclette Polymultipliée*, referred to as the "Poly" for short. The Poly events made a major contribution to French bicycle and derailleur development. It was a combined bicycle race and technical

Left: Fig. 8.8. 14-year-old Lyli Herse (left) after winning the 1949 Poly de Chanteloup mixed tandem race with her tandem partner Prestat (right). She continued winning the tandem events for many years after. Lyli's father, René Herse is second from the left.

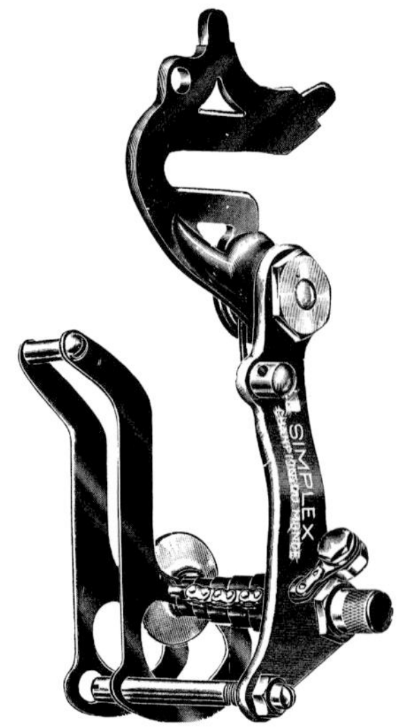


Above: Fig. 8.9. The 1946 Simplex Champion du Monde rear derailleur was a continuation of the pre-war model. It was a pretty grim plunger-type single-pulley model with a 10-tooth capacity

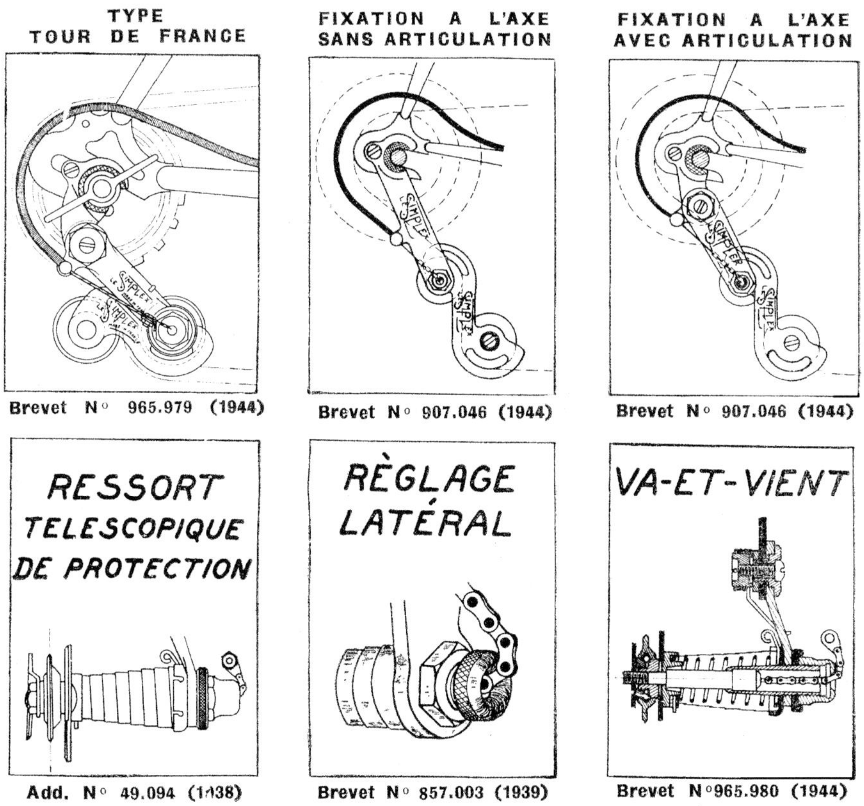


Left: Fig. 8.10. Lucien Juy kept busy during the occupation of France. In 1944, he filed several patents for the design of plunger-type cable-operated two-pulley rear derailleurs with two spring-loaded pivots—one at the mounting point, the other one at the cage. The three lower drawings illustrate how a plunger-type derailleur moves the cage in or out.

competition for bicycles with multispeed gears. Poly events were held in Chanteloup in May and in Lyon in September. The mountainous courses were typically about 200 km (120 miles) long. There were separate races for professionals, randonneurs, and mixed tandem pairs (a man and a woman each). These prestigious events ranked next to the major races in importance. Everybody who was anybody in the French bicycle industry attended, and the results were widely publicized.

René Herse was a premier builder of top-quality custom bicycles, and his daughter Lysiane, or Lilly, was his secret weapon. She won the tandem event from 1946 to 1956 behind a succession of captains. She trained early in the morning and then worked a full day as the bookkeeper and assembler in the family business.

## Simplex

Simplex dominated the derailleur market in this period, much as Shimano does today. Simplex made more derailleurs than everyone else combined. From pre-war days, Simplex derailleurs were light, simple, and cheap. The racing models were also limited in capacity and rather more flimsy than the touring models.

Jan Heine tested a 1946 Simplex Tour de France derailleur on a 1951 Follis racing bicycle with 49/46 chainwheels and a 14–22 five-speed freewheel. He found that it shifted very smoothly with some overshifting required in all gears.

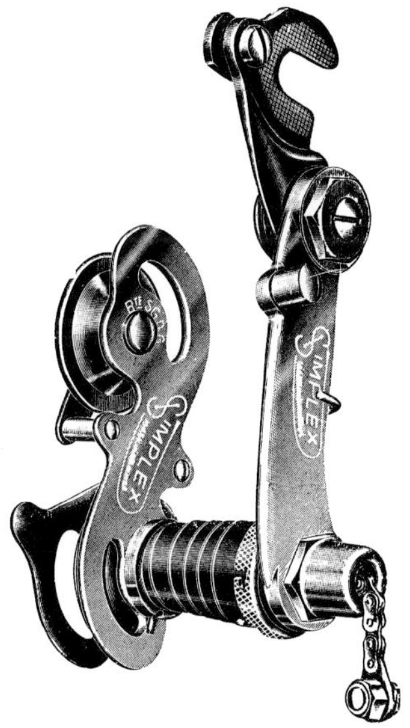

Above: Fig. 8.12. In 1947, Simplex added a second pulley to the Champion du Monde to improve the chain-wrap on the small sprocket.

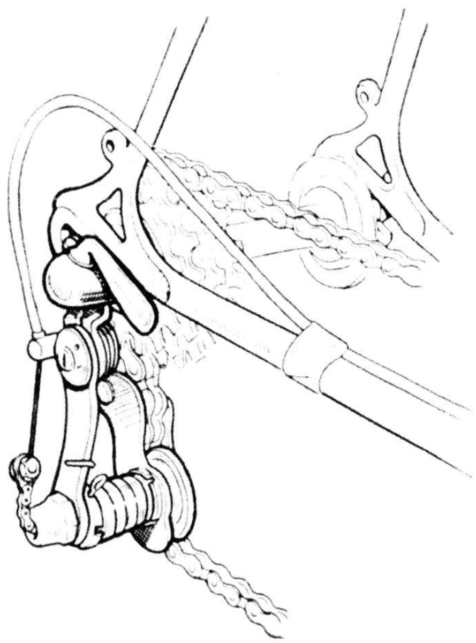

Right: Fig. 8.13. The 1947 Simplex Tour de France, showing how the second pulley gave slightly better chain wrap.

Above: Fig. 8.11. Exploded view of the 1957 Simplex Tour de France rear derailleur, showing the two springs. The spring at the top of the arm swings back the whole derailleur. The second spring rotates the cage, so that the second pulley increases the chain wrap around the sprocket. The inner coil spring and the outer *puttee* spring push the cage toward the larger sprockets.

Above: Fig. 8.14. The modified 1948 version was called the Tour de France after Robic's 1947 victory. The total capacity was 16 teeth. It could handle a 14–24 freewheel or a 10-tooth chainwheel difference. Variations of this model were made for a decade.

## One Pulley Versus Two Pulleys

I have received numerous letters and e-mails from people who fondly remember fondly the 1950s Simplex Tour de France rear derailleurs. I call this process "gilding the old oaken bucket." I think that the truth is about as follows:

The racing public thought that derailleur pulleys caused friction, so they preferred single-pulley derailleurs, like the tension arm Super Champion or the pre-war Simplex Champion du Monde. However, single-pulley derailleurs have limited capacity and do not shift as well as properly designed double-pulley derailleurs.

Simplex figured out a way to give the racers what they wanted: a "single chain bend" rear derailleur that looked like a single-pulley model. At the same time, Simplex made a modest improvement to the capacity and the shifting performance. They added a second pulley with a cage and a spring to press against the back of the chain between the lower jockey pulley and the sprocket (see Figs. 8.9 and 8.10). This Tour de France model was a huge marketing success, and Simplex sold more of them than any manufacturer sold of any other model up to then.

Viewed with hindsight, the Tour de France was a retrograde derailleur. It had all of the basic

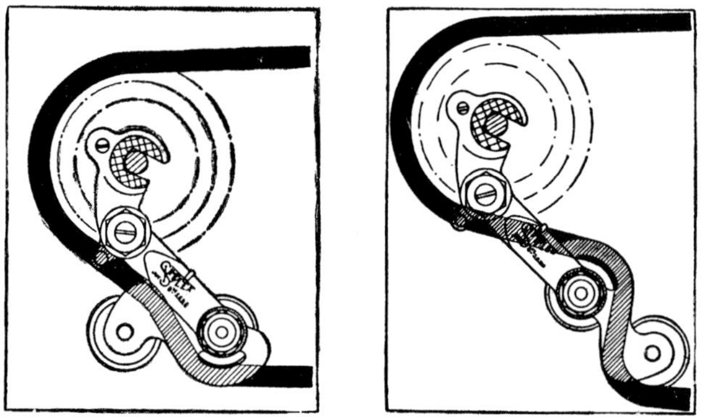

**Montage correct** **Montage incorrect**

Left: Fig. 8.15 A & B. Simplex included this installation drawing to prevent mechanics from installing the derailleur in what seemed to be a more logical fashion. Note how the *"Montage incorrect"* would actually provide more chain wrap.

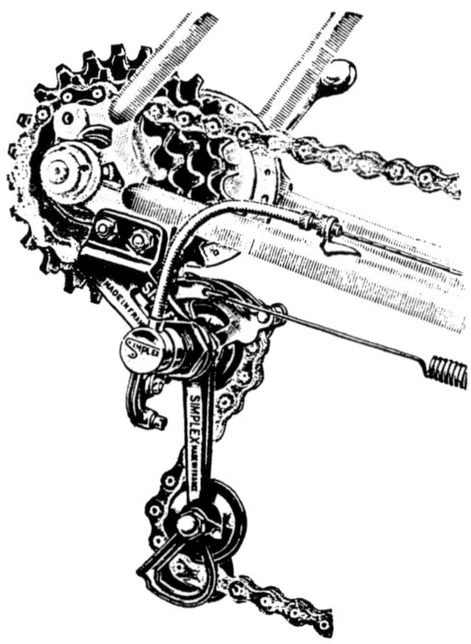

Above: Fig. 8.16. The 1952 Simplex Grand Tourisme could handle a 5-speed freewheel using $^{3}/_{32}$-inch chain. The total capacity was 37 teeth. It could handle a 13–28 freewheel and up to a 22 teeth difference on the chainwheels. Note the bell-crank. The Rigidex was a similar model.

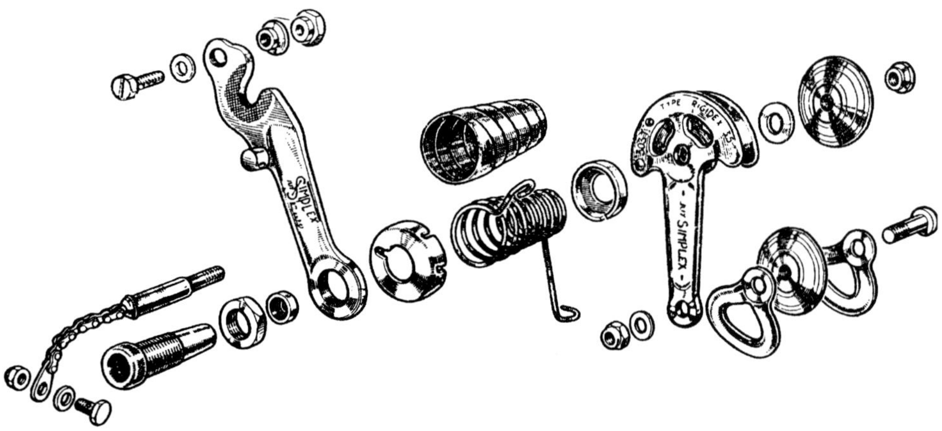

Left: Fig. 8.17. Exploded view of the 1956 Simplex Touriste 303.

elements of a modern rear derailleur: a jockey pulley, a tension pulley, two springs, and a cage; but it gave up capacity and shifting performance. Look at right-hand detail in Fig. 8.15 on the facing page. Simplex called this the "double chain bend." They knew how to build a better rear derailleur for the same money, but they gave the public what it wanted.

When I first encountered old Simplex Tour de France derailleurs in the 1970s, I said to myself. "What a stupid design. It's upside down with the tension pulley on top". With the passage of time, I now believe that Simplex provided what the market wanted. Simplex made racing derailleurs for the racers, and touring derailleurs for the tourists. The racers thought that bending the chain around pulleys added friction, so they got a single-pulley design.

Above: Fig. 8.18. The Simplex 1956 Touriste 303 was a less expensive pull-chain touring rear derailleur. It could handle a 14–28 freewheel and up to a 20-tooth difference between chainwheels.

The first post-war Simplex racing derailleur, the Champion du Monde, was a reissue of the pre-war version. The first Paris Bicycle Show was held in September 1946, and Lucien Juy was ready with a range of Simplex derailleurs. Simplex made a significant change to their racing derailleur, the Champion du Monde Model 46. They added a second pulley. In 1947, it was renamed the Tour de France after Robic's 1947 victory in the Tour de France.

Above: Fig. 8.19. The 1946 Simplex front derailleur had a turning lever to move the cage.

Below: Fig. 8.20. The 1947 Simplex front derailleur had a lower pivot and a notched plate for the lever.

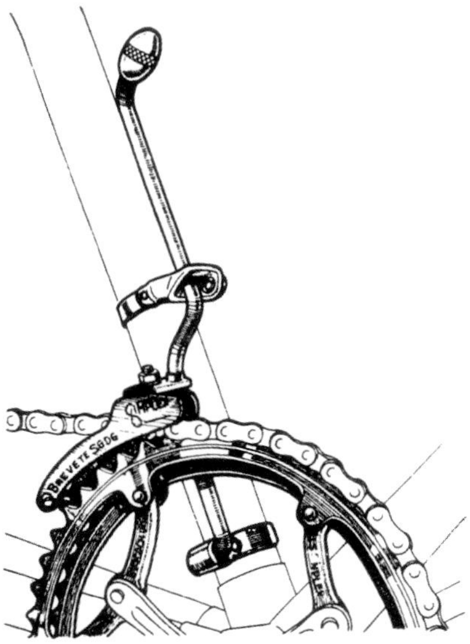

Above: Fig. 8.21. The 1949 Simplex Competition lever-operated front derailleur, available until 1970, was the definitive model. It could cope with a 52/28 double chainwheel. British cyclists called it a "clanger" because of its noisy shifts. By 1960, all of Simplex' competitors were selling low-priced cable-operated front derailleurs instead.

Above: Fig. 8. 22. The 1952 Simplex JUY 39 was a cable-operated plunger type. It was suitable for wide-range double chainwheels (up to 50/28).

The Tour de France was a pull-chain plunger-type model in the pre-war Simplex tradition. Even though it had two pulleys, it was a single-pulley design with slightly more capacity. Single-pulley derailleurs are inherently limited in capacity. Simplex' solution was to provide a cage with a second pulley to increase the chain wrap.

Above: Fig. 8.23. The 1950 Simplex Tour de France Tension rear derailleur. The total capacity was 16 teeth, and the maximum freewheel was 13–26. It had two cables. The second cable adjusted the chain tension. Only one cable came from the shift lever. The second cable was attached forward of the casing.

Below: Fig. 8.24. The 1951 Simplex JUY 51 Tension rear derailleur connected the second cable forward of the casing. Note the improved cage shape, with the pivot between the pulleys, which gave more chain wrap and increased the total capacity to 20 teeth.

At the 1946 Milan Show, Simplex boasted that 146 of 195 derailleur bicycles on display used Simplex derailleurs. At the 1949 Paris Show, the count of rear derailleurs was as follows:

| | |
|---|---|
| Simplex: | 519 |
| Cyclo: | 131 |
| Huret: | 53 |
| Super Champion: | 23 |
| others: | 25 |

Simplex' production figures are a measure of the increasing demand for derailleur bicycles. Simplex made a million derailleurs in 1948 and 1.5 million in 1949.

The patented features of Simplex' racing derailleurs were the two spring-loaded pivots. The top spring swung the whole derailleur back to provide chain tension. The bottom spring rotated the cage so that the second pulley pushed the chain gently forward against the freewheel sprockets. It was advertised as a single chain bend, even though it had two pulleys. Vélocio's aversion to pulley friction was still being felt.

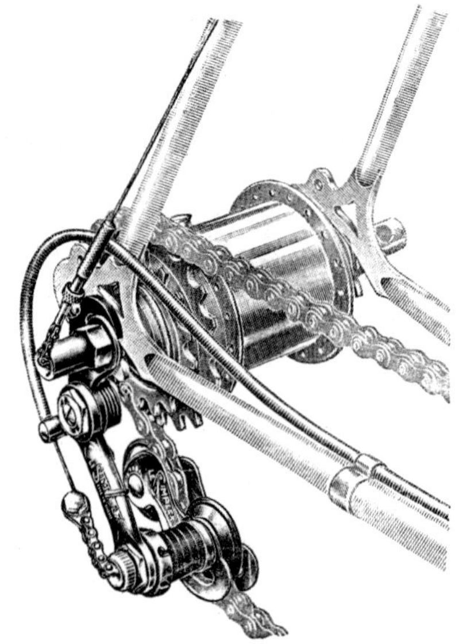

Above: Fig. 8.25. This 1953 hybrid gearing set-up combined a Simplex Tour de France rear derailleur with two sprockets and a Sturmey-Archer three-speed hub gear.

Not too many of these old Simplex derailleurs have survived, because they were light and flimsy. However it was as good as most of the competitive rear derailleurs at the time and it was inexpensive.

Simplex also made touring derailleurs. They looked more conventional, with an upper jockey

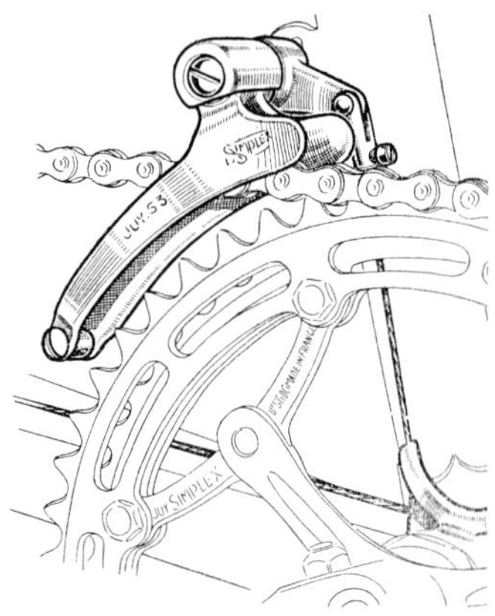

Above: Fig. 8.26. The 1952 Simplex JUY 53 cable-operated helically guided front derailleur was available until 1956.

Right: Fig. 8.27. Simplex dual shift levers, introduced in 1950, for front and rear derailleurs.

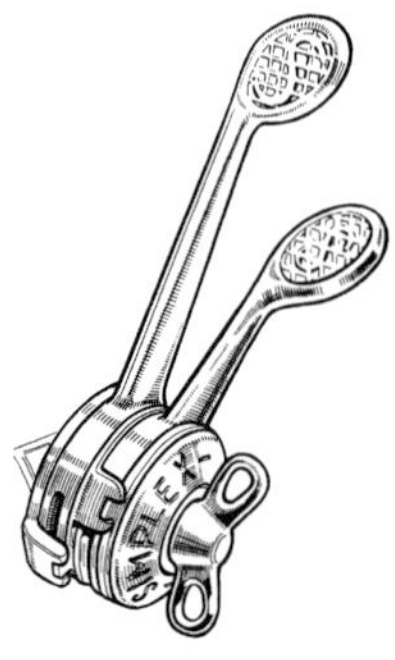

pulley and a lower tension pulley and a cage shaped to help shifting. The key difference was that the less expensive models had direct action: The cable pulled the jockey pulley outward from the larger sprockets to the smaller, and the spring pushed it inward onto the larger sprockets. The better models used a bell-crank instead of the pull-chain to reverse the action. This provided more positive shifting and cost a little more. The bell-crank models were called Grand Tourisme or Rigidex. The bell-crank had first appeared on the 1936 Simplex Route 36. A major reason that racers did not use the better-shifting touring derailleurs was that chain was wrapped around the sprockets, so it took a little longer to change the rear wheel.

Since Vélocio's time, racers had believed that tension in the lower run of chain was bad. Around 1949, special rear derailleurs were provided to alleviate this presumed problem. Huret provided this feature first. It must have seemed important at the time, because both Simplex and Cyclo copied Huret's design. However, there was no practical benefit, and eventually it disappeared.

Simplex, Cyclo, and Huret all made similar pull-chain-operated plunger-type racing rear derailleurs. Simplex used their patented spring-loaded top pivot. Lucien Juy was a skillful salesman. Simplex had the financial muscle to sponsor most of the teams in the major French races, including the Tour de France. Their derailleurs were reasonably reliable, and the racers and mechanics were used to them.

Simplex derailleurs were on the winning bicycles in the Tour de France in 1947, 1949, and 1950. Fausto Coppi's 1949 win with a Simplex led Campagnolo to introduce their Gran Sport model, which was similar to the Simplex design. Louison Bobet used Huret derailleurs to win the Championship of France in 1950 and 1951 and the Tour de France in 1953 and 1954. Fausto Coppi used a Campagnolo derailleur to win the 1952 Tour de France. Lucien Juy realized that he had to make a major technical change to keep competitive. This resulted in the Simplex LJ 543 rear derailleur, which will be described in the next chapter.

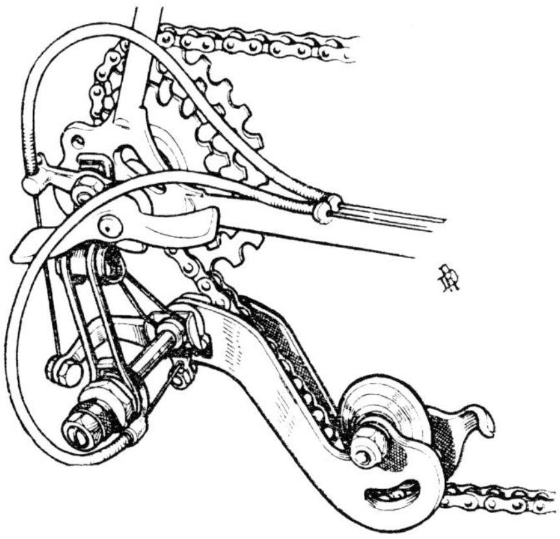

Above: Fig. 8.28. The 1946 Huret Course Competition racing rear derailleur had a single pulley, which swung backward to tension the chain. Limited capacity is inherent in this design. There were two models, with either one cable or two cables. The second cable tensioned the chain.

Simplex changed derailleur names frequently. Identifying and dating Simplex derailleurs vexes today's historians and collectors. Simplex sometimes tacked a number onto the name to denote the year of first manufacture. Dozens of slightly different rear derailleurs were called Champion du Monde or Tour de France.

## Huret

André Huret was an old bicycle racer. He started the Huret company in 1920 to make wing nuts, and the first Huret derailleur was made in 1930. Lucien Weiss won the 1932 Paris–Reims race with a Huret derailleur. Huret and Simplex both made a full range of touring and racing derailleurs, but Huret was the more innovative company. Huret copied Simplex or Cyclo only when necessary. Huret had made a single pulley derailleur in 1942, which was discontinued after the war.

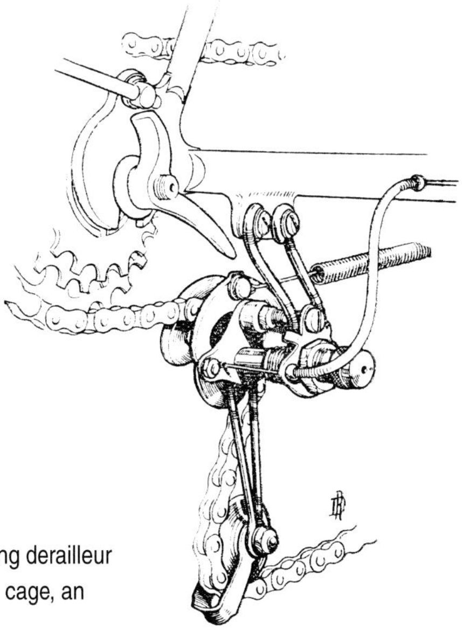

Right: Fig. 8.29. The 1946 Huret Route Touriste touring derailleur looks more conventional by today's standards, with a cage, an upper jockey pulley, and a lower tension pulley.

Above: Fig. 8.30. The 1942 single-pulley Huret rear derailleur was cast from a material referred to as "Aluvac." The compression spring was inside the cylinder. It was not made after the war.

Huret started post-war production in 1945 with two "piano wire" rear derailleurs that were made largely by hand. The attachment brackets were tempered steel rods, which combined strength and flexibility. It was impossible to buckle the rear dropout in a crash. By 1946, Huret had five employees, and in 1947, the factory went from hand production to an assembly line. Huret had two sons, Roger and Jacques. Roger took over the marketing department, while Jacques took over the technical department.

In 1949, Huret introduced the two-pulley Competition rear derailleur. The cage design was a significant improvement. In 1950, Huret brought out the Competition gruppo (*gruppo* being the Italian term for a group of matching components that has become common usage in the US;

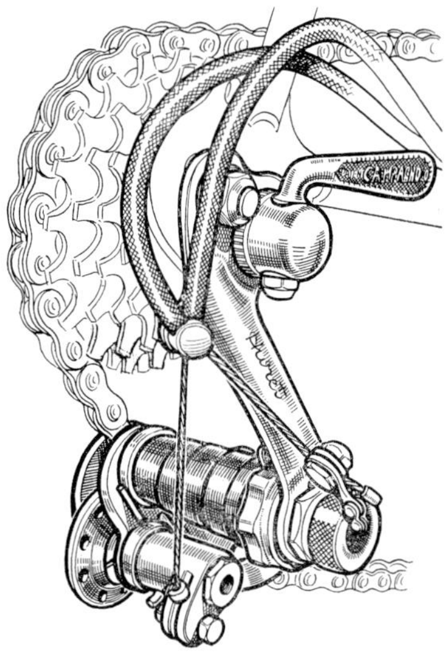

Right: Fig. 8.33. The 1953 Huret Louison Bobet racing rear derailleur was a deluxe version of the Competition, with a second cable to adjust the cage torsion spring.

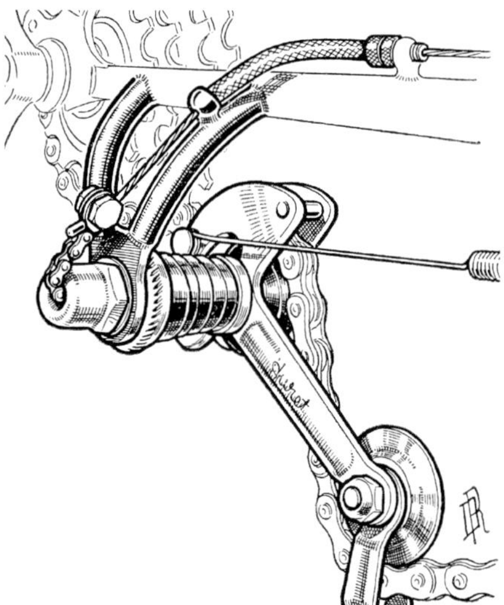

Above: Fig. 8.31. The 1950 Huret Randonneur touring rear derailleur used a single cable and a pull-chain rather than two cables and a helicoid like the Cyclo. It had an external spring. There were models for braze-on and for bolt-on mounting. The total capacity was 30 teeth: 10 teeth on the freewheel and 20 teeth on the chainwheels.

Right: Fig. 8.32. The 1950 Huret Competition gruppo comprised a rear derailleur, a front derailleur, and dual shift levers. The shift levers for both derailleurs were on the right side of the downtube.

commonly known as *group-set* in the UK), consisting of a racing rear derailleur, a helical-operated front derailleur, and dual shift levers. The rear derailleur was a plunger-type pull-chain design without a spring-loaded top pivot. The cage design was a significant improvement over the Simplex Tour de France because the

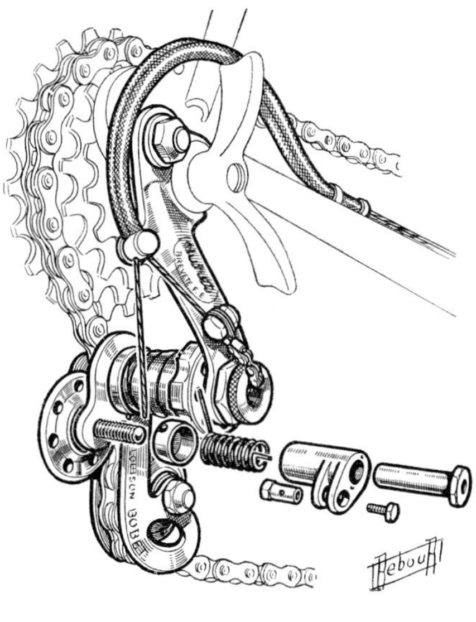

Above: Fig. 8.33. Partially exploded view of the 1953 two-cable Louison Bobet rear derailleur, showing how the second cable adjusted the

Right: Fig. 8.35. The 1952 Huret lever-operated front derailleur had a helical operator and a return spring.

jockey pulley was closer to the cage pivot. The total capacity was 22 teeth: 8 teeth on the freewheel and 14 teeth on the chainwheels. The derailleur pivoted on its own mounting bolt behind the rear wheel axle, allowing easy removal of the rear wheel.

Louison Bobet used Huret derailleurs for most of his racing career. He won a French Championship, a World Championship and the 1953, 1954, and 1955 Tour de France races for Huret. Then he switched to Campagnolo, though Huret continued to sell their Louison Bobet model.

In recent years, Jan Heine tested a Huret Randonneur derailleur on a 1952 René Herse

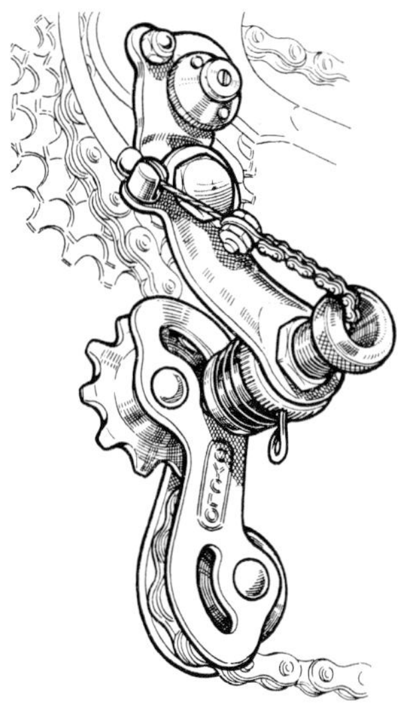

Above: Fig. 8.36. The 1950 French Cyclo Sport rear derailleur copied the Huret Competition. In 1953, Cyclo made a two-cable version of the Sport. The second cable adjusted the chain tension.

Right: Fig. 8.37. The 1950 French Cyclo Route was a less expensive version of the Standard. It used a pull-chain to avoid the need for a double cable. Note the curved slot for the spring attachment, which increased chain tension as the jockey arm moved back.

bicycle. Compared to a Cyclo or a Nivex, he found it to be noisier and required more finesse to engage the right gear. The shift lever also took more force, because it had to work against a return spring. The similar Simplex touring derailleurs would be prone to similar problems.

## French Cyclo

There was a large demand for bicycles in France during World War II. The Cyclo factory north of St.-Étienne expanded from 90 employees to 167. Derailleur production reached 2,000 per day, and freewheel production was even greater. Albert Raimond, who owned Cyclo, openly supported the French Resistance, and he was imprisoned and maltreated by the Gestapo. He emerged sick and diminished by the ordeal. At the end of the war, he returned to Cyclo, but he lacked the spirit to develop new products to keep up with the competition.

For nearly forty years, the Cyclo Standard was the best rear

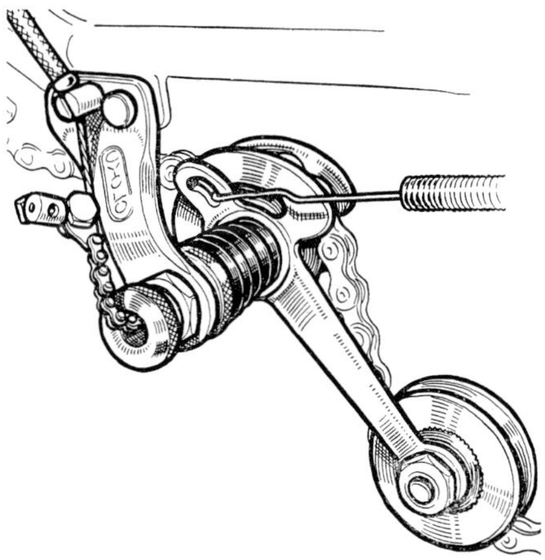

derailleur available for touring bicycles. It used two cables, a helical actuator, an external spring, and (usually) a brazed-on mounting. It was powerful, reliable, and could handle wide-range touring gearing. Hundreds are still in use on vintage bicycles, and they still shift very well. It was available for either clamp-on or braze-on mounting on the chainstay. The helicoid operator required an external spring (often two springs) to tension the chain. The double cable had to be adjusted as the cable stretched; however, double cable derailleurs are inherently self-centering. Flanged tension pulleys were available as an option. On later models, chain guides were added to the top of the cage. By the mid 1950s, the Cyclo Standard rear derailleur faced cut-rate competition from Simplex and Huret.

Jan Heine tested a Cyclo Standard rear derailleur on a 1947 René Herse tandem. He found that it required a decisive pull on the lever, but gave immediate positive, silent shifts, even uphill on the tandem. No over-shifting or fine-tuning was required.

The French Cyclo factory equipment was obsolete and worn out, but Raimond could not attract the investment needed to modernize, and his sons Jean and Ado took over the operation of the factory. Cyclo's main customers were bicycle tourists, and this market declined.

Cyclo produced less expensive versions of the Standard. The Cyclo Route used a pull-chain instead of the dual cable and helicoid operator. However, in the lower-priced utility market, Simplex' Route model was preferred by the bike shop mechanics because it was cheap and easy to repair. Who cared if it lacked Cyclo's quality and durability? Cyclo's continuous cables needed care and adjustment, which worked against their wide acceptance. In the 1950s, French Cyclo's racing derailleurs offered nothing unique. They were copies of Huret designs.

Raimond died suddenly in 1953. Faced with a declining market for touring components, the sons decided to concentrate on freewheels. Derailleurs became a sideline. Production of the Cyclo Standard, which started in 1924, ceased around 1962.

Cyclo had a British marketing branch in Birmingham, run by Louis Camillis. In 1933, Raimond and Camillis had founded the Cyclo Gear Co. Ltd. in partnership to make Cyclo components in Britain. In 1949, they became two separate companies sharing the name Cyclo. It is difficult to separate French Cyclo from British Cyclo Gear Co. Ltd. They were two separate companies making slightly different components. The British Cyclo Gear Co. Ltd. is covered under Britain elsewhere in this chapter.

## Super Champion

Oscar Egg's company, Super Champion, assembled derailleurs from components made by subcontractors. In 1945 or 1946, production of the 1939 models restarted in the factory in France.

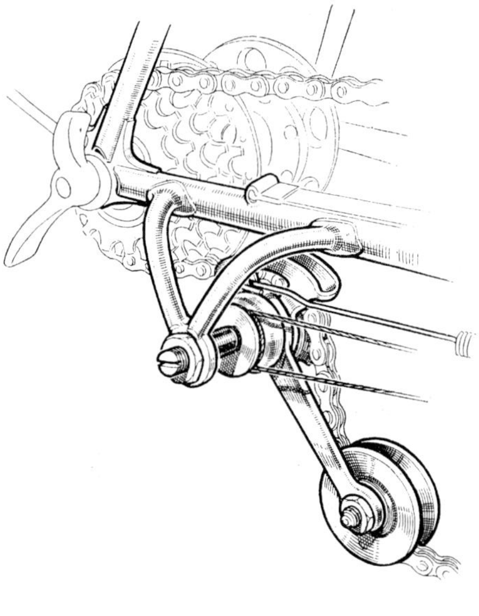

Left: Fig. 8.38. Dating back to 1924, the Cyclo Standard was simply the best touring rear derailleur available for nearly forty years. This 1950s Touriste version has the flanged tension pulley and a cage around the jockey pulley.

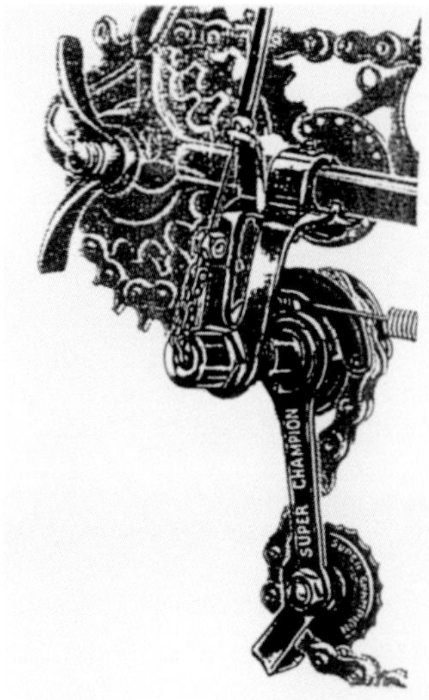


Right: Fig. 8.39. The 1946 Super Champion Sport 46 was the post-war version of the Grand Sport, with the same 30-tooth capacity.

These were the shifting-fork Super Champion Tour de France model, the double-pulley Grand Sport, and the single-pulley Champion du Monde. They were called Super Champion on the Continent and Constrictor Osgear in Britain.

Fork-type derailleurs were tricky to shift, and the tension arm limited the capacity. By 1950, they were out of fashion. Oscar Egg added copies of the prevailing Cyclo, Huret, or Simplex models. The Mondial copied the single-pulley Simplex Champion du Monde. The Sport was a replica of the Huret Randonneur. In fact, Huret made it for Super Champion. With no new products, Super Champion did not survive this decade as a derailleur manufacturer.

## Other French Derailleur Makers

About 20 companies made derailleurs in France during this decade. Most of them made inferior or copycat products that simply disappeared in the heat of competition.

Most of the French companies that made derailleurs before World War II started up again after the war. It was a seller's market for a year or so—until competition set in and the shakeout began. Most of them copied Huret, Simplex, or Super Champion, but a few had novel features.

Half a century later, Rebour's illustrations and comments in *Le Cycle*, and a few derailleurs in the hands of collectors, are all that survives for cycling historians. A number of these devices are shown in the accompanying illustrations.

Nivex, and Spirax, and JIC were the exceptions. In the opinion of those who used them, they were probably the best rear derailleurs of their day. They did not survive because they were expensive and lacked the marketing power to compete with Simplex and Huret.

### Nivex

Nivex made their first parallelogram rear derailleurs in 1938, ten years before the Campagnolo Gran Sport. Many who have used these derailleurs on period bikes agree that these were probably the finest derailleurs ever made, always shifting predictably and quietly.

Alex Singer, one of the major French custom builders, had a special relationship with Nivex. Singer

Above: Fig. 8.41. The Nivex rear derailleur, seen from the left side with the wheel removed. Note how the chain has been shifted onto the chainrest, installed on the fixed portion of the ingeniously split rear spindle.

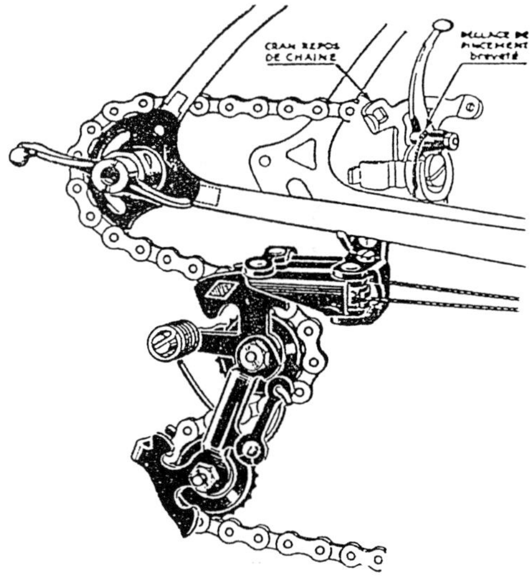

Right: Fig. 8.42. The Nivex horizontal-parallelogram rear derailleur with 2-cable operation was usually installed with a Nivex split-axle rear wheel.

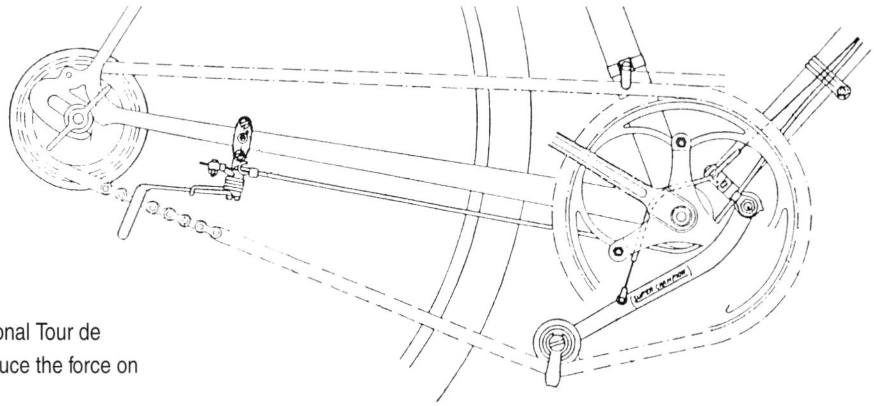

Right: Fig. 8.40. The 1939 Super Champion Professional Tour de France rear derailleur included a second cable to reduce the force on the tension arm.

bicycles could be ordered with the Nivex split-axle rear hub or with the Nivex chain holder on the right dropout to facilitate wheel removal.

Jan Heine tested a Nivex rear derailleur on a 1962 Singer with 46/32 chainwheels and 14–22 four-speed freewheel (plus chain-rest). Its' shifting was a revelation. A slight tug on the shift lever produced an instantaneous shift with no overshifting under all circumstances. He rates it as the best touring derailleur ever created.

Above: Fig. 8.43. The Nivex shift lever, showing the position for placing the chain on the chainrest for wheel removal or installation.

Below: Fig. 8.44. This Nivex rear derailleur is installed on a 1950 Singer tandem bicycle.

### Spirax

This was another innovative company that deserved to survive. Invented in 146 by L. Bon, this rear derailleur was a technical *tour de force*. It used a single cable, a helicoid operator like Cyclo, and an internal coil spring with a large flat cross-section. The chain tension remained constant as the cage rotated. It could handle a 6-sprocket freewheel. The capacity was 38 teeth: 14 teeth on the freewheel (14–28) and 24 teeth on the chainwheels (52/28). Spirax also made a cable-operated front derailleur with a helical operator.

Ernest Czuka, of Singer, says that Spirax did not use top-quality materials, and their derailleurs did not wear well. On the other hand, Jan Heine tested a Spirax derailleur on a 1953 René Herse randonneur bike. He reported that it shifted astonishingly well—comparable to today's Hyperglide—and this with standard 1950s chain and sprockets!

Left: Fig. 8.45. The 1950 Spirax rear derailleur was suitable for racing or touring. The capacity was 38 teeth, 14 teeth on the freewheel (14–28) and 24 teeth on the chainwheels (52/28).

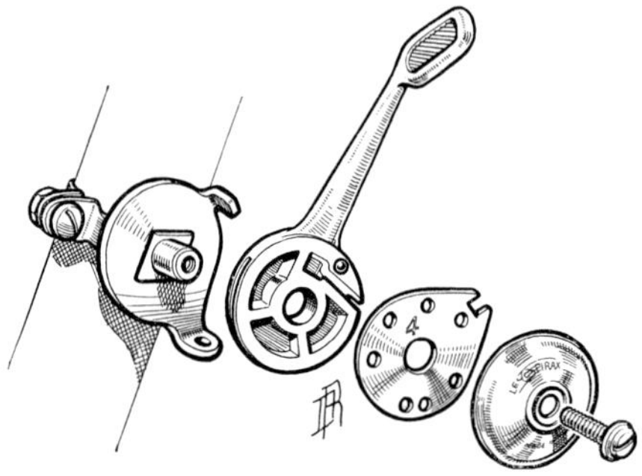

Above: Fig. 8.46. The 1950 Spirax detented shift lever probably index-shifted quite well.

Top right: Fig. 8.47. Spirax derailleur installed on a 1953 René Herse randonneur bike.

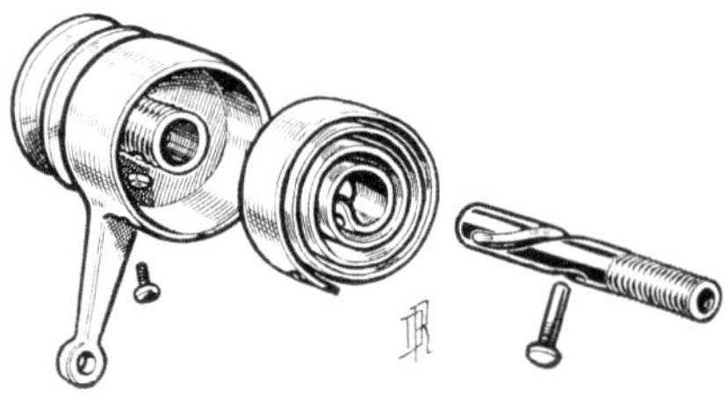

Bottom right: Fig. 8.48. Exploded view of the 1950 Spirax rear derailleur's mechanism. The cable rotated the shifter pulley, which moved the derailleur cage against the fixed helical groove.

## And More…

JIC, PGR, RAS, Super Inax, and Vita-Lux were small French derailleur makers that survived for a while after World War II. Except for JIC, I have found no Daniel Rebour drawings of their products in *Le Cycle*, although there are some of his drawings from other

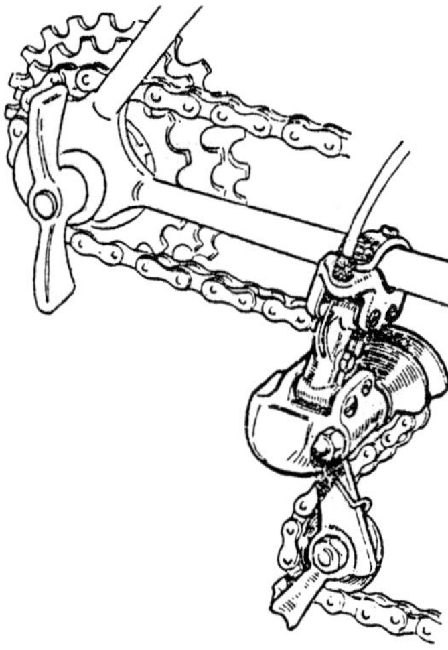

Above: Fig. 8.49. The 1946 JIC Tourisme was a double-pulley parallelogram rear derailleur that was clamped to the chainstay.

Rght: Fig. 8.50. The 1950 Super-Leader rear derailleur had two sliding rods and a single pulley.

Bottom center: Fig. 8.51. The 1946 Chaluret 93 was a single-pulley plunger-type pull-chain rear derailleur with an unusual cage shape.

Below: Fig. 8.52. The 1911 Le Chemineau rear derailleur survived until 1958. Joanny Panel continued to install it on bicycles that he made for the local and regional market.

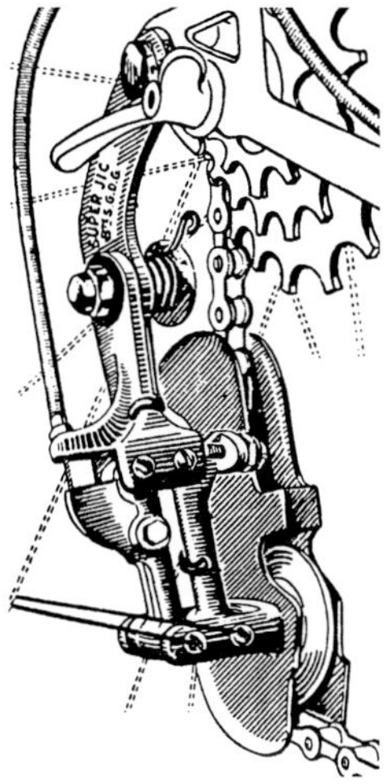

Above:: Fig. 8.53. The 1946 JIC Course was a 3-speed parallelogram rear derailleur made of aluminum alloy. It had a single pulley and a spring-loaded top pivot.

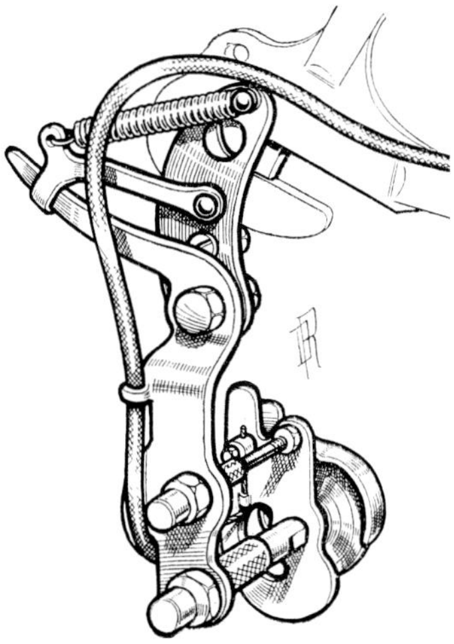

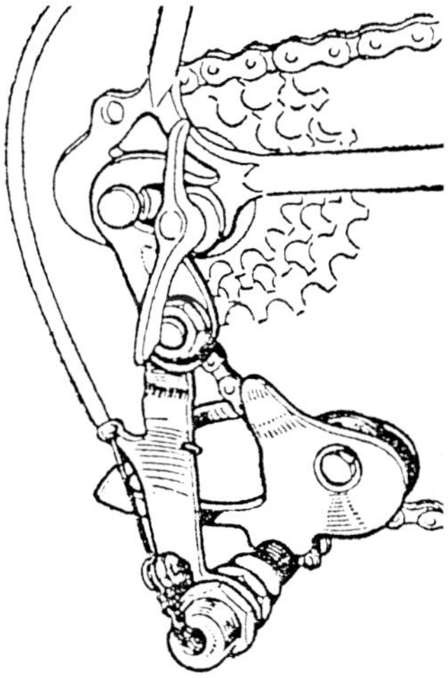

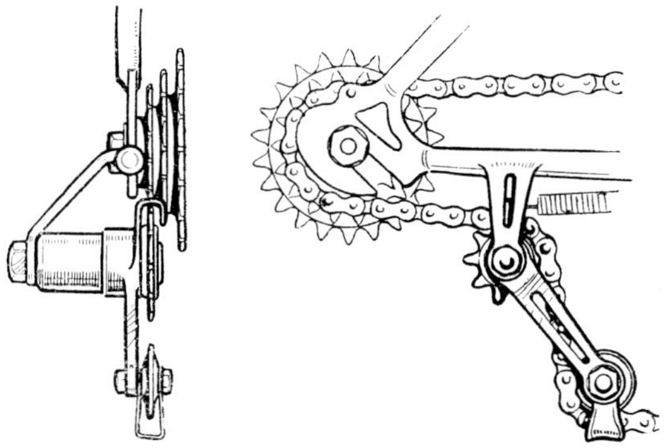

Abpve: Fig. 8 54. The 1947 Renalb-Lux rear derailleur was not cable-operated. Instead it was shifted by back-pedaling.

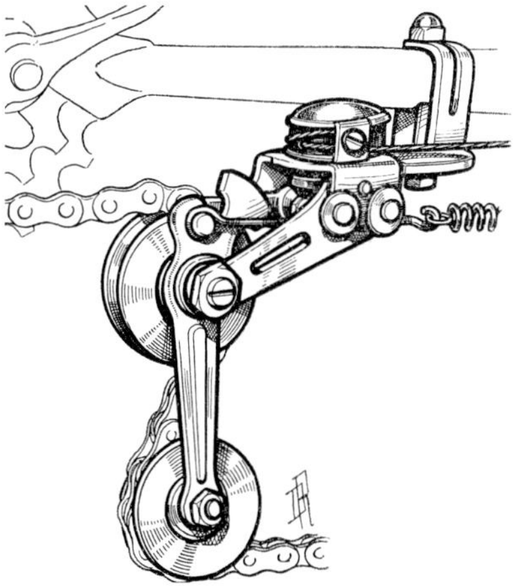

Above: Fig. 8.55. The 1949 CMP Le Samson rear derailleur was an inexpensive utility model with a rotary shifting action.

Below: Fig. 8.56. The 1946 New Lewis touring rear derailleur was an interesting design, and it probably shifted quite well.

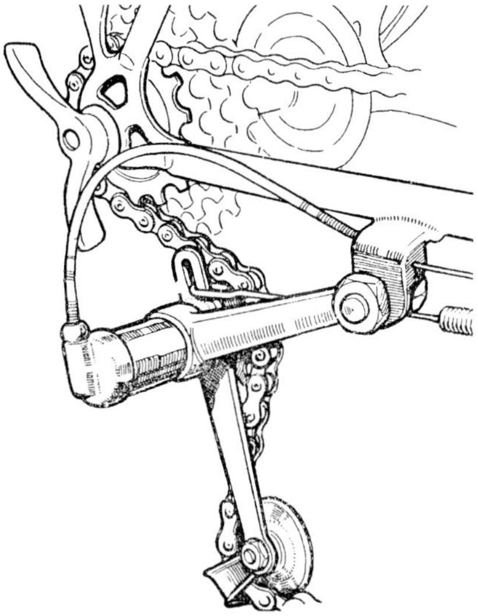

sources. I do not know exactly when JIC made their first parallelogram rear derailleur, except that it was before World War II.

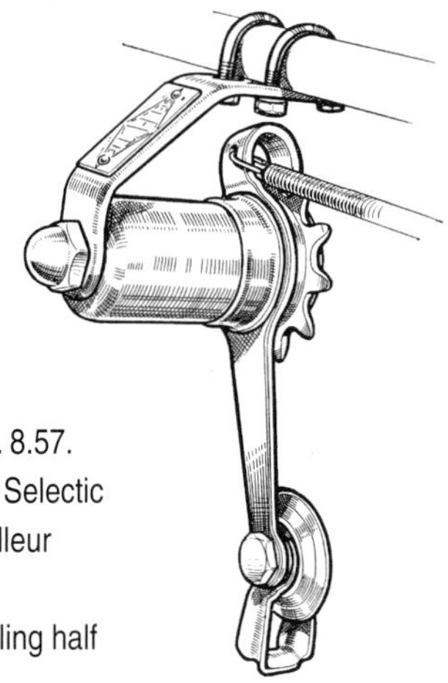

Right: Fig. 8.57. The 1951 Selectic rear derailleur shifted by backpedaling half a turn.

Left: Fig. 8.58 The 1951 Morel rear derailleur had two pulleys. A rubber bellows protected the sliding rod.

Right: Fig. 8.59. The Manufrance was a copy of the double-cable Cyclo Standard with different cosmetics. It was made from 1939 until 1959.

### Front Derailleurs

There was not a large market for front derailleurs because most derailleur bicycles had only a single chainwheel. Some riders of double chainwheel bikes simply moved the chain across the chainwheels by hand or with the heel of their right foot. Finally, many of the French custom bicycle makers preferred to make their own front derailleurs. However, by the 1950s all of the major makers were marketing front derailleurs as well.

Above: Fig. 8.60. The 1952–1956 Spirax front derailleur was suitable for triple chainwheels.

Below: Fig. 8.61. Alex Singer custom-built this 35-gram front derailleur in the early 1950s.

### Bar-End Shifters

In 1946, Jacques Souhart made the first *pass-vitesses* ("many speeds"), the name used in France to describe bar-end shifters. By 1949, they were being used in the Tour de France. In 1949, British Cyclo introduced their Benelux bar-end shifters. In 1951, Campagnolo also made bar-end shift levers, and they were used on Koblet's Tour de France winning bicycle.

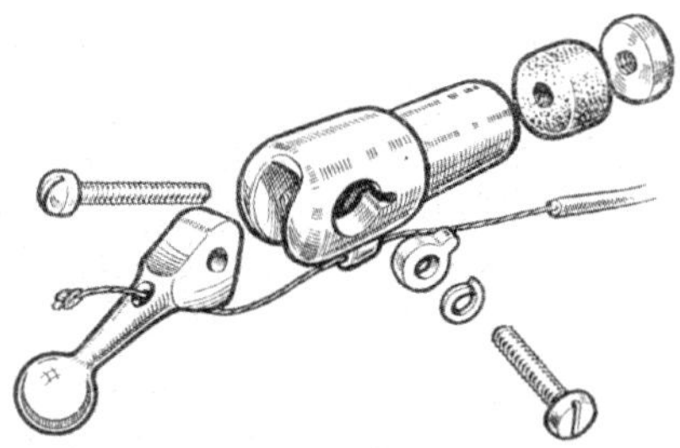

Above: Fig. 8.62. The 1946 Souhart Pass-vitesses bar-end shift levers were popular with both tourists and racers.

Below: Fig. 8.63. In 1950, Souhart made a parallelogram front derailleur that moved upward as it moved outward. It was controlled by a detented (i.e., indexed) bar-end shift lever.

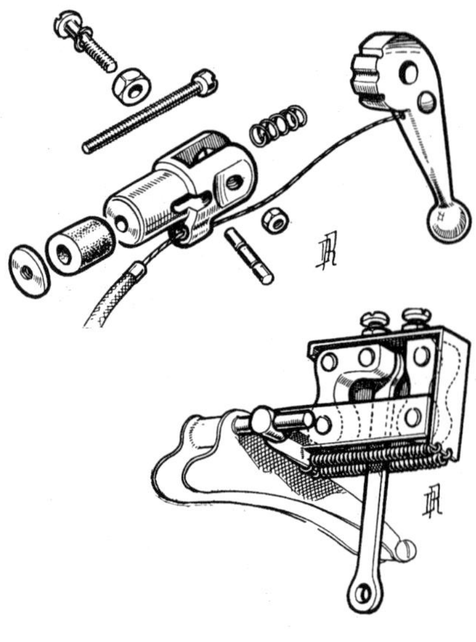

## Italy

The Italian bicycling industry was centered around Milan, which suffered the greatest bomb damage during the war. Italian bicycle manufacture was characterized by numerous small and medium-sized companies. Most of them were back in business by early 1946. The shortage of tires ended when the Pirelli factory resumed production and concentrated on bicycle tires. By the end of 1946, Italy was making more than a million bicycles a year, many for export. Italian bicycle usage was similar to that in France, but the emphasis was on utility single-speeds and racing models. Compared to France, there were fewer randonneurs and almost no bicycle tourists.

Carrying over from the pre-war period, most of the Italian makers made fork-type rear derailleurs without jockey pulleys. The tension pulley was often under the bottom bracket. Many used lever-operators to avoid cables. It was as if there was a technological wall at the Alps. The Italian makers copied other Italian makers, and the French makers copied other French makers.

Arduino, Cardinalli, Gasparetto, Gipiemme, Gambatto, Monviso, and Supercambio all made similar plunger-type single-pulley rear derailleurs.

The market told a more realistic story. After each Milan bicycle show, Simplex ran an advertisement in *Le Cycle*, pointing out

Right: Fig. 8.65. Gino Bartali, shifting his Corsa derailleur on the Stelvio Pass in the 1948 Tour de France. Bartali was a master of the Corsa and reportedly could shift while sprinting flat-out. That story was probably reporter's hyperbole.

Above: Fig. 8.64. The 1946 Campagnolo Corsa 1001 rear derailleur had two shift levers, and the derailing fork was on the top run of the chain, so that it required back-pedaling to shift.

Left: Fig. 8.66. Author Frank Berto, wearing his Russian national team cycling shorts, shifting the Campagnolo Corsa while pedaling around a parking lot. After 10 minutes of practice, he finally completed a shift and smiled triumphantly. The bike had about an inch of toeclip overlap, which steepened the learning curve.

that 60 to 80 percent of the Italian bicycles at the Milan show were equipped with Simplex derailleurs. Lucien Juy built a Simplex factory in Italy in the late 1940s, which provided heavy competition in the low-priced market.

### Campagnolo

At the end of the war, Campagnolo was a small company whose main product was the quick-release. During the 1930s, Campagnolo had established a reputation for meticulous workmanship. Everything was made by hand.

Campagnolo continued to make small numbers of their bizarre 1933 automatic quick-release rear derailleur, which was called the Corsa after the war. It was expensive to build and sold in the hundreds, while Simplex sold hundreds of thousands.

The Corsa had some things to commend it: Compared to its contemporaries amongst racing rear derailleurs, the Corsa did not have any pulleys, eliminating all pulley friction, and it was unaffected by mud. There was no loose chain on the return run, and no cable to break. It was silent and unbreakable.

On the down side though, it was expensive, and the special toothed rear dropouts had to be accurately mounted by the framebuilder. It required special sprockets with the front corner of the teeth chamfered to assist backpedal shifting. Because of the unique hub and dropouts, the entire team had to use the Corsa, so that they could switch wheels.

It was also quite difficult to shift gears. First, the rider reached back and turned the upper lever to unlock the rear wheel quick-release. Then he pedaled backward and turned the lower lever to move the derailing fork against the chain, until the chain jumped to the next sprocket. The wheel moved backward by its own weight on upshifts. On downshifts, the chain pulled the wheel forward. Then he pedaled forward lightly so that the

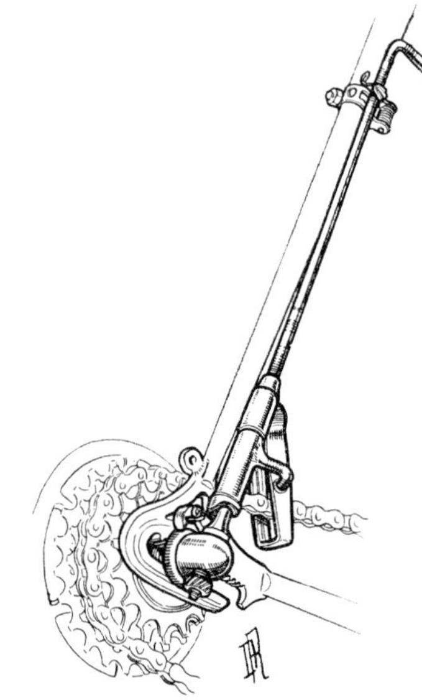

Left: Fig. 8.68. The 1949 Campagnolo Paris–Roubaix rear derailleur had a single lever, but the derailing fork was still on top, requiring backpedaling to shift.

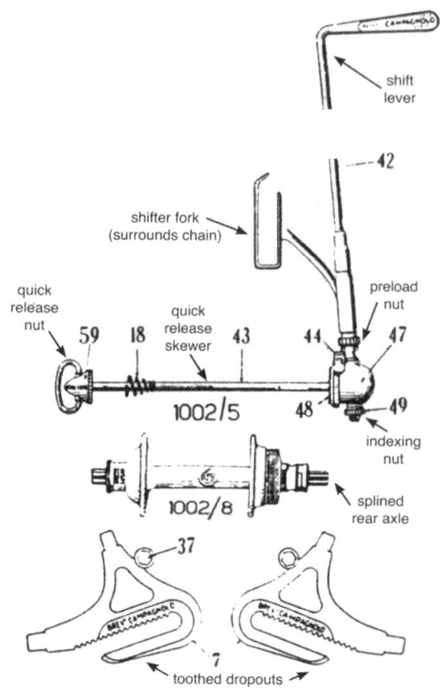


Below: Fig. 8.69. The mechanism that opened the quick-release and then moved the shifter fork. In the other direction, it tensioned the chain and then closed the quick-release.

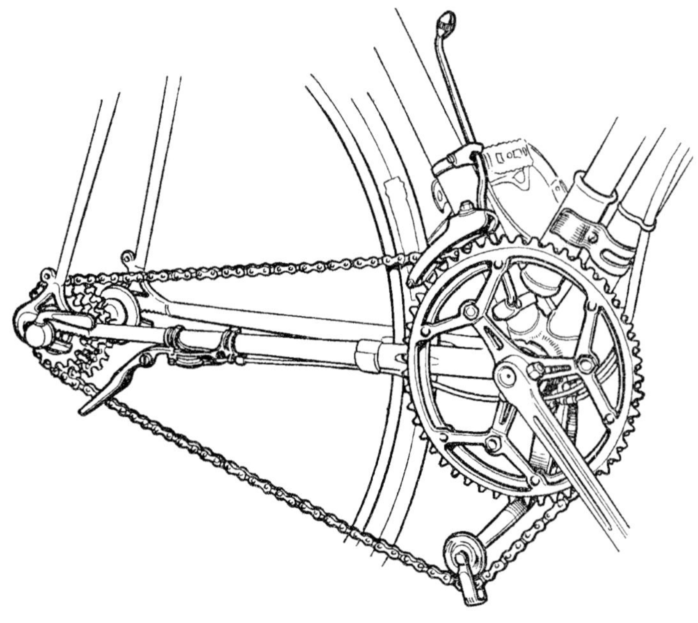

Left: Fig. 8.67. The 1949 Cervino derailleurs were the final development of the derailing-fork-and-tension-arm design. Note the front derailleur and the half-step chainwheels.

chain was not stretched too tight. Finally, he turned the upper lever back to lock the quick-release.

It was particularly difficult to shift while climbing. Because of the slow, cumbersome shift, the racer had to pull out of the peloton, shift, and then sprint to catch up.

Jan Heine tested a Campagnolo Corsa derailleur on a 1950s Learco Guerra bicycle with 16–22 four-sprocket freewheel. He found shifting cumbersome but not too difficult. The chain moves easily to the next cog while backpedaling.

Campagnolo made several versions of their Mozzi Calibrati rear derailleur. The first was called Corsa (Course). The second was called Paris–Roubaix, after Fausto Coppi's breakaway victory in the 1950 Paris–Roubaix. The two-lever-operation was replaced by a single lever, which unlocked the quick-release in the first 1/8 turn and then moved the derailing fork. It included an adjustment to tension the chain. Operation was easier, but the derailing fork was still on top, requiring back-pedaling to shift.

Two later versions came out in the early 1950s. They probably did not get past the prototype stage. The derailing fork was moved to the lower chain. When the quick-release was unlocked, the rear wheel was loose in the dropouts. With the derailing fork on the lower chain, any hang-up during the shift could shoot the wheel out the front. Note the stop. These prototypes did not seem to have had a name. Perhaps the Italian racer's expletives did not translate well. All versions required a high degree of concentration and coordination to shift. Nevertheless, the top Italian teams used Campagnolo's Corsa and Paris–Roubaix derailleur, and Campagnolo continued to make the Paris–Roubaix into the mid-1950s.

When Campagnolo introduced the Gran Sport in 1950, it took over the high-priced Italian market. The other Italian derailleur makers of single-pulley rear derailleurs simply could not compete. Their products cost more and did not perform as well. The Gran Sport put them out of their misery, and they disappeared from the market.

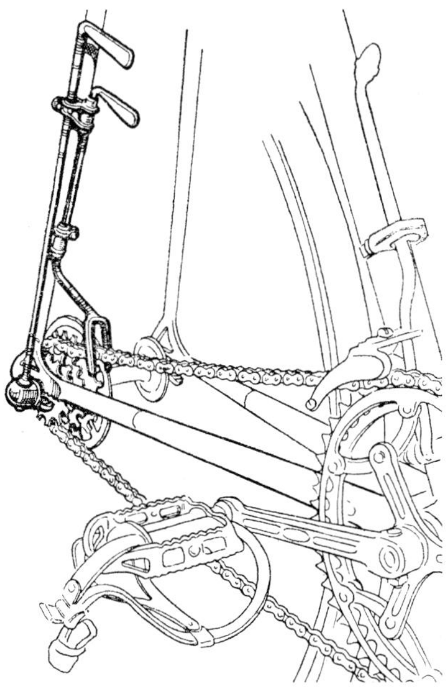

Above: Fig. 8.70. Some of the Italian racers in the 1948 Tour de France added a front derailleur to the Campagnolo Corsa. This looks like a half-step chainwheel. Imagine the drill for a double shift—pedaling backward for the rear shift and pedaling forward for the front shift.

Right: Fig. 8.71. Lucien Juy supervising the installation of Simplex derailleurs on Fausto Coppi's Bianchi in 1949. Tragalla, Bianchi's *Directeur Sportif*, shrugs his shoulders.

### The Passing of the Flag

The two Italian greats, Gino Bartali and Fausto Coppi, were the best cyclists in the world for a decade. They each won major races with the Corsa gear (or in spite of it). Because the lower part of the chain remained taut, it was a good derailleur for rough courses like Paris–Roubaix. Bartali won the 1948 Tour de France and Coppi won the 1950 Paris–Roubaix with the Corsa.

Campagnolo strongly supported Italian racing teams. Coppi rode for the Bianchi team until he started his own company. Bartali rode for Legnano. Both Bianchi and Legnano were underwritten by Campagnolo and were supposed to use Campagnolo components.

Lucien Juy paid Coppi handsomely to use Simplex for the

1949 Tour de France, and Coppi went on to win the race in 1949. The rest of the Bianchi team stuck with Campagnolo, so Coppi could not switch rear wheels with his teammates.

It got worse for Campagnolo when Gino Bartali became enamored with the Cervino rear derailleur and bought an interest in the Vittoria company, which made the Cervino. The race results reported in the advertising pages of *Le Cycle* are confusing: Simplex advertised that Coppi won the Tour de France with Simplex, while Campagnolo advertised that Coppi won the Grand Prize of the Mountains with Campagnolo.

Coppi and Bartali were near saints to the Italian racing fans. Coppi's Simplex and Bartali's Cervino were severe blows to Tullio Campagnolo's pride. He decided to develop a rear derailleur that did not have so many disadvantages. In less than three months, Campagnolo developed a prototype parallelogram rear derailleur, the Gran Sport, which was shown at the December, 1949 Milan Show.

The Gran Sport is often described as the first parallelogram rear derailleur, but it was not. Most writers suggest that Campagnolo copied the Gran Sport from the 1938 Nivex or the JIC. However, I believe that the real inventor of Campagnolo's Gran Sport parallelogram rear derailleur was probably Francesco Ghiggini. He obtained Italian patents on the design in 1937, 1938, and 1941. He sold the rights to his patents to Campagnolo in 1951. The story of the birth of the Gran Sport is conjecture, because neither Tullio Campagnolo nor his son Valentino, ever provided a proper history of the company.

With the Gran Sport, Campagnolo took the parallelogram rear derailleur design, refined it, and made it a commercial success.

Right: Fig. 8.72. Bartali leads Coppi over the Izoard mountain pass in the 1949 Tour de France. The pack is nowhere in sight; neither is Campagnolo. Note the Cervino on Bartali's bike and the Simplex on Coppi's bike. Coppi won the race, beating Bartali by 10 minutes.

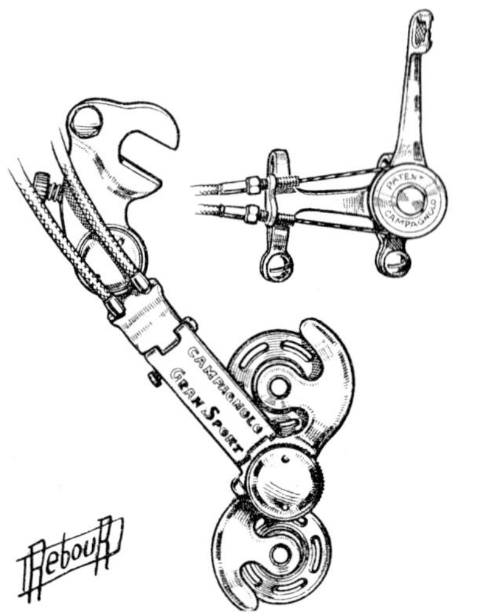

Above: Fig. 8.74. The 1950 Campagnolo Gran Sport parallelogram rear derailleur and shift lever were first shown at the 1949 Milan show. This two-cable model was probably a prototype for testing. It never appeared in a catalog.

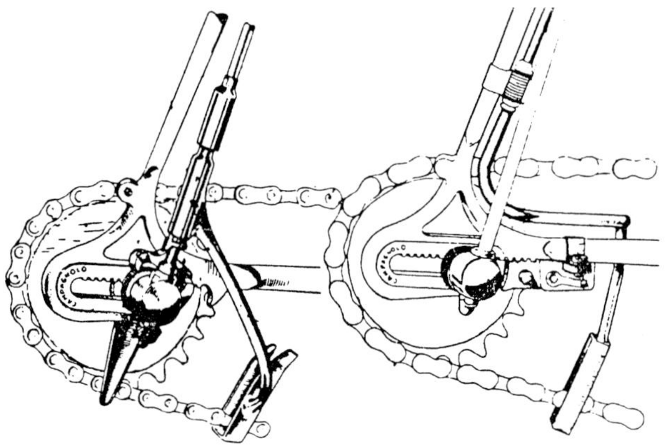

Left: Fig. 8.73. About 1951, Campagnolo made these two prototypes. Both had the derailing fork on the lower chain The one on the right had two levers.

With hindsight, it might be suggested that Coppi and Bartali did Campagnolo a great favor in 1949 by forcing him to abandon the Corsa and Paris–Roubaix gears and introduce the Gran Sport.

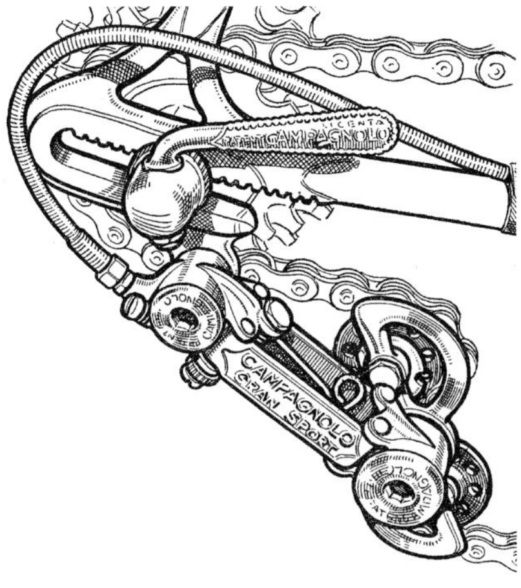

Above: Fig. 8.75. The definitive 1951 Campagnolo Gran Sport used a single cable and a return spring. This one is mounted on Coppi's 1951 Tour de France bicycle, which still had the

Campagnolo showed a Gran Sport prototype at the Milan show in the fall of 1949. This was the two-cable derailleur shown in Fig. 8.76 on page 169. Daniel Rebour wrote:

> We noted with surprise that Campagnolo had placed, among its collection, a French-style derailleur. We may be wrong, but we were under the impression that Campagnolo was staking out a position rather than presenting an actual product.

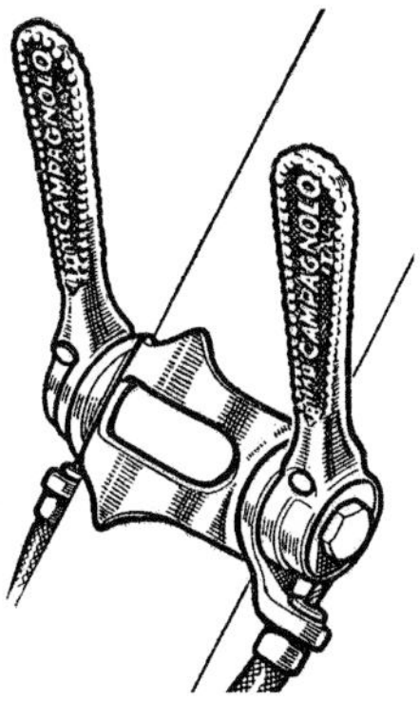

Right: Fig. 8.76. The 1952 Campagnolo downtube shift levers were made for 33 years with only minor cosmetic changes.

A year later, Campagnolo introduced the definitive Gran Sport with a single cable and a return spring. It was the archetype for

Above: Fig. 8.77. The Campagnolo Gran Sport cable-operated plunger-type front derailleur was simple and rugged.

Below: Fig. 8.78. The standard production Gran Sport appeared in the 1953 Tour de France and was made for the next ten years with only minor changes.

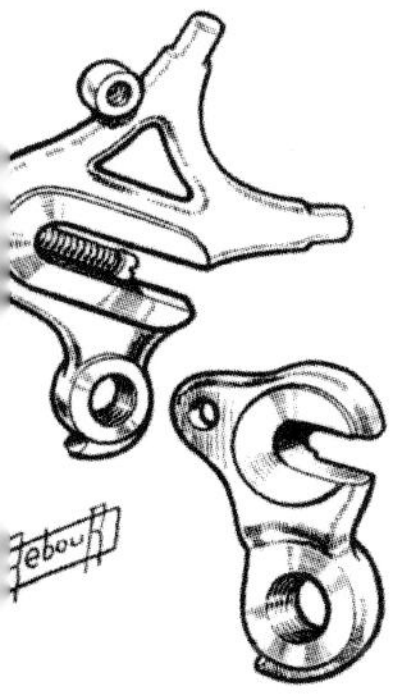

...pout with hanger and ...o-fit hanger.

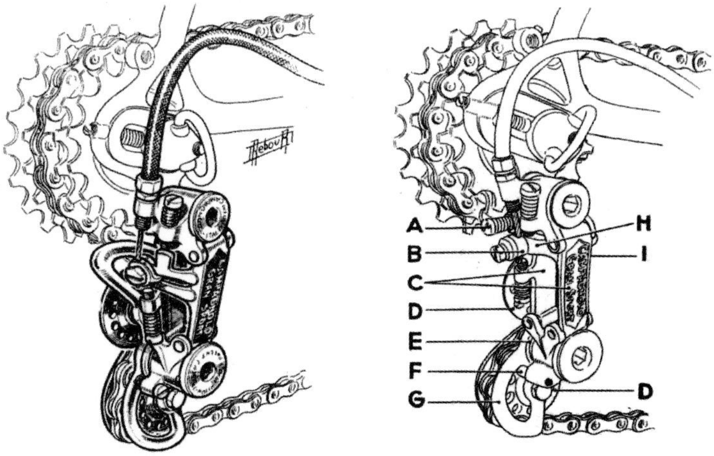


*On the left, the new "GRAN SPORT" ["Extra"] derailleur [model 52-1]. Note the longer arm for the shifter cable attachment, new cable clamp, larger pulley cage and large, drilled pulleys. On the right, the old model [51-1], with the changes as shown:*
*A: Screw for adjusting derailleur angle deleted.*
*B: Cable clamp with hole replaced by pinch bolt [and groove].*
*C: Parts made from aluminum bronze replaced with stamped [cast?] steel.*
*D: 30 mm pulleys with 1/8" balls replaced with 35 mm pulleys with 5/32" balls.*
*E: Now 6 holes for the end of the chain tension spring [to adjust chain tension].*
*F: Limit screw for pulley movement easily removeable.*
*G: Outer flange [of pulley cage] extended on top.*
*H: Arm for shifter cable attachment is 5 mm longer*
*I: Spring for lateral movement (invisible in the drawing) much stronger and more supple.*

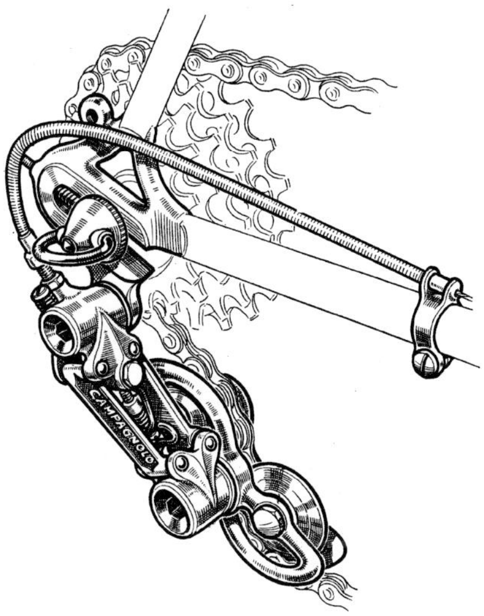

Left: Fig. 8.79. In the 21 June 1952 issue of *Le Cycle*, Daniel Rebour explained the differences between the 1951 Gran Sport (right) and the 1952 Gran Sport Extra (left). The translation is by Jan Heine.

Campagnolo's superlative line of single-pivot parallelogram rear derailleurs for the next 30 years.

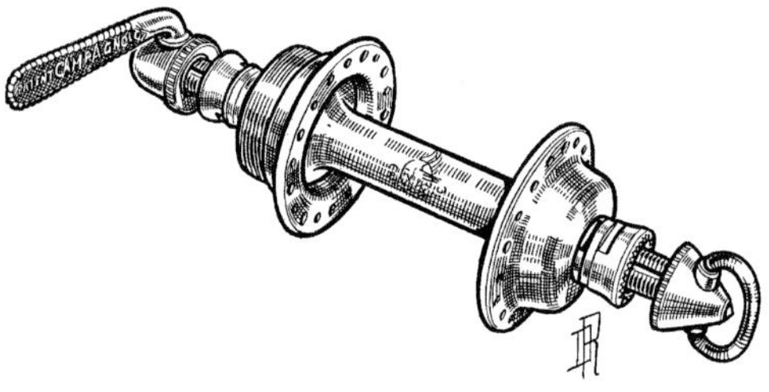

Above: Fig. 8.80. The first Campagnolo conventional rear hub, introduced 1952, was a step toward the first Campagnolo gruppo.

Below: Fig. 8.81. The 1954 Campagnolo Sport single-pulley rear derailleur had limited capacity, like all other single-pulley derailleurs, and could only be used with narrow-range freewheels. However, there was a market prejudice against double pulleys, so Campagnolo, Simplex, and Huret made products to meet market demand.

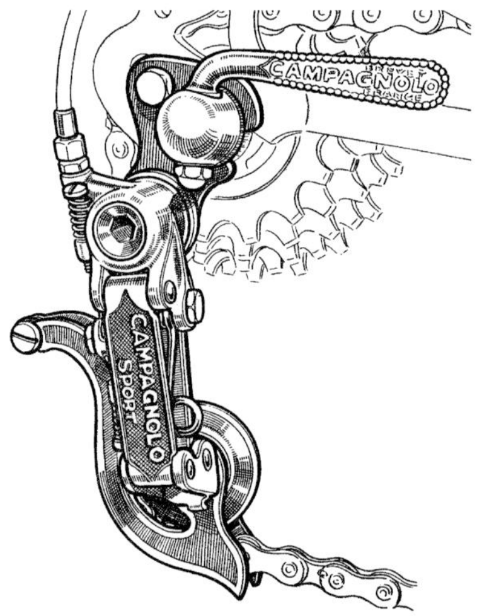

It generated a hundred different copies.

Campagnolo collectors thrive on minutiae, and the Gran Sport gives them lots to feed on. There were running changes to the 1950, 1951, 1952, and 1953 models of the Gran Sport. There were at least four versions, with changes in cable-clamp bolt, the limit screws, the attachment bolt, plain pulleys or pulleys with 9, 12, or 16 holes, different parallelogram arms, and different cage shapes. And at one point the parallelogram material was changed from bronze to steel.

## Vittoria

Vittoria's Turin factory was a major derailleur maker before the war. The Nieddu brothers, Tommaso and Amadeo, owned Vittoria. Both served in the Italian army in World War II. Vittoria was slow to rebuild after the war. The brothers quarreled and Tommaso struck out on his own.

Amadeo was left in charge of the original Vittoria company, but he was better at public relations than at designing new derailleurs. He introduced a Simplex copy, a modified Campagnolo copy, and the Duplex. The 1951 Vittoria Duplex rear derailleur had a two-pulley cage to perform the shifting. Then the pulleys were latched up out of the way and the wheel was moved back manually in the dropout. This was only marginally better than turning the rear wheel, but it satisfied the fixation against pulley friction. It could not compete with the Gran Sport and it was the end of the line for Vittoria.

## Cervino

Tommaso and his son Mario were more successful. In 1948, they introduced the Cervino (deer) rear derailleur. It was an improvement over both the Vittoria Margherita and the Super Champion. The tension pulley still pivoted from the bottom bracket, but the cable-operated derailing fork was on the lower run of chain so that it shifted while pedaling forward.

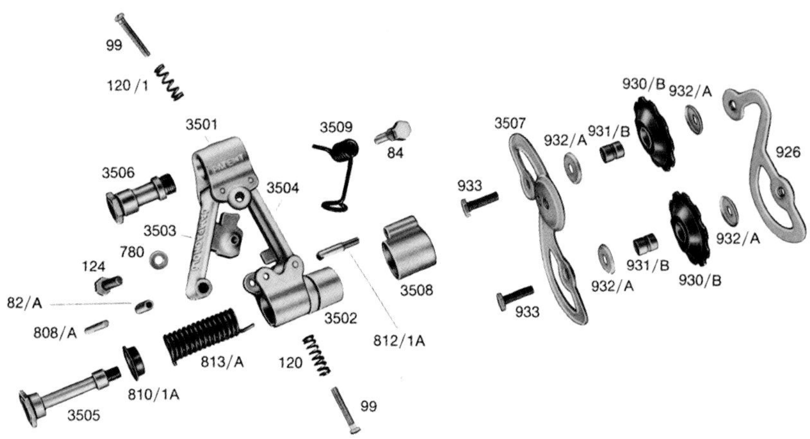


Left: Fig. 8.82. Exploded view of the Campagnolo Gran Sport rear derailleur. It remained essentially unchanged for some 10 years.

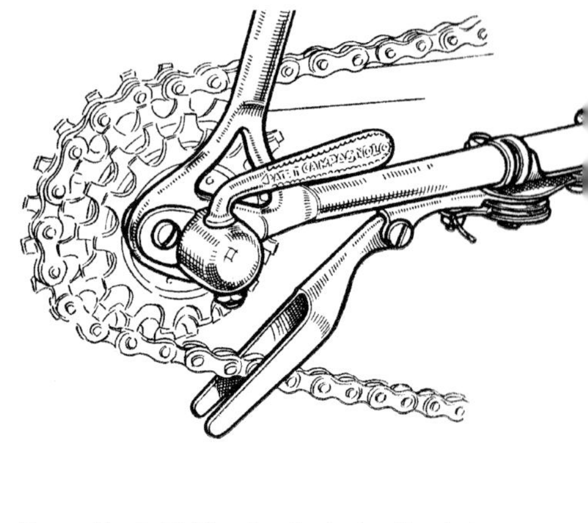

Above: Fig. 8.83. The Cervino's derailing fork was on the lower run of the chain.

Bartali had won the 1938 Tour de France with a Vittoria Margherita, and the 1948 Tour de France with a Campagnolo Corsa. Subsequently, he started his own bicycle company, and became enamored with the Cervino. He bought into Tommaso Nieddu's company, and became a director. He installed a Cervino on his 1949 Tour de France bicycle, which he hoped to ride to victory. However, It was not to be: he was at the end of his long career, and Coppi won the Tour using either Simplex or Campagnolo derailleurs, with Bartali was second, 10 minutes behind.

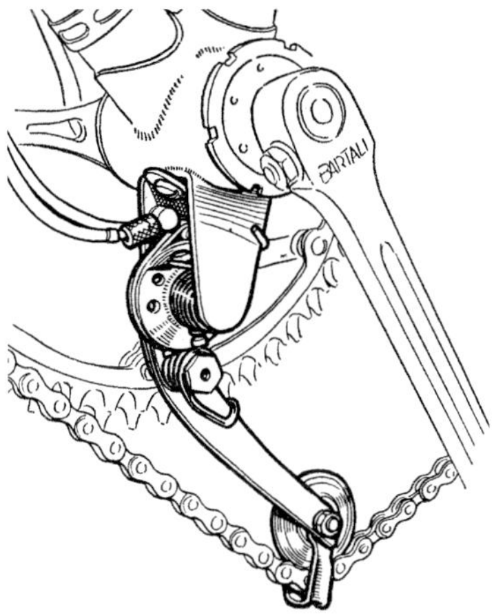


Above: Fig. 8.84. The Cervino's take-up pulley on the bottom bracket was less susceptible to damage when going over a curb, and the chain tension was adjustable.

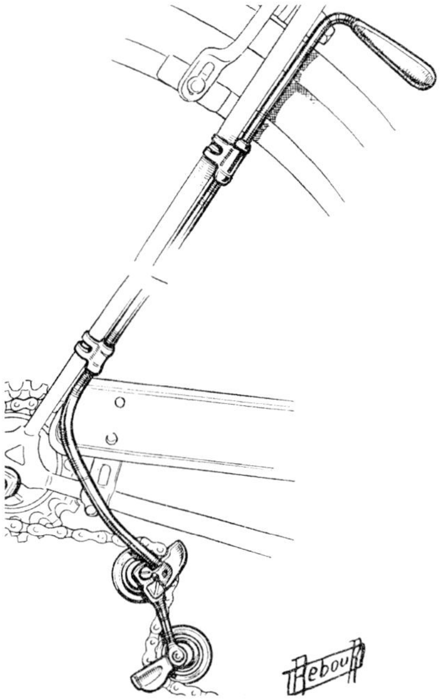

Above: Fig. 8.85. The 1949 Gambatto rear derailleur was a rod-operated two-pulley model.

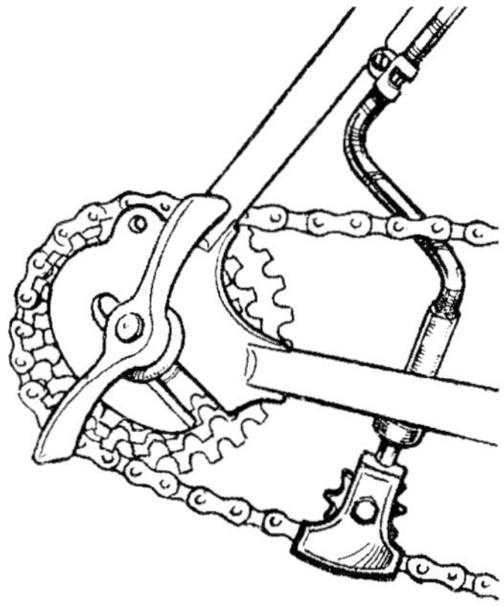

Left: Fig. 8.86. The 1947 Cardinalli Sport rear derailleur had a single spring-loaded pulley that moved down when it was shifted to the larger sprockets.

### Other Italian Derailleurs

The Italian situation was similar to the French. There were numerous small derailleur companies in addition to Campagnolo, Cervino, and Vittoria, but Simplex' Italian factory in Milan made more derailleurs than all the others combined. Most Italian companies used rods rather than cables to operate the rear derail-

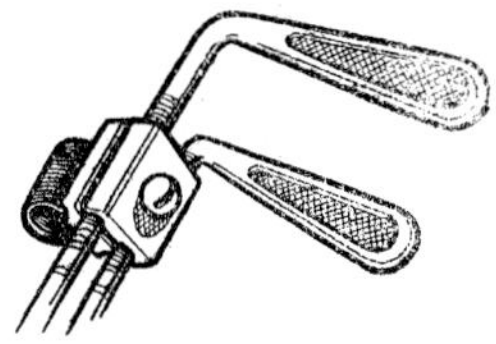

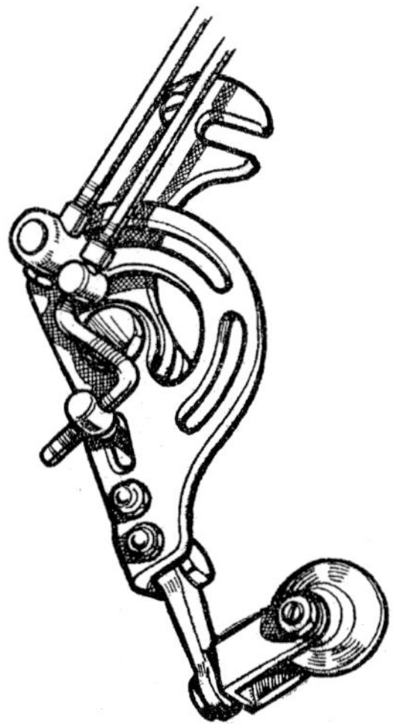

Left: Fig. 8.89. The 1948 Monviso was another rod-operated single-pulley rear derailleur. The jockey pulley moved back and forth to tension the chain.

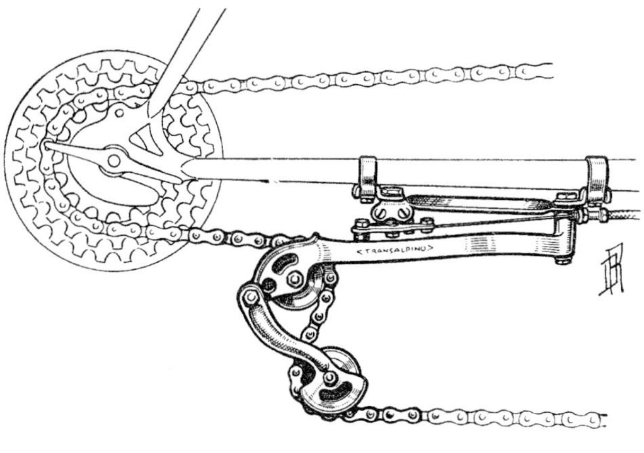

Left: Fig. 8.87. The 1951 Transalpino derailleur had a horizontal parallelogram like the Nivex.

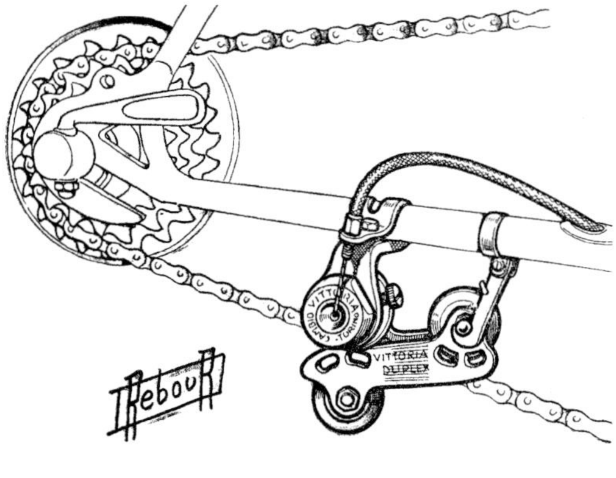


Right: Fig. 8.88. The pulleys on the 1951 Vittoria Duplex could be latched up to reduce friction. The fronts of the sprocket teeth are cut away to improve shifting.

leur. The small companies could not compete with the Campagnolo Gran Sport for quality or with the Simplex Tour de France for price. By 1954, they were all out of business, and Rebour's illustrations shown here are all that remains of them.

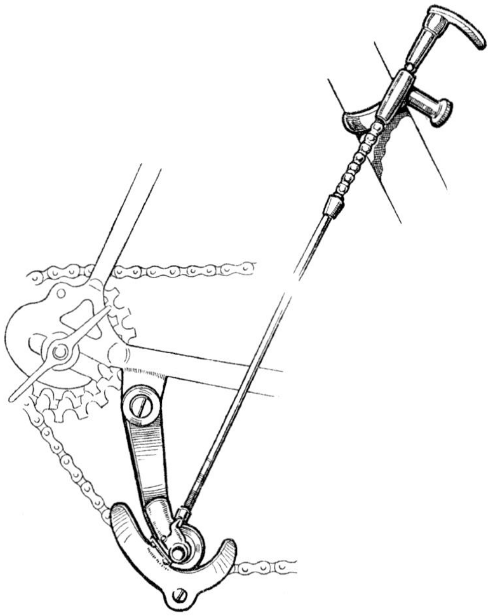

Above: Fig. 8.91. The 1948 Arduino was a single-pulley rod-operated rear derailleur.

Below: Fig. 8.92. The 1949 Gipiemme rear derailleur was a Campagnolo Corsa copy with a single lever and the derailing fork on the lower chain run.

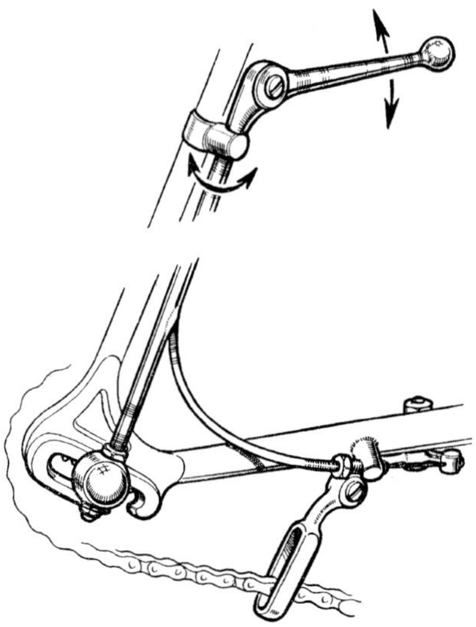

The Transalpino was an exception. It had been designed in he 1930s by Giovanni Mari, a hatter

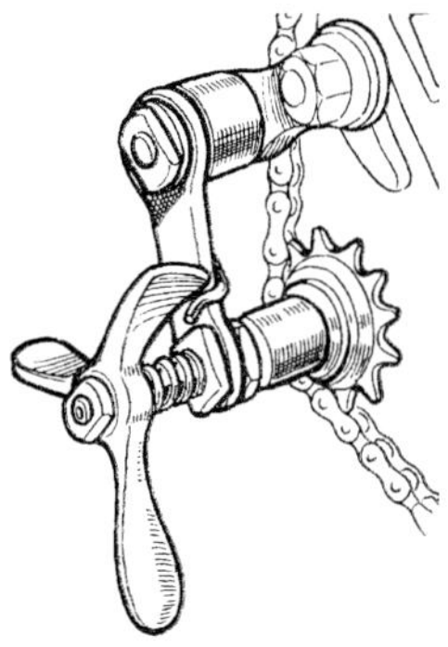

Above: Fig. 8.90. The 1946 API Sport rear derailleur had a foot-operated three-prong "foot-wheel" to screw the single pulley back and forth

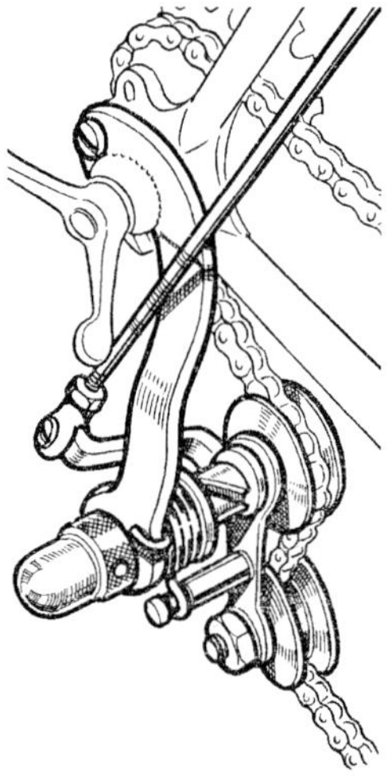

Right: Fig. 8.93. The 1948 Meccan Cambio rear derailleur was a rod-operated copy of the Cyclo Standard design.

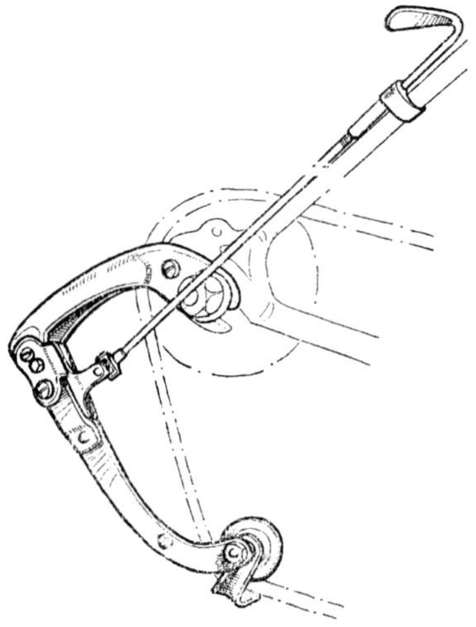

Above: Fig. 8.94. The 1948 Gasparetto rear derailleur was another Campagnolo Corsa copy, with the derailing fork on the lower chain run.

from Casalbuttano who was also a gifted inventor. The Rebour drawing (Fig. 8.87) suggests that it would have shifted positively and reliably, much like the Nivex.

## Great Britain

At the end of the war, there were more than 10 million bicyclists in Britain, and most of them used their bicycles for personal transportation. Britain had more than 40 small and medium-sized bicycle makers and a significant number of derailleur makers. Encouraged by government incentives, bicycle production grew from 2 million in 1946 to 3.5 million in 1949. The export market was helped by the 1949, devaluation of the pound from $4.03 to $2.80. In 1949, 2.2 million bicycles were exported, two thirds of the total production. Raleigh made more than 750,000 bicycles in 1949 and the majority were exported. The British bicycle

Below: Fig. 8.95. The 1946 Supercambio was a rod-operated single-pulley rear derailleur. It is easy to see why this design provided less chain wrap on the smaller sprockets.

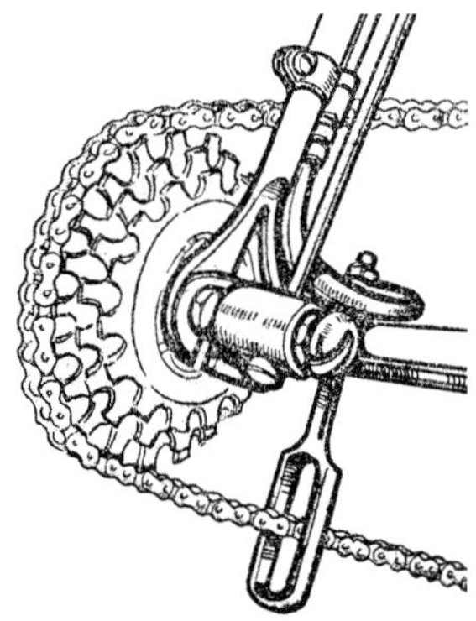

industry thrived because of the export market.

The Olympic Games were held in London in 1948, and the first post-war Earls Court Cycle and Motorcycle Show, at which 40 British bicycle makers exhibited, was held the same year. The three largest companies were the combines run by BSA, Raleigh, and Tube Investments (TI). More than a dozen different British rear derailleurs were on display, made by BSA, Constrictor-Osgear, British Cyclo, Hercules, Overhill, Phillips, Resilion, and Tri-Velox. Nearly one half of the 1,100 bicycles exhibited had gearing. Raleigh's bicycles used only Sturmey-Archer hub gears, but many of the other makers showed both derailleur and hub-geared bicycles. The show included 145 derailleur bicycles.

In this critical period, Britain fell behind the technical progress that was taking place in France and Italy. Britain was the major exporter of lightweight bicycles, but they were mostly what the Americans called "English Racers," i.e., upright roadsters with three-speed hubs.

In the 1940s, the British League of Racing Cyclists (BLRC) organized massed-start road races. The National Cyclist's Union (NCU) responded with massed-start road races of their own in the 1950s. Most of the racers used derailleurs on their bicycles. There was a surge in the use of bicycles for recreation and for bicycle touring, and many of these bicycles also used derailleurs.

As the economy picked up, the public sought to make up for the deprivation of the wartime years. Motorcycles and automobiles were high on the shopping list. As automobile production and road building increased, Britain developed a conscious anti-bicycle bias. Bicycling was seen as a working-class pastime. The total mileage traveled by British cyclists dropped steadily from a peak of 12 billion miles in 1952 to 2 billion miles when it bottomed out in 1974.

Faced with a declining home market, the British bicycle industry could only shrink and consolidate. Raleigh was the exception. They made more than a million bicycles in 1951 and they increased the size of the

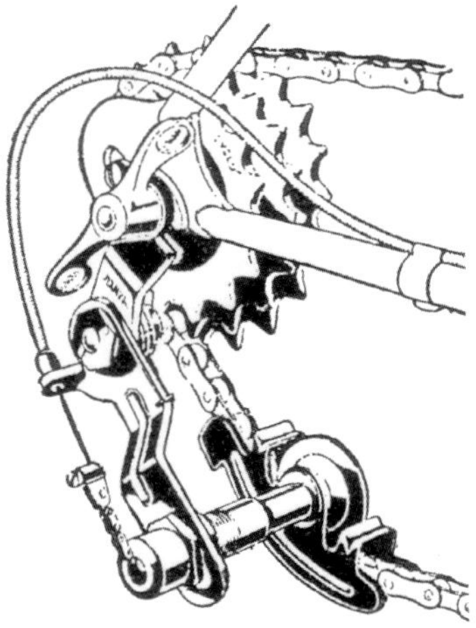

Left: Fig. 8.97. The 1947 British Cyclo Super Olympic rear derailleur was an improved version of the 1946 British Cyclo Olympic, which evolved from the pre-war British Cyclo Poly. It was called the Super Olympic to celebrate the 1948 Olympic Games in London. With only a single pulley, it was no competition to the Simplex Tour de France.

Below: Fig. 8.98. For 1948–1950, the British Cyclo Super Olympic Professional got an improved cage.

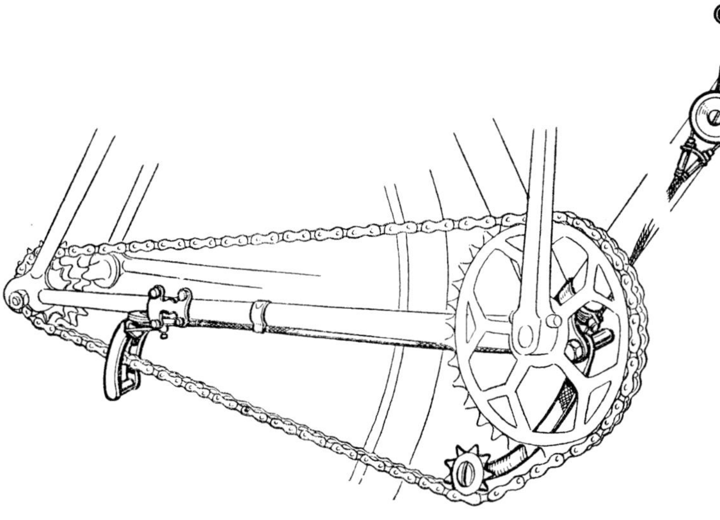

Left: Fig. 8.96. The pre-war British Cyclo Ace rear derailleur remained in production until about 1954. It was a copy of the Osgear Super Champion.

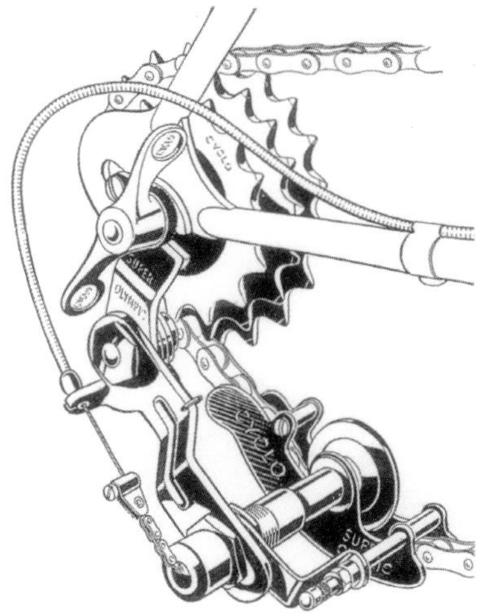

Nottingham factory and their capacity to mass-produce inexpensive middle-weight bicycles. With 8,000 employees, Raleigh Nottingham was not just the largest bicycle factory in the world, according to the *Guiness Book of World Records*, it was the largest factory of any kind.

The smaller companies could not compete, and one by one they went to the wall, while Raleigh picked up the pieces. In 1954, they absorbed two rival groups, Triumph and Three Spires.

To compete with derailleurs in the early 1950s, Sturmey-Archer offered 14 different hub gears—with 3 or 4 speeds, alloy or steel hub shell, narrow, medium, or wide-range, with or without hub brake. The rarest model was the narrow-range ASC with fixed gear, which is now a coveted collector's item.

## British Derailleur Makers

Louis Camillis, the UK agent for Cyclo, had established a separate

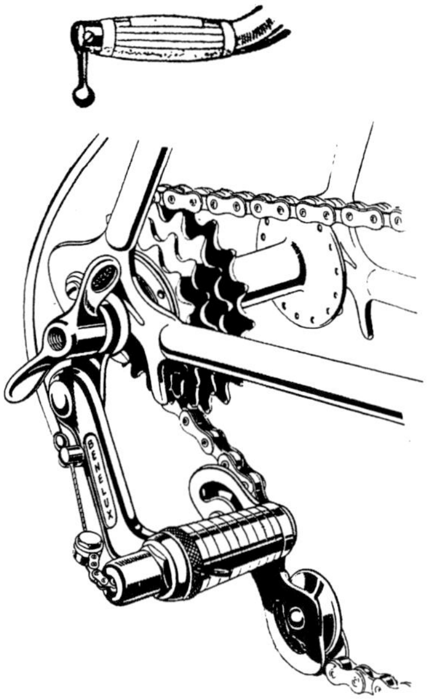

Above: Fig. 8.99. The 1950 British Cyclo Benelux was a pull-chain plunger-type racing derailleur. It could handle a 5-speed 13–23 freewheel. Note the bar-end shifter

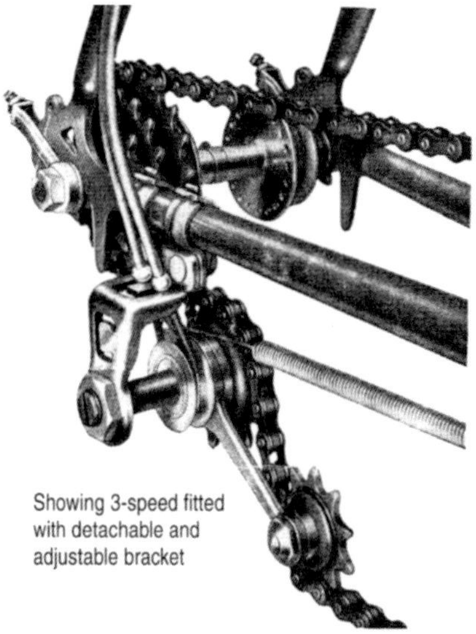


Right: Fig. 8.100. The 1946 Cyclo Touring was made in both the French and the British Cycle factories.

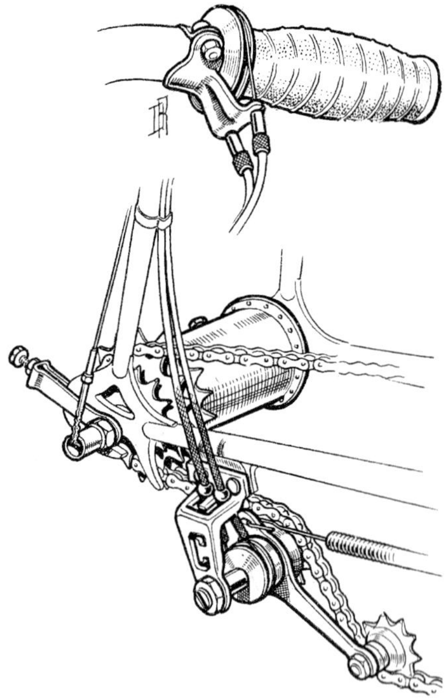

Above: Fig. 8.103. A 1949 British Cyclo Benelux derailleur combined with a Sturmey-Archer hub gave nine speeds. British Cyclo made hybrids like this for thirty years, but they never sold well. Note the twist-grip shifter.

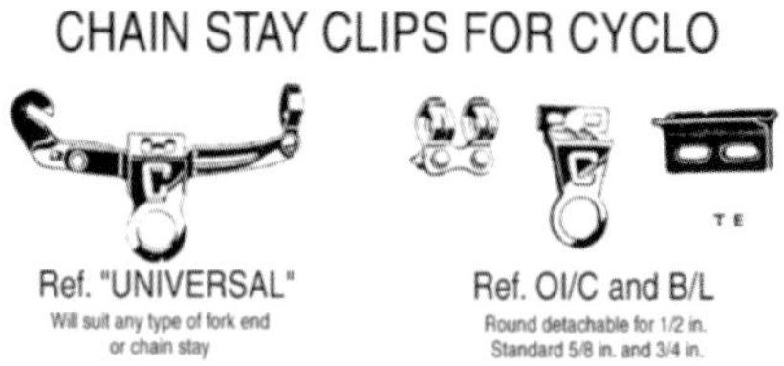


Below: Fig. 8.101. This 1949 BSA rear derailleur was only used on BSA bicycles. It shifted poorly because of the single-pulley.

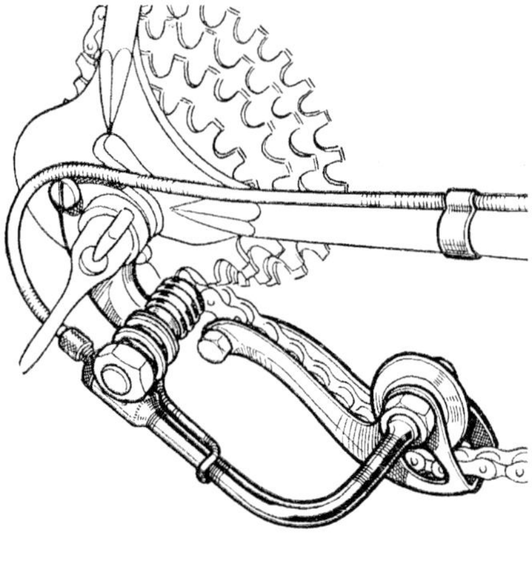

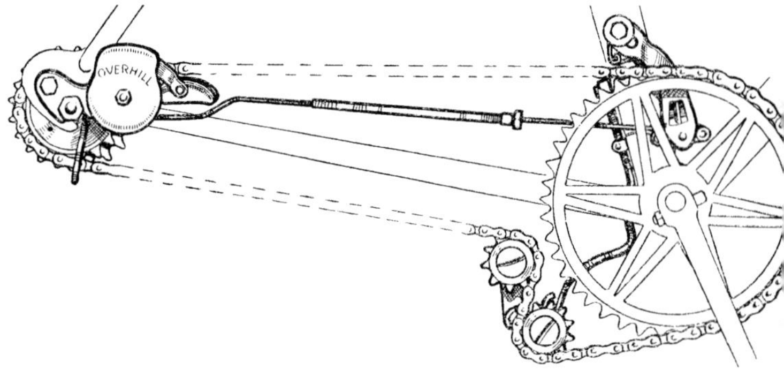


Right: Fig. 8.102. The 1949 2-speed Overhill rear derailleur combined a derailing fork, low capacity, and an exposed take-up arm. It shifted by back-pedaling. It's easy to see why it did not survive for long.

company in 1933, the Cyclo Gear Company Ltd. I refer to this venture as British Cyclo. They made derailleurs, hubs, freewheels, chainwheels, and bicycle tools. Before World War II, most of the derailleurs had been copies of French Cyclo, but with British screw-threads.

Following World War II, the two Cyclo companies went in quite different directions. After the death of Albert Raimond in 1953, French Cyclo made freewheels only, whilst British Cyclo became the largest derailleur company in Britain and introduced several competitive new derailleurs.

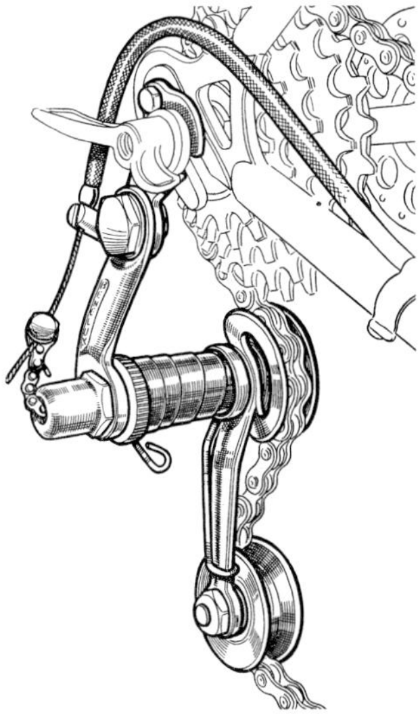

Above: Fig. 8.105. On the 1954 Benelux Tourist rear derailleur, British Cyclo substituted an internal coil spring and a pull chain for the helical operator. A caged jockey pulley allowed 4 or 5 speeds with a 3/32-in. chain. It could handle a 13–28 freewheel and a 20-tooth difference between the chainwheels.

Right: Fig. 8.108. This 1949 Constrictor rear derailleur prototype had an ingenious planetary mechanism to maintain parallelism, but it still had only one pulley.

Left: Fig. 8.106 A and B. The 1952 British Cyclo Perry backpedaling (coaster) brake conversion added 3-speeds to a coaster-brake hub. It was restricted to a 5-tooth difference between the sprockets. Notice the mechanism to wrap up the chain when braking (see lower image).

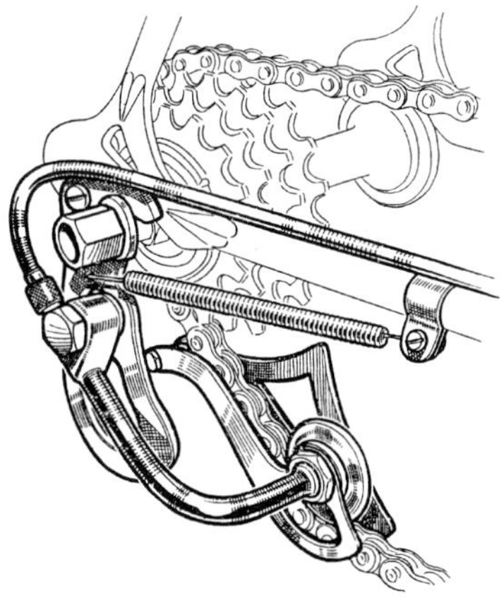

Above: Fig. 8.109. For 1952, the BSA rear derailleur got an external spring and a bigger, clumsier cage than the 1949 version.

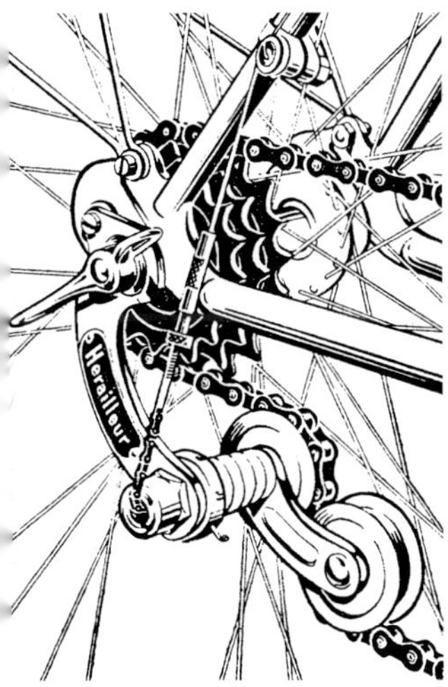


Left: Fig. 8.104. The 1949 Hercules Herailleur was a rugged little plunger-type pull-chain rear derailleur. The handlebar-mounted shift lever was detented (i.e., indexed).

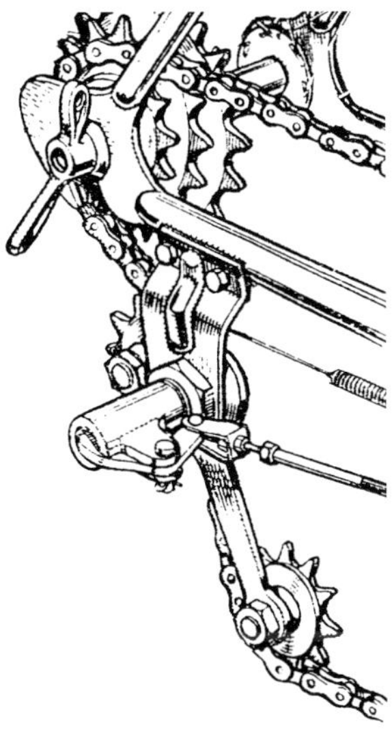

Above: Fig. 8.107. During the period 1948–1950, TriVelox continued their pre-war sliding sprocket derailleur. Then they switched to conventional plunger-type rear derailleurs, using a bell-crank instead of a pull-chain.

Below: Fig. 8.110. Starting in 1954, Phillips made this rear derailleur for their bicycles with the required chainstay mounting. This was also called the Resilion.

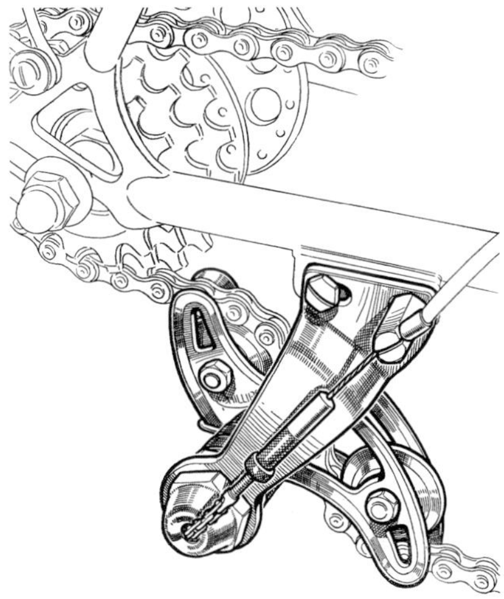

Cyclo Gear Company Ltd. was the largest British derailleur maker, and they probably had more than half of the market. However, it was not a large market because of the competition from three-speed hubs and from Simplex and Huret. There were at least eight other derailleur companies at the start of this decade.

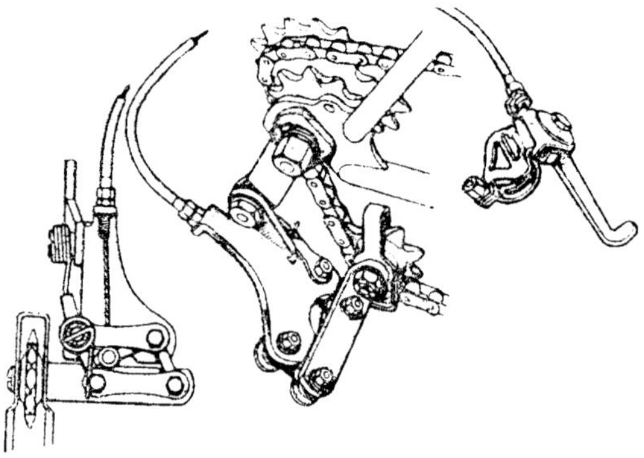

Above: Fig. 8.111. The 1947 Conloy Osgear rear derailleur had a parallelogram, but with only one pulley, its total capacity was only about 6 teeth.

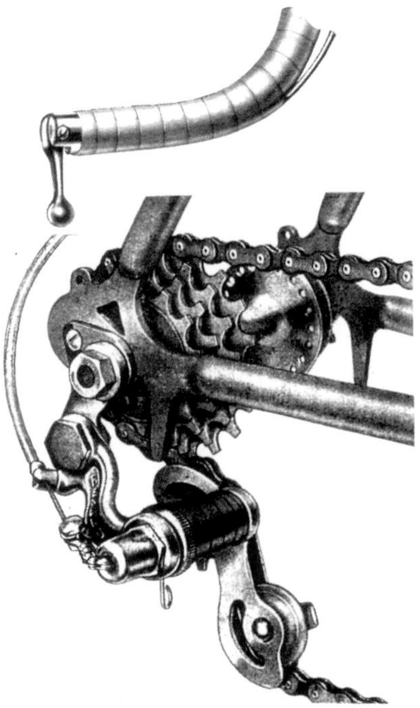

Above: Fig. 8.112. Between 1952 and 1958, the British Cyclo Benelux and Benelux Sport had minor cosmetic changes. Note the bar-end shifters on both of these derailleurs.

Only British Cyclo and Resilion survived to 1955.The Constrictor Tyre Company (Conloy) distributed Oscar Egg derailleurs in Britain. They also made their own models.

## Germany and Austria

After World War II, the German bicycle industry started from scratch. There was little bicycle racing. Bicycles provided essential transportation. Gearing was not a priority. Getting a good pair of tires was much more important. The pre-war racing component companies started up again after the war. In the *Third Reich*, only German-made components were allowed, but even immediately after the war, only Fichtel & Sachs and, to a lesser extent, Wissner derailleurs were available.

In the early 1950s, a modest German derailleur production developed. As in France, Italy, and Britain, there were many small manufacturers. Fichtel & Sachs-Amsler and Simplex-PWB were the largest brands, but Altenburger, Honer, Kreis, and Renak were also significant. Kreis patented a 2-speed automatic derailleur in 1951. Alda, Bismarck, Magura, Optima, Rabeneick, Rasant, Reinhold, Tectoran, Wissner, and Velo were minor derailleur makers.

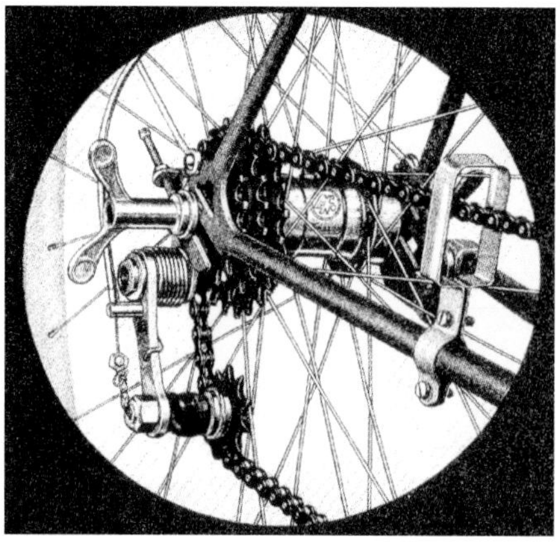

Above: Fig. 8.113. The 1950 single-pulley Fichtel & Sachs rear derailleurs were usually combined with coaster brake hubs, which limited their capacity to about 8 teeth between the sprockets. Note the cage to catch the slack chain when braking.

Below: Fig. 8.114. The 1950 Renak was a copy of the Huret Tour de France. Renak stood for Reichenbacher Naben- und Kupplungswerke. It was an F&S plant set up in Eastern Germany during World War II.

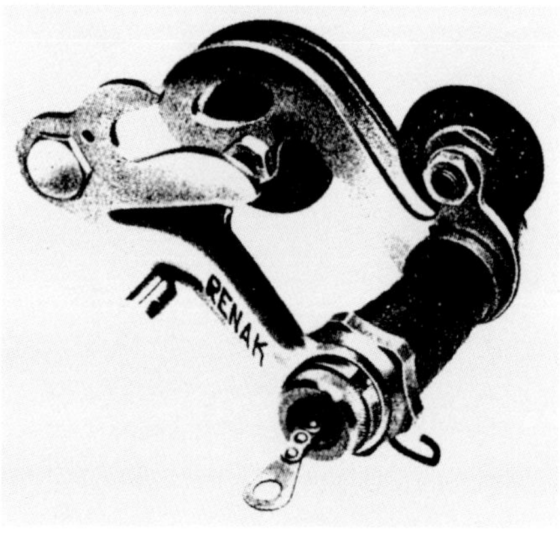


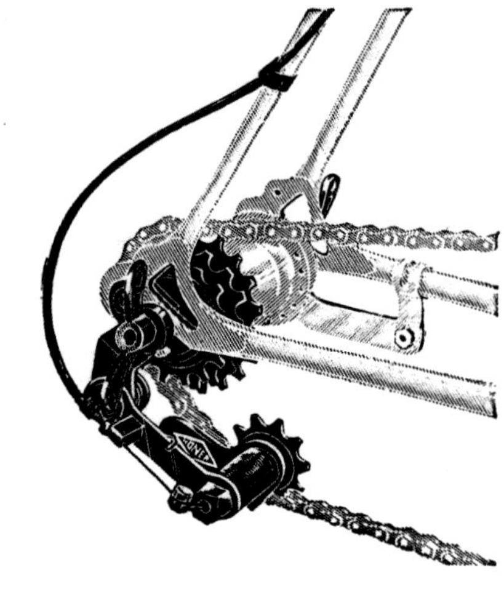

Above: Fig. 8.115. The 1951 Honer was a single-pulley F&S copy. Like most other German derailleurs, it was combined with a coaster brake.

The situation was similar in Austria after the war, where the major domestic derailleur makers were Busto, Capo, and Suwe. Significant competition soon developed from Simplex and Huret, and later from Campagnolo.

### Altenburger

Altenburger restarted after World War II, making aluminum alloy rims for bicycles and motorcycles. In 1950, Karl Altenburger bought the patent for the horizontally pivoting rear derailleur from Nieddu, Mariani, and Blaser. The first derailleurs were marked, "Altenburger/ Brev. To. Nieddu." Heinz Müller, a famous German bicycle racer, was also a skilled toolmaker, and he revised the first model to incorporate a horizontal parallelogram. He successfully raced the revised versions. It was sold in the UK as the GB Altenburger. A front derailleur was added in 1951. Production ended in about 1960.

Above: Fig. 8.116. The 1954 Austrian Suwe rear derailleur had a horizontal parallelogram.

## United States

After Pearl Harbor, U.S. automobile production ceased, and gasoline was rationed. Demand for bicycles soared, and in April 1942, sales were restricted to defense workers. This market for transportation bicycles ceased when the war ended. U.S. bicycle production went from 1.7 million in 1946 to 2.8 million in 1948. Then it declined until it reached about to 1.5 million in 1954.

The standard U.S. bicycle in the post-war decade was a single-speed balloon-tired roadster with a coaster brake. These machines were used almost exclusively by children or teenagers too young to get a driver's license. Bendix and New Departure made back-pedal coaster brakes with two speeds.

In 1946, there was a mini-boom in lightweight bicycles as the returning GIs remembered the bicycles they had encountered in Europe. That year, 50,000 bicycles were imported, mostly by ex-servicemen, but imports declined to less than 20,000 by 1949. Interest picked up again at the start of the 1950s, as bicycle imports began a steady growth to 250,000 in 1952 and 600,000 in 1953.

Most of the imports were English 3-speed bikes (350,000 of the 600,000 bicycles imported in 1953 came from Britain), but there was also a small market for French and Italian derailleur bicycles.

Schwinn was the only significant maker of quality bicycles in the U.S. in this decade. President Frank W. Schwinn believed that there was a market for lightweight adult bicycles, and he constantly promoted the idea. In the early

Below: Fig. 8.118. The Bendix 2-speed coaster-brake hub came in two versions. One gave a 32% reduction, and the other one gave a 47% increase. They were shifted by momentary back-pedaling. For m any years, it remained America's most popular gearing option.

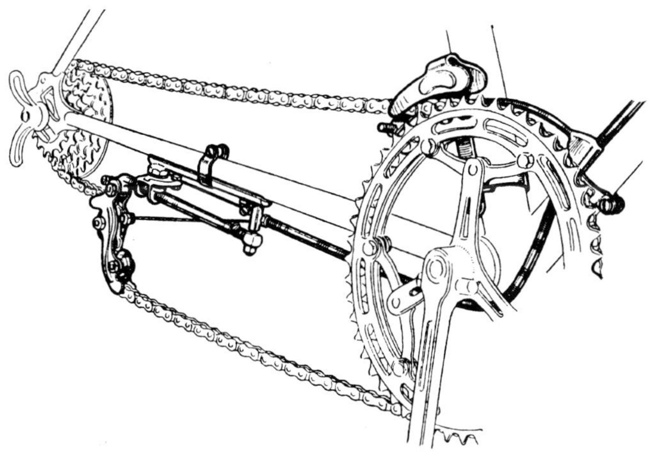

Left: Fig. 8.117. The 1950 Altenburger rear derailleur had a small tube that swivelled horizontally to move the jockey pulley in or out. It was available in three versions, called Universal, Touriste, and Champion respectively.

1950s, Schwinn sold the World Traveler, fully equipped with caliper brakes, lights, luggage bag, and a Sturmey-Archer three-speed hub, but it was heavier and more expensive than the bikes imported from Britain and France.

For the connoisseur, Schwinn made the Paramount road racing bike, which was custom-built to the buyer's specifications. In 1954, Schwinn introduced middleweight bicycles with 26×1¾-inch tires. The vast majority were single-speed bikes with coaster brakes, but the top models had three-speed hubs.

Throughout the 1940s, U.S. racers had used fixed-wheel bikes. There was no prohibition of gearing, but most racers felt that a fixed gear encouraged a better cadence—and derailleurs were hard to come by. By the early 1950s, the experience of the European racers crossed the Atlantic and derailleur-equipped bikes took over the races.

Right: Fig. 8.119. The 1949 Schwinn Black Phantom was every American 15-year-old's dream. Fifty years later, Schwinn was selling Black Phantom replicas for $3,000.

## Japan

Shikanosuke Maeda had started making SunTour 8.8.8. freewheels in 1912, and Shozaburo Shimano had started making Shimano 3.3.3. freewheels in 1922. Both companies were in Sakai, and during World War II, both were taken over by Japan's military government to make armaments.

The Japanese bicycle industry was centered in Osaka and Sakai, the port city on the south side of Osaka harbor. Osaka was heavily bombed in World War II. In June 1945, a B-29 air raid destroyed

Left: Fig. 8.120. The 1960 Schwinn World Traveler with a 3-speed hub represented the limit of public acceptance of gearing in the U.S. during this period.

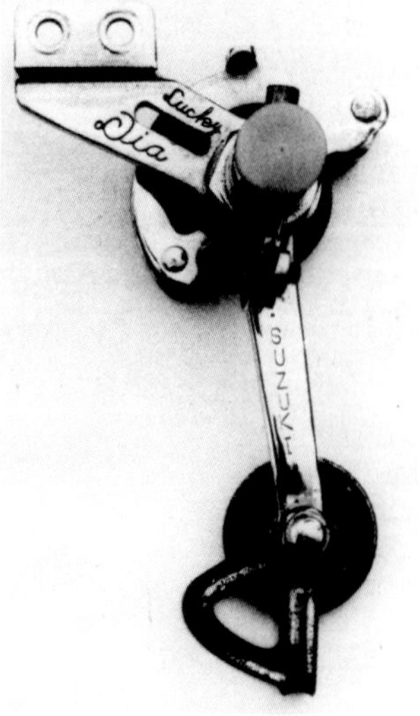

Right: Fig. 8.121. This circa 1950 DIA Lucky touring derailleur, made by Suzuki, was a copy of the Simplex Route.

Below: *Table 8.2. Derailleur time line and model overview, 1924–1954, Part A* *(continued on facing page)*

| | 1924–1929 | 1930–1934 | 1935–1939 | WW-II | 1945–1949 | 1950–1954 |
|---|---|---|---|---|---|---|
| **Cyclo – France** | **Albert Raimond** | | **1885–1953** | | | |
| Standard 2-, 3-, 4-speed | –+———— | ——+——— | ———— | | ———— | ———— |
| Touriste 5-speed | | | — | - | ———— | |
| Dural 5-speed | | | ——— | | | |
| Witmy | | – | ———— | | | |
| Poly | | | ———— | | | |
| Super Cyclo | | | ———— | | | |
| Sport, Sport 2-cable | — | -------+– | | | | |

| | 1924–1929 | 1930–1934 | 1935–1939 | WW-II | 1945–1949 | 1950–1954 |
|---|---|---|---|---|---|---|
| **Cyclo – Britain** | **Louis Camillis** | | **1889–1987** | | | |
| Standard 2-, 3-, 4-speed | | ——— | ——— | | ——— | ——— |
| Dural | | | | | — | ——— |
| Witmy, Star | | — | ———+——— | | — | |
| Super Cyclo | | — | | | | |
| Sinclo | | | ——— | | | |
| Oppy | | | ——— | | ——— | |
| Ace | | | ——— | | ——— | ——— |
| Olympic, Olympic Pro | | | —— | | ——— | |
| Benelux 50, 52, 53 | | | | | | ——+—+—— |
| Benelux Sport | | | | | | ——— |
| Benelux Tourist | | | | | | — |
| | | | | | | |
| **Huret** | **André Huret** | | **1890–1946** | | | |
| Course | | ——— | ——— | | | |
| Aluvac | | | | — | | |
| Course Competition | | | | | ——— | |
| Route Touriste | | | | | ——— | |
| Competition | | | | | — | ——— |
| Randonneur | | | | | ——— | |
| Louison Bobet | | | | | | ——— |
| | | | | | | |
| **Simplex** | **Lucien Juy** | | **1899–1976** | | | |
| Route 2-speed | —— | ——— | | | | |
| Route 3-speed | | ——— | | | | |
| Type Anglais 33 | | —— | | | | |
| Tour de France 34 | | — | — | | | |
| Champion 35 | | | —— | | | |
| Champion de France 35, 36 | | | —— | | | |
| Route 35 Route Leger 36 | | | —+—— | | | |
| Super Simplex, Selection | | | ——— | | | |
| Champion du Monde 46 | | | | | ——— | |
| Tour de France 48 | | | | | — | |
| Tour de France 49, 50, 52 | | | | | — | ——+——— |
| Juy Tension Model 50, 51 | | | | | | —+——— |
| Grand Tourisme | | | | | | ——— |
| Juy 543 | | | | | | — |
| | | | | | | |
| **Super Champion** | **Oscar Egg** | | **1890 – 1961** | | | |
| Champion | | — | | | | |
| Super Champion | | —— | ——— | | ——— | |
| Course, Professional | | | ——— | | | |
| Route | | | — | | | |
| Grand Sport | | | —— | | ——— | |
| Champion de Monde 39 | | | — | | ——— | |
| Mondial | | | | | | ——— |
| Sport | | | | | | ——— |
| | | | | | | |
| **Campagnolo** | **Tullio Campagnolo** | | **1901–1983** | | | |
| Cambio Bacchetta | | | ——— | | | |
| Corsa 1001 | | | | | ——— | |
| Paris-Roubaix 1002 | | | | | — | — |
| Gran Sport 1012 | | | | | | ——— |
| Gran Sport Extra | | | | | | — |
| Sport 1013 | | | | | | —— |

## Author's Note:

After adding the names of the six founders of the six major European derailleur makers, I was struck with these thoughts: All six founders were born near the end of the 19th century. All six started their companies around 1930. All six had a son (or sons) to whom they turned over their companies. Five of the sons (grandsons in the case of Huret) were unable to save their companies. Valentino Campagnolo was the exception, and though he went through a difficult period in the 1980s, he was able to save the company and its reputation—due largely to the loyalty of its customers.

the Shimano factory, while a July air raid burned out most of the Maeda factory. Both companies slowly were rebuilt after the war.

By January 1946, Maeda's 58 employees were producing freewheels and agricultural components at pre-war levels. Shimano resumed manufacture of freewheels and hubs in November 1946. They also made frames and complete bicycles in 1947, because that gave them priority for raw materials. They stopped making bicycles in 1954 and concentrated on freewheels, hubs and spokes. Shimano did not make their first derailleur until 1956.

Japan's bicycle population fell from 10 million to 6 million during the war. Many of the survivors used rope for tires. There was a huge demand for basic single-speed transportation bicycles. However, there was also a small market for derailleur bicycles, and there were at least six derailleur makers. Most of Japan's pre-war experience with multi-speed bicycles had been with British Sturmey-Archer and BSA three-speed hub gears.

Kayaba Industries made the first post-war Japanese derailleur and applied for a patent. In 1947, Sanko made a 3-speed derailleur based on the two-cable British Cyclo. This model became also known as the Fuji Speed-X Gear. Neither of these early derailleurs caught on in the Japanese market.

In 1949, Junzo Kawai visited Europe and brought back samples of freewheels and derailleurs. He concluded that Maeda should make derailleurs and multi-speed freewheels. By 1950, Maeda had 122 employees and reported annual sales of about 50 million yen.

In 1951, a series of Japan versus U.S. bicycle races was held in Japan. Several French racing bicycles with Simplex Tour de France derailleurs were brought into Japan for the races. Shortly afterward, the Sanko Company (Sanko-sha) came out with a copy of the Simplex Tour de France. By 1952, the Japanese component makers were familiar with Campagnolo, Simplex, Huret, and Cyclo components.

The early Japanese derailleur makers included Maeda SunTour, Cheruvino, Dia/D.N.B./Dynamic, and Sanko. In the early 1950s, they all made copies of the pull-chain Simplex Tour de France or the Huret Competition models.

Above: Fig. 8.122. Early Maeda SunTour pull-chain rear derailleur. It appears to be a copy of a French Cyclo derailleur of the period. A wide-range touring version was also available.

Left: Fig. 8.123. This circa 1950 Sanko SR-1 racing derailleur was a Huret copy.

## Cycling Magazines

France had two first-rate cycling magazines. *Le Cycle* and *Le Cycliste*. *Le Cycle* was published twice monthly. It covered both the bicycle and motorcycle industry, but it was widely read by bicycle enthusiasts because of its strong technical content. Daniel Rebour was both technical editor and illustrator. Many of his illustrations are featured in this book.

*Le Cycliste* was published bi-monthly. It was *the* magazine for bicycle tourists, founded by Vélocio in 1887, and still going strong half a century later.

Britain had two weekly magazines, *The Bicycle* and *Cycling*, and, for Cyclists' Touring Club members, its monthly, the *CTC Gazette*.

In the U.S., the trade publication *American Bicyclist and Motorcyclist* was still around, but there were no consumer magazines.

# 9 Slow Growth: 1955–1964

BY 1955, bicycles in Europe were losing the battle to mopeds, scooters, motorcycles, and small cars. The Suez Canal crisis of 1956 caused an oil shortage and gasoline rationing in many European countries, which created a brief resurgence of interest in transportation bicycles.

In 1959, Alex Moulton filed for a British patent for a small-wheeled bicycle with 16-inch wheels, high-pressure tires, front and rear suspension, and a step-through "unisex" frame. When the machine was finally introduced in 1962, it created a bicycle revival in Britain for both recreation and transportation.

Of course, small-wheel bikes were not new, but Moulton's concept was unique: None of the earlier small-wheel bikes—nor those copied after the Moulton's introduction—had the high-pressure tires and suspension that made this machine so efficient, fast, and comfortable to ride.

In the U.S., a trend to use bicycles for recreation led to a steady growth in imports of lightweight bicycles. The American public started to develop an interest in exercise, and there were the first glimmerings of awareness that automobiles caused pollution and congestion. In 1960, Schwinn introduced the Varsity, the first derailleur bicycle designed for American teenagers.

In France, Italy, and Britain, the small derailleur makers became extinct. By the end of 1964, only Simplex, Huret, Campagnolo, and British Cyclo had survived.

Japanese companies started to export children's bikes. Shimano and SunTour made their first major sales in the export market.

Above: Fig. 9.1. Jacques Anquetil, shown here with Tullio Campagnolo (left), won the Tour de France five times during this decade.

## Bicycle Racing

Jacques Anquetil dominated the Tour de France in this decade, winning it five times. He won in 1957, 1961, and 1962 on Helyett bicycles with Simplex derailleurs. In 1963, he switched teams and won in 1963 and 1964 on Gitane bicycles with full Campagnolo gruppos.

In 1955, Jacques Anquetil raised the unpaced one-hour record to 46.16 km (28.69 miles). However, the record was not officially recognized because he refused to submit to a drug test. In 1962, Anquetil again raised the record to 47.49 km (29.51 miles), which was recognized by the UCI.

In the early 1960s, European racers switched from "half-step" gearing (something like a 52/48 crankset with a 14–24 freewheel) to "crossover" gearing (something like a 52/44 crankset with a 13–19 freewheel). The overall gear range was about the same, but the shifting sequence was simpler. The larger front difference affected the shifting ability of all of the 1950s racing rear derailleurs. A new generation of front and rear racing derailleurs was developed to cope with the wider chainwheel differences.

Since I first wrote this, I have been told that racers did not purposely set up "half-step" gearing. The smaller inner chainring was provided merely to increase the range with a slightly higher gear. at each end

I have since this was first written been told that racers did not purposely set up "half-step" gearing. The smaller chainwheel was added to increase the range with a slightly lower and slightly higher gear at each end, and they did not use double shifting to get intermediate gears.

Meanwhile in 1962, José Meiffret, now 49 years old, raised his own motor-paced world speed record to 204 km/h (127.4 mph). His bike had a 180-tooth TA chainwheel and a 15-tooth sprocket. I suppose someone had alerted him to the relative inefficiency of smaller sprockets (the same ratio could have been achieved using a chainring with "only" 120 teeth and a 10-tooth sprocket). At the record speed his cadence would have been 187 RPM.

Above: Fig. 9.2. José Meiffret with his 1962 paced speed record bike. It had a 180-tooth chainwheel and a 15-tooth sprocket.

## France

Domestic sales of French bicycles grew very slowly from 725,000 in 1955 to 820,000 in 1964. By 1959, the membership in La Fédération Française de Cyclotourisme had dropped from 14,000 members to 5,000. Interest in bicycle racing also declined, but compared to touring, it was in better health with 23,000 registered racers.

*Table 9.1. Tour de France winners, 1955–1964*

| Year | Rider (Country) | Bicycle | Derailleur |
|---|---|---|---|
| 1955 | Louison Bobet (France) | Louison Bobet (Mercier) | Huret |
| 1956 | Roger Walkowiak (France) | Geminiani | Campagnolo |
| 1957 | Jacques Anquetil (France) | Helyett | Simplex |
| 1958 | Charly Gaul (Luxembourg) | Learco Guerra | Campagnolo |
| 1959 | Federico Bahamontès (Spain) | Coppi | Campagnolo |
| 1960 | Gastone Nencini (Italy) | No brand | Campagnolo |
| 1961 | Jacques Anquetil (France) | Helyett | Simplex |
| 1962 | Jacques Anquetil (France) | Helyett | Simplex |
| 1963 | Jacques Anquetil (France) | Gitane | Campagnolo |
| 1964 | Jacques Anquetil (France) | Gitane | Campagnolo |

The racers were more motivated. They cycled for pride, glory, and money. The tourists rode for recreation and enjoyment of nature. They were not interested in the intense training required for serious competition.

Bicycle racing was a young man's sport, whereas bicycle touring could be practiced until the elderly rider's balance became uncertain. Bicycle touring was both an art and a philosophy of life.

These two different worlds could never be united in a single French organization. They cordially scorned each other, but the old racers would merge into the pelotons of bicycle tourists, and the tourists would shout themselves hoarse at the time trials, which in France were competitions for bicycles with fenders. Everyone became reconciled during the Tour de France to cheer the champions, especially if they were French champions.

In France, the major bicycle and component companies survived by becoming larger, while the smaller companies disappeared. All of the small French derailleur makers had disappeared by 1955. The large bicycle companies probably survived because they also made cyclomoteurs (mopeds), vélomoteurs (mini-motorcycles), motorcycles, and scooters. As bicycles lost market share, most of the development effort went into motor-powered two-wheelers.

By 1955, Simplex, Huret, and Cyclo were the only surviving French derailleur makers. They all made similar plunger-type pull-chain rear derailleurs. By 1964, this design had become extinct. In 1955, most front derailleurs were lever-operated. By 1964, they were all cable-operated, except for Simplex, who still offered both types.

Cyclo made the splendid Cyclo Standard touring derailleur almost unchanged through the decade. Huret and Simplex made simpler wide-range derailleurs for the touring market. By 1964, Huret had developed the Allvit, and Simplex had developed the Prestige, their definitive rear derailleurs, which they would continue making for the next twenty years. Cyclo pretty much abandoned the derailleur market to concentrate on freewheels.

Left: Fig. 9.3. This 1956 Peugeot advertisement in *Le Cycle* shows nothing but *"moteurs,"* reflecting a common trend during this decade when Europeans got motorized in droves. Just the same, this (temporary) switch of emphasis for the trade is what saved many of the bicycle manufacturers and bike shops from extinction.

Above: Fig. 9.4. The Competition/Tour de France, introduced in 1950, was Huret's basic pull-chain plunger-type rear derailleur until 1958, when it was replaced by the Allvit.

## Huret

From its beginning, Huret was more innovative than Simplex, but Lucien Juy of Simplex was the better salesman. From 1950 onward, Huret offset the cage pivot in front of the jockey pulley. In my opinion this was a better design than the otherwise similar Simplex Tour de France, because it provided more chain wrap. During the 1950s, all of the competitors adopted Huret's cage design.

Louison Bobet was a national hero. He had won the Tour de France for the third time on a Louison Bobet bicycle using Huret's Louison Bobet model rear derailleur. Two months later

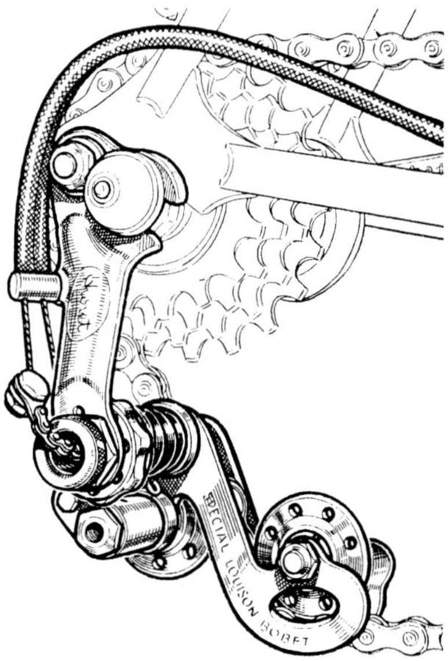



Above: Fig. 9.5. The 1954 top-of-the-line Huret Special Louison Bobet rear derailleur had two cables. The second cable controlled the chain tension.

Below: Fig. 9.6. The 1955 Huret lever-operated front derailleur used a helix to move the cage.

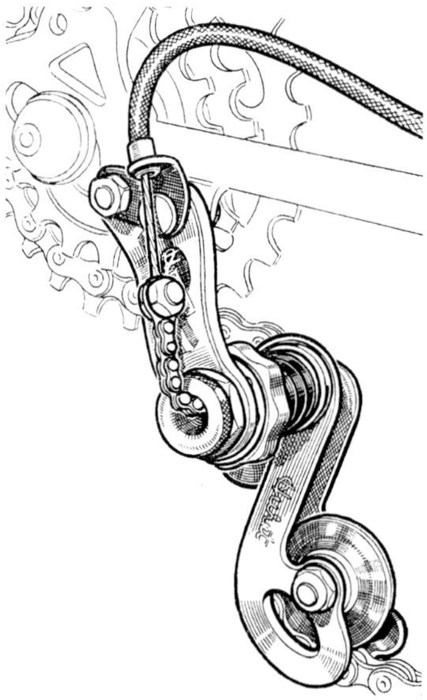

Above: Fig. 9.7. The 1958 Huret Monobloc rear derailleur required a two-bolt braze-on mounting on the chainstay.

Below: Fig. 9.8. Huret Allvit exploded view. The significant change from the first Allvit was the steel guard. The bike could fall on the right side without damage to the derailleur. The mechanism was still flimsy, but now it was well protected.

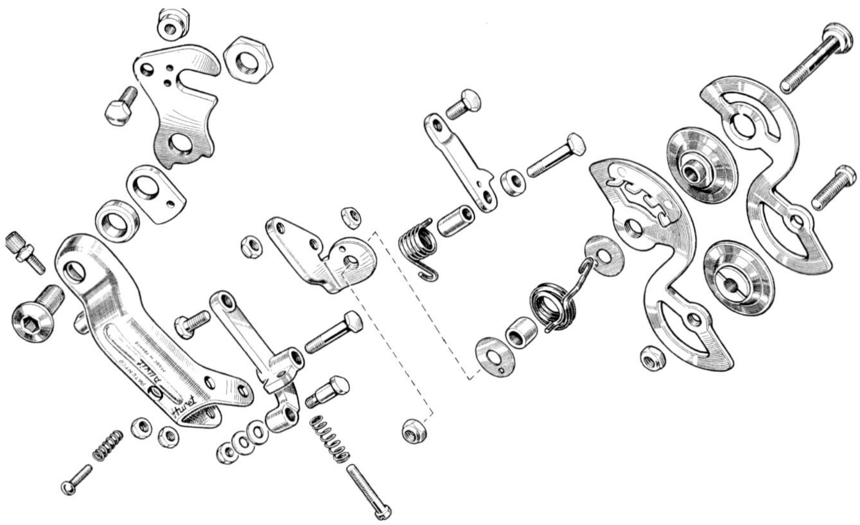

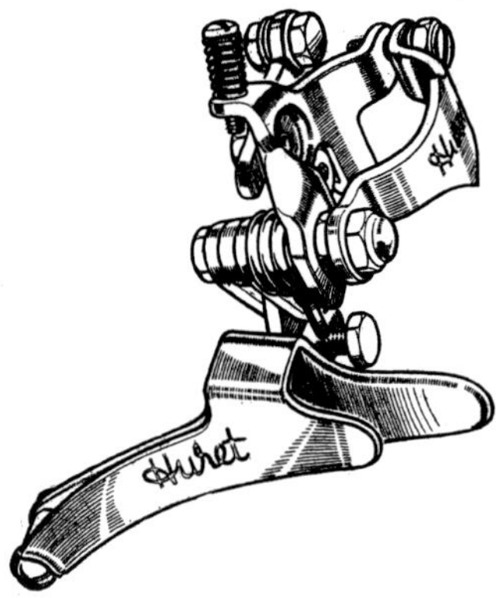

Above: Fig. 9.9. The 1961 Huret Allvit cable-operated parallelogram front derailleur shifted better than the plunger-type Simplex.

Below: Fig. 9.10. The first version of the Huret Allvit rear derailleur, introduced in 1958, did not have a guard to protect the parallelogram.

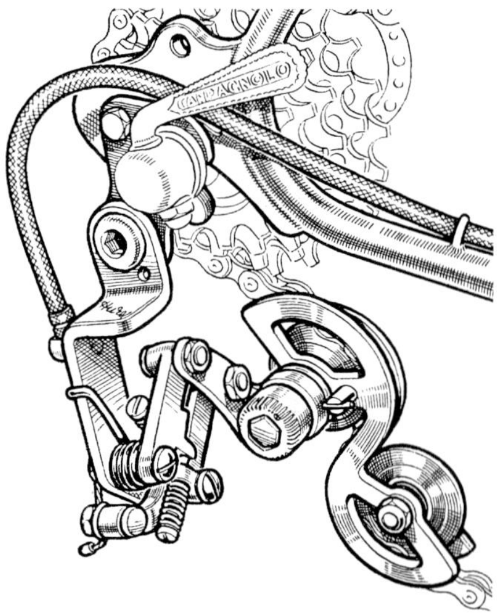

he announced that Louison Bobet bicycles would be equipped with Campagnolo derailleurs in 1956.

In 1958, Huret introduced the brilliantly engineered Allvit rear derailleur. Allvit derailleurs were used by several racing teams during this period, although they never won a major race.

Jan Heine tested a long-cage 1958 Huret Allvit on a René Herse bicycle with wide-range gearing (52/26 chainwheels and a 15–30 freewheel). He found that it shifted reliably but "with a clunk." It was not as smooth as the Nivex and Spirax derailleurs of the 1950s that he had tested.

André Huret died in 1964, at the age of 73. He had turned company operations over to his sons, Roger and Jacques. A new factory was built in Chepy. Five million Allvits had been sold by 1965.

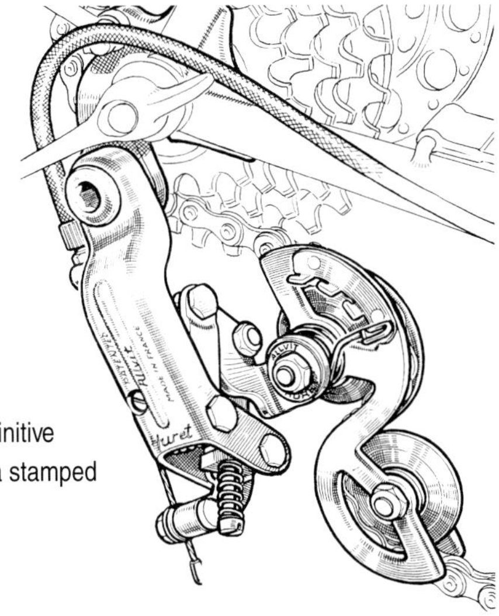

Right: Fig. 9.11. Introduced in 1961, this was the definitive Huret Allvit. The mechanism was now protected by a stamped steel guard.

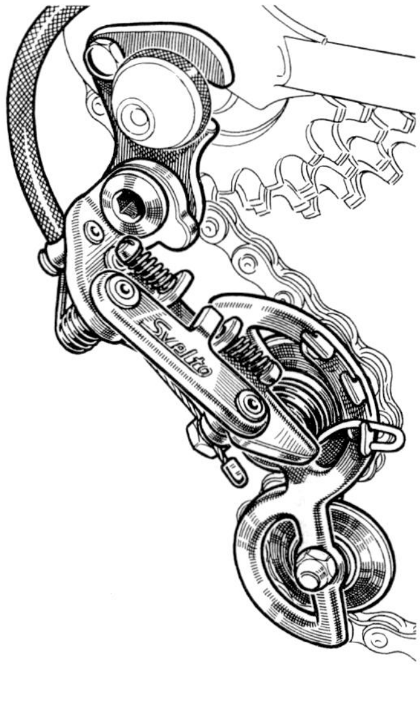


Above: Fig. 9.15. The 1963 Huret Svelto was an economy parallelogram rear derailleur that replaced the Tour de France model.

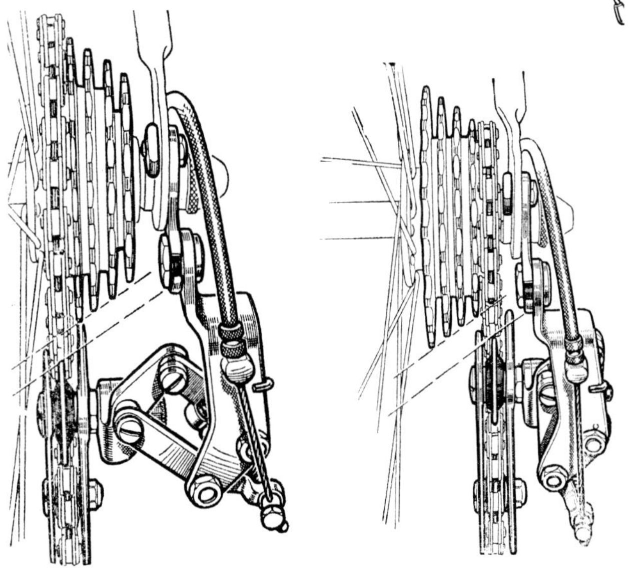

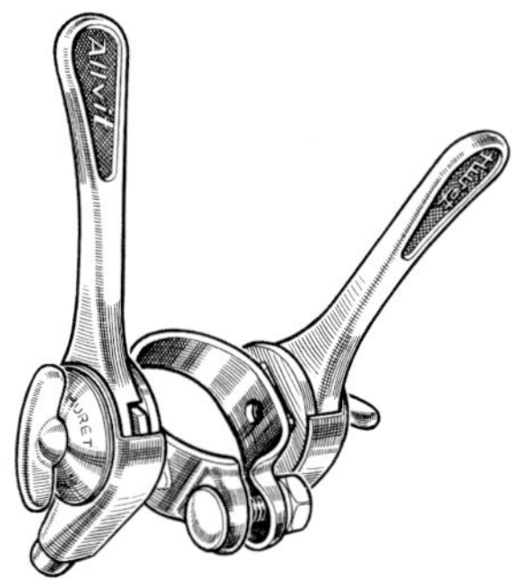

Right: Fig. 9.14. Huret's 1962 downtube shift levers had a larger-diameter cable-drum because Huret derailleurs required more cable movement than Simplex parallelogram derailleurs.

Above: Fig. 9.12 A and B. The parallelogram of the Huret Allvit moved the jockey pulley downward as it moved inward to provide a constant chain gap and full chain wrap around the sprockets. Huret patented this design, so nobody could copy it. On the down side, the Allvit was a bit flimsy, and the pivots would get sticky unless they were regularly lubricated.

Right: Fig. 9.13. This 1960s long-cage Huret Allvit Super Touring rear derailleur is mounted on a René Herse Démontable bicycle (which had a frame that could be divided into two parts to allow transporting the bike in the trunk of a car).

### Simplex

Simplex made a nearly complete gruppo of derailleurs, shift levers, cranksets, freewheels, seatposts, and quick-release hubs in this decade. They also made speedometers for bicycles and mopeds. The patented design of Simplex derailleurs, with two spring-loaded pivots, was better than that of the competitors. This feature is an essential part of modern rear derailleurs.

However, somewhere between 1955 and 1964, Simplex lost its way. They continued to make obsolete designs long after Huret and Campagnolo moved on to improved models. The low-priced Simplex Tour de France rear derailleur and the lever-operated Simplex Competition front derailleur were the best-selling derailleurs on the market from 1946 until 1958 because they performed adequately and they were cheap. Simplex kept churning them out.

Simplex did not have a top-quality derailleur for professional racers to compete with the rugged, reliable Campagnolo Gran Sport. It was time for Lucien Juy to react, and at the end of 1954, he introduced the Simplex LJ 543. In a typical Simplex publicity operation, it was tested with great ceremony on the Col du Galibier by Kübler, Bahamontes, and Dotto. In 1955, riders using the LJ 543 won the World Championship, Paris–Roubaix, and the Grand Prize of the Mountains in the Tour de France. It was still a pull-chain plunger-type Simplex, but it looked new. An alloy cover hid the pull-chain and the sliding rod. Although no one realized it

Above: Fig. 9.18. The JUY 59 rear derailleur replaced the JUY 51 in 1959. It had Simplex' two spring-loaded pivots and conventional cage geometry, with a total capacity of 20 teeth.

Below: Fig. 9.19. The 1960 JUY Record 60 was the last Simplex plunger-type rear derailleur. It was a slight modification of the JUY 543, without the second tensioning cable.

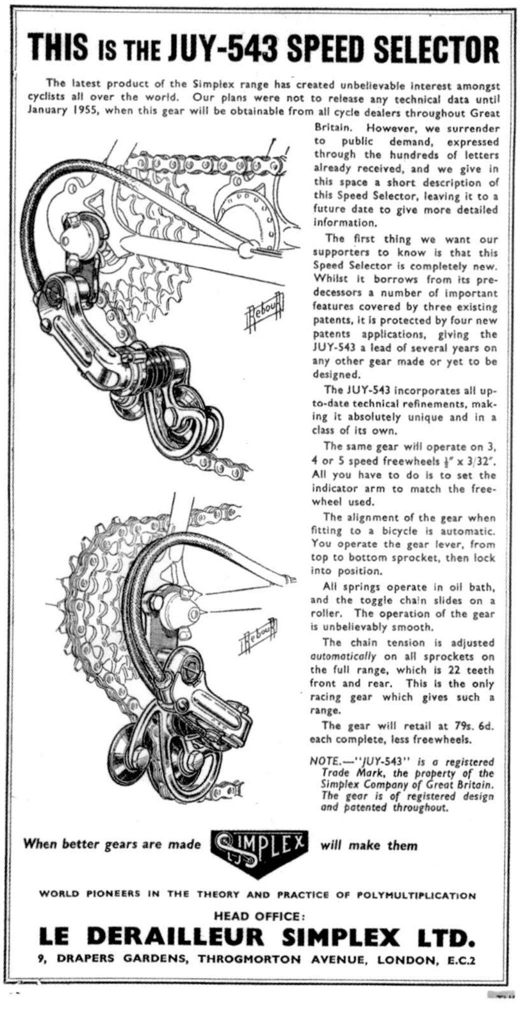


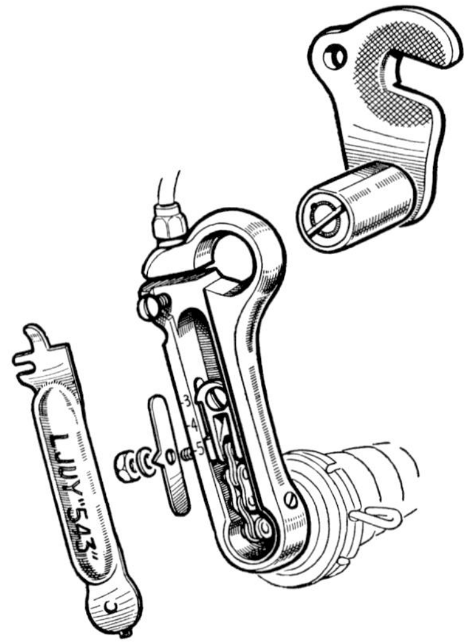


Above: Fig. 9.16. Partially exploded view of the 1954 Simplex 543, showing the pull-chain mechanism that could be set for 3, 4, or 5 speeds.

Left: Fig. 9.17. The 1954 Simplex JUY 543 was the ultimate pull-chain plunger-type rear derailleur. Many collectors believe it was the finest Simplex derailleur ever made. It could handle a 12-tooth difference on the freewheel, plus 10 teeth between the chainwheels. It used two cables. The second cable controlled the chain tension.

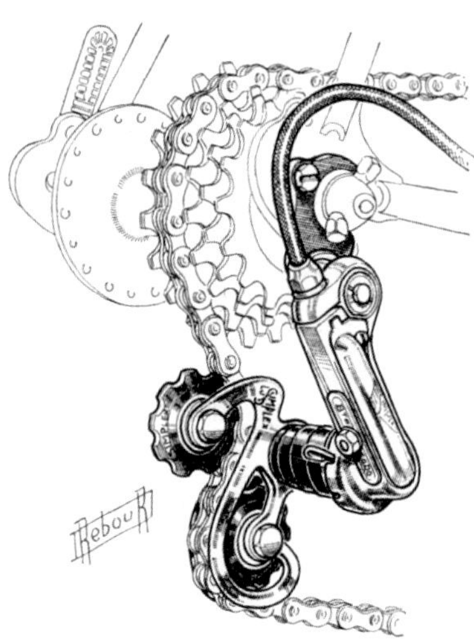

at the time, the LJ 543 was probably the peak of Simplex' innovation in derailleur development.

The 1961 Simplex Juy Export 61 was a first-class derailleur design that shifted as well as the Huret Allvit, and better than the Campagnolo Gran Sport. It had two spring-loaded pivots and a proper cage design with the jockey pulley on top. It maintained a fairly constant chain gap and shifted with a light, quiet action. The total capacity was 21 teeth.

A year later, in 1962, Simplex converted their production line to Delrin, billed as an "Industrial Resin" (read: plastic). Simplex had a long-term working relationship with Peugeot, who did much of the engineering for the conversion to Delrin.

Simplex named their flagship rear derailleur the Prestige. It continued the design of the Export 61; however, it was flexible and flimsy, and it did not wear well. It could well have been called the "No-Prestige."

With 20–20 hindsight, Lucien Juy's decision to make Simplex derailleurs from Delrin doomed the company. No matter what they did afterward, the Simplex name had become synonymous with cheap plastic trash.

Above: Fig. 9.20. The Simplex JUY Export 61 was Simplex' first parallelogram rear derailleur. This was not a Campagnolo Gran Sport copy. It was a superior design. The JUY Export 61 Luxe was the premium model with revised pivots.

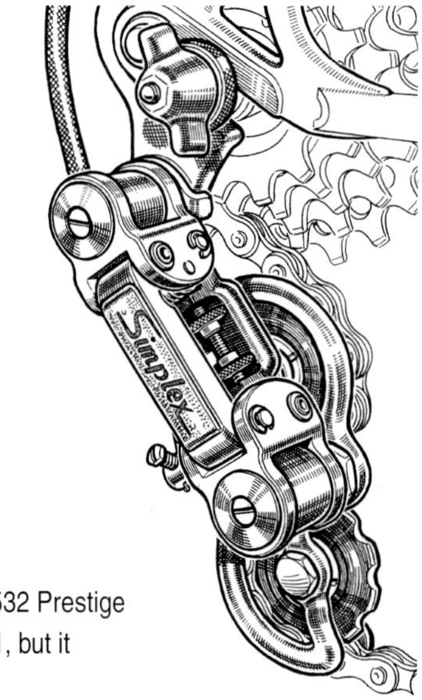


Right: Fig. 9.21. The parallelogram of the 1962 Simplex 532 Prestige rear derailleur continued the design of the JUY Export 61, but it was—unfortunately—made of Delrin plastic.

### Simplex Front Derailleurs

Simplex made the Competition lever-operated front derailleur from 1949 to 1962 with minor cosmetic changes. It was their least expensive model. By 1960, all of the competitors had shifted to cable-operated front derailleurs.

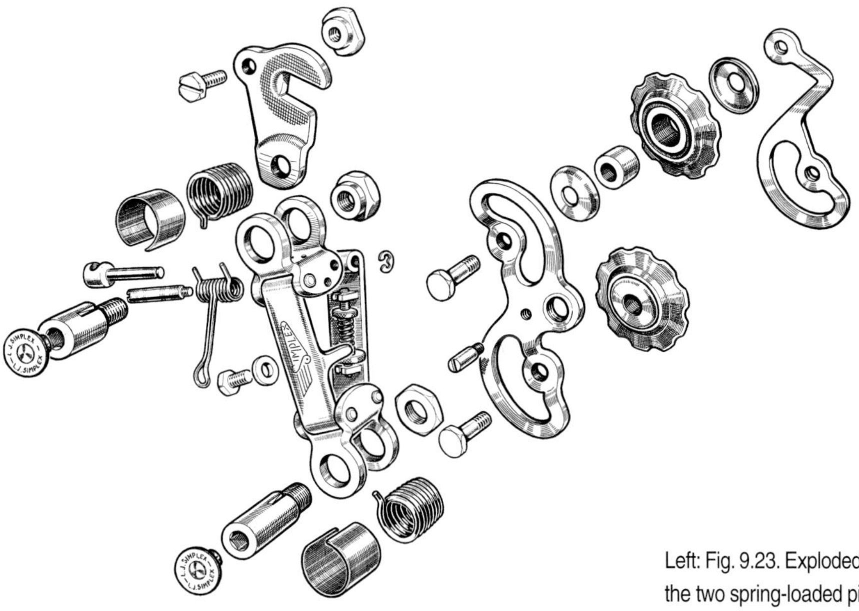

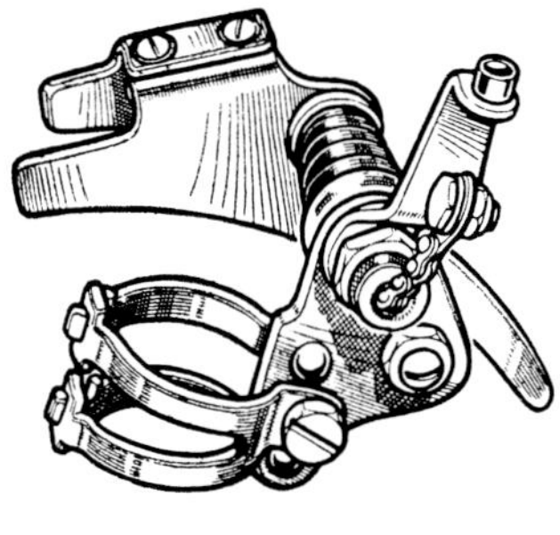

Above: Fig. 9.22. The Simplex 1956 JUY-56 was a cable-operated plunger-type front derailleur. It was based on the 1939 Touriste front derailleur, but the cage width could be adjusted for either 1/8-in. or 3/32-in. wide chains.

Left: Fig. 9.23. Exploded view of the 1961 Simplex JUY Export 61. This shows the two spring-loaded pivots and the cage with an offset pivot.

Simplex made a series of different cable-operated, plunger-type front derailleurs in this decade. The 1958 LJ-23 was a copy of the Campagnolo Gran Sport. The first models had an alloy body, but that was replaced by Delrin plastic in 1961. Simplex made plunger-type front derailleurs for twenty years. They looked cheap and shifted poorly compared to the Huret and Campagnolo parallelogram front derailleurs. The continued use of this dated design reinforced Simplex' "cheap plastic" reputation.

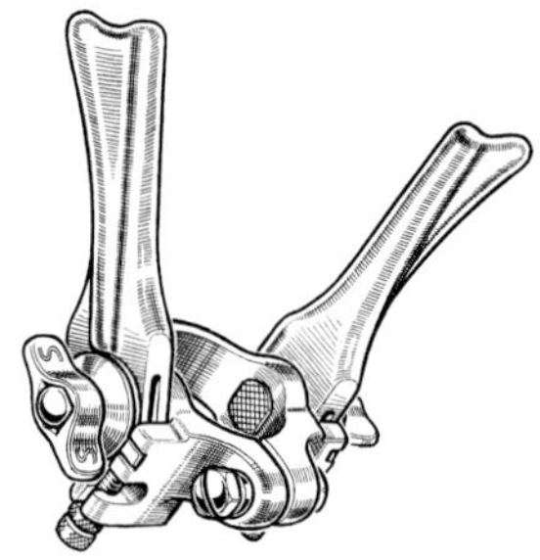

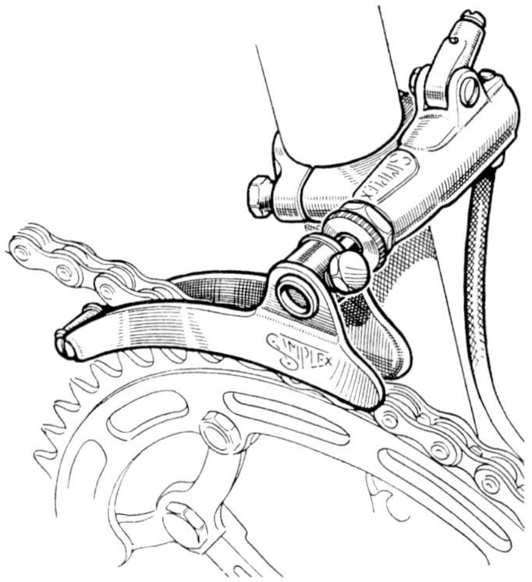

Top right: Fig. 9.26. Whereas the 1961 JUY Export control levers were made of alloy, with rubber covers, these Simplex shift levers, first Introduced in 1963, were made of plastic—and ugly compared to Huret's and Campagnolo's.

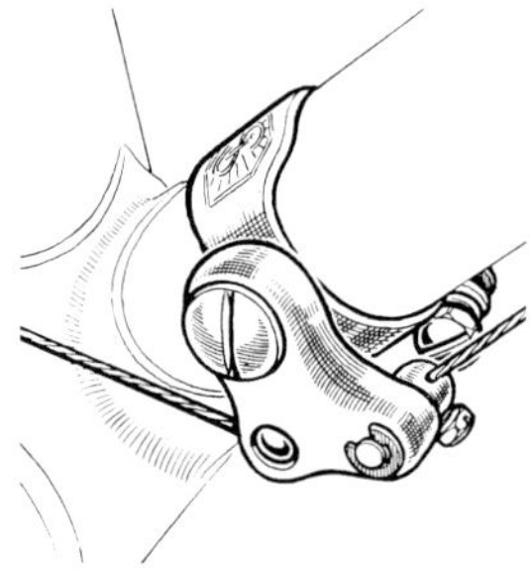

Lower right: Fig. 9.27. The 1962 Simplex *Demulitiplicateur* allowed the shift lever to have more cable movement than the derailleur.

Above: Fig. 9.24. The 1957 Simplex front derailleur had a cam to move the cage. Simplex' Italian factory made it for a year.

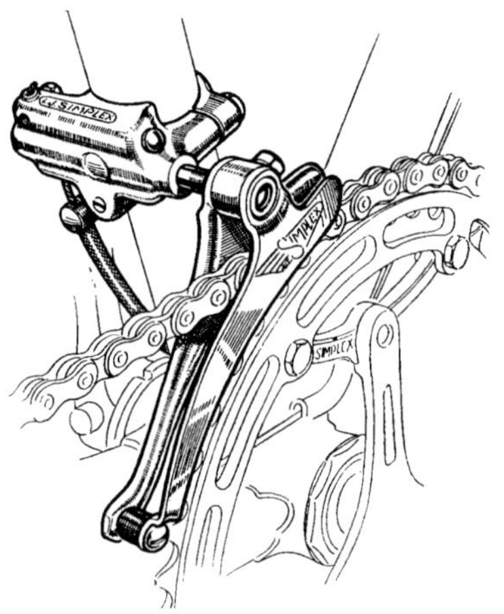

Right: Fig. 9.25. The 1958 LJ 23 cable-operated plunger-type front derailleur was a copy of the Campagnolo Gran Sport. It had an alloy body. In 1962, it was replaced by the AV 223 with a Delrin body.

Below: Fig. 9.28. Half-step gearing, shown on the Heyett machine with which Jacques Anquetil won the 1957 Tour de France. Note the 4-tooth difference between the 52-tooth large chainwheel and the 48-tooth small chainwheel and the 26-tooth (or 24-tooth) large sprocket on the freewheel. Note also the Simplex 543 rear derailleur, the lever-operated Simplex front derailleur, and the Stronglight crankset with TA chainwheels.

## French Cyclo

From 1955 to 1962, Cyclo continued to make both racing and touring derailleurs. When Albert Raimond died suddenly in 1953, his sons decided to concentrate on making quality freewheels. Derailleur production was secondary, and there was little innovation. Cyclo kept making the old favorites as long as they could be sold. The Cyclo Standard touring rear derailleur changed little from 1939 to 1960. It had two cables, a helical actuator, and one or two external springs. In France, it was usually installed with a braze-on mount. The tension pulley changed slightly over the years.

The Nivex and the Cyclo Standard were the best touring derailleur available. French Cyclo's pull-chain plunger-type rear derailleurs were clearly dated and provided no competition for the Huret Allvit or the Simplex Prestige. It was the end of the road, and French Cyclo derailleur production ceased sometime after 1962.

## Nivex

Nivex ceased derailleur production some time during this decade. However, Alex Singer, who had always maintained a special relationship with Nivex, took over their inventory and spare parts. Thus, for many years after, Nivex derailleurs, arguably the finest "old-school" derailleurs ever made, could still be had—but only on a Singer bike.

Jan Heine tested a Nivex rear derailleur on a 1962 Singer with 46/32 chainwheels and 14-22 four-speed freewheel (plus chainrest). Its shifting was a revelation: A slight tug on the shift lever produced and instantaneous shift without need for overshifting under all circumstances. He rates it the best touring derailleur ever made.

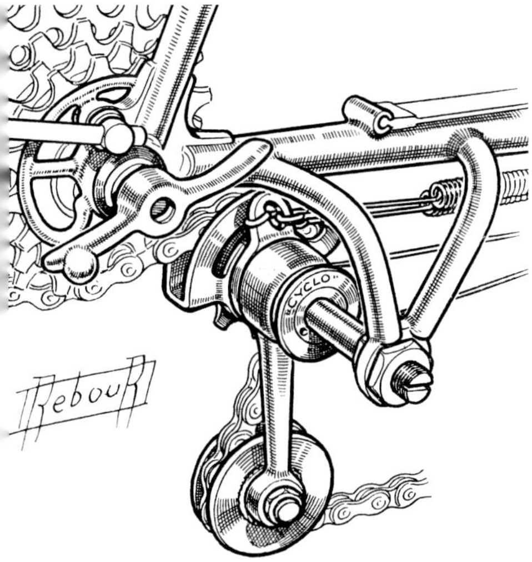


Above: Fig. 9.29. In the late 1950s, the Cyclo Touriste derailleur had a cage around the jockey pulley and a flanged tension pulley. Five- and six-speed versions were available. The 6-speed version shown here used the additional space for a circular chain rest.

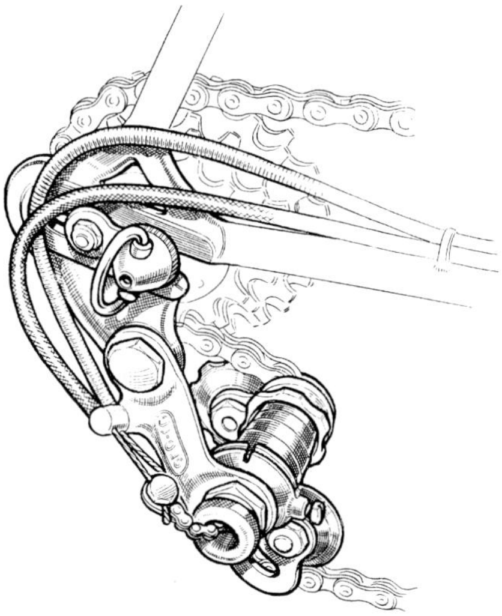

Left: Fig. 9.31. The 1955 French Cyclo racing rear derailleur had two cables. One shifted the jockey pulley, and the other adjusted the chain tension. Huret and Simplex also offered derailleurs with this feature.

Below: Fig. 9.32. The 1962 French Cyclo racing rear derailleur was an inexpensive Huret copy.

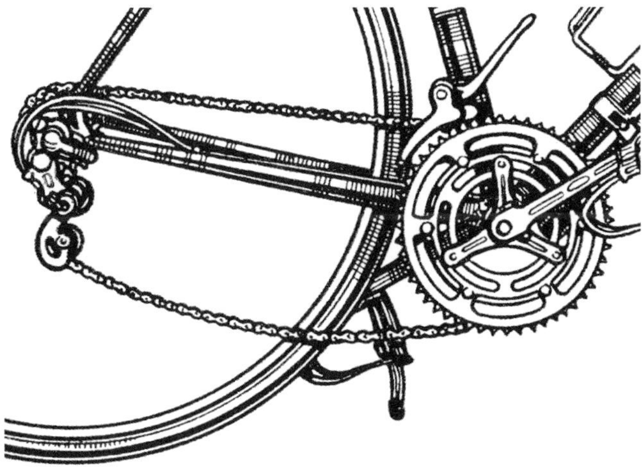

Left: Fig. 9.30. This detail of a 1956 Tour de France Mercier shows how the second cable worked: It allowed chain-tension to be reduced until the chain hung loose. Although some riders thought that would reduce friction, the basic idea was to increase tension on the small sprockets.

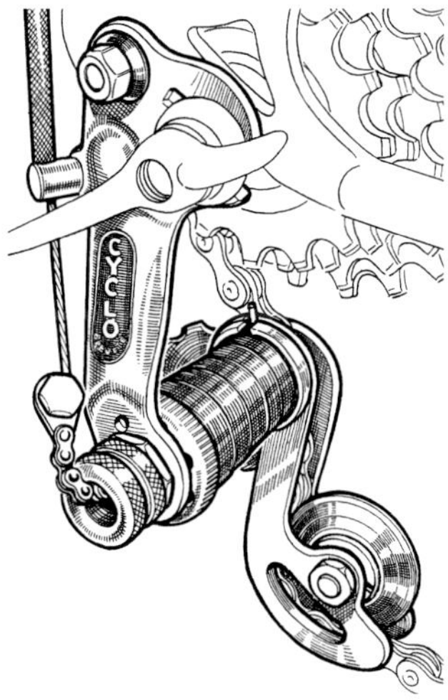

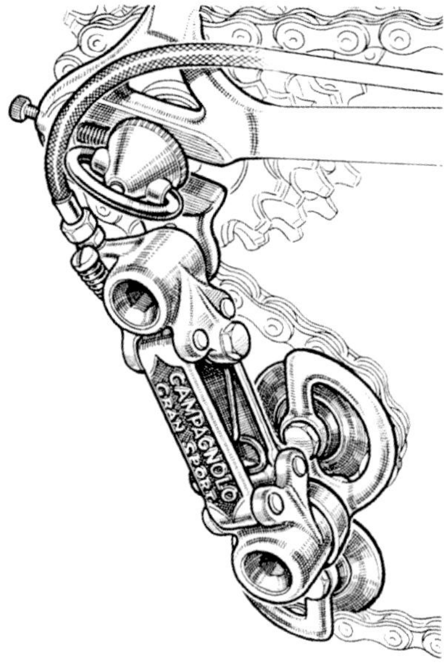


Above: Fig. 9.33. The 1951 Campagnolo Gran Sport rear derailleur was made for 13 years without significant changes.

## Italy

Italy's bicycle population grew from 5 million in 1938 to 8 million in 1965, despite the growing popularity of motorcycles and motor scooters. Italian bicycle production grew from 380,000 in 1956 to 900,000 in 1965. Italian bicycle exports grew from 17,000 bicycles in 1953 to 134,000 in 1962. All of the small Italian derailleur makers were out of business by 1955. Only Campagnolo and Gian Robert survived.

### Campagnolo

Tullio Campagnolo loved expensive top-quality racing components, and Campagnolo dominated the top-of-the-line racing market. Tullio kept trying to widen the company's customer base by introducing components to use with medium-priced racing bicycles and touring bicycles. However, his heart was not in it when he designed such components, and these were less successful.

Jan Heine recently tested a 1963 Campagnolo Record rear derailleur on a 1956 Cinelli Supercorsa bike with a 14–28 tooth five-speed freewheel. He found that it shifted well on the larger sprockets but required considerable overshifting to shift onto the 14-tooth sprocket.

The lower-priced 1963 Campagnolo Sportsman rear derailleur was rated for a 13–32 freewheel and a triple crankset, but its capacity did not match these specifications. No one tested derailleur capacity in those days, certainly not

Below: Fig. 9.34. A page from a 1960s Alex Singer catalog, showing the components that could be ordered, including Nivex rear derailleur and shifter, and Singer's own ultra-lightweight front derailleur—available either lever- or cable-operated.

in Italy, where bicycle tourists were almost non-existent.

In 1956, Campagnolo introduced a whole range of other components with the Record name: seatpost, headset, and pedals. When the Record crankset was introduced in 1959, it helped to put Campagnolo well on their way toward a nearly

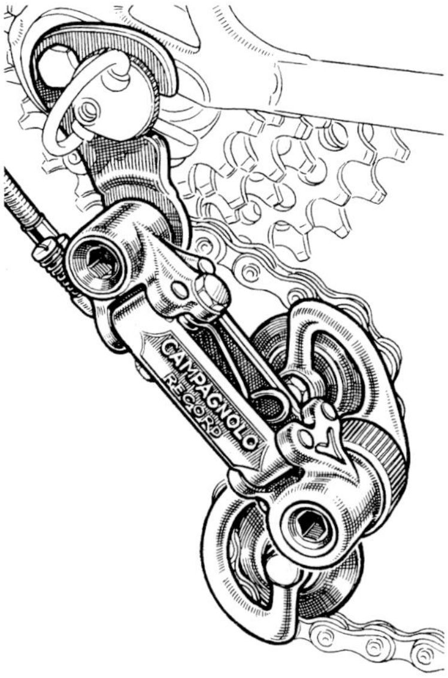


Above: Fig. 9.35. The 1963 Campagnolo Record rear derailleur was made of steel and chrome-plated bronze. The significant difference between the Gran Sport and the Record was the location of the cage pivot. The cage was revised to move the pivot further forward and closer to the jockey pulley, so that front shifts had less effect on the chain gap. The Record had a bit more capacity. It could handle a 28-tooth rear sprocket, but it was unhappy if the chainwheel difference was more than about 6 teeth.

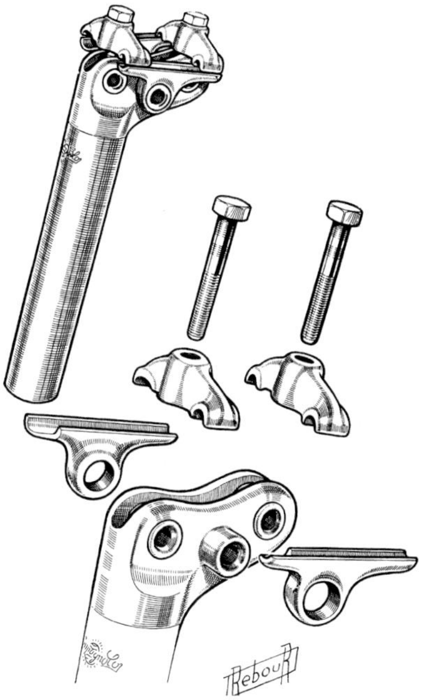


Right: Fig. 9.36. Exploded view of the 1956 Campagnolo micro-adjustable seatpost.

Above: Fig. 9.37. Campagnolo's Valentino rear derailleur, introduced in 1964, was an economy copy of the Gran Sport, with a stamped steel parallelogram.

Below: Fig. 9.38. Crossover gearing on the Gitane which Jacques Anquetil rode to victory in the 1963 Tour de France. Note the 10-tooth difference between the 52-tooth large chainwheel and the 42-tooth small chainwheel and the smaller large sprocket on the freewheel. Even though this is a French bicycle, it now had a Campagnolo parallelogram rear derailleur, a Campagnolo cable-operated front derailleur, and a forged alloy Campagnolo crankset.

complete *gruppo* of top-quality components.

A *gruppo* (Italian for group, and usually known as a group-set in Britain) is a matched set of components from a single maker. Gruppos are an effective marketing tool. In later years, Shimano and SunTour started offering complete gruppos at various price levels. The French makers formed marketing associations, like FREXA, which offered components selected from different makers that were supposed to work well together. This was a less effective marketing tool.

When the Record rear derailleur replaced the Gran Sport in 1963, Campagnolo had a nearly complete Record gruppo, consisting of front and rear derailleurs, shift levers, hubs, crankset, pedals, headset, and seatpost with matching cosmetics. *Tout Campagnolo* (all Campagnolo) became the hallmark of a top-quality bicycle.

At first, the French teams used Simplex or Huret, and the Italian teams used Campagnolo. But by the end of the decade, Campagnolo was the equipment of choice because of their reputation for reliability and performance. Campagnolo also had a significant PR budget for race team support.

The 1964 Valentino rear derailleur totally lacked the Campagnolo polish. If you were Tullio Campagnolo's son Valentino, wouldn't you get an inferiority complex over this name?

Above: Fig. 9.39. The 1962 Campagnolo Sportsman rear derailleur was a lower-priced wide-range version of the Record.

### Gian Robert

For manufacturers offering lower-priced bicycles, there was a significant market for imitators to make something that looked like Campagnolo Record but cost less. Gian Robert was one of the companies that made Campagnolo copies. In 1963, Gian Robert took Simplex' Delrin mistake to the next level and converted to polypropylene plastic bodies and parallelograms.

## Great Britain

British bicycle production peaked at 3.6 million in 1955, then declined to 2.3 million in 1960, and 1.9 million in 1965. Exports fell from the peak of 2.4 million in 1955 to 1.5 million in 1960 to less than a million in 1965. In the late

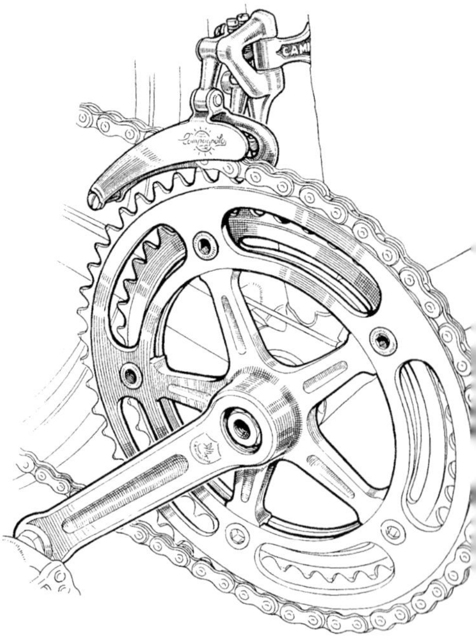

Above: Fig. 9.41. The 1960 Campagnolo Record cable-operated parallelogram front derailleur was a classic design. Twenty years later, most of the parts were still interchangeable. It is shown here with the new Campagnolo Record crankset on a 1962 Tour de France bicycle. Note that by 1962, the racers had switched to crossover gearing with a larger chainwheel difference.

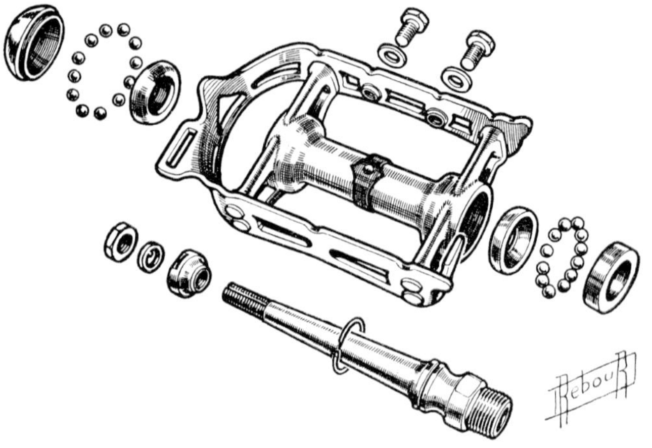

Right: Fig. 9.40. Exploded view of the 1956 Campagnolo racing pedal. Note the bearing seal on the crank-side bearing. Both the pedal and the seatpost sired dozens of copies.

1950s, exports were strongly encouraged by the government and aided by the devalued pound. Under serious price competition, the quality of mass-produced British bicycles deteriorated. At the same time, many overseas markets became saturated, and in many third-world countries, local bicycle companies took over the national markets.

Above: Fig. 9.42. This 1962 Gian Robert rear derailleur was one of the dozens of copies of the Campagnolo Gran Sport/Record.

Below: Fig. 9.43. Starting in 1964, the next generation of Gian Robert rear derailleur used polypropylene plastic for the body and the parallelogram

Domestic bicycle sales peaked at 1.4 million in 1950, and they were still at 1.2 million in 1955. The Suez crisis and expensive gasoline kept sales up until 1957. As in most other European countries, the trend in transportation during this period was that commuters abandoned bicycle transportation and bought scooters, mopeds, motorcycles and minicars. Domestic sales declined from 810,000 in 1960 and 730,000 in 1965. Cycling to work became a working-class "cloth cap" symbol.

### Raleigh

Raleigh lost their way in the 1960s. In 1957, Raleigh absorbed BSA, wich included Sunbeam and New Hudson. By that time, domestic sales were down to about 800,.000 bicycles per year, and Raleigh had about half of the domestic market. The other major maker was British Cycle Corp. (BCC), which was owned by Tube Investments (TI). Both Raleigh and BCC had considerable excess capacity and neither was very profitable.

In 1960, Raleigh and BCC merged to become TI-Raleigh. BCC's brands Hercules, Norman, Phillips, and Sun became part of the merged company and their plants were all shut down. TI-Raleigh still had more than 7,000 employees and enough factory capacity to make 1.5 million bicycles per year. They had 75 percent of the British market—but it was a shrinking market.

Right: Fig. 9.44. The 1955 Trent Sports was the first derailleur-equipped Raleigh bicycle. It was offered with a 4-speed British Cyclo Benelux derailleur as an alternative to the Sturmey-Archer hub gear. Also note the Sturmey-Archer hub-generator lighting system.

Over the years, TI-Raleigh absorbed Armstrong, BSA, Gazelle, Hercules, Humber, New Hudson, James, Norman, Phillips, Robin Hood, Rudge, Sun, Sunbeam, Three Spire, and Triumph. After the takeover, the nameplate typically survived for a few years as a badge on a Raleigh-made bicycle.

Carlton Cycles was purchased in 1960. Carlton had been a custom frame builder, making quality racing bikes in its small factory in Worksop. From 1960 to 1973, Raleigh's top-of-the-line derailleur bicycles were made in the Carlton factory and were sold under both the Carlton and Raleigh brand names.

Many of the smaller British bicycle companies had been the main sources of innovation. By the end of the 1960s, only privately-owned Dawes, Holdsworthy, and Moulton provided significant competition to TI-Raleigh.

By the mid-1950s, only the buyers of utility bicycles were satisfied with three-speed hubs. Buyers of medium- and high-priced bicycles demanded derailleurs-equipped bikes. Raleigh made their first 4-speed derailleur models for the British market in 1955. By 1957, there were four derailleur models, including an 8-speed. By the end of this decade, Raleigh offered either British Cyclo Benelux or Campagnolo derailleurs on many models.

When the TI-Raleigh merger took place in 1960, Raleigh was exporting about 60,000 bikes a year to the franchised dealers in the U.S. These were good-quality roadsters or middleweights with hub gears or derailleurs. The export derailleur models usually had "alpine" gearing, with 52/40 chainrings and 14–28 freewheels.

Before the merger, BCC had developed a major market in the U.S. for inexpensive middleweight roadsters with private-labels to the American gaspipe bicycle companies like Huffy, department stores like Sears, and automotive supply stores. TI-Raleigh took over the BCC market and in 1963 they shipped 800,000 bicycles to the U.S. Only 60,000 of those were badged Raleighs and sold through franchised dealers.

These private-label mass-market sales to the U.S. kept the factory busy, but they were unprofitable. The wholesale price of such bicycles was reduced to as little as $18. In 1963, TI-Raleigh decided to get out of the U.S. mass-market and to concentrate on Raleigh brand-name sales through franchised dealers.

### Moulton

Alex Moulton was born in 1920. He worked for the Bristol Aeroplane Company during World War II. After the war, he completed his engineering degree and founded his own company to

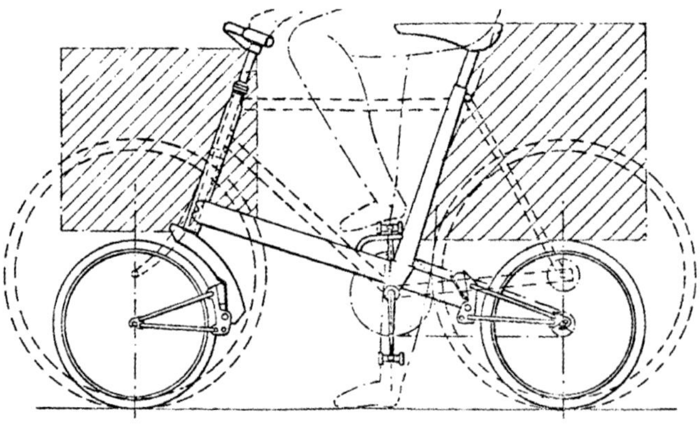

Above: Fig. 9.45. In 1959, Alex Moulton obtained a patent on a bicycle with small wheels, high-pressure tires, front and rear suspension, and a step-through "unisex" frame.

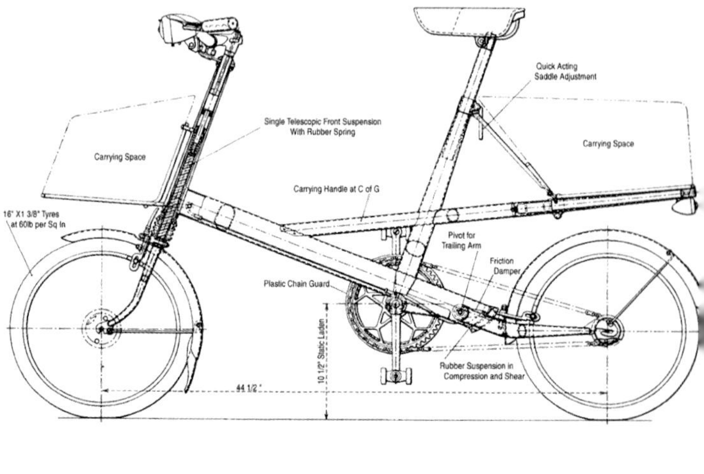


Right: Fig. 9.46. The 1962 Series I Moulton was available with six different gearing options: a single-speed coaster brake, a 2-speed F&S Duomatic, a 3-speed or 4-speed Sturmey-Archer hub gear, a 2 X 4 Cyclo/Sturmey-Archer hybrid, or a Cyclo P-2 derailleur.

perform research and development of rubber spring elements. His Hydrolastic suspension system was a vital part of the British Motor Corp. (BMC) mini-cars.

At the age of 35, he became interested in bicycle design. He became convinced that the basic design of the classic safety bicycle would be improved by using smaller wheels and high-pressure tires. The resulting harsh ride would be compensated by front and rear suspension. This was the basis of all subsequent Moulton bicycles. Moulton had a good working relationship with Dunlop, who agreed to develop a suitable high-pressure tire. A number of prototype Moultons were built and tested.

Moulton offered to license his final design to Raleigh, but after much delay, it was rejected at the last minute because of the "NIH (i.e., not invented here) factor." He then built a factory on the family estate in Bradford-on-Avon. The Series I Moulton was announced in November 1962.

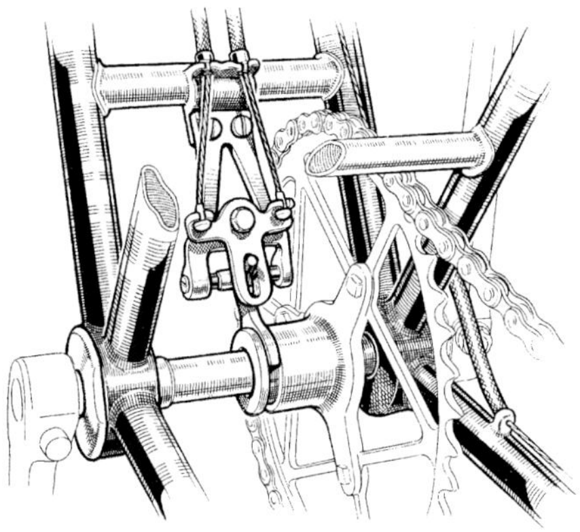

Above: Fig. 9.47. The 1956 Elswick-Hopper Vampire Convincible bicycle with a movable chainwheel was the last hurrah of the British obsession with chain alignment. After shifting on the rear, the chainwheel was moved sideways to line up with the freewheel. This was a prototype to cope with a very short chainstay length.

## Components

Of the 429 bicycles exhibited at the 1955 London Bicycle Show, 170 had derailleurs and 148 had 3- or 4-speed hubs. Only one of the 30 Raleigh bicycles had derailleurs. By 1955, except for Cyclo, all the (small) British derailleur companies had disappeared, as had happened in France with the smaller derailleur manufacturers.

The death of the British component companies in the 1950s and 1960s deserves some discussion. Why did the French and Italian companies survive so much longer? There were a number of reasons. There was a lower level of British public interest in professional bicycle racing. There

Right: Fig. 9.48. The 1956 British Cyclo Benelux Mark 12 rear derailleur automatically increased the chain tension on the smaller sprockets, which increased the total capacity to 23 teeth. Note how the single cable is attached to the pull chain and also loops around a drum to tension the take-up spring, avoiding the need for a second cable. There was an adjustment screw on the top pivot to adjust the chain gap.

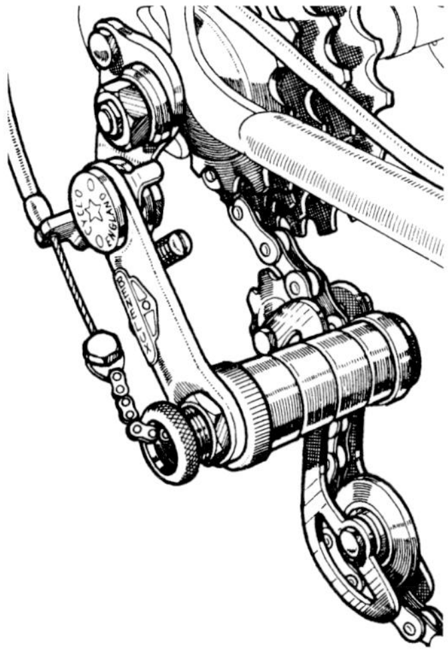

Above: Fig. 9.49. The 1957 British Cyclo Benelux Mark 7 was the Mark 12 without the chain-tensioner. It was a conventional plunger-type pull-chain rear derailleur. It was rated for a 13-24 freewheel and a four-tooth chainwheel difference, giving a total capacity of 15 teeth.

Below: Fig. 9.50. The1958 British Cyclo Benelux Mark 8 Tourist was a development of the 1954 Benelux Tourist, with a knurled adjuster for the pull chain. It was rated for a 13–28 freewheel and a 20 tooth chainwheel difference. By 1959, plunger-type pull-chain derailleurs were obsolete.

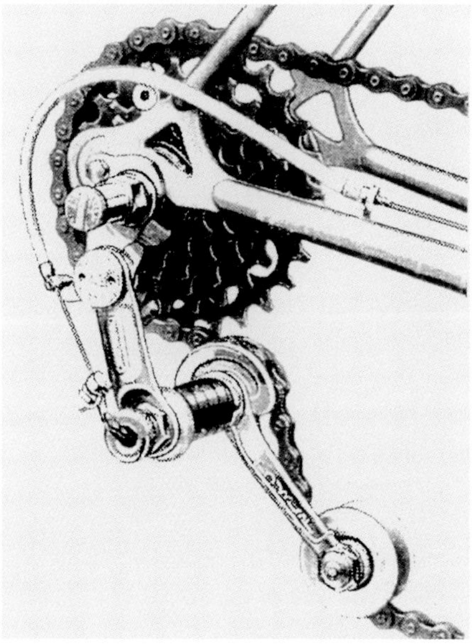

was no British equivalent of the Tour de France or the Giro d'Italia. From the 1910s to the 1950s, there was a long-standing fixation on chain alignment in Britain, which led to the preference for three-speed hubs. Raleigh had 75% of the British bicycle market, and since they owned Sturmey-Archer, they had a strong motive to encourage hub gears.

Eventually, market pressures prevailed and, starting in 1955, Raleigh catalogs showed derailleur-equipped bicycles using British Cyclo-Benelux derailleurs as an alternative to Sturmey-Archer hubs.

In my opinion, the dominance of TI-Raleigh was the overwhelming reason for the demise of the British component companies.

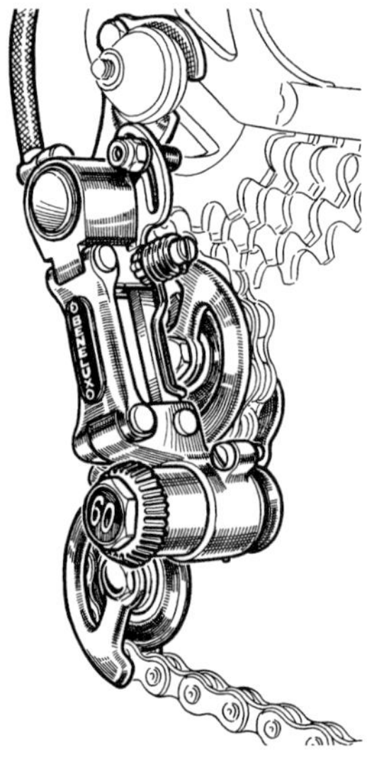

Above: Fig. 9.51. The 1960 British Cyclo Benelux Super 60 parallelogram rear derailleur was inspired by Campagnolo, but it had unique features, including adjustable spring tension for the parallelogram and an angle adjustment.

When the bicycle market declined in the mid-1960s, TI-Raleigh was the only significant component buyer, and they used their monopoly buying power to drive down prices. One by one, the British component companies went out of business. Dunlop stopped making bicycle tires rather than cave in to TI-Raleigh's price demands. Two British derailleur makers, British Cyclo, and Resilion, survived until 1955. By 1965, only British Cyclo remained.

### British Cyclo

By 1955, Louis Camillis largely owned the Cyclo Gear Company in Birmingham, which made freewheels, chainwheels, derailleurs, downtube and bar-end shifters, hubs, small parts and tools. They made the Cyclo Standard touring derailleur and various pull-chain racing rear derailleurs. They added parallelogram derailleurs in 1960.

Over the years, the Cyclo Gear Company called their derailleurs both Benelux and Cyclo. When French Cyclo switched emphasis to freewheels in 1962, the Cyclo Gear Company had been designing its own products for almost three decades. derailleur production in Birmingham suffered from foreign competition and ceased in the late 1960s.

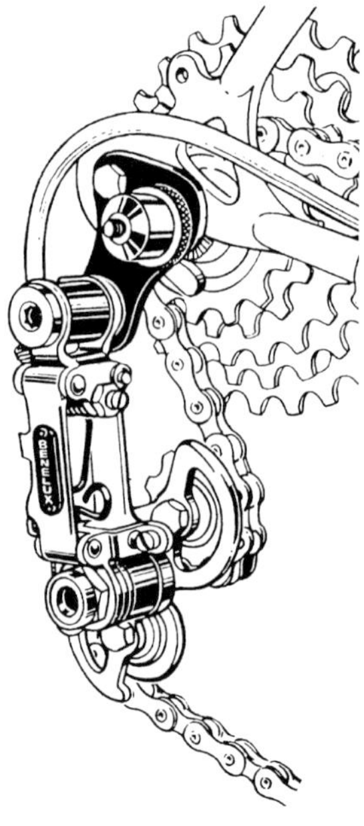

Right: Fig. 9.52. The1963 British Cyclo Benelux Sport P2 rear derailleur was an economy version of the Super 60. This was the last Benelux deraileur.

## Germany and Austria

By 1955, bicycles carried a low status in Germany. The 1956 guidelines for road building stated: "Mopeds, bicycles and pedestrians on the road are to be regarded as laterally moving obstacles that diminish the effective width of the road."

Simplex, Huret, and Campagnolo were the major suppliers of

Above: Fig. 9.53. The 1958 Resilion Crimson Star rear derailleur featured red nylon pulleys for quiet operation. Otherwise, it was a close copy of the British Cyclo Mark 7. This was the last Resilion deraileur.

gear-train components to the German and Austrian markets. Fichtel & Sachs (F&S) and Altenburger in West Germany, and Renak and Optima in East Germany were domestic competitors. F&S's main products were hub gears with backpedal coaster brakes, which was what the German market demanded. The F&S Duomatic two-speed hub shifted by back-pedaling briefly (whereas backpedaling further applied the brake). The F&S Automatic was a two-speed hub that automatically shifted into the higher gear when a certain pedaling speed was exceeded, and it too was combined with a back-pedaling brake. F&S made a small number of derailleurs, but they were usually combined with a F&S hub with a back-pedaling brake.

## Spain

Spain was isolated after the war. The two major Spanish bicycle companies, BH Cycles, in Vittoria, and Zeu,s in Vizcaya, made almost everything in-house. Zeus made a complete line of components. Their derailleurs were such faithful copies of Campagnolo models that even the parts (and part numbers) were interchangeable. Another Spanish company, Triplex, made poor-quality Campagnolo copies.

## United States

U.S. bicycle production was a little under 2 million from 1955 to 1958. It was about 2.5 million from 1959 to 1961, 3 million for 1962, and 4 million in 1964. Children's bicycles made up the majority of the U.S. bicycle market until the 1970s. Adult bicycles were a sideline.

Bicycle imports peaked at 1.2 million in 1955 and 1956. Most of these were middleweight (30-lb.) roadsters that were sold at auto supply stores and department stores. Raleigh made bicycles for Huffy and Western Auto. Hercules made bicycles for Roadmaster. Imports dropped to

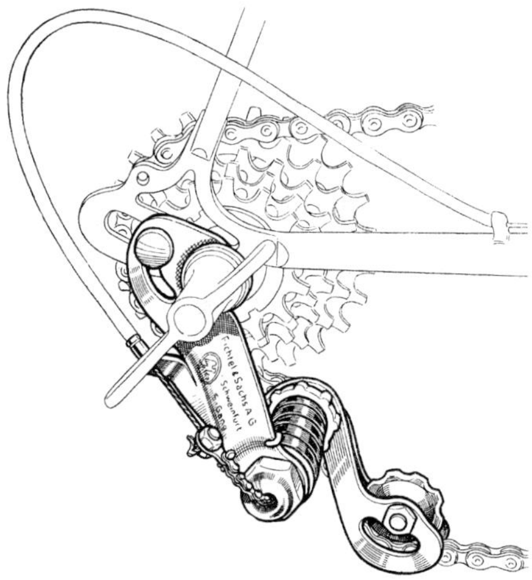

Right: Fig. 9.56. The 1955 Altenburger Champion rear derailleur had a horizontal parallelogram to maintain a constant chain gap. This was an effective design that was used by the German teams in the Tour de France. It was sold in Britain as the GB Altenburger.

Above: Fig. 9.54. The 1958 Fichtel & Sachs rear derailleur was similar to the Huret Competition, but the angle of the arm was adjustable. This Rebour drawing shows the 5-speed model used by the Swiss-German team in the 1958 Tour de France. They still used wing nuts instead of quick-releases on the wheel axles.

Right: Fig. 9.55. In 1958, The French Cycling Federation had this special bicycle made for U.S. President Eisenhower, whose physician, Dr. Paul Dudley White, had recommended cycling as a form of exercise.

700,000 in 1957 and slowly built back up to 1.3 million in 1963.

Bruce Epperson quotes a fascinating story that illustrates the "Law of Unintended Consequences": The Westfield Manufacturing Company, the last remnant of the Pope bicycle empire, was sold to the management for $4 million in 1961. It had a worn-out high-cost factory. Westfield's president Norman Clarke recalls the story:

> I needed all of the help that I could get. I got it a few months later. Western Auto, who ran a nationwide chain of auto parts stores, also sold appliances and other hard goods. They were the largest retailer of Raleigh bicycles in America.
>
> Western Auto was told by Raleigh that they would no longer sell them English lightweight geared bikes, 3-speeds and so on. Western Auto had catalogs out, prices out, everything ready for distribution and Raleigh pulled out, just like that.
>
> Western Auto approached Schwinn asking for replacements. Frank W. Schwinn, who had successfully pulled out of the made-to-brand mass-market after a twenty-year struggle, refused.

They next turned to Westfield, which Clarke was in the process of renaming the Columbia Manufacturing Company. Clarke knew from his own sources how much Western Auto was paying for Raleigh's bicycles. He also knew that he could not possibly match the price because Raleigh owned Sturmey-Archer, who made the bike's three-speed hubs and sold them to the parent company at below-market prices. However, he could make it—just barely—if he could use Shimano hub gears, which he could get for half the price of the English product.

> We were familiar with Shimano's product and had been testing them for a while. They were just as good.
>
> Appalled at the thought of using Japanese parts, Western Auto refused.
>
> Okay, a week goes by. Western Auto came down with their executives. They said that we must have 175,000 units to fulfill our existing commitments. We'll use them if you will stand behind them.
>
> I said we would. I hopped on a plane the next day to Japan.

He was greeted at the airport by a brass band, and spent the next three weeks being entertained quite fabulously. After meeting with 40 different manufacturers,

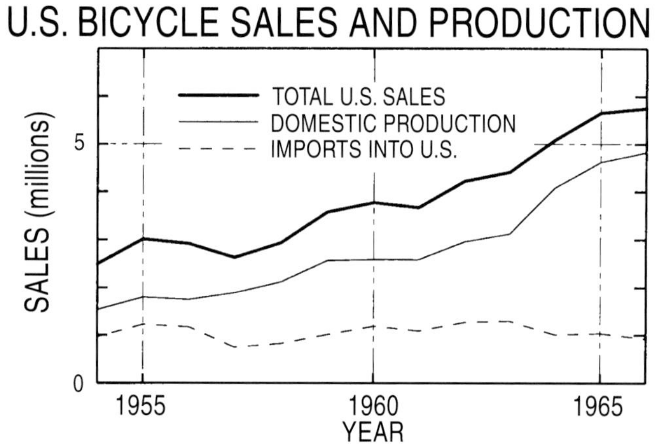


Above: Fig. 9.57. Total U.S. bicycle sales, domestic bicycle production, and bicycle imports, 1954–1965. (Source of data: *American Bicyclist and Motorcyclist*, December 1979.)

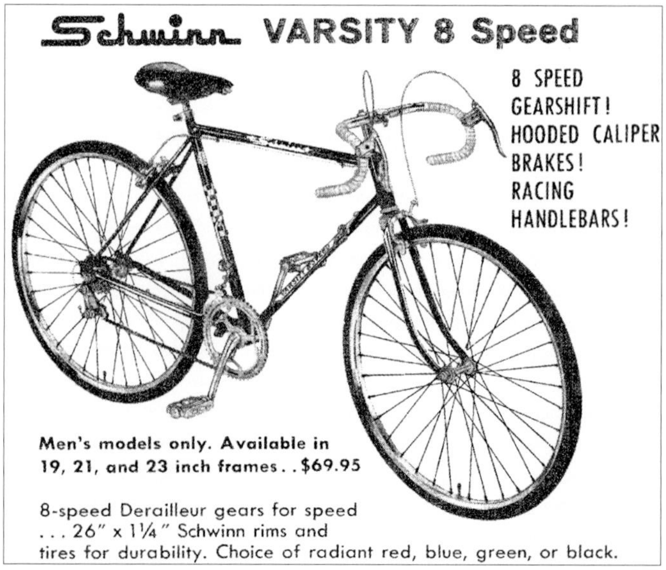


Right: Fig. 9.58. Introduced in 1960, the 8-speed Schwinn Varsity was the breakthrough derailleur bicycle in the U.S. It proved that American teenagers could handle derailleurs. This page from a 1960 Schwinn brochure shows the first Varsity, with a Simplex Tour de France rear derailleur and a Simplex Competition front derailleur.

Clarke left orders for some $9 million worth of parts—Shimano three-speed hubs, Honda chains, Araya rims, and so on. "We broke even, but it gave us volume and we needed that."

More importantly, Clarke's 1963 buying spree opened the American bicycle market to Asian component makers. Every one of the eight American bicycle makers went back the next year and started buying Japanese parts. In no time, they were selling $100 million worth of parts in the U.S.

In 1955, bicycle racing and recreational riding on lightweight derailleur bicycles were clustered in about a dozen cities. These cities could support one or two specialty bicycle stores that imported bicycles from Britain, France, or Italy. Boston, New York, Philadelphia, Chicago, Los Angeles, San Francisco, and San Diego had active bicycle clubs that sponsored club rides and club races.

A small number of enthusiasts, like Fred DeLong, Phyllis Harmon, Keith Kingbay, and Dr. Clifford Graves spread the gospel of lightweight geared bicycles for recreation and touring. Racing clubs thrived, especially in Los Angeles and San Francisco. America sent a cycling team to the 1960 Olympics in Rome.

Starting in 1960, there was a groundswell of opposition to the pollution and congestion caused by fast-growing automobile usage. The Department of the Interior offered $2 million for bicycle paths, to be matched by local funds. The American Youth Hostels (AYH) provided a bicycle touring program, which was not limited to young people. Year by year, more people embarked on cross-country bicycle tours.

Schwinn's president, Frank W. Schwinn, believed in lightweight bicycles. In the mid-1950s, he tested the market for derailleur bicycles by making 1,000 drop-handlebar 8-speed Continentals. They did not sell well, and the test was not a success, but they tried again in 1960 with the Varsity.

For the next few years, more Schwinn Varsities were sold in the U.S. than all other derailleur bikes combined. Schwinn added additional derailleur models. By 1964, Schwinn dealers carried six different derailleur bicycles, from the $70 Varsity to the $237 Campagnolo-equipped Paramount.

Schwinn's network of franchised dealers spread the Varsity's success, and soon 10-speed bicycles were found throughout America. This led the low-end mass-manufacturers like Huffy, Columbia, Murray, and Ross to introduce their own derailleur bicycles. Many of these bicycles were 3-speed middleweights, but by 1965, there was a considerable market for 10-speeds.

Keith Kingbay was the Service Manager for Schwinn in the company's glory days. In Kingbay's own words, this is the story of the birth of the Varsity and how it got to be equipped with Huret Allvit derailleurs:

> The Varsity was totally my baby. In 1959, I worked up a parts list and built a prototype and sold the old man (Frank W. Schwinn) on it. We agreed to go ahead with the project just to show me that I was wrong. Schwinn had made several unsuccessful attempts at derailleur bicycles previously, including a 2-speed on a New Departure coaster brake.

Right: Fig. 9.59. Someone installed a Huret Allvit derailleur and added a spoke protector on this 1960 Varsity. It would take another year before Schwinn offered the Varsity with the Huret Allvit derailleur.

The bike was developed to offset the small numbers of foreign derailleur bikes coming in, mainly through the West Coast and at quite high prices. This was at a time when we were just getting the dealers to accept hand brakes and three-speed hubs.

When I told Bill Stoeffhaas, the general manager and Number Two to F.W. Schwinn, his comment was, "I wonder how many of these SOBs we'll have to eat?"

Bob Wilson of Wilson Sales in Los Angeles called that morning and Bill told him what we were projecting. Bob agreed to buy the entire trial lot of 500 Varsities. Stoeffhaas called and told me to double the trial order. I talked him out of it, saying that we should tread carefully.

Some time later, the leaders of FREXA (the French component combine), came to Chicago to visit Schwinn. We were now a potent factor and all of the imported parts except the derailleurs were bought from FREXA. I approved all orders for imported parts so they visited me. The group consisted of Pierre Maillard, Roger and Jacques Huret, Dominick Le Bec and Jean Nollet of FREXA. We were using more derailleurs than had ever been imagined. After the usual niceties, Roger Huret, who spoke impeccable English, asked me why we did not use his Huret Allvit derailleur. I answered because it cost $1.37 more than the Simplex derailleur that we were using.

That night, I entertained the group at the famous Stockyard Inn, a truly American restaurant where steak was king. I suggested that they select the wine but was told that we should have an American meal with American drinks. I ordered dry Beefeater martinis on the rocks for everyone. After the third drink, Roger Huret leaned over and said, "Mr. Kingbay, we will equal Simplex' price."

The next morning, I told the old man and he insisted that I confirm the deal when everyone was sober. That was Schwinn's policy. That afternoon, I went to lunch with the Huret brothers and we worked out the arrangements. That agreement was the breakthrough for Huret in America, and Lucian Juy of Simplex never forgave me.

Right: Fig. 9.60. The 1964 Schwinn Super Deluxe Sting-Ray had a 2-speed coaster-brake rear hub.

Left: Fig. 9.61. The Schwinn Corvette, introduced in 1961, was a middleweight bicycle with a 5-speed Huret Allvit geartrain. It sold so poorly that some had to be retrofitted with coaster brakes. Some people argue that it was the first mountain bike.

The first 1960 Schwinn Varsity was an 8-speed with a Simplex Tour de France rear derailleur and a lever-operated Simplex front derailleur on the downtube. Most later models were 10-speeds

The Varsity was designed for the average inexperienced young American cyclist accustomed to a balloon-tire bike. It allowed them to get acquainted with true cycling. The bicycling business cannot survive on elitist cyclists. There were only a few elitists in the pre-Varsity days.

If you did not have a dealer nearby, you could buy a lightweight bicycle by mail-order. Gene Portuesi started the Cyclo-Pedia mail-order company in 1937. In 1964, he was selling British Dunelt 3-speeds (built by Raleigh) for $45. The Louison Bobet Pro Racer with a Reynolds 531 frame and a Campagnolo Record gruppo was $190.

In 1964, the few surviving members of the League of American Wheelmen (LAW) met in Chicago. They agreed to reorganize into a national organization. Keith Kingbay of Schwinn provided financial support and an office. Fred DeLong and Phyllis Harmon were early volunteers. Soon, the LAW had multiple chapters and a monthly magazine, and bicycling once again had an advocacy voice in the U.S.

The period ended with the "high-riser," a phenomenon that started in Southern California in 1962. Kids made "pig bikes" by adding a banana seat and longhorn handlebars to a 20-inch-wheel children's bike. Schwinn jumped on the idea and introduced the Sting-Ray in late 1963. They had sold 45,000 by the year's end. By 1964, the competition was also offering high-riser bicycles, and these other companies took a major share of the market.

## Japan

In Japan, there was a significant domestic market for bicycles for urban transportation. In 1959, about 20 million bicycles were in use in Japan, and 3 million bicycles were sold. Lightweight sport bicycles became popular in the mid-1950s, providing a market for derailleurs and freewheels. The flurry of demand for derailleur bicycles lasted only a few years, and then the Japanese sport bike buyers switched to three-speed hubs. There were about six Japanese derailleur makers in 1955. All but SunTour and Shimano disappeared during this decade.

### Maeda SunTour

By 1955, Maeda SunTour had 177 employees and sales of about 90 million yen ($250,000). In late 1956, the first SunTour rear derailleur, the 8.8.8. SunTour Wide, was made for Maeda by the Iwai Seisakusho company.

From the very beginning, there was confusion over the correct spelling of the company's brand name. Most of the early SunTour derailleurs were stamped "Sun-Tour" (one word with a capital "T" in the middle). Sometimes they were stamped "Suntour"

Above: Fig. 9.63. Maeda's first SunTour derailleur, the 8.8.8. Wide, introduced in 1956. It was a beefed-up copy of a French pull-chain plunger-type touring derailleur.

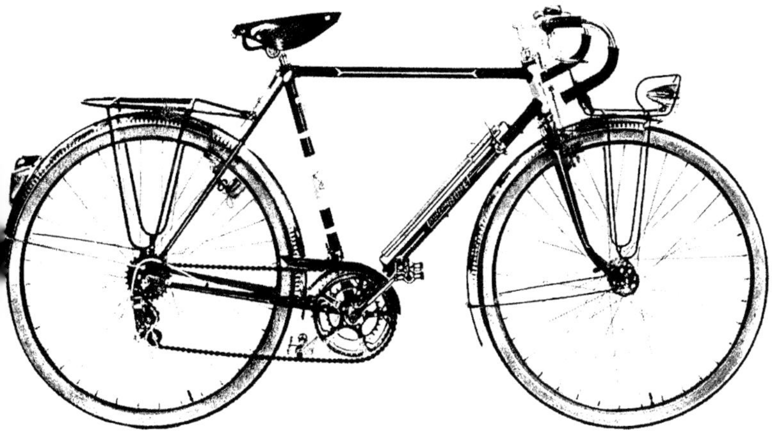

Left: Fig. 9.62. In 1964, mail-order company Cyclo-Pedia sold this fully equipped French Louison Bobet Cyclotouriste for $90. It had a 15-speed Huret Allvit gear-train.

(with a lower-case "t"), and sometimes there was a space between "Sun" and "Tour" (two words).

When Iwai Seisakusho went bankrupt in 1958, Maeda started making SunTour derailleurs in their own factory. Sales of mopeds boomed in the late 1950s, and Maeda supplied moped components. By 1960, Maeda had 156 employees and reported sales of 120 million yen ($330,000). From the middle 1960s to 1980, Maeda also sold SunTour three-speed hubs, which I've been told were Sturmey-Archer AW hubs with a SunTour logo on the shell.

Nobuo Ozaki, the head of product development at Maeda, invented the slant parallelogram rear derailleur in 1964 and Maeda obtained worldwide patents on the design. For the next 20 years, SunTour produced technically su-

Left: Fig. 9.64. The first SunTour Skitter of 1960 was an inexpensive stamped steel parallelogram rear derailleur similar to the Huret Svelto. Some say Huret made it.

Right: Fig. 9.66. The 1964 Grand Prix was SunTour's first slant parallelogram rear derailleur. The distance between the pulleys could be adjusted to match the freewheel.

Below: Fig. 9.65. Exploded view of the 1958 Shimano 3.3.3. three-speed hub, made using Shimano's cold-forging process. It was smaller, stronger, and lighter than the competition from Sturmey-Archer and Fichtel & Sach. It was the product with which Shimano broke into the U.S. market.

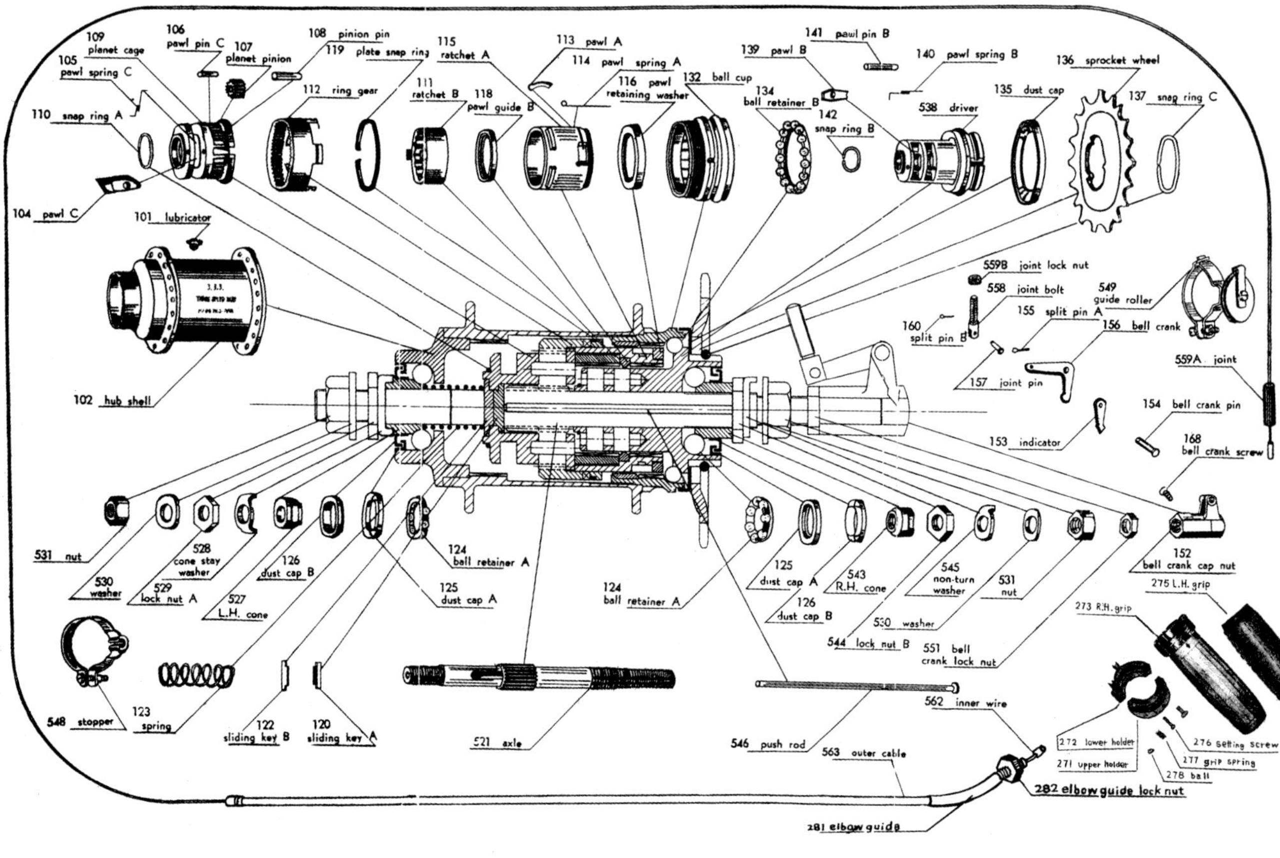

perior derailleurs. SunTour's slant parallelogram broke the image of Japanese components as being cheap copies of European components.

SunTour had good patent attorneys, and they vigorously defended the slant parallelogram design. All of the major derailleur makers incorporated slant parallelograms once the patent expired in 1984.

The timing of the 1964 Grand Prix rear derailleur was ideal because the bicycle market, both worldwide and in Japan, was turning from three-speed hub gears to derailleurs. The bicycle races at the 1964 Olympics in Tokyo provided great publicity. SunTour became Japan's largest derailleur manufacturer.

### Shimano

In 1956, Shimano made a pull-chain rear derailleur for two years until the Japanese market for sports bicycles switched back to three-speed hubs.

When the first Japanese derailleur "bubble" popped, Shimano had gone back to freewheels and hubs. In 1958, they developed an excellent 3-speed hub. Shimano built a new factory to mass-produce 3-speed hubs and aggressively sold them worldwide. Columbia was their first major U.S. customer. Shimano developed proficiency in cold forging aluminum alloys.

The Japanese market shifted back to derailleurs in 1965. The Japanese market went from three-speed hubs to derailleurs, back to three-speed hubs, and back to derailleurs within a single decade. No wonder Shimano and SunTour learned to react to market swings!

## Cycling Magazines

*Le Cycle* switched from twice monthly to monthly in 1960. Daniel Rebour became editor-in-chief in 1963. The *Northern California Cycling Association Newsletter* started publication in 1962. The name was changed to the *American Cycling Newsletter* in 1964. This was the first nationwide consumer-oriented bicycling magazine in the U.S. since the 1900s.

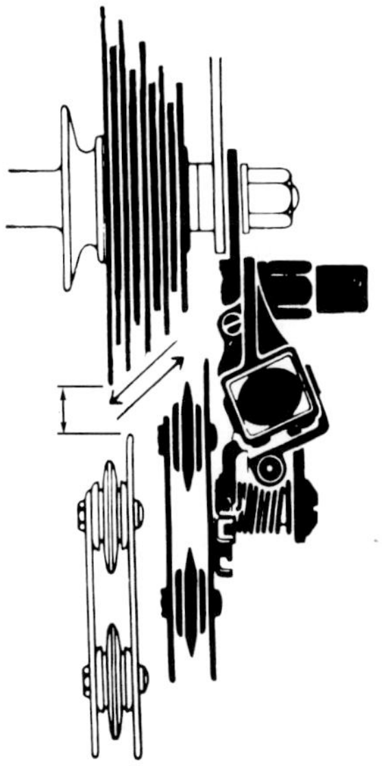

Left: Fig. 9.67. SunTour's 1964 slant parallelogram was a significant invention. The jockey pulley tracked the angle of the freewheel sprockets and maintained a nearly constant distance ("chain gap") between the jockey pulley and the sprockets.

Above: Fig. 9.68. The 1964 Shimano Archery parallelogram rear derailleur was an inexpensive copy of the Campagnolo Gran Sport.

Left: Fig. 9.69. The 1956 Shimano 3.3.3. plunger-type pull-chain rear derailleur was based on similar Simplex, Huret, and Cyclo models.

Right: Fig. 9.70. The origins of *Bicycling* magazine can be traced back to this humble publication, the *Newsletter* of the Northern California Cycling Association.

10.

# The "Great American Bike Boom": 1965–1974

THIS DECADE began with the bicycle industry in a slump and ended with what I refer to as the "Great American Bike Boom." At the end of the decade, America, rather than Europe, had become the biggest market for derailleur bicycles, and the largest makers of derailleur bicycles and components were in Asia instead of in Europe.

Even before the introduction of the Moulton in 1962, small-wheeled utility bicycles had become popular in Europe and saved the declining bicycle market. Sixteen different makers exhibited small-wheel folding bicycles at the Paris Show in 1966.

In 1965, Simplex and Huret in France, and Campagnolo in Italy were the dominant derailleur suppliers. Simplex "owned" the low-priced market and competed strongly in the mid-priced market. Huret had a major share of the mid-priced market, while Campagnolo dominated the top-quality market. A Campagnolo gruppo defined a quality bicycle.

Left: Fig. 10.1. This was the time of the "ten-speed," and this is the way some were advertised to the public.

In 1965, Shimano and SunTour were satisfying the Japanese derailleur demand and had started to seek export markets.

The Great American Bike Boom, from 1971 to 1973, totally changed the world's bicycle marketplace. The three major

European derailleur makers—Campagnolo, Huret, and Simplex—were unable to supply the bike boom demand. This allowed a second wave of makers to enter the market. Shimano and SunTour had expanded their factories and were quick to fill the void. Altenburger, Belleri, Gian Robert, Triplex, and Zeus picked up the leftovers. Shimano and SunTour became significant competitors in low-priced markets everywhere, and by 1972, Shimano claimed that they were the world's largest derailleur makers.

Ron Kitching and André Bertin had an agreement to use the name Milremo as a private-label for components that Kitching sold in Britain and Bertin sold in France. In the mid-1970s, Ron Kitching became the agent for SunTour components and Bertin became the agent for Shimano components. This was the beginning of the Japanese invasion of Europe.

Prices of bicycles and components increased dramatically in 1973 and 1974 because of the bicycle shortage and the inflation caused by the increased oil price.

## France

In 1965, Peugeot and Motobécane, the two largest French bicycle makers, were concentrating on mopeds and small motorcycles, and the French component makers were also concentrating on motorized two-wheelers. French moped sales went from 520,000 in 1953 to 1,200,000, in 1968, and this seemed to be the market trend.

Then the bicycle market turned around, and French bicycle production increased from about 900,000 in 1965 to about 2,400,000 in 1974. Domestic sales increased from 800,000 to 1,900,000, and exports increased from 100,000 to 850,000 over the same period. About 65 percent of the French exports were adult bicycles. French bicycle exports to the U.S. increased from 20,000 per year before the bike boom to a peak of 500,000 in 1973.

In the early 1970s, Simplex opened a new factory, and Huret doubled the capacity of their Picardy factories. Peugeot and Motobécane, the giants of the French industry, were churning out low-priced 10-speeds for export, and Bertin, Gitane, and Lejeune expanded their production facilities. Exchange rates for exports to the U.S. were more favorable for France than for Japan, but when the bike boom began in late 1971, the French bicycle makers and component companies were operating at nearly full capacity.

By 1972, the U.S. represented about 60 percent of the export market for lightweight bicycles. In April 1972, Daniel Rebour editorialized about the need to maintain the quality of French bicycle exports to the U.S. He warned against exactly what would happen over the next few years. Japan became noted for quality, and France became noted for fussy components and fragile bicycles, especially at the low end of the market.

In 1973, at the peak of the bike boom, Daniel Rebour again bewailed France's bicycle export problems. He was particularly

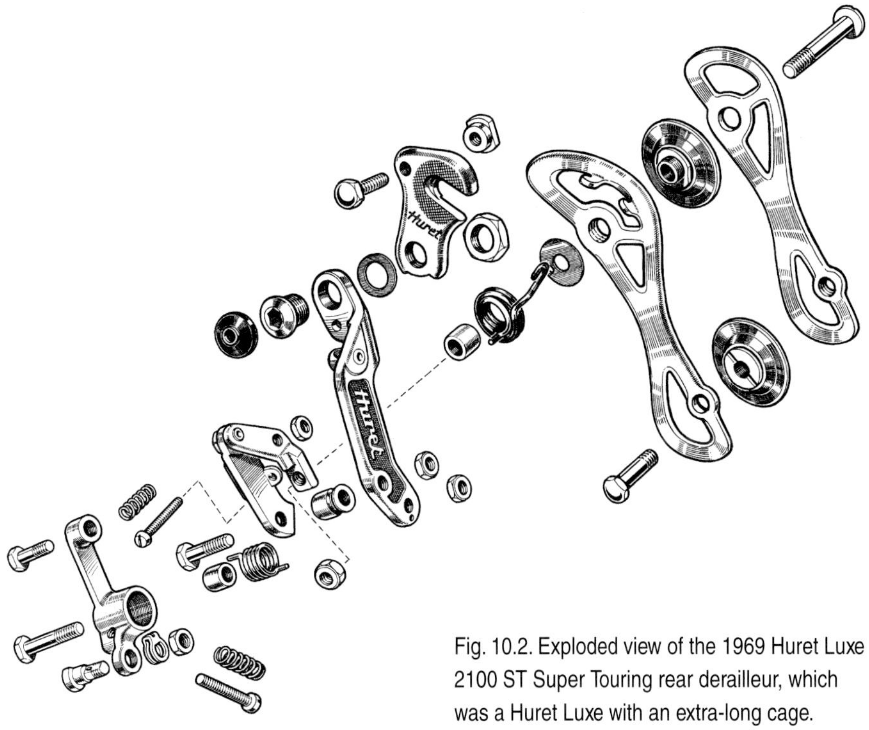


Fig. 10.2. Exploded view of the 1969 Huret Luxe 2100 ST Super Touring rear derailleur, which was a Huret Luxe with an extra-long cage.

concerned about French bicycles with Japanese components. When French bicycle makers could not get French components on time, they used Japanese components. Rebour feared that the loss of the French image would cause French bicycle exports to decline a few years later.

The French component industry realized that the bicycle makers wanted component gruppos rather than mismatched components from different small companies. In 1973, Atax, Huret, Idéale, Mafac, Nervar, Nervex, Pivo, Poutrait-Morin, Sedis, Simplex, Stronglight, Super Champion, and Velox formed the trade promotion association STC (Selection Tricolore Course).

There were two problems with French components in this decade. The first was that the French component makers, including Simplex and Huret, could not meet the rapidly growing demand. The second was much more serious. The equivalent Japanese components were cheaper, and they usually worked better.

## Huret

Huret entered the decade with the Allvit and Svelto rear derailleurs and the 600 front derailleur. These changed little over the decade. In 1965, Daniel Rebour reported that Huret had sold 5 million Allvits since its introduction in 1958. By 1965, Huret was producing more than 100,000 Allvits a month.

In 1967, Huret introduced the lower-priced Luxe rear derailleur and the 700 front derailleur and shift levers. The Luxe rear derailleur was a lighter, prettier version of the Allvit. It was rated for a 13 to 20 six-sprocket freewheel and 36/53 chainwheel. The 700 front derailleur had an enclosed spring.

In 1969, the Huret Allvit received new cosmetics and a new cage design, with two steps for chain tension adjustment instead of four.

By 1972, there were three versions of the Luxe: the narrow-range Luxe 2100C, the medium-range Luxe 2100T, and the wide-range Luxe Super Touring 2100ST. The 2100ST was not a good touring derailleur because its specifications were broader than its real capacity. It worked best with a narrow-range freewheel and a wide range between the chainwheels. The long-cage Luxes were too flimsy for wide-range freewheels.

In 1972, Huret introduced the elegant Jubilee front and rear de-

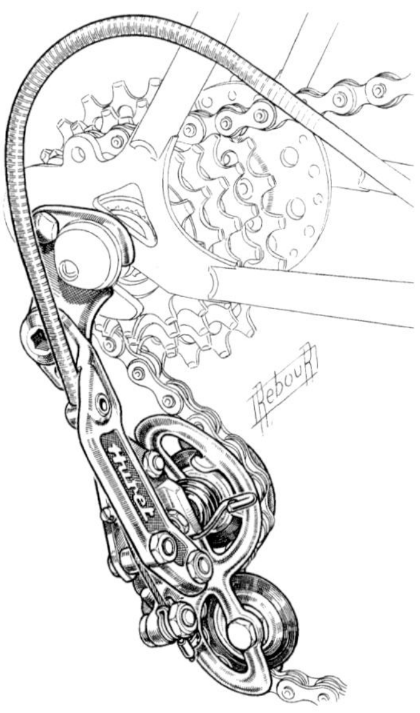

Above: Fig. 10.3. The 1967 Huret Luxe rear derailleur was a lighter, prettier version of the Allvit—without the shroud. It was rated for a 13–20 freewheel and 53/36 chainwheels.

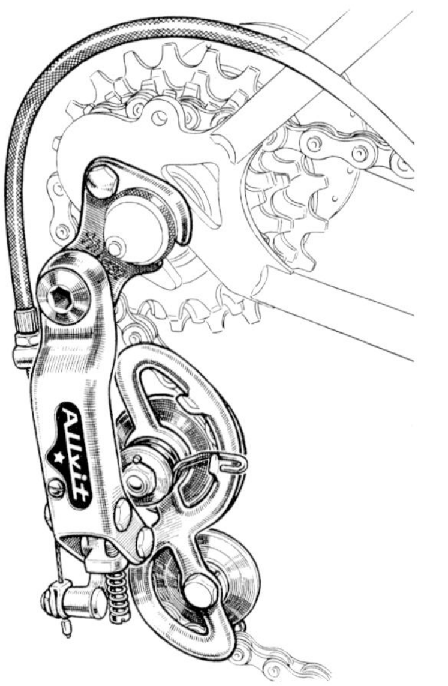


Right: Fig. 10.4. The 1969 Huret Allvit got new cosmetics and a new cage with two, instead of four, steps for chain-tension adjustment.

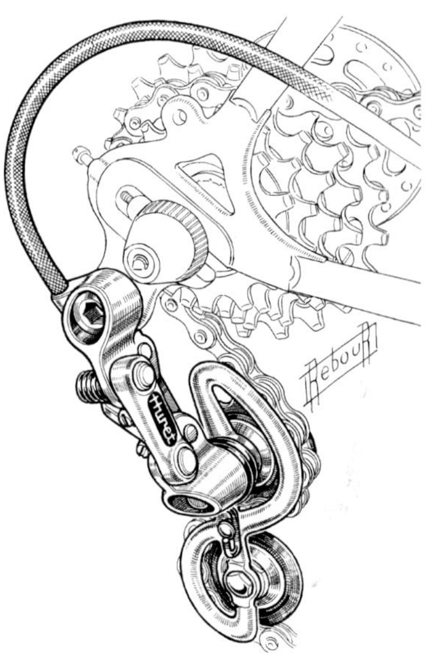

Above: Fig. 10.5. At 133 g (4 oz.) the 1972 Huret Jubilee was one of the lightest rear derailleurs ever made. It shifted quite well if you did not push its capacity.

Facing page: Fig. 10.6. The 1970 Huret range of derailleurs and shifters.

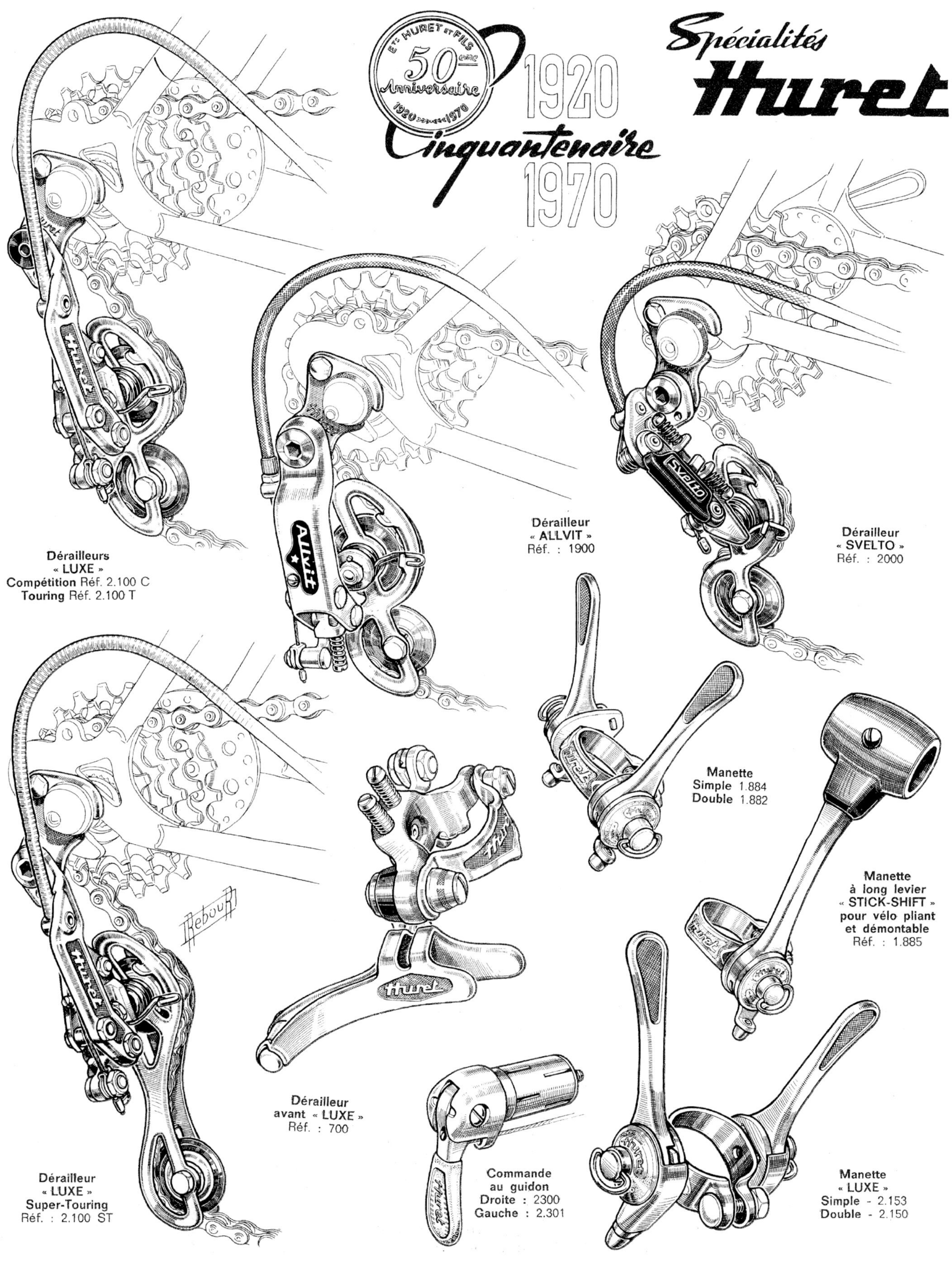
Spécialités
Huret
Ets HURET et FILS
50ème
Anniversaire
1920 1970
1920
Cinquantenaire
1970
Dérailleurs
« LUXE »
Compétition Réf. 2.100 C
Touring Réf. 2.100 T
Dérailleur
« ALLVIT »
Réf. : 1900
Dérailleur
« SVELTO »
Réf. : 2000
Manette
Simple 1.884
Double 1.882
Manette
à long levier
« STICK-SHIFT »
pour vélo pliant
et démontable
Réf. : 1.885
Dérailleur
avant « LUXE »
Réf. : 700
Dérailleur
« LUXE »
Super-Touring
Réf. : 2.100 ST
Commande
au guidon
Droite : 2300
Gauche : 2.301
Manette
« LUXE »
Simple - 2.153
Double - 2.150

railleurs. This was the "lightness" decade, when everything except water bottles and pumps was drilled out. At 133 g (4 oz.), the Huret Jubilee was one of the lightest rear derailleurs ever made. It shifted quite well if you did not push its capacity. The rear Jubilee was a conventional two-pulley parallelogram design. The true capacity limit was about a 24-tooth sprocket. Huret made a long-cage touring version of the Jubilee for a few years, but it was really too delicate. The final versions of the Jubilee had a drilled-out cage. Today, surviving Jubilees are prized by collectors.

There is a bit of controversy over the weight of the Jubilee rear derailleur. Huret listed weights from 133 to 148 g. The drilled-cage versions were actually heavier (it had to be beefed up to make up for the loss of strength due to the holes...). My Jubilee weighs 132 g, including the mounting bolt. The 80 g Huret Jubilee front derailleur was too light for its own good. You had to shift carefully or the cage would bend.

The 1973 Huret Super Allvit had a longer cage and it could wrap up lots of chain. It had a bit more capacity than the Touring Jubilee. However, by this time, the SunTour VGT was a much better touring derailleur for less money.

In 1974, Huret introduced the Challenger rear derailleur, which shared the cage and the geometry of the Jubilee but used a less expensive body and parallelogram. The cage pivot of the Challenger rear derailleur could be set in two positions—for either a 24-tooth or a 28-tooth large sprocket. The total capacity was about 30 teeth. Huret would use this stamped body for the next eight years. The 1982 Challenger II included an angle adjustment screw.

Another interesting derailleur introduced by Huret during in 1972 was the Duopar, which had a double parallelogram system to cover a large range with fairly

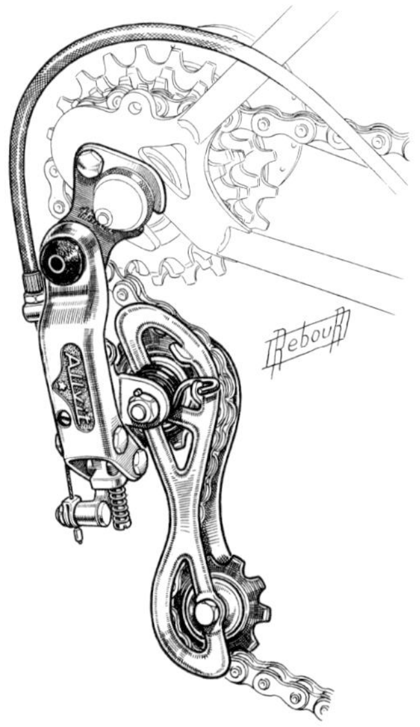

Above: Fig. 10.7. The 1973 Huret Super Allvit had a bit more capacity than the Touring Jubilee, and it could wrap up lots of chain.

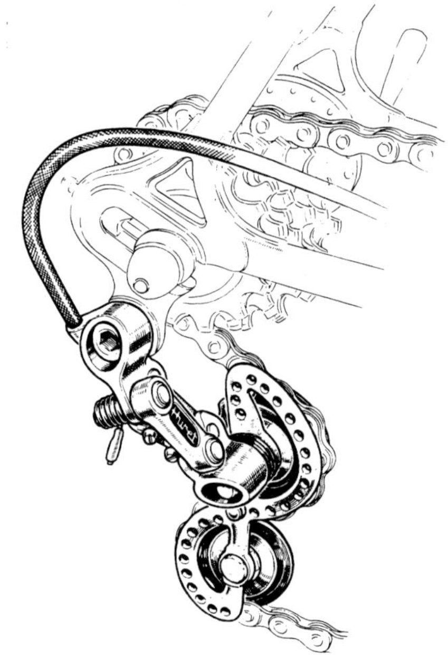

Right: Fig. 10.8. This late (1987) model Huret Jubilee had a drilled-out cage, but it had to be beefed-up to make up for the resulting loss of strength, ironically making it *heavier* than the non-drilled-out version.

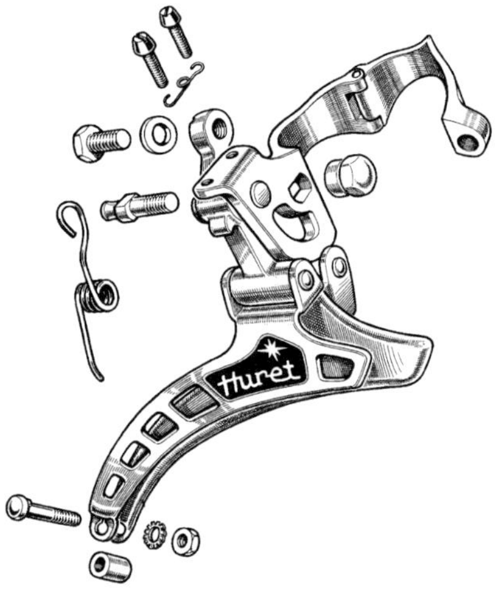


Above: Fig. 10.9. The 80 g (3-oz.) 1972 Huret Jubilee front derailleur was too light for its own good. You had to shift carefully or the cage would bend.

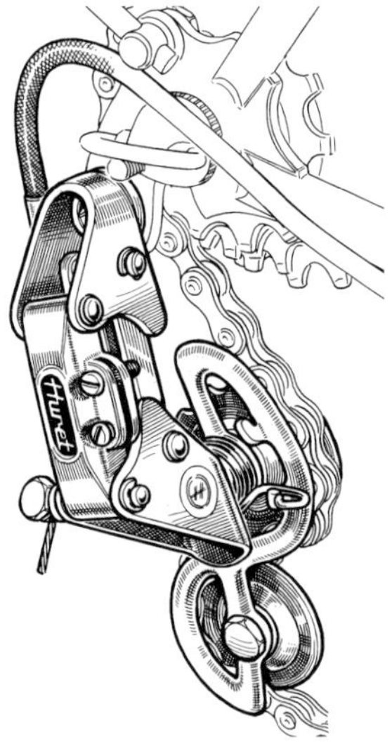

Above: Fig. 10.10. The cage pivot of the 1974 Huret Challenger rear derailleur could be set in two positions, for either a 24-tooth or a 28-tooth large sprocket. The total capacity was about 30 teeth.

constant chain gap. This would remain the best wide-range derailleur available until Shimano and SunTour introduced their mountain-bike specific derailleurs in the mid-1980s.

Jan Heine tested a Huret Duopar on three different French randonneur bicycles, and found that their performance very much depended on the hanger and the resulting chain gap. Mounted on a touring hanger, resulting in a large chain gap, shifting was "appalling" and required much overshifting.

Frank Berto's own experience has been much better, and he has successfully used Duopars on all his touring bikes until index shifting came along. Of course, he knew about the significance of chain gap and minimized it by selecting the appropriate hanger design.

By the 1970s, Huret and Simplex had abandoned much of the standardization that had benefitted the French bicycle industry in the 1940s and 1950s. Nowhere was this more obvious than in the derailleur mounting on the rear dropout. If you bought a custom-built French bicycle, the frame builder first had to decide on the derailleur, and then select the dropout to match.

A Huret Jubilee or a Challenger rear derailleur required the unique Huret rear dropout, which had the angle adjustment tab at

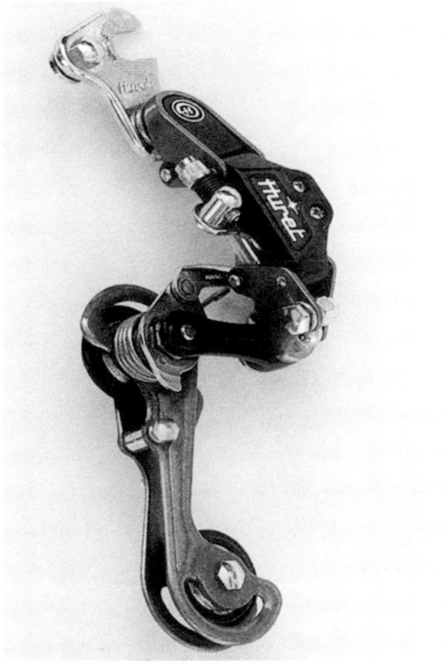

Above: Fig. 10.11. The first Huret Duopar rear derailleur, introduced in 1972, showing its double-parallelogram mechanism to handle ultra-wide-range gearing.

Above: Fig. 10.13. Huret's Jubilee shifters were as elegant, light, and drilled-out as the derailleurs themselves.

Left: Fig. 10.12. The 1972 Duopar front derailleur, which was identical to the Success.

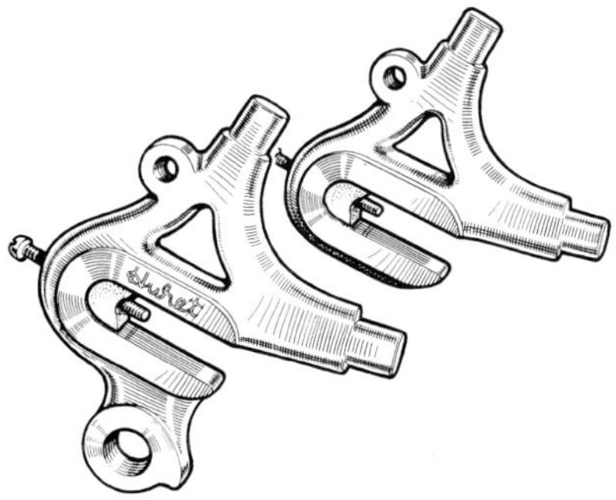

Above: Fig. 10.14. Huret's Jubilee and Challenger rear derailleurs required the unique Huret rear dropout. Notice the angle adjustment tab at the 3-o'clock position. If your bicycle had one of these "bastard" dropouts, you could only use a Huret derailleur.

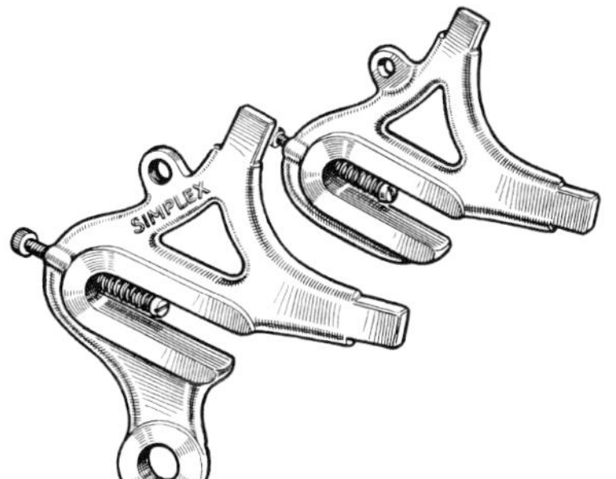


Above: Fig. 10.15. A Simplex rear derailleur required one of the unique Simplex rear dropouts. Notice the absence of an angle-adjustment tab. Some, but not all, Simplex dropouts had a non-threaded derailleur bolt hole.

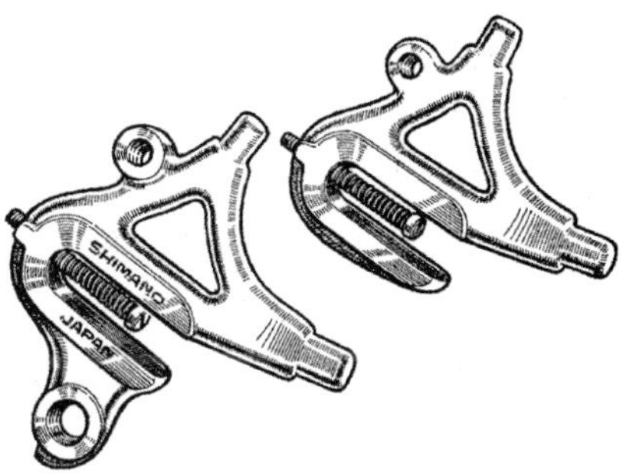


Above: Fig. 10.16. Shimano and everyone else used the standard Campagnolo-pattern rear dropout. Notice the angle adjustment tab at the 7 o'clock position.

the 3-o'clock position. If your bicycle had one of these "bastard" dropouts, you had to use a Huret derailleur.

A Simplex rear derailleur required one of the unique Simplex rear dropouts. Notice the lack of an angle adjustment tab. Some, but not all, Simplex dropouts had an non-threaded derailleur bolt hole

Shimano, SunTour, and everyone else had standardized on the Campagnolo dropout dimensions, which had become the *de facto* standard. The unique Huret and Simplex dropouts provided excellent reasons *not* to use their derailleurs.

Below: Figs. 10.17, 10.18, and 10.19. In 1965, Simplex had three rear derailleurs, from left to right: Prestige 537, Prestige Criterium, and Prestige Cadet. The Cadet was a cheap derailleur for 2- or 3-sprocket freewheels. With only one pulley, it could not wrap up much chain.

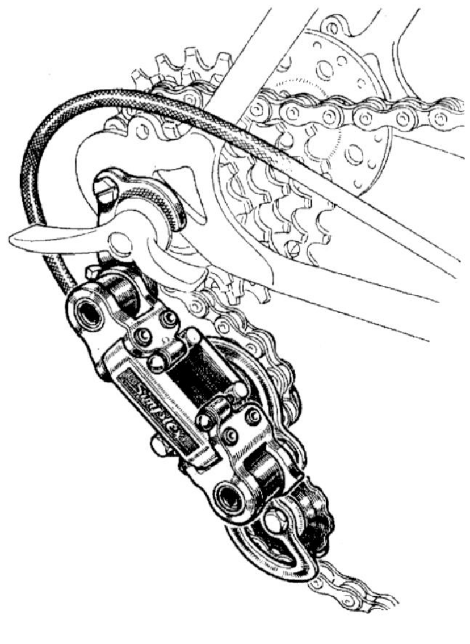

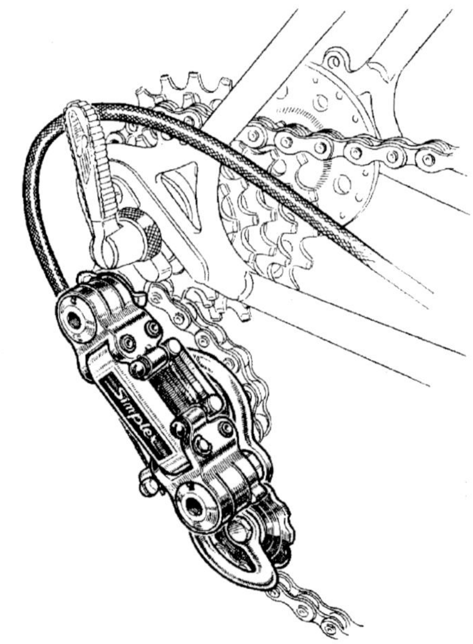

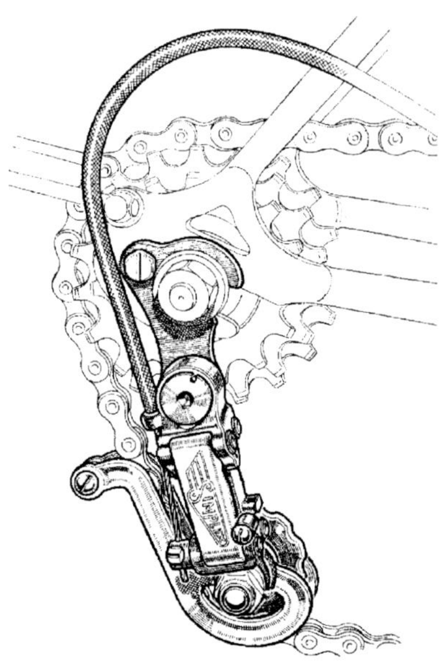

### Simplex

Simplex entered the decade with the Prestige front and rear derailleurs, with bodies and parallelograms made of Delrin plastic. They were really too flexible for reliable shifting. Simplex emphasized that Delrin could be twisted and would spring back without breaking. However, any twisting made the derailleur shift poorly, especially with wear. In 1965, Simplex changed the limit stops from knurled wheels inside the parallelogram to screws on the top and bottom housings.

In 1965, Simplex had three rear derailleurs: Prestige 537, Prestige Criterium, and Prestige Cadet. The Cadet was a cheap derailleur for 2- or 3-sprocket freewheels. With only one pulley, it could not wrap up much chain: its total capacity was only 6 teeth.

Over the years, Simplex made a series of changes to beef up the Prestige. In 1966, the Prestige-Criterium 537 rear derailleur got metal side plates to stiffen up the plastic parallelograms. By 1967, all Simplex rear derailleurs had metal side plates.

In early 1967, Kevin O'Donovan, of Carlton (a Raleigh subsidiary by

Above: Fig. 10.20. Kevin O'Donovan, of Carlton, using a crescent wrench to twist a plastic Simplex Prestige derailleur 90 degrees. Sure, it bent and didn't break, but unfortunately, he drew the wrong conclusion from this test.

then), and Frank Donnelly, of Raleigh, visited Lucien and Henry Juy at the Simplex factory in Dijon. Kevin O'Donovan uses a crescent wrench to twist a plastic Simplex Prestige derailleur 90°. It sprung back, and they thought that flexibility was a benefit, so Raleigh confirmed their decision to use Simplex derailleurs on lower-priced bicycles, formerly equipped with British Cyclo derailleurs.

Above: Fig. 10.21. In 1966, the Simplex Prestige-Criterium 537 rear derailleur got metal side plates to stiffen up the plastic parallelograms.

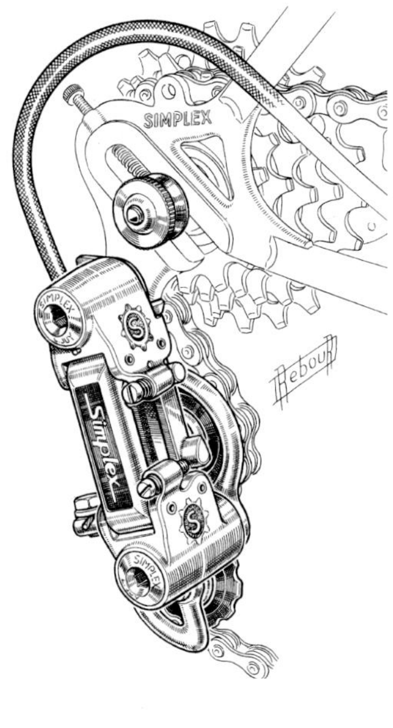


Below: Fig. 10.22. The 1973 Simplex Super LJ rear derailleur had alloy housings and parallelograms.

Simplex' inconsistent designations of model names and numbers have plagued bicycle historians anc collectors to this day. In 1965, the Criterium was the lower-priced model with a separate derailleur hanger. The Prestige P (for Professional) was a beefier, higher-priced model designed for mounting on a Simplex dropout. Both models were called 537. In 1966, the Prestige-Criterium became the better model.

The 1970 Simplex Super Competition parallelogram front derailleur was a good derailleur and clearly overdue. The Super Competition front derailleur only came with the expensive Simplex Criterium ensemble. The plunger-type Prestige front derailleurs lasted for five more years in the low-priced gruppos, giving an obsolete look and poorer front shifting to Simplex-equipped bicycles.

By 1970, there was some order to Simplex' names. Rear derailleurs and shift levers came in good (Criterium) and cheap (Prestige) versions. Front derailleurs came in good (Super Competition) and cheap (Criterium) versions. These were combined in three ensembles. The good rear derailleur, with good shift levers and good front derailleur made up the Super Competition ensemble. Good, good, and cheap made up the Criterium ensemble. Cheap, cheap, and cheap made up the Prestige ensemble. The Prestige rear derailleur was available with or without a derailleur hanger.

At the 1972 Paris show, Simplex introduced Super LJ front and rear derailleurs and shift levers. The Super LJ rear derailleur had an alloy housing and parallelogram. Simplex finally made a first-class "industrial strength" derailleur. They were expensive and shifted well, but Simplex had ruined its reputation with "cheap plastic." Consequently, as good as it was, most people who could afford an expensive bicycle opted for Campagnolo instead of Super LJ.

Above: Fig. 10.23. The 1970 Simplex Super Competition front derailleur was a good parallelogram-type front derailleur.

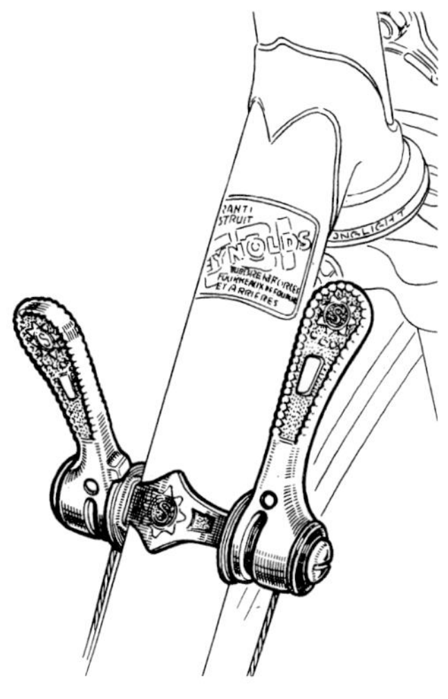

Right: Fig. 10.24. The 1973 Simplex Retrofriction shift levers used a ratcheting balancing spring.

Jan Heine tested a 1973 Simplex Super LJ on a 1999 "retro" Rivendell road bike. He found that the shifting performance varied with the freewheel and chain. Using a standard six-speed freewheel and chain, it shifted adequately, at best. However, with a more flexible Sedisport 7-speed chain it gave precise, silent shifts with some overshifting.

The 1973 Simplex Retrofriction shift levers used a ratcheting balancing spring, giving it a response similar to that of the SunTour Power-Ratchet. These worked so nicely that you often saw them on fully Campagnolo-equipped bikes. For touring bikes, Simplex also offered Retrofriction bar-end shifters.

In the 1940s and 1950s, the French bicycle trade had benefitted from the joint industry program that had standardized bicycle dimensions. This was the source of those "French threads" that plague today's bicycle restorers. By the 1970s, the French companies were operating in a vacuum. The Japanese had standardized on English threads and Campagnolo's key dimensions. These became the *de facto* industry standard. Only Huret and Simplex continued using their separate non-standard dimensions.

In 1973, Simplex moved from the original site in Dijon to a new factory in the industrial zone of St.-Appolinaire. Simplex had 200 employees in 1973. It was their bad fortune that the Prestige was the most popular derailleur on the low-end "garbage bikes" sold during the bike boom. Indeed, by 1975, a Simplex Prestige derailleur was the hallmark of a "garbage" 10-speed.

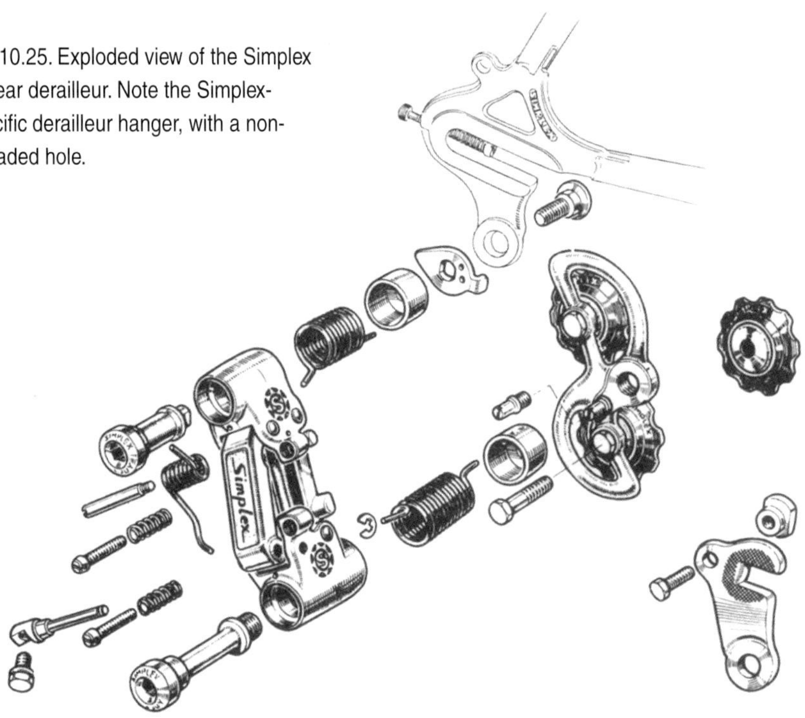


Fig. 10.25. Exploded view of the Simplex LJ rear derailleur. Note the Simplex-specific derailleur hanger, with a non-threaded hole.

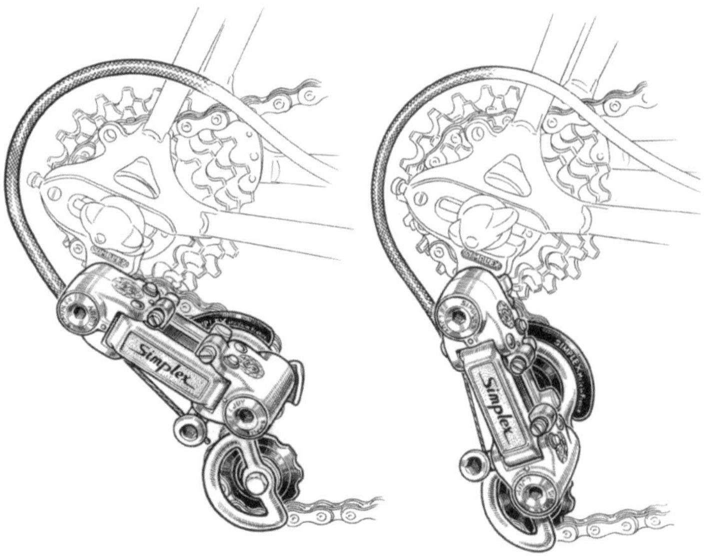


Left: Fig. 10.26 A and B. The Simplex LJ, as it was shifted in the front from the large (left) to the small chainring (right). The cage swings back and the jockey pulley remains in about the same position, keeping the chain gap constant.

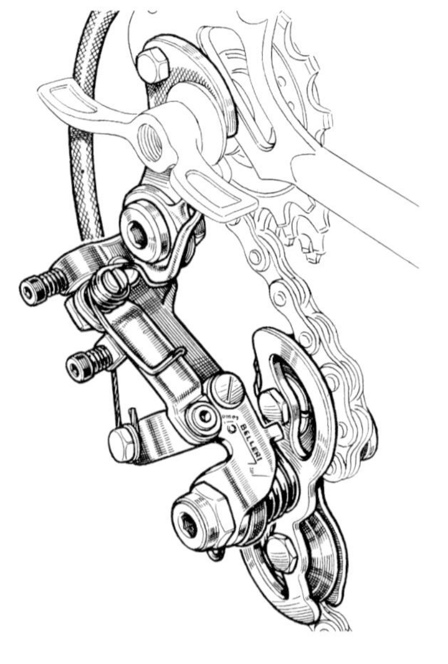

Above: Fig. 10.27. The 1972 Belleri rear derailleur was a copy of the cheap Campagnolo Velox.

### Belleri

The Belleri company made handlebars, stems, and an inexpensive line of stamped steel derailleurs. Often their products were labeled "Belri," rather than "Belleri." They disappeared around 1974.

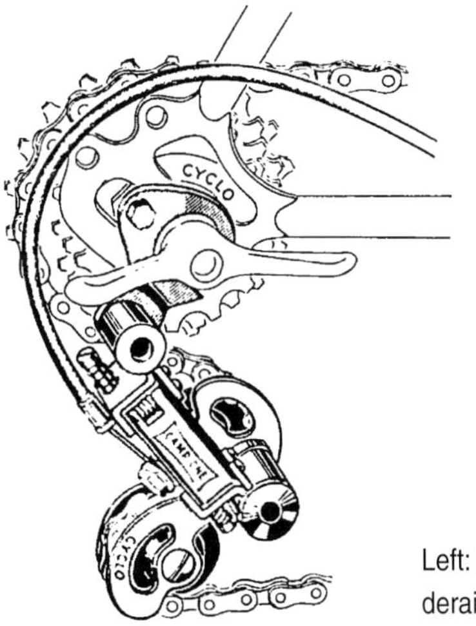


### Cyclo

French Cyclo stopped making the classic Standard touring derailleur in the early 1960s. They showed derailleurs in their catalog as late as 1975. The Campione and the Witmy were apparently Gian Robert derailleurs with Cyclo nameplates. The Campione rear derailleur was made of alloy and the Witmy was made of polyamide plastic. The two derailleurs were different, but Cyclo used the same advertising plate block with a different name on the parallelogram.

Left: Fig.10.28. The 1974/1975 French Cyclo Campione rear derailleur was made of aluminum alloy, while the similar Witmy was made of polyamide plastic.

## Bicycle Racing

In 1965, the typical Tour de France bicycle weighed around 10 kg (22 lbs.). The big chainwheel grew from 52 to 53 or 54 teeth. Most bicycles used a six-sprocket freewheel with a 13-tooth small sprocket. 170 and 172.5 mm

Left: Fig. 10.29. Eddy Merckx won the Tour five times in this decade. By 1965, racers used Campagnolo almost exclusively.

Above: Fig. 10.30. Eddy Merckx, photographed at a 1972 bicycle trade show with Daniel Rebour, who is holding a Merckx-Kessels frame made of ultra-light Reynolds 753 tubing.

*Table 10.1. Tour de France winners, 1965–1974*

| Year | Rider (Country) | Bicycle | Derailleur |
|---|---|---|---|
| 1965 | Felice Gimondi (Italy) | Magni | Campagnolo |
| 1966 | Lucien Aimar (France) | Geminiani | Campagnolo |
| 1967 | Roger Pingeon (France) | Peugeot | Simplex |
| 1968 | Jan Janssen (Holland) | Lejeune | Campagnolo |
| 1969 | Eddy Merckx (Belgium) | Merckx (Kessels) | Campagnolo |
| 1970 | Eddy Merckx (Belgium) | Merckx (Kessels) | Campagnolo |
| 1971 | Eddy Merckx (Belgium) | Merckx (Kessels) | Campagnolo |
| 1972 | Eddy Merckx (Belgium) | Merckx (Kessels) | Campagnolo |
| 1973 | Luis Ocaña (Spain) | Motobécane | Campagnolo |
| 1974 | Eddy Merckx (Belgium) | Merckx (Kessels) | Campagnolo |

cranks were used for the level stages, and 175 and 180 mm cranks in the mountains.

By 1965, Campagnolo was the equipment of choice for professional racers. They would use Huret or Simplex only if the team were sponsored by those manufacturers.

### The Hour Record

The Union Cycliste Internationale (UCI) was the governing body for the one-hour record. They maintained firm standards for the classic track bicycle. The records from the turn of the century to 1972 reflected the athletic prowess of the rider, rather than the skill of the aerodynamic designer. The one-hour record crept up very slowly. In 1967, Ferdinand Bracke raised it to 48.09 km (29.9 miles). In 1972, Eddy Merckx set a new record at 49.43 km (30.7 miles).

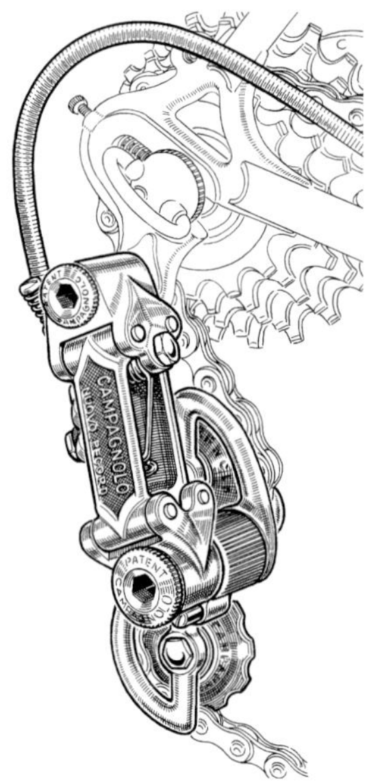

Above: Fig. 10.31. The 1967 Campagnolo Nuovo Record rear derailleur was made of cold-forged aluminum alloy.

This was the last "classic" hour record. After this, the riders and designers developed riding positions and aerodynamic changes to skirt the UCI rules and improve their performances.

## Italy

Italian bicycle production went from 900,000 bicycles in 1965 to 1,6 million bicycles in 1968. The domestic Italian market consisted of a large number of transportation bicycles and a smaller number of lightweight pure racing bicycles. There were almost no touring bicycles, and motor scooters were taking over as two-wheeled transportation due to the tax laws, which were different from those in France, where they benefitted mopeds more.

Right: Fig. 10.32. The Campagnolo Valentino Extra rear derailleur was made from 1969 until 1985. The cage shape was changed in the version sold during the last few years. This illustration is from the 1974 catalog.

Below: Fig. 10.33. The 1974 Campagnolo Nuovo Gran Sport was a homely alloy-and-steel copy of the Nuovo Record, and it shifted the same.

### Campagnolo

For the first four years of this period, Campagnolo continued with the Record, Valentino Super, and the Nuevo Sport models. In the fall of 1967, Campagnolo introduced the Nuovo Record rear derailleur. It was made of cold-forged aluminum alloy, which resulted in a significant weight-saving compared with the bronze Record. The cosmetics were also improved The Nuovo Record was made without change for fifteen years and it set the standard for derailleur quality and reliability.

The Nuovo Record did not shift precisely, but users got accustomed to it. Indeed, many books described the Campagnolo overshift-and-recenter drill as the standard shifting technique.

Facing page: Fig. 10.34. This page from a 1971 issue of *Le Cycle* shows the complete Campagnolo Record gruppo, except for the brakes and the headset. It shows the old Record (1020) rear derailleur. The Nuovo Record (1020/a) rear derailleur had been available for two years, but the French distributor used his old Rebour printing plate.

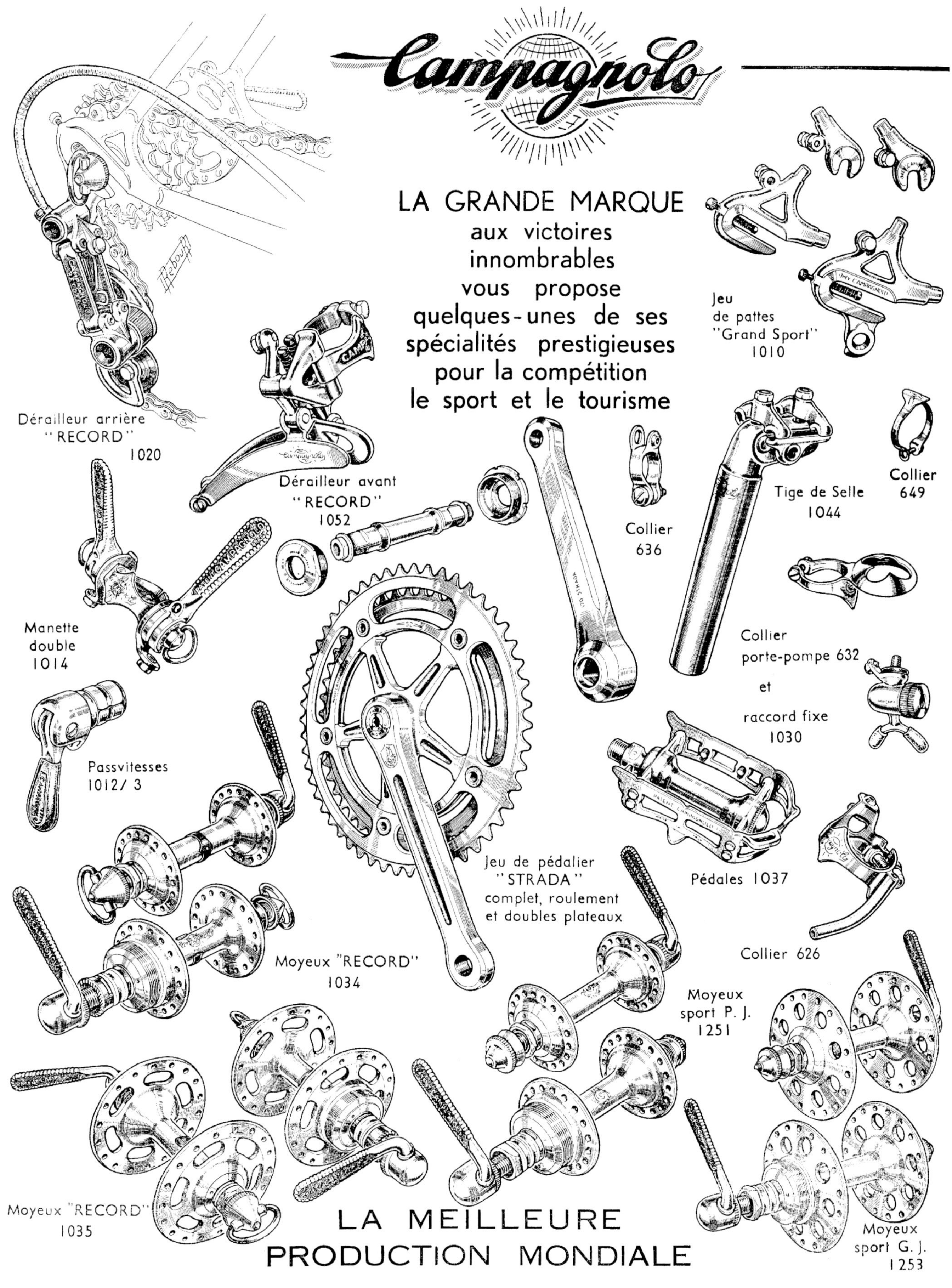
Campagnolo
LA GRANDE MARQUE
aux victoires
innombrables
vous propose
quelques-unes de ses
spécialités prestigieuses
pour la compétition
le sport et le tourisme
Jeu
de pattes
"Grand Sport"
1010
Dérailleur arrière
"RECORD"
1020
Dérailleur avant
"RECORD"
1052
Collier
636
Tige de Selle
1044
Collier
649
Manette
double
1014
Collier
porte-pompe 632
et
raccord fixe
1030
Passvitesses
1012/ 3
Pédales 1037
Jeu de pédalier
"STRADA"
complet, roulement
et doubles plateaux
Collier 626
Moyeux "RECORD"
1034
Moyeux
sport P. J.
1251
Moyeux "RECORD"
1035
LA MEILLEURE
PRODUCTION MONDIALE
Moyeux
sport G. J.
1253

If imitation is the sincerest form of flattery, then the Nuovo Record set the record. Over the next fifteen years, it was copied by dozens of makers. Many buyers preferred a pretty Nuovo Record copy to a lower-priced Campagnolo rear derailleur.

Jan Heine recently tested a 1968 Campagnolo Nuevo Record on a 1972 Cinelli road bike with 52/42 chainwheels and a 13–22 tooth six-speed freewheel. He found it shifted well on the larger sprockets, but as the chain gap increases on the smaller sprockets, it took much overshifting to engage those higher gears. It was reluctant to shift onto the 13-tooth sprocket. My ow experience with the Nuovo Record has been very much the same.

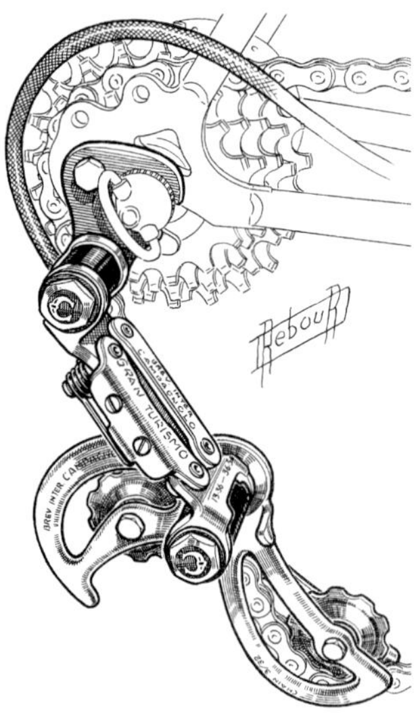

Above: Fig. 10.35. The 1970 Campagnolo Gran Turismo rear derailleur had bad cage geometry, and it shifted terribly, but it was so strong that when the chain jammed, it bent the bike rather than the derailleur.

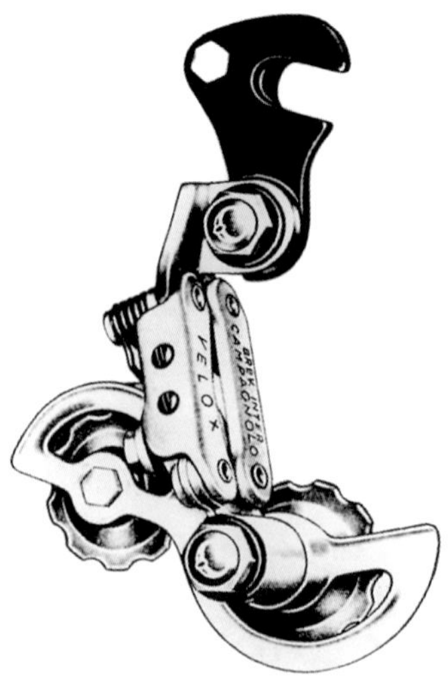


Right: Fig. 10.36. The 1972 Campagnolo Velox rear derailleur had an undersized parallelogram.

The 1969 Campagnolo Valentino Extra rear derailleur used the stamped steel parallelogram of the Velox. Valentino derailleurs looked cheap and they did not shift as well as the Nuovo Record.

Campagnolo expanded their focus to include touring derailleurs. At the 1970 Paris show, they introduced the Gran Turismo touring rear derailleur. The Gran Turismo had bad cage geometry and it shifted terribly, but it was so strong that when the chain jammed, it bent the bike rather than the derailleur. It was arguably the worst rear derailleur to carry Campagnolo's logo. In 1975, I wrote the truth about it. The Gran Turismo was my litmus test for bicycle writers. If a writer praised it, it meant that either he had never used it or he was lying, and the rest of his opinions were suspect.

The 1972 Campagnolo Velox rear derailleur had an undersized parallelogram. In 1974, the Nuovo Gran Sport replaced the Velox.

In 1974, the Rally touring rear derailleur replaced the Gran Turismo. It was similar to the Shimano Titlist. It had two spring-loaded pivots, and the key dimensions were nearly identical. I became unpopular with the Campagnolo fraternity when I reported my test results. The ten-dollar SunTour VGT shifted better than the more expensive Shimano Titlist or Campagnolo Rally.

Jan Heine tested a Campagnolo Rally 3450 on a 1990 Mercian bicycle with a five-sprocket 14–28 Regina freewheel and standard chain—the kind of components likely to be used in 1974. It shifted very poorly, the worst of any derailleur he has tested so far. It required overshifting and still missed many shifts. This agrees with my own experience. But, interestingly, it shifted surprisingly well

Above: Fig. 10.37. The Campagnolo Rally touring rear derailleur, 1974–1982, was similar to the Shimano Titlist. It had two spring-loaded pivots, and the key dimensions were nearly identical.

when used with a 7-sprocket 13–26 Dura-Ace freewheel and a narrow Sedisport chain.

Dating Campagnolo derailleurs can be difficult. Campagnolo and its distributors were frugal with both catalogs and advertising plate blocks. They would often run an advertisement that showed obsolete items. fig. 10.30 shows an advertisement from a 1971 issue of *Le Cycle*, showing the complete Campagnolo Record gruppo except for the brakes and the headset. It shows the old Record (1020) rear derailleur. The Nuovo Record (1020/a) rear derailleur had been available for two years, but the French distributor used his old Rebour plate. Or perhaps he still had a warehouse full of old Records.

### Gian Robert

Gian Robert made copies of the Campagnolo Gran Sport in the 1960s. By 1970, they were making copies of the Simplex Prestige in both alloy and plastic, either Delrin, polypropylene, or polyamide, depending on which catalog you read.

## Great Britain

There was an economic recovery in Britain in the early 1960s. People could afford to buy mopeds and motorcycles and cars—and gasoline to run them. Riding a bicycle for transportation became

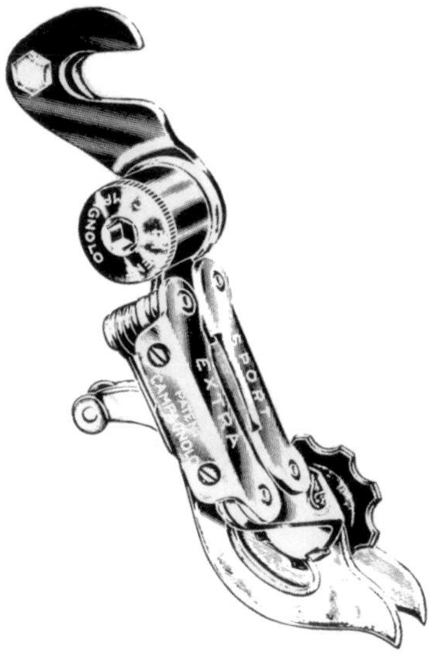

Above: Fig. 10.39. The 1967 Valentino front derailleur.

Left: Fig. 10.38. Campagnolo made single-pulley derailleurs from 1953 to 1969. This is the 1969 Sport Extra.

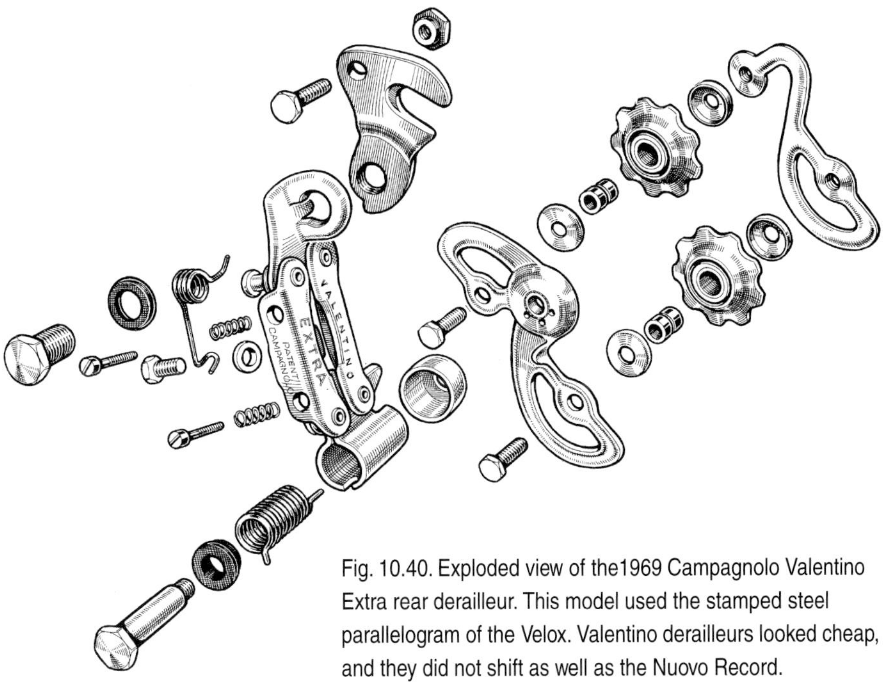

Fig. 10.40. Exploded view of the1969 Campagnolo Valentino Extra rear derailleur. This model used the stamped steel parallelogram of the Velox. Valentino derailleurs looked cheap, and they did not shift as well as the Nuovo Record.

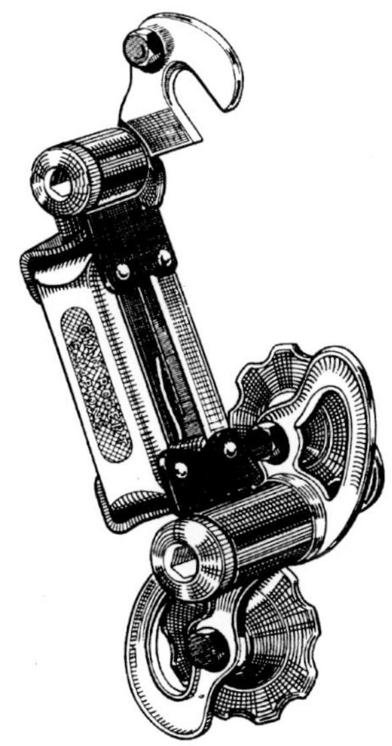

Fig. 10.41. The 1970 Gian Robert Campione Sport rear derailleur was a Simplex copy.

the mark of someone who could not afford anything better.

Starting in 1962, the Moulton small-wheel bicycle sparked a revival of interest in recreational cycling and in short-range urban riding. Raleigh, Dawes, Royal Enfield, and many others made small-wheel bikes—but without suspension. The 1965 Dawes Kingpin was one of the better Moulton copies. It used a Sturmey-Archer three-speed hub. These small-wheel bikes raised domestic sales from 730,000 in 1965 to 1.1 million in 1975.

Total British bicycle production (domestic and export) dropped from 1.9 million in 1965 to 1.6 million in 1970. Aided by the Great American Bike boom, it was back up to 2 million in 1975. Exports were 950,000 in 1965, 1 million in 1970, and 1.1 million in 1975.

### TI-Raleigh

The 1960 takeover by Tube Investments gave TI-Raleigh 75% of the British bicycle market—but it was a declining market. Raleigh had turned down the opportunity to make the Moulton in the early 1960s. When small-wheel bicycles caught hold, Raleigh produced their own small-wheeled bikes: the RSW-16, with 16-in. wheels; the RSW-14, with 14-inch wheels; and finally the Twenty with 20-inch wheels. The small-wheel Raleighs were heavy, and

Above: Fig. 10.42. The 1965 Dawes Kingpin was one of the better Moulton copies from the mid-1960s, though without suspension. It was equipped with a Sturmey-Archer three-speed hub.

Above: Fig. 10.43. Introduced in 1965, the folding Raleigh RSW-16 Compact copied the Moulton look, but it lacked Moulton's easy-rolling high-pressure tires and suspension.

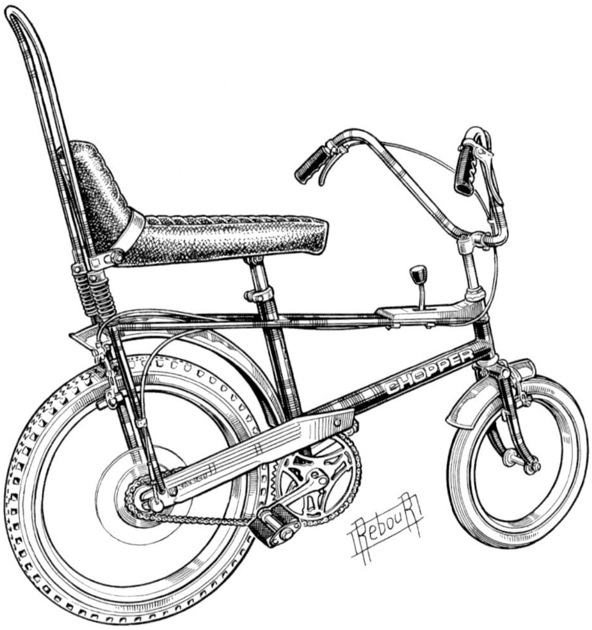


Right: Fig. 10.44. The 1970 Raleigh Chopper looked like a drag racer, and it usually came with a Sturmey-Archer three-speed hub operated by a stick shift installed on the top tube.

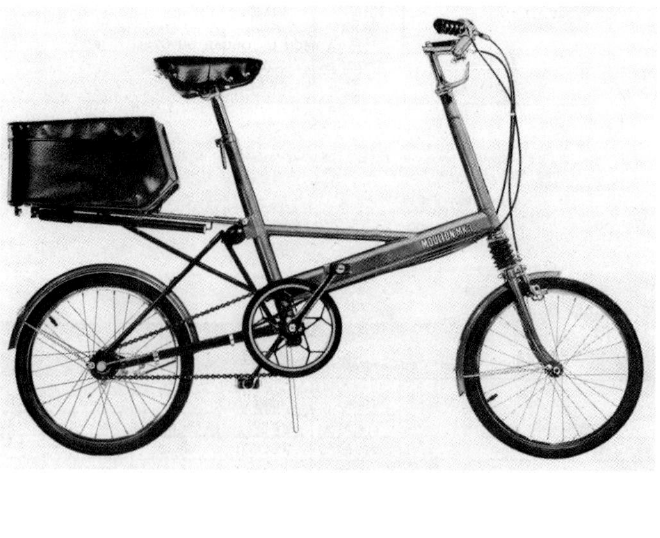

Left: Fig. 10.45. The 1970 Raleigh Moulton III had a three-speed Sturmey-Archer hub.

they used low-pressure balloon tires, but they were inexpensive and they sold well. Raleigh's UK sales went from 280,000 in 1965 to 530,000 in 1974.

In 1967, Raleigh bought the rights to produce the Moulton bicycle. Alex Moulton continued as a consultant. This resulted in the 1970 Moulton III. Raleigh made Moulton suspension bikes and non-suspension RSW-16s and RSW-20s.

The Raleigh Chopper outsold them all. Raleigh had moved too slowly, and missed most of the high-riser fad in America. However, in 1970, using a frame that owed a lot to the RSW-16, they produced the Chopper, which looked like a drag racer. It usually came with a Sturmey-Archer three-speed hub and a stick shifter. It was a huge seller, and over the next 10 years, 1.5 million were sold in Britain and on the continent. It established that there *was* a market for an expensive children's bicycle.

However, Raleigh moved too slowly, and consequently missed the emerging high-rise fad in America, though Raleigh had an extensive dealer network in the USA. For five years, high-rise bikes made up three-quarters of the U.S. market, but Raleigh dealers had nothing to sell. One wonders whether Raleigh USA failed to send a message or TI-Raleigh wasn't listening.

Raleigh was a vertically integrated company. They made most of their components in-house, often with unique Raleigh threads and dimensions. Everything that was not made in-house was bought from low bidders. By the end of this decade, the TI-Raleigh monolith had laid waste to the surviving innovative British bicycle and component makers.

Something similar happened to British motorcycles, British cars and, alas, even to British Meccano (something similar to what Americans know as an Erector-Set—but technically more sophisticated). Every "Meccano boy" dreamed of owning a large enough Meccano set to build Supermodel No. 4, the Giant Block-Setting Crane. In 1979, when Meccano went bankrupt, I purchased a second-hand 1927 No. 7 set, the one of my boyhood dreams, and built the Giant Block-Setting Crane.

Right: Fig. 10.46. The 1965 Raleigh Record for the British market offered a choice of four different gears: 5- or 10-speed Cyclo Benelux P2, and 5- or 10-speed Campagnolo Gran Sport.

Left: Fig. 10.47. The 1965 Raleigh Record for export to the U.S. came with Huret Allvit derailleurs and "Alpine" gearing, i.e., 52/40 chainwheels and a 14–28 freewheel.

Right: Fig. 10.48. Author Frank Berto's pride: The Giant Block-Setting Crane by Meccano—another British company that bit the dust during this period.

The Meccano system has proven useful in other ways too: I frequently use it to build prototypes of component designs or testing equipment. See Chapter 14 for an example, used to develop a rear derailleur design for White Industries.

### Raleigh's U.S. Export Sales

Raleigh's program to sell through franchised dealers succeeded. Raleigh had 1,400 franchised dealers in the U.S., second only to Schwinn. Raleigh's U.S. dealers had a problem in the early 1960s because Raleigh did not make a high-riser bicycle, and these made up more than half of the U.S. market.

Nevertheless, sales of Raleigh bicycles in the U.S. increased year by year until they exceeded 200,000 in 1968, and more than half a million in 1973, at the peak of the bike boom. To supply the U.S. bike boom, Raleigh made Record and Grand Prix 10-speed bicycles in Nottingham, Worksop (Carlton), Holland (Gazelle), Raleigh-Ireland, and possibly even Malaysia.

Raleigh's quality-control problems got worse during the bike boom. Raleigh was mass-producing and exporting poorly built bicycles. As the boom peaked, buyers and bike stores could compare the quality bicycles from Japan with the shoddy bicycles from Nottingham. The word spread, and soon nobody wanted a Raleigh.

### Moulton

More than 1,000 Moultons a week were sold in 1965. By 1965, they were available with a single-speed coaster brake, a 2-speed F&S Duomatic, a 3-speed or 4-speed Sturmey-Archer hub gear, 2 x 4 hybrid gearing, using a Cyclo converter, or 6 speeds with a Cyclo P-2 derailleur.

Above: Fig. 10.49 A and B. The 1966 Moulton Speedsix used a British Cyclo P2 derailleur with an 11–21 freewheel.

Below: Fig. 10.50. The 1966 Moulton Mini, with 14-in. wheels, was available with various gearing options.

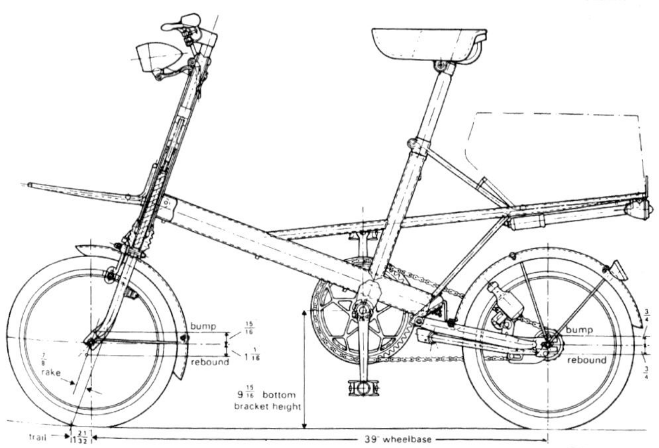

Above: Fig. 10.51. The 1966 Moulton S (Speed) was sold as a frame kit, made of Reynolds 531 tubing. This one had been built up with a Huret Allvit rear derailleur.

Moulton Bicycles Ltd. became the second-biggest bicycle maker in Britain. Raleigh's introduction of the RSW-16 in 1965 triggered a price war between Moulton and Raleigh. To compete, the Bradford-on-Avon factory closed, and all Moultons were made by the British Motor Corporation (BMC).

In 1966, Moulton introduced the 7/8-scale Mini range with a shorter wheelbase and 14-in. wheels, intended for women. Over the years, it became available with a broad selection of gearing options.

The late-1966 Moulton Speedsix used a British Cyclo Benelux P2 derailleur mounted backward and a 6-speed freewheel with an 11-tooth small sprocket. It was the first British 6-speed freewheel. In 1967, Moulton patented a 5-speed freewheel that accepted a nine-tooth sprocket. The S (Speed) was sold as a frame kit, which the buyer equipped to order.

By 1967, sales were down to 750 per week, and Moulton was operating at a loss. In 1967, Alex Moulton sold Moulton Bicycles Ltd. to Raleigh and returned to his consulting business. Raleigh was one of his clients, and this led to a series of Raleigh Moultons. The contract with Raleigh ended in 1974.

### Lambert-Viscount

The Lambert Aerospace bicycle was as poor as the Moulton was brilliant. Three models, priced from $130 to $150, were introduced with great fanfare in 1973. They advertised, "Lambert Aerospace Ultralite components, including sealed bearing hubs, sealed bearing bottom bracket, needle bearing pedals, and 15-speed alloy crank assembly." The brazed lugless frame used "Lambert 1027 alloy steel tubing to aircraft specifications." The Grand Prix weighed 22¾ lbs. (10 kg) and the Grand Prix Professional weighed 20 lbs. (9 kg). The original Lamberts used Simplex derailleurs (which should have sounded the alarm bell). Later Viscounts used Shimano derailleurs.

It sounded too good to be true, and it was. The Lambert cut too many corners. It had great specifications and poor quality and reliability. The poorly made Lambert components used nonstandard threads, and when they failed, it was difficult to get replacements. Lambert quality control was atrocious, even by Raleigh standards. There were frame and fork failures, and soon Lambert went bankrupt.

In 1974, Trusty Manufacturing bought the factory and the tooling. The bike reappeared in 1975 as the Viscount Aerospace Pro, selling for $240. It weighed 24 lbs. (10.8 kg) with tubular tires. In 1976, Yamaha (of motorcycle and piano fame) took over. They made six Viscount models, priced from $140 to $270. The top three models used the Lambert-design cast aluminum fork. The forks suffered fatigue failures and had to be recalled, which was the end of the Aerospace bicycle.

### Cyclo-Benelux

The 1962 Sport P2 was the last Cyclo-Benelux rear derailleur. This model was used on the Moulton Speedsix. By 1970, Cyclo-Benelux had no market among the British bicycle makers, and they went out of the derailleur business. The surviving Crown Gear Works continued to make bicycle tools and a hybrid derailleur-sprocket kit to convert three-speed hubs to six speeds.

Right: Fig. 10.52. Too many corners were cut on the 1973 Lambert Aerospace Pro. On paper, it had great specifications, but it was of poor quality and reliability.

## Other European Countries

The story of derailleur bicycles in Spain is largely the story of Zeus. Don Nicolas de Arregui had started Zeus Industrial in 1926, in Eibar in northern Spain. Initially, Zeus made simple components, like axles and spindles.

According to the 1973 Zeus catalog, Zeus made the first parallelogram rear derailleur in approximately 1932, and the prototype was shown at the company's 50th-anniversary dealer meeting. I suspect that this is a myth, because I've never seen either this derailleur or a photograph of it.

Zeus fell on hard times from the end of the Spanish Civil War through World War II. Following the war, Zeus reverted to making simple components. In the 1960s, materials and financing were available to broaden the range. By 1965, Zeus was making and exporting a nearly complete line of components, including brakes, pedals, cranksets, hubs, seatposts, and derailleurs.

By 1973, Zeus had 200 employees. Zeus was unique in that they made both frames and components. They sold complete bicycles using entirely Zeus components. Zeus bicycles were imported into the U.S. during the bike boom.

The 1970 Zeus gruppos—Criterium, Alfa 72, and Alfa Junior—were all copies of Campagnolo. Zeus did not have a high reputation for reliability.

Triplex was another Spanish company. Their 1970 Triplex Sport rear derailleur was one of the poorer copies of the Campagnolo Nuovo Record.

In 1965, West Germany produced 1.1 million bicycles. 205,000 bicycles were exported,

Above: Fig. 10.53. The 1970 Zeus Criterium rear derailleur was a faithful copy of the Campagnolo Record.

Left: Fig. 10.54. Although Zeus also made some fine Campy replicas, their 1970 Alfa Junior was nothing to be proud of.

Above: Fig. 10.55. The Polygraph Tectoron KS-01 was an East-German copy of the Campagnolo Nuevo Record.

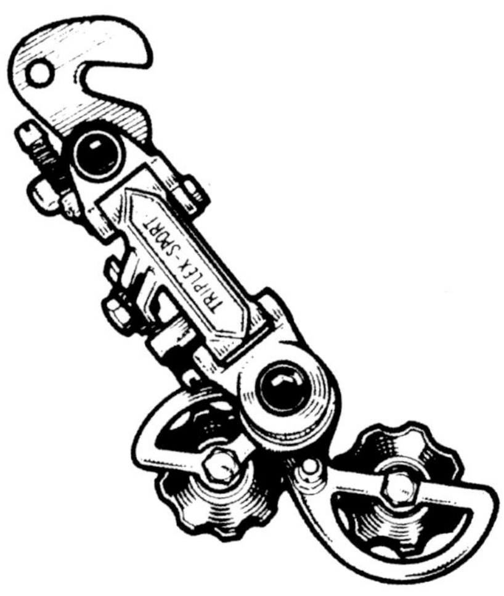


Left: Fig. 10.56. The 1970 Triplex Sport rear derailleur from Spain was one of the poorer copies of the Campagnolo Nuovo Record.

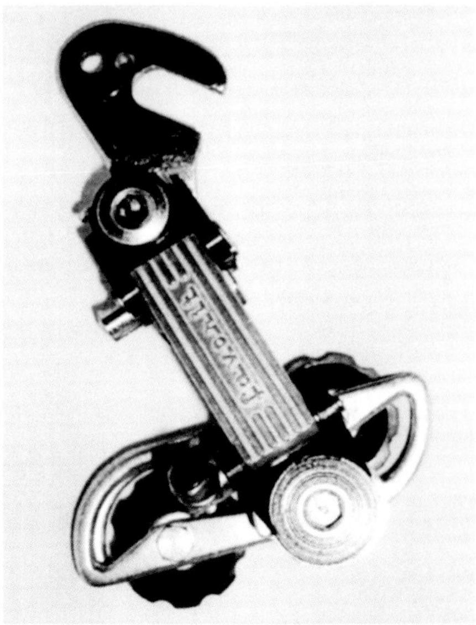

Right: Fig. 10.57. The Czech Favorit was one of many Campagnolo Record copies.

and 105,000 bicycles were imported. In 1968, West Germany produced about 2 million bicycles.

Sachs, still called Fichtel & Sachs at the time, was the only significant derailleur maker for the German market, but their main products were Torpedo coaster brakes and hub gears.

Meanwhile, Austria, East Germany, and Czechoslovakia also had some bicycle component manufacturers making derailleurs—with some exceptions, these were mainly copies of Campagnolo products.

## United States

In 1965, the majority of bicycles sold were for 12- to 14-year-olds. From 1965 to 1970, high-riser bikes were the hot items. America's total (domestic and imported) bicycle sales rose slowly from 5.5 million in 1965 to 7 million in 1970, which was the beginning of the Great American Bike Boom. Sales then took off, reaching 9 million in 1971, 12 million in 1972, 15 million in 1973, and 14 million in 1974. Then the boom collapsed, and sales were halved in 1975.

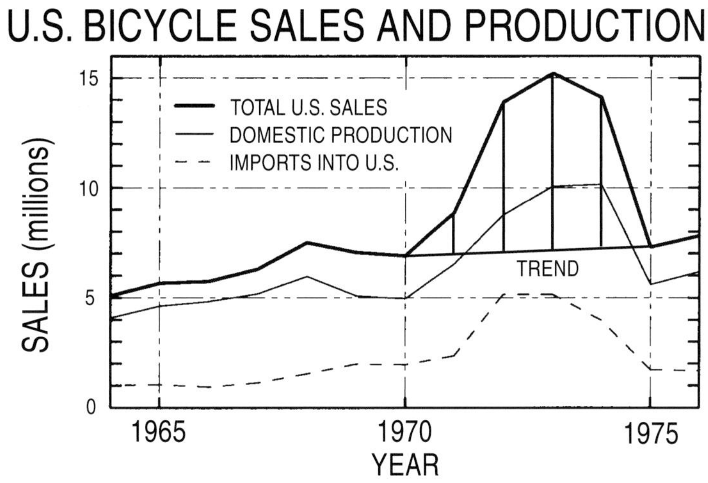


Right: Fig. 10.58. Graph of U.S. domestic bicycle production, bicycle imports, and total bicycle sales, 1964 to 1975. (Source of data: *Schwinn Reporter*, February 1978.)

The sales numbers are deceptive because they include children's bicycles and high-risers, and those numbers were relatively constant. The number of adult bikes increased by a factor of 20. By 1975, an extra 20 million adult bikes had been sold in the U.S., and this permanently changed the marketplace.

U.S. domestic bicycle production grew slowly from 4.5 million in 1965 to 5 million in 1970. It then soared to 10.2 million in 1974, and plummeted back to 5.5 million in 1975.

Imports were 1 million in 1965, 2 million in 1970, 5 million in 1973, 4 million in 1974, and then fell to 1.7 million in 1975.

### Adult Recreational Bicycles

Most U.S. bike stores did not know how to sell or service 10-speed adult bicycles. However, a small band of dedicated pioneers laid the groundwork for public

Left: Fig. 10.59. The 1963 Schwinn American was a typical middleweight bicycle of the period. Of course, that's "middleweight" by American standards of the day—some 40-odd pounds, or nearly 20 kg.

Fig. 10.60. Chuck Harris' wide-range Beech Hill front and rear derailleurs.

acceptance of bicycles as adult recreational vehicles rather than children's toys. Dr. Paul Dudley White, Eisenhower's personal physician, preached the gospel of regular exercise. The American Youth Hostels (AYH) organized multi-day bicycle tours and trained cyclists to lead bicycle tours. Dr. Clifford Graves founded the International Bicycle Touring Society.

In 1964, Phyllis Harmon, Keith Kingbay, and Fred DeLong had breathed new life into the moribund League of American Wheelmen (LAW), and by 1966, thy had 500 members. In 1970, the Great Eastern Rally (GEAR) and the Great Western Bicycle Rally (GWBR) each attracted more than 500 participants.

In 1965, you could buy a Raleigh Record, with Huret derailleurs, 52/40-tooth chainwheels, and a 14–28 freewheel, for $95. A Raleigh Carlton, with a Reynolds 531 frame, a Williams crankset, and Huret derailleurs, cost $106.

Chuck Harris was very active in the AYH. He was a gearing experimenter, and in the 1960s he made his own sprockets and chainwheels. His favorite setup used a 13-17-26-34-42 freewheel combined with 68/44/26 chainwheels. He built his own front and rear derailleurs to handle his ultra-wide-range gearing, called Beech Hill. He was the first American derailleur maker; starting in 1958, he produced some 40 Beech Hill derailleurs.

American women became interested in recreational cycling and in bicycle racing, where they often did better than the men in international competition. In 1969, Audrey McElmury won the women's World's Championship in Brno, Czechoslovakia, making her the first American to win a world cycling championship since 1912.

## Schwinn

In the early 1960s, the standard American bicycle was still a single-speed, coaster-brake, balloon-tired "tank" that weighed about 50 lbs. and was marketed exclusively to kids. It was styled to look like a motorcycle to appeal to the juvenile market. In the 1950s, Schwinn converted the market from heavyweight bicycles with $2\frac{1}{8}$-inch wide tires (heavy tanks) to middleweight bicycles with $1\frac{3}{4}$ inch wide tires (medium-weight tanks). The Schwinn American came in two wheel sizes: 24-inch for 11-year olds and 26-inch for teenagers. It was available with a 2-speed coaster-brake hub for an extra $10.

In 1965, the majority of U.S. bicycles were cheap, heavy "gas pipe" bikes, with coaster brake, and balloon tires roadsters, aimed at 12- to 15-year-olds. "Gaspipe" is a pejorative description of the thick-wall steel tubing used to make the welded frames of those American bicycles. They were churned out by Huffman (Huffy), Columbia, Murray-Ohio, American Machine and Foundry (AMF), and Chain Bike (Ross) to sell for less than $50 in department stores and mail-order catalogs. By 1965, these makers also imported 3-speed, 5-speed, and 10-speed middleweight for adults.

Schwinn was the dominant American bicycle maker, with

Left: Fig. 10.61. One of the first titanium-frame bikes, the 1974 Teledyne Titan was made in the U.S. using large-diameter thin-wall titanium tubing.

Right: Fig. 10.62. The 1974 Tokheim Gear-Maker was a 5-speed expanding-sprocket rear hub with a 3-to-1 range.

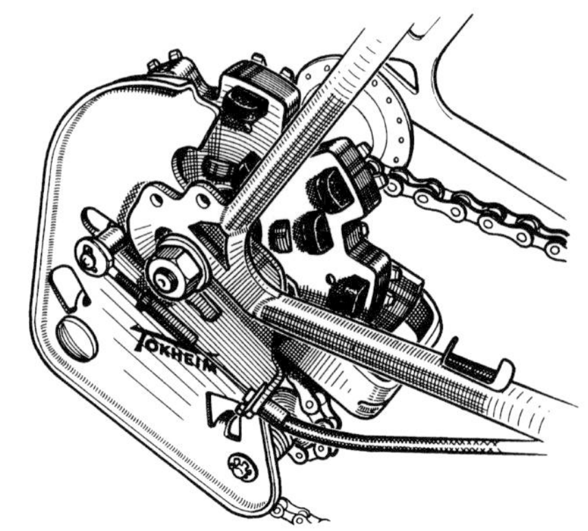

about 20% of the U.S. market. The 2,000 employees at Schwinn's Chicago factory could make about 1.5 million bicycles a year. The vast majority were for children. Schwinn competed with the low-priced gaspipe bicycles, but their bikes were better. Schwinn assembled frames by a technique called flash-welding. Their quality was better, and their bikes were sold only through franchised bike stores. The 1960 Schwinn Varsity had proved that there was a market for inexpensive American-made 10-speed derailleur bicycles. Schwinn dealers had a derailleur demonstrator to help overcome doubts about these new-fangled devices.

Schwinn expanded the range of derailleur models each year. In 1965, the Schwinn dealers had $70 Varsities, $90 Continentals, $100 Sierras, and $130 Superiors. The pride of the store was the $237 Paramount. All used Huret Allvit derailleurs, except the Paramount, which was equipped with Campagnolo Nuovo Record components.

By the early 1970s, the ancient Schwinn factory in Chicago could not supply the dealer demand. In 1972, Schwinn went to Japan and signed contracts to import Schwinn-Approved World bicycles from Bridgestone and National (Panasonic). Yoshi Shimano was the translator for the negotiations. Not surprisingly, the bikes used Shimano components. The Shimano derailleurs were labeled Schwinn-Approved. With lugged carbon steel frames and Shimano derailleurs, the World imports were lighter and better shifting than the more expensive Chicago Varsity. Schwinn's Japanese imports upgraded the image of the Japanese bicycle industry in the U.S.

By 1975, Schwinn was offering nine different drop-handlebar derailleur bicycles, ranging from the Schwinn-Approved World Traveler at $130, and ending with the Paramount at $578. The Varsity was $133.

## Bicycle Imports

By 1965, there was a small but significant market in the United States for imported British, French, and Italian lightweight bicycles. About 200 bike shops stocked imported bicycles. Raleigh was the dominant British import and was developing a network of franchised dealers. However, Falcon and Dawes were also available. Peugeot, Motobécane, and Gitane were the leading French brands. Atala, Bianchi, and Frejus were the leading Italian brands.

If you wanted an imported bicycle, you could get one, but you might have to drive to a big city. The really enthusiastic racer or bicycle tourist would have his bicycle made to order by a special-

Above: Fig. 10.63. The 1974 Schwinn Varsity had Huret Allvit derailleurs and a steel crankset. It sold for $120 and weighed 40 lbs. (18 kg).

Right: Fig. 10.64. The 1964 Schwinn Sting-Ray, with banana seat, sissy bar, and longhorn handlebars.

ist custom-builder in Britain, France, or Italy, although a few American frame builders had also entered the fray.

The major U.S. bicycle makers also imported bicycles with three-speed hubs and 10-speed derailleurs. Raleigh made the majority of these bicycles. The 1965 Huffy Carlton was a typical imported 10-speed with Allvit derailleurs and a Stronglight crankset.

By 1969, most cities had a number of bicycle shops that carried imported lightweight bicycles with either three-speed hubs or derailleur gears. Even as the market for high-riser bicycles peaked, the demand for imported lightweight bicycles was slowly growing, and more bicycle shops added imported bicycles to their offerings. With their large network of franchised dealers, Raleigh was the largest supplier of lightweight bicycles. By 1970, most Raleighs were 10-speed racers rather than 3-speed roadsters.

By the early 1970s, most cities had at least one specialist bicycle shop that catered to the racers, and imported custom frames from specialist builders like Bob Jackson, Ron Cooper, Claud Butler, Harry Quinn, Hetchins, Mercian, Woodrup, Masi, Colnago or Cinelli. These specialized shops advertised in *Bicycling!* magazine on pages titled "America's Leading Bike Stores." The market for lightweight bicycles grew steadily at a rate of roughly 20% per year between 1967 and 1971. Then it exploded with the bike boom.

### American Products

Some innovative American makers ventured into bicycle manufacture. The 1974 Teledyne Titan was made in the U.S. from large-diameter thin-wall titanium tubing. It was a commercial success, but the necked-down section of the downtube created a stress riser that failed after a few years.

Left: Fig. 10.65. The 1969 Schwinn Orange Krate was the ultimate Sting-Ray. It was equipped with a 5-speed Huret Allvit derailleur and a Huret Stick-Shift on the top tube.

Introduced in 1974 the Tokheim Gear-maker was a 5-speed expanding sprocket rear hub with a range of 3 to 1. It was made of polyamide plastic. Tokheim marketed the Gear-maker for about five years to the makers of low-price bicycles. Then it disappeared.

### High-Riser Bicycles

The high-riser bicycle story began in San Diego in 1963, when kids started taking used 20-inch-wheel juvenile bikes and equipping them with "longhorn" handlebars, "banana" seats, and "sissy" bars to support the rear of those long seats. The San Diego kids called them "pig bikes." Many parents thought that they looked stupid, which made them even more attractive to their owners. The fad soon spread north to Los Angeles.

Schwinn's West Coast dealers noticed the booming sales of these exotic parts and reported the activity back to Schwinn's Chicago headquarters. Al Fritz, an engineer in Schwinn's management, recognized the potential and convinced Schwinn to make a trial batch of high-riser bicycles, complete with banana seats, sissy bars, and longhorn handlebars.

Schwinn started making Sting-Rays in mid-1963, and 45,000 were sold by the year's end. It was a huge success, and Schwinn's name Sting-Ray became the generic name for the breed.

The next year, the rest of the domestic industry jumped aboard the bandwagon, and the high-riser bike moved east into the mainstream. In the process, it picked up a rear derailleur and hand brakes. The derailleur "stick-shift" lever was modeled after the hot-rodder's "four-on-the-floor." It was kid brother's version of big brother's hot rod.

Sales of high-riser bikes increased by about a million bikes per year for five years, until they peaked in 1968 at 5.7 million. In 1970, almost three-quarters of the bicycles sold in the U.S. were high-risers. They continued to be a significant part of the market through 1974. The marketers did not neglect the girl's market: There also were high-riser "Fair Ladies" and "Slick Chicks." The Schwinn Orange Krate was the ultimate Sting-Ray. It was equipped with a 5-speed Huret Allvit derailleur and a Huret Stick-Shift.

## The Bike Boom

The Great American Bike Boom was the most significant event since the first Bicycle Boom at the end of the 19th century. After the Great American Bike Boom, America, rather than Europe, was the main market for quality derailleur bicycles. The center of gravity of the world's bicycle industry shifted from the Atlantic to the Pacific.

British and French bicycle makers, such as Raleigh, Peugeot, Motobécane, and Gitane had dominated the quality-bicycle market for decades. After 1974, Japanese bicycle makers, like Fuji, Bridgestone, Panasonic, and Takara, were the major exporters.

The high-riser bicycle replaced the middleweight bicycle in the 1960s. This was not particularly important—just another strange American fad. The change in the teenage mind-set was important. In the early 1960s, only "nerds" with steel-rimmed glasses rode bicycles to high school. In the late 1960s, millions of American teenagers rode their high-riser bikes to school because they were "cool." It was the "in" thing to do. With just a bit of practice, you could "pop a wheelie" and ride on the rear wheel indefinitely.

For the first time since the beginning of the century, young Americans kept riding bicycles past their teens, even though they had a driver's license. The more than 20 million high-riser bikes that were sold in the eight years preceding the bike boom created the millions of youthful riders who were the main buyers of 10-speeds during the bike boom. In 1978, Schwinn commissioned an exhaustive study of the bike boom. The study concluded:

> This dramatic increase in "adult" style 26- and 27-inch wheel multi-speed bicycles is attributable in large part to "youth" riders in the 13-to-17-year-old segment of the population, graduating from high-riser bicycles to sophisticated derailleur-equipped lightweight bicycles.

So the convulsion in the worldwide bicycle industry that

*Table 10.2. List of exporting countries and bicycle makes imported into the U.S. during the bike boom.*

| Country | Bicycle Makes Exported to U.S. |
|---|---|
| Austria | Steyr |
| Belgium | Flandria, Superia |
| Brazil | Caloi |
| Canada | CCM |
| Denmark | BKC |
| France | Bertin, Cazenave, Chaplait, Corso, Delcroix, Ficelle, Flandria, Follis, Gitane, Hosteller, Jeunet, Joannou, Lejeune, Liberia, Mercier, Michard, Motobécane, Paris Sport, Peugeot, St.Etienne, Star France, Stella, Sutter, Vélo Sport, Urago |
| Germany | American Arrow, Kalkhoff |
| Holland | Ancona, Batavus, Eroba, Gazelle, Magneet, Pegasus |
| Italy | Atala, Bianchi, Bottechia, Bugatti, Chiorda, Cortina, Doniiselli, Frejus, Fiorelli, Garlatti, Italvega, Legnano, Maserati, Moretti, Vanquer, Veloce |
| Japan | Azuki, Bridgestone-Kabuki, Centurion, Chimo, Fuji, Itoh, Nishiki, Panasonic, Sekai, Sekine, Takara |
| Mexico | Windsor |
| Norway | DBS |
| Spain | Torval, Zeus |
| Sweden | Crescent |
| Switzerland | Mondia |
| UK | Dawes, Falcon, Holdsworth, Lambert, Mercian, Moulton, Raleigh |

resulted from the Great American Bike Boom was caused by U.S. teenage buyers and their high-riser bicycles. I call it the "Great American Bike Boom" to distinguish it from the 1890s bike boom described in Chapter 3.

In 1972, 1973, and 1974, more bicycles than automobiles were sold in the U.S. America became the biggest market for derailleur bicycles, and remained the dominant force until this day.

The bike boom happened because young people decided that riding 10-speeds was a neat thing to do. However, the timing was also right. The 1960s and early 1970s were periods of unprecedented wealth for young Americans. The baby-boomers grew up and got their first jobs.

The same social and economic forces that allowed young people to buy expensive high-fidelity equipment, camping gear, cameras, and other consumer goodies unleashed this huge group of baby boomers onto the bicycle market. It was an easy sell, because the lightweight 10-speed bicycle was fast, fun, and exotic—and a lot cheaper than a good hi-fi set.

The oil shortage did not cause the bike boom. The Arab-Israel war began in October 1973, and the lines at the gas stations came at the end of 1973, when the bike boom was two-thirds over and bike sales were winding down for the winter. The best that can be said about the energy crunch was that public concern over the oil crisis kept the bike boom going for an extra year.

### The Bike Boom, Year-by-Year

**1970:** U.S. bicycle sales decreased slightly to less than 7 million from 7.5 million in 1968. Of the 7 million bicycles sold, 5.5 million were for children. About 12% of total sales (820,000) were derailleur-equipped adult lightweight bicycles. About 60% of the derailleur lightweights were imported, and 40% were domestic. The term "lightweight" is relative, because most of the domestic bikes were Schwinn Varsities, which weighed more than 40 lbs.

**1971:** The bike boom began. Nearly 9 million bicycles were sold. This jump in bicycle demand was a complete surprise, especially the trebled sales of 10-speeds. By the middle of 1971, the U.S. bicycle makers had orders in hand that exceeded their entire production capacity for the year. Schwinn dealers were on allocation from May 1971 through 1974. The market breakdown included 4.1 million high-riser bikes and 3.7 million lightweights. Sales of imported lightweight bicycles tripled. 10-speed bicycles were in short supply, and everything that was available quickly sold at a handsome profit. This attracted newcomers to import, distribute, and sell 10-speed bicycles.

Above: Fig. 10.66. The 1974 Peugeot U0-8 had Simplex Prestige derailleurs.

Right: Fig. 10.67. The 1974 Gitane Gypsy was another French 10-speed of the period. It came with whatever derailleurs were available to the manufacturer at the time.

**1972:** Nearly 14 million bicycles were sold, amounting to 66 bicycles per 1,000 population. The domestic U.S. manufacturers worked overtime to produce 8.75 million bicycles. Sales of imported 10-speeds tripled again, and imports took 37% of the market. Cycling was "in," and TV commercials for all kinds of products showed adults on bicycles. The sources of the 5 million imported bikes were as follows: Japan 24%, Taiwan 14%, Austria 12%, Britain 11%, Germany 10%, France 10%, Italy 7%, and Holland 2%. Sales of adult balloon-tired bicycles fell to about 0.5 million. Sales of lightweight bicycles exploded to 8 million. Between 1970 and 1972, the U.S. market for lightweight adult bicycles grew 40-fold, from 200,000 to 8,000,000.

**1973:** The boom peaked, and more than 15 million bicycles were sold, amounting to 72 bicycles per 1,000 population. Domestic U.S. production increased to 10.1 million. More than 5 million imported bicycles were sold in 1973, the same number as in 1972. Overseas factories were at capacity or they were limited by component availability. The import market share dropped back to 34%.

**1974:** Sales decreased slightly to 14.1 million, amounting to 67 bicycles per 1,000 population. U.S. domestic production increased to 10.2 million, but less than 4 million imported bicycles were sold. The market demanded cheaper bicycles: the best-sellers were inexpensive domestic "gaspipe" bicycles with 26x1⅜-in. rather than the 27x1¼-in. tires used on lightweight bikes.

**1975:** The bike boom collapsed, and total bicycle sales fell to just over 7 million. This included 2.2 million high-risers and BMX bikes. Domestic sales dropped to 55% of the 1974 peak. The importers took a blood bath when sales slumped to one-third of their 1973 level. Many importers and bike stores had built up their inventories over the winter, expecting to profit from the rapidly inflating prices. When the bubble burst in 1975, the industry was up to the saddle in inventory and down to the pedals in bank balance.

### Bike-Boom Bicycles

At the peak of the bike boom, the demand for derailleur 10-speeds greatly exceeded the supply. Anything with a derailleur could be sold. The American makers of gaspipe bicycles slapped on

Above: Fig. 10.68. The 1974 Columbia Super X was made by bolting a derailleur onto a coaster-brake frame. See the text for an evaluation of its quality.

Right: Fig. 10.69. The 1968 Huffy Sportster was equipped with the Shimano Lark W 5-speed rear derailleur, controlled from a twistgrip shifter.

derailleurs, hand brakes, and narrower tires to cobble together a $75 ten-speed. Some of these were truly horrid.

In the worst cases, the rear derailleur was added to a frame designed for a coaster brake, making rear wheel removal almost impossible. Gaspipe bicycles were sold by department stores or auto supply stores, often only partly assembled in the carton. Even if the bikes were assembled, the stores did not have mechanics capable of adjusting or repairing them.

The 1974 Columbia Super X was made by bolting a derailleur onto a coaster brake frame. It sold for $85 and was best used for a one-way trip to the nearest dump.

The 1974 Raleigh Record DL130 was a typical $120 British 10-speed. It had Simplex Prestige derailleurs, a cottered steel crankset with 49/40 chainwheels, a 14–28 freewheel, high-flange alloy nutted hubs, and Weinmann centerpull brakes. 2030 carbon steel unbutted tubing was used and it weighed 31 lbs. (14.1 kg).

The 1974 Peugeot UO-8 was a typical $140 French 10-speed. It had Simplex Prestige derailleurs, a cottered steel crankset with 52/40 chainwheels, a 14–28 Atom freewheel, Normandy alloy quick-release, high-flange hubs, and Mafac Racer centerpull brakes. Many UO-8s came with 52/36 chainwheels. Lightweight unbutted tubing kept the weight down to 28 lbs. (12.7 kg).

The 1974 Fuji Road Racer was a typical $140 Japanese 10-speed. It had SunTour GT and Spirt derailleurs, a Sugino alloy cotterless crankset with 50/40 chainwheels, a SunTour 14–34 freewheel, Sunshine alloy high-flange hubs with quick-releases, and Dia-Compe centerpull brakes. The alloy crankset saved two pounds, but the Fuji still weighed more than the Peugeot. The heavier frame could take more abuse. Though the gear spread was much wider, every writer commented on the smooth shifting SunTour derailleurs.

Another example of a French bike-boom bike was the Gitane Gypsy. Gitane used Huret, Simplex, or SunTour derailleurs—whatever they could get. The gearing was stock alpine: 52/40 chainwheels and a 14–28-tooth freewheel. This was a low-priced model that came with wing nuts instead of quick-release hubs.

The 1974 Schwinn Varsity sold for $112. It had Huret Allvit derailleurs, an "Ashtabula style" (i.e. one-piece) steel crankset with 52/39 chainwheels, a Maillard 14–28 freewheel, Normandy alloy nutted hubs, and Weinmann sidepull brakes. It weighed around 40 lbs. (18 kg), due to its heavy thick-wall welded steel frame.

The 14 million imported bicycles that came into the U.S. dur-

Above: Fig. 10.70. The 1974 Raleigh Record DL130 was a typical $120 British-made 10-speed with Simplex Prestige derailleurs.

Right: Fig. 10.71. The 1974 Fuji Road Racer had SunTour GT and Spirt derailleurs.

ing the bike boom also included some pretty grim examples. Prices and profits were up. In the early 1970s, Raleigh was the major importer. Motobécane, Peugeot, and Gitane had smaller dealer networks. When these traditional suppliers ran out of bicycles in 1971 and 1972, bicycle makers from around the world rushed in to fill the void. The old magazine advertisements showed the brands and countries listed in table 10.2.

When the bubble burst in 1975, most of these names disappeared from the U.S. market. Now they show up at garage sales, and some people assume that the foreign name must make them valuable.

During the bike boom, there was a price break at about $125 for a 10-speed bicycle. If you paid less, you got a poor bargain. If you paid a bit more, you could get a fairly decent bike.

## Japan

Japanese bicycle production increased from 3 million bicycles in 1965, to 4.5 million in 1970, to a peak of 9.5 million in 1974. Production was down to 7.5 million in 1975. Japanese bicycle exports increased from 450,000 bicycles in 1965, to 1,07 million in 1970, to a peak of 1,55 million in 1972, and then declined to 1,4 million in 1975.

The Japanese bicycle industry recognized the export growth potential of the lightweight bicycle market. The Japan Trade Center exhibited Japanese components at the 1970 Amsterdam trade shows, and they maintained bicycle showrooms in Los Angeles and New York in 1970 and 1971. Shimano and SunTour were serious competitors in the export market starting in the late 1960s.

By 1972, Japan was exporting more than a million bicycles a year to the U.S. Japanese-made bicycles, such as American Eagle (which was renamed Nishiki), Campania, Centurion, Fuji, and Sekai, took a significant share of the U.S. market. Some of these Japanese imports were inexpensive children's bikes, but Japanese adult 10-speed bicycles acquired a reputation for quality and value.

Japanese 10-speed bicycles used SunTour or Shimano derailleurs. When the bike boom exploded, the Japanese makers had two advantages over Huret, Simplex, and Campagnolo: They had concentrated on the low-priced market, and they had designed their derailleurs to match the Alpine gearing that was popular on low-priced U.S. 10-speed bicycles. Alpine gearing used a 52/40 chainwheel and a 14-17-20-24-28 freewheel. The vast majority of the low-priced 10-speeds sold in the U.S. during the bike boom had this gearing. The Huret Allvit and Svelto, the Simplex Prestige, and the Campagnolo Velox and Valentino rear derailleurs were designed for typical European gearing, which used smaller freewheel sprockets. The Simplex Prestige plunger-type front derailleur was obsolete, and it shifted poorly.

### SunTour

In this decade, SunTour was more innovative than Shimano and had technically superior designs. Their key advantage was the 1964 invention of their slant parallelogram, which kept the chain gap constant throughout the range.

Until about 1968, SunTour derailleurs were only sold in the domestic Japanese market. The 1964 Grand Prix and Super Competition models were rarely exported. In 1965, Huret built Svelto rear derailleurs labeled Skitter for SunTour. SunTour had to wait for the Japanese bicycle makers to develop an export market. In 1965, SunTour built a second factory at Mihara.

The 1968 SunTour Perfect freewheels used three splined and two

Right: Fig. 10.72. The 1968 SunTour Perfect freewheel had three splined and two screwed sprockets.

screwed-on sprockets. They offered a complete range of sprockets from 14 to 34 teeth. They were reliable and inexpensive.

The breakthrough 1969 SunTour V rear derailleur was made of aluminum alloy. The slant parallelogram made it the best-shifting derailleur available.

In 1969, SunTour made a detented (i.e., indexed) shift lever, the Five-Speed Click, and a combined freewheel-plus-rear hub, the Unit Hub, which was available with 1, 3, or 4 sprockets. These innovations preceded Shimano's SIS and Freehub by more than a decade. Neither of them found a market, and they were abandoned.

1970 was the turning point for SunTour. The company changed its name from Maeda Iron Works to Maeda Industries and adopted the SunTour brand name for all of its products. SunTour President Junzo Kawai made extensive studies, and concluded that it was time to enter the international market. They moved rapidly, and by the end of 1970, SunTour was the dominant Japanese derailleur maker, with steel models for the low-priced market and alloy models for the medium-priced market. The 1970 SunTour Compe V front derailleur was the mid-priced version, with an alloy body and a steel cage.

By 1973, the V line included three rear derailleurs: the narrow-range racing V, the mid-range VT, and the wide-range touring VGT. Unlike Huret and Simplex, SunTour fine-tuned the cage length and the pivot location of the different models to match the target gear range. The VGT touring rear derailleur had a wrap-up capacity of 36 teeth. Unlike French and Italian touring derailleurs, the VGT was quite happy with a 34-tooth sprocket. It was the best shifting wide-range derailleur available in 1973, and it only cost $10.

Thanks to their slant parallelogram, SunTour rear derailleurs shifted precisely and reliably. SunTour derailleurs were inexpensive, especially on the replacement market. It was hard for snobbish bike writers to accept that an $8 SunTour shifted better than a $40 Campagnolo. Campag-

Above: Fig. 10.73. The 1969 SunTour Unit-Hub combined the hub and the freewheel. It was available with 1, 3, or 4 sprockets.

Right: Fig. 10.74. The1969 SunTour V rear derailleur was made of aluminum alloy. The slant parallelogram made it the best-shifting derailleur available.

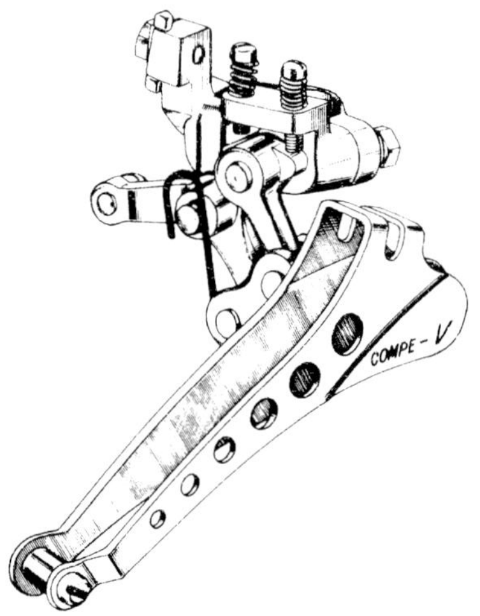


Above: Fig. 10.75. The 1970 SunTour Compe V front derailleur was the mid-priced version with an alloy body and a steel cage.

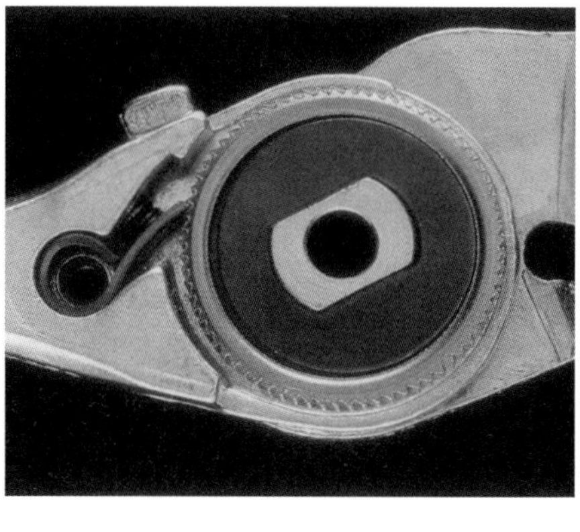

Right: Fig. 10.76. SunTour's 1973 Power-Ratchet mechanism provided a "sweet," light feel. They lasted right through indexed shifting for front derailleurs.

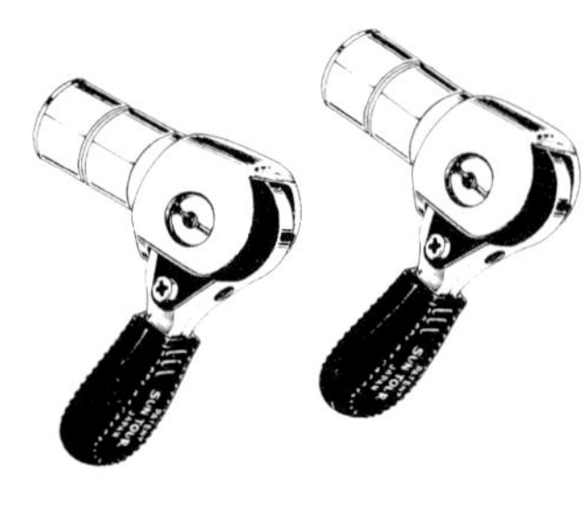

Right: Fig. 10.77. SunTour Bar-Con handlebar-end shifters with Power-Ratchet.

nolo's main advantage was better bearings, which lasted longer. They would shift poorly forever!

SunTour's technical innovation extended to their front derailleurs, which worked backward compared to those of the other brands. You pulled on the lever to shift from the large to the small chainwheel. The return spring shifted from the small to the large chainwheel. This was a better design. First, both levers gave you a lower gear when you pulled backward. Second, down shifts were more severe as you dropped to the smaller chainwheel under load. I used to write that the SunTours worked correctly, and that all of the other front derailleurs worked backward.

SunTour's Power Ratchet shift levers included a small ratchet wheel that released the friction from the lever when pulling on the cable. This Power Ratchet mechanism provided a lovely light feel. The same ratchet was used in SunTour's Bar-Con handlebar-end shift levers. They were the nicest bar-end shifters available, and they remained in production until indexed shifting took over. SunTour sold more Bar-Cons than all other bar-end shift levers combined. Mixing a SunTour Bar-Con with somebody else's derailleurs was one of those neat touches that said this bike builder knows his business.

### Shimano

Shimano made their first inroads into the U.S. market in 1963, when they sold three-speed hubs to Columbia. In 1965, they provided Schwinn with freewheels with bearing seals after Maillard refused to modify their design. Prior to 1965, Shimano concentrated on spokes, freewheels, and three-speed hubs. They became the world's largest maker of three-speed hubs. In 1965, both the domestic and export markets turned from three-speed hubs to derailleurs, and Shimano re-entered the market with their Archery rear derailleur.

By 1967, Shimano had developed the Sky Lark rear derailleur, which soon became the Lark. The Shimano Lark rear derailleur was an improvement of the Simplex Prestige. The parallelogram was more nearly horizontal, and

Left: Fig. 10.80. At the 1973 New York Cycle Show, Mitsuharo Hontani, VP of Shimano America, presents the new Dura-Ace gruppo to the bike trade.

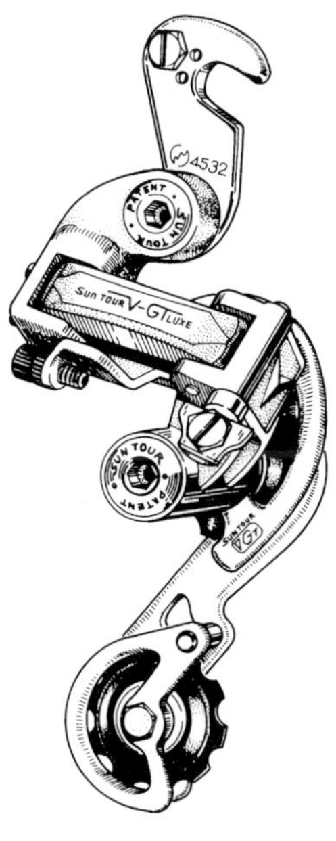

Left: Fig. 10.78. The 1973 SunTour VGT touring rear derailleur had a chain wrap-up capacity of 36 teeth. Unlike French and Italian touring derailleurs, the VGT was quite happy with a 34-tooth sprocket. It was the best-shifting wide-range derailleur available at the time, and it only cost $10.

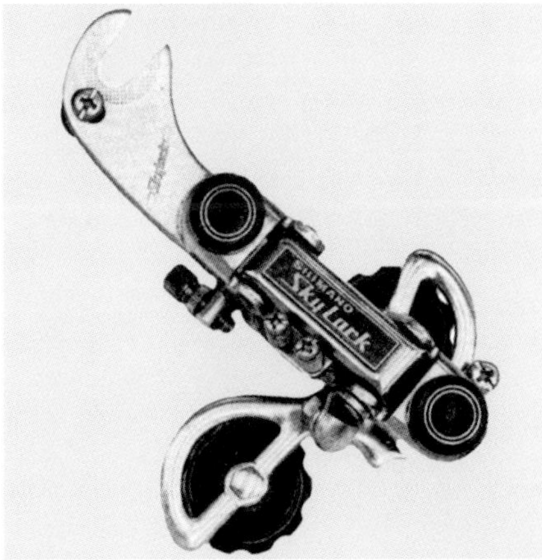

Right: Fig. 10.79. The 1970 Shimano Sky Lark rear derailleur was an improved copy of the Simplex Prestige.

the two spring-loaded pivots matched the cage geometry better. As a result, the Lark shifted more positively than the Prestige, and its steel construction was much more reliable.

The Lark was followed shortly by the Eagle, which had a steel guard to protect the derailleur if the bike fell on its side. This was a feature of the Huret Allvit and a requirement for Schwinn-Approved derailleurs.

Shimano became a major supplier of three-speed hubs to the American gaspipe bicycle makers. They established the Shimano American Corp. in New York in 1965 and Shimano Sales Corp. in California in 1974. Shimano used their American contacts to enter the developing market for inexpensive 5- and 10-speed bicycles. Shimano's advertising emphasized the top-of-the-line Dura-Ace front and Crane rear derailleurs, but it was the Lark that established Shimano's reputation in the bike boom.

Shimano had the combination of financial muscle, solid engineering, marketing skill, and large factories. By 1968, they had more than 1,000 employees and they passed SunTour in derailleur production. Two years later, they were challenging Simplex. Shimano moved their main production to a new factory in Shimonoseki, in western Japan in 1972.

By 1974, Shimano offered the widest range of derailleurs of any maker. There were seven different price/quality levels, with racing and touring (GS) rear derailleurs at each level. From the top, the rear derailleur names were Crane, Titlist, Tourney, 500, Eagle, Eagle SPO, and Lark.

The racing models had a total capacity of 28 teeth and could handle a 28-tooth sprocket. The touring models had a total capacity of 34 teeth and could handle a 34-tooth sprocket. The more expensive models had less steel, more alloy, and a finer finish. The 1974 Shimano Titlist

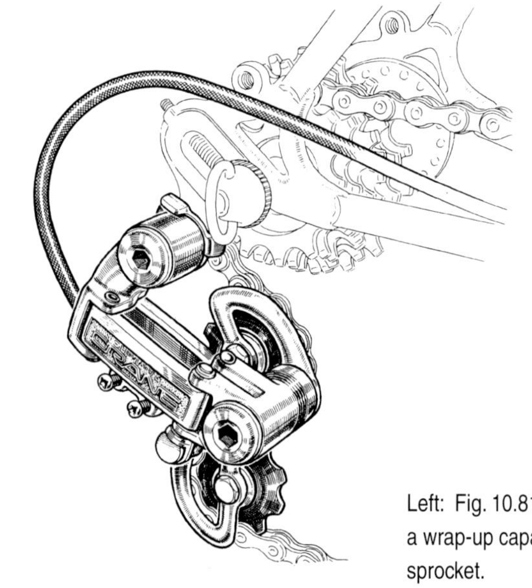

Left: Fig. 10.81. The 1973 Shimano Crane racing derailleur had a wrap-up capacity of 28 teeth and could handle a 28-tooth sprocket.

Below: Fig. 10.82. The 1974 Shimano Titlist GS touring rear derailleur (lower detail, with large cage) had a wrap-up capacity of 34 teeth. The largest sprocket it could handle was 34 teeth. It shifted well, though not as well as the SunTour VGT.

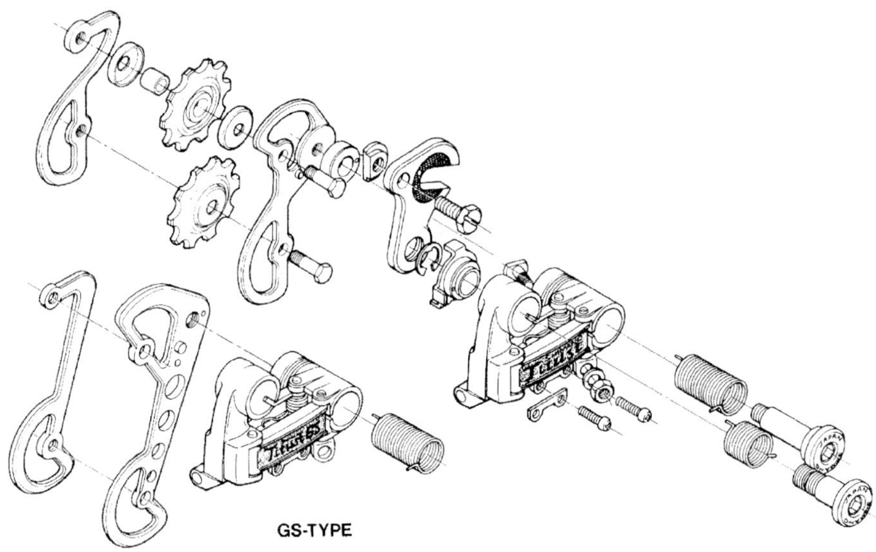


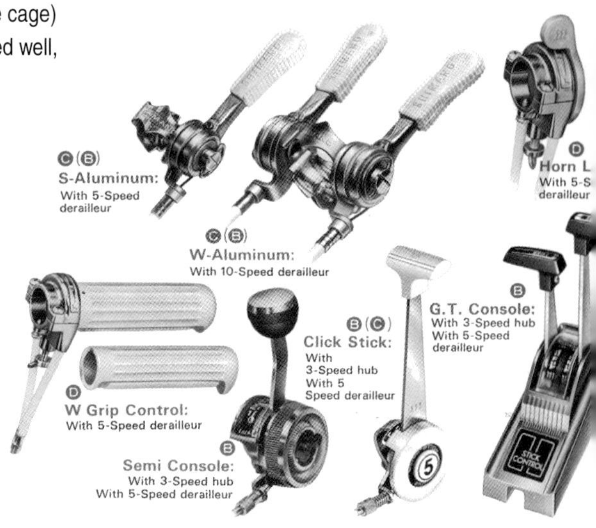


Above: Fig. 10.83. In addition to conventional shift levers, Shimano made a wide range of controls for their derailleurs, including twistgrip, twisting, and "console shifters" for high-rise bikes.

GS shifted well, but not as well as the SunTour VGT. The bicycle companies liked Shimano's large range of nearly identical models, because it allowed them to differentiate between similar bikes at slightly different price points.

Like Simplex, all Shimano rear derailleurs shared a common design using two spring-loaded pivots. Shimano had a nearly horizontal parallelogram. They heavily advertised their Servo-Pantograph design and claimed it was a technical breakthrough.

Shimano's front derailleurs used a standard parallelogram design. Both Shimano and SunTour realized that front derailleurs had to be rigid.

Shimano had 1,200 employees in 1973, and boasted that they were the world's largest maker of derailleurs. Shimano's policy was that 10% of their work force must be graduate engineers and that 10% of their employees must work in research and development. By the early 1970s, these two Shimano policies put them in the forefront of bicycle innovations. In 1974, Shimano matched Campagnolo with a complete Dura-Ace gruppo: hubs, crankset, brakes, freewheels, shift levers, and derailleurs.

## Cycling Magazines

At the end of 1969 Jean Billion, the owner and publisher of *Le Cycle*, decided to cease publication. Daniel Rebour and René Geslin, a noted bicycle collector, managed to keep *Le Cycle* going. At the end of 1973, *Le Cycliste* merged with *Le Cycle*. Daniel Rebour retired as editor at the end of 1973, but his illustrations continued until *Le Cycle* ceased publication with the May 1974 issue.

The *American Cycling Newsletter* (née *Northern California Cycling Association Newsletter*) changed its name, first to *American Cycling*, and then, in 1970, to *Bicycling!* Dr. Clifford Graves was the touring correspondent and Fred DeLong was the technical editor.

*Bike World* first appeared in 1972. It had a funky low-key style, because unpaid enthusiasts wrote most of the articles. Joe Kossack was its technical editor. My first article appeared in the August 1973 issue of *Bike World*, describing how to get lower gearing on inexpensive bicycles. I described five different wide-range derailleurs and concluded that the SunTour VGT was the clear winner.

Above: Fig. 10.84. *Bicycling!* was published from Northern California, and it featured many articles on touring, as well as technical articles.

*To enjoy cycling in hilly areas, you really need some good low gears.*

*by Frank Berto*

GETTING INTO REALLY LOW GEARS

The San Francisco Bay area where I live is famous for its hills and biking there and in other hilly areas is more fun when you can pedal up, rather than walk.

The typical 10-speed bike with "alpine" gears is not geared low enough for the beginning weekend cyclist who is not yet in top condition. If you insist on wide-range gearing, some bike stores offer two solutions: get in better shape, or buy a very expensive 15-speed touring bicycle. But there is a third solution.

For around $10, you can modify some 10-speeds for wide range touring. The changes can be done by the dealer or by an intelligent do-it-yourselfer. There are three requirements:

- a wide-range touring derailleur
- a wide-range freewheel
- a front inner chainwheel that uses the full capacity of the derailleur.

TOURING DERAILLEURS

The key item is the rear derailleur. It sets an absolute limit on the gear spread. The following table gives specifications of five wide-range touring derailleurs:

| Make & Model | Approx. Cost | Capacity | Max. Rear Sprocket | Weight |
|---|---|---|---|---|
| Campagnolo Gran Tourismo | $18-$25 | 43 teeth | 36-tooth | 16.5 oz |
| Suntour GT or VGT (alloy) | $8-$10 | 40 teeth | 34-tooth | 12 oz |
| Shimano Eagle GS or GTO | $9-$10 | 38 teeth | 34-tooth | 14.1 oz |
| Huret Luxe Super Touring | $14-$16 | 41 teeth | 30-tooth | 10.5 oz |
| Simplex Prestige | $6-$8 | 37 teeth | 28-tooth | 9.2 oz |

The Suntour derailleur is the best wide-range derailleur. Pick a bike with a Suntour derailleur if you plan to install wide ratio gearing. The slanted pantograph of the Suntour makes it particularly smooth-shifting on wide-range freewheels.

The Campagnolo Gran Tourismo has three teeth more capacity than the Suntour GT. It costs twice as much, weighs more and shifts less smoothly You usually find it only on top-line touring bikes.

The Shimano Eagle has two teeth less capacity and shifts less smoothly than the Suntour. The Eagle can also handle wide-ratio (34-tooth) freewheels. The Eagle is a second choice if you can't find a Suntour-equipped bike.

The Huret Luxe Super Touring and the Simplex Prestige are poorer choices, because they can't handle the ultra-wide range freewheels. The Schwinn Super Sport uses a Huret Allvit derailleur with an extra long cage. It has the same ultra-wide range limitations as the Huret Luxe Super Touring.

If your bike has a low capacity derailleur, your best bet is to change to a Suntour. This will cost $9.00 to do yourself, or about $12.00 at a bike shop.

DERAILLEUR CAPACITY

This specifies the amount of chain that the derailleur must wrap up when set at the smallest chainwheel-freewheel combination. The derailleur unwinds all of this chain when you shift to the largest chainwheel-freewheel combination. For example, the Suntour has a capacity of 40 teeth. If you want to make full use of this capacity, the best gear combination is:

| Front Chainwheel | |
|---|---|
| Large | 52 teeth |
| Small | 32 teeth |
| Difference: | 20 teeth |

| Rear Freewheel | |
|---|---|
| Largest | 34 teeth |
| Smallest | 14 teeth |
| Difference: | 20 teeth |

The total tooth difference from the widest 52 to 34 combination to the narrowest 32 to 14 combination is 40 teeth. This gearing uses the entire capacity of the Suntour derailleur and that's as far as you can go. It really can handle 40 teeth.

There are other combinations of chainwheels and freewheels that add up to 40 teeth, but 52T-32T chainwheel, 34T-14T freewheel turns out to be a very practical choice for an inexpensive bike. Note that when you're in the lowest gear with the 32-tooth chainwheel and the 34-tooth freewheel, your gearing is less than one to one. Each turn of the crank gives less than one turn of the wheels. And that, children, is what makes it easy for us old people to pedal up hills!

MAXIMUM SPROCKETS

In addition to wrapping up chain, the derailleur has to derail the chain from sprocket to sprocket on the freewheel. The pantograph must be big enough to cover the spread from the smallest to the largest freewheel sprocket. The Campagnolo and the two Japanese derailleurs can handle the widest range freewheels (14T-34T) that are readily available.

The Huret and Simplex derailleurs will not shift smoothly to freewheel sprockkets larger than 30 teeth. To use their full chain-wrapping capacity requires very small (26- or 28-tooth) front chainwheels. Chainwheels this small can't be mounted on inexpensive bikes without changing the crankset.

Finally, a touring derailleur needs to be solidly engineered to supply the large forces required to derail the chain over the large steps on wide-range freewheels. All three of the recommended derailleurs are good beefy desings.

WIDE-RANGE FREEWHEELS

The easiest way to widen a bicycle's gear ratio is to install larger sprockets on the freewheel. Maeda and Shimano, the

–28–

Bike World Magazine

Right: Figs. 10.85 and 10.86. *Bike World* was also published in Northern California. This is the issue in which author Frank Berto's first article about wide-range gearing appeared.

# 11 The Dawn of Mountain Biking: 1975–1984

THE GREAT AMERICAN Bike Boom ended in 1973 for imported bikes, and in 1974 for domestic bikes. Total U.S. bicycle sales fell from 14 million in 1974 to 7 million in 1975. It took a while for everyone to accept that the boom was really over.

Many importers and bike stores had built up inventories, expecting to profit from the rapidly inflating prices. There was a huge backlog of unsold bicycles and components in the distribution system, and the whole bicycle industry slumped. When the bubble burst in 1975, the industry was over the saddle in inventory and under the pedals in bank balance.

Left: Fig. 11.1. "Pre-industrial" mountain bikes getting ready for the May 1977 Repack race on Mount Tamalpais, in Marin County, California. At the time, the winning bikes had derailleur gears and a mix of drum brakes and cantilever brakes.

However, the U.S. had become the largest market for derailleur bicycles, and henceforth developments would go from the U.S. to the rest of the world. The ten-fold increase in the price of crude oil in 1974 had created a worldwide recession. The U.S. suffered from unemployment, inflation, and interest rates all over 10%. This provided a wonderful stimulus for bicycling in western Europe and the U.S. For a few years, bicycle commuting and bicycle shopping became socially acceptable for middle-class people.

These new middle-class riders demanded better bicycles and were prepared to pay for them. Annual cycling mileage increased dramatically. The increased number of bicycles on the road got the attention of motorists and eventually of politicians and highway departments.

BMX (Bicycle Moto-Cross) appeared on the U.S. scene at the end of the bike boom. The American bicycle market gradually recovered, but the sales slump in 1982/1983 was the second part of the one-two punch. Bike sales fell, except for mountain bikes. Competition was fierce and prices were down in spite of inflation. Taiwanese bicycles became competitive in the medium-price range.

Japanese companies gradually took over the component market. By 1983, SunTour equipped more than half of the medium-price derailleur bikes sold in the U.S., and Shimano supplied almost all of the rest. A few Huret Duopars were used on top-quality touring bikes, and a few Campagnolo Nuovo Records were used on top-quality road racing bikes.

Huret was the first of the old time component makers to go, merging with Sachs in 1982.

A new generation of copycat derailleur makers entered the market place, but they did not thrive. Most of the European newcomers, Belleri, Galli, Mavic, Roto, Triplex, and Zeus, copied Campagnolo. Gipiemme and Ofmega copied Simplex. The Taiwanese newcomers, Falcon, Joytech, Long-Yih, and SunRace, copied Shimano. None of the new European companies had any impact in the crucial U.S. market, but Taiwanese components were sometimes specified on economy 10-speeds.

## United States

The U.S. bicycle market changed continuously. It was necessary to follow the trends (or fads) to survive. Large bicycle makers like Schwinn, Raleigh, Motobécane, and Peugeot lacked the necessary flexibility and they fell on hard times. Bicycle prices more than doubled in this decade of high inflation rates.

The mountain bike was the most significant change of the decade. Mountain bike sales grew from about two hundred in 1979 to about a million in 1984. The user-friendly mountain bike introduced a whole new population of bicycle riders to the market place.

Bicycle touring became popular in the U.S. Loaded-touring bicycles were developed to satisfy the demand. Bicycle racing and triathlons also grew steadily.

Sales of imported bicycles (mostly 10-speeds) dropped from 5 million in 1973 to 4 million in 1974 to 2 million in 1975. With

Below: Fig. 11.2. Graph of U.S. total bicycle sales, domestic bicycle production, and bicycle imports,1974–1986. (Source of data: Bicycle Manufacturers' Association of America.)

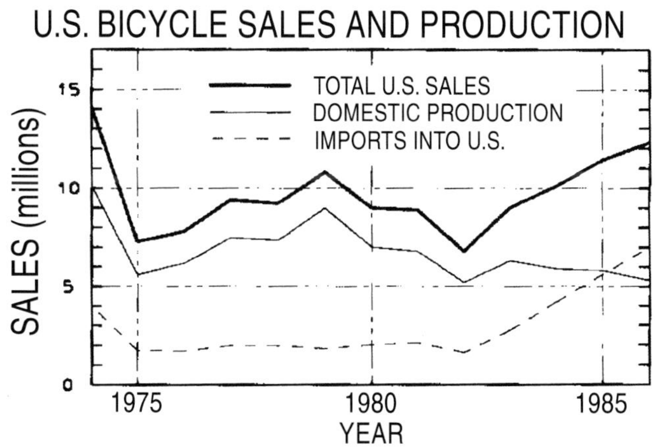

Above: Fig. 11.3. Introduced in 1979, the first production Ritchey MountainBikes sold for $1,300. Tom Ritchey built the frames. Gary Fisher and Charlie Kelly assembled them into complete bikes, and sold them to customers in Marin County, California. These sales established that there was a significant market for mountain bikes.

full warehouses, the importers of little-known European brands like Atala, Batavus, Cazenave, Crescent, Falcon, and Stella, ran advertisements in *Bicycling* looking for dealers. Unfortunately, the dealers were also suffering from lack of sales, and many bike shops closed their doors.

The 1975 slump was especially hard on the French and Italian component makers. They had received huge orders from the American builders of low-priced bicycles. When the boom ended, these buyers stopped buying for a few years while they sold off their component inventories. In the early 1970s, the American fad for high-riser bikes moved to Europe, pumping up European sales for a while.

By 1976, the market had recovered a bit for domestic bikes and the strong businesses survived. 1977, 1978, and 1979 were good years and domestic production exceeded the best pre-boom years. 1980, 1981 and 1982 were bad years. Total bicycle sales were under 7 million. Then the mountain bike took over and the market turned up again.

The market for imported bicycles rattled along between 2 million and 2.5 million from 1975 to 1982. By and large, these were good quality bicycles made by reliable companies and sold by reliable importers.

U.S. buyers had became much more knowledgeable about lightweight bicycles. The 15 million or so new adult cyclists with new 10-speeds had a major influence. Many read the cycling magazines and learned about gearing, tubing, and frame geometry. It was no longer necessary to order custom-built bicycles from Britain, France, or Italy. There was a market for a growing number of American custom bike builders. By 1977, *Bicycling* listed 21 U.S. custom builders.

## Schwinn

Schwinn, the quality American bicycle maker, floundered in this decade. Schwinn's ancient Chicago factory could only make bicycles with heavy flash-welded frames, such as the Varsity and the Continental. However, the market demanded lighter lugged-frame bicycles. In 1974, Schwinn went to Japan and negotiated with Bridgestone and Panasonic for lugged-frame bicycles. The imported Japanese Schwinn bicycles were lighter and less expensive than the Chicago Schwinns.

Yoshi Shimano, the youngest brother, was in charge of Shimano's U.S. sales. He was Schwinn's negotiator and translator at these negotiations. Not surprisingly, the Japanese-built Schwinn bicycles came equipped with Shimano components. The Shimano derailleurs were labeled "Schwinn-Approved."

In March 1980, Schwinn's Chicago workers voted to join the United Auto Workers. The UAW was unable to negotiate a satisfactory contract with Schwinn, and in September there was a four-month strike. The workers prevailed, but it was the end of the line for the old Chicago factory. During the strike, Schwinn increased their imports from

Left: Fig. 11.4. The 1984 Diamond Back Formula One was the BMX bike that your son really wanted.

Right: Fig. 11.5. The 1977 Schwinn Mag Scrambler was modeled after BMX bikes, but real BMX bikes did not have chain guards and reflectors.

Bridgestone and Panasonic in Japan, and from Giant in Taiwan. The imports were more profitable than Schwinn's Chicago-made bicycles. The Chicago factory closed at the end of 1983.

Schwinn built a new factory in Greenville, Mississippi, which opened in 1981. It was poorly located, and it was never profitable in its ten-year life.

Throughout the market's ups and downs, there was a steady demand for children's and teenagers' bicycles. BMX racing developed at the end of the bike boom. BMX bikes were small-wheeled single-speeds that teenagers raced over obstacle courses. The early BMX models were old Schwinn Sting-Rays that were stripped of derailleurs and other non-essential parts. BMX bikes sold well, but Schwinn missed most of the BMX market. Red Line, Mongoose, and Diamond Back were the major suppliers. BMX sales grew to 1.75 million bicycles by 1977. BMX sales peaked in 1982 at around 2 million.

## Trek

Trek started making bicycle frames in 1976 in Waterloo, Wisconsin. By 1980, they were building 6,000 hand-brazed lugged frames per year on a 15-man production line. Trek started making complete bicycles in 1981. In 1982, the Trek 957 was their top-of-the-line model, with a Columbus double-butted frame and a Campagnolo Super Record gruppo. Their advertising emphasized high-end bicycles and advanced frame technology. By 1984, Trek had 130 employees, and they were offering road bikes like the $460 Trek 560, with a lugged frame made of Reynolds 501 tubing with SunTour Cyclone derailleurs. Trek also had four loaded-touring bikes and three mountain bikes. By 1984, they were making 60,000 bicycles per year. However, they expanded too rapidly, and lost money for the next two years.

## Cannondale

In 1977, Cannondale started making panniers and bags in a factory in Stanford, Connecticut. In 1983, Cannondale made their first bicycles using oversized aluminum frame tubes. The early Cannondales had problems with warped tubes and defective welds, but these were quickly corrected. The 1985 Cannondale SR900 was their top racing bike. It weighed 22.5 lbs. and had a Campagnolo Nuovo Record geartrain. Like Trek, they concentrated on the high-priced market, and they learned to follow the trends.

Right: Fig. 11.6. The 1982 Trek 957 was a high-end bike that was equipped with Campagnolo Super Record components.

Left: Fig. 11.7. The 1985 $1,000 Cannondale SR900 was equipped with Campagnolo Nuovo Record derailleurs.

### Raleigh Cycle Company of America

TI-Raleigh had serious problems with the fast-changing and trendy U.S. market because they operated on a "Father knows best" basis. Worse still, Raleigh bicycles were poorly assembled, and it took a conscientious dealer several hours to inspect and rebuild each Raleigh to fix the shoddy Nottingham workmanship. Meanwhile, Fujis and Panasonic bikes came out of the box perfectly assembled, with the cables greased and cut to length. They could be assembled in ten minutes.

TI-Raleigh was five years late on high-riser bicycles. They had lost their reputation for quality during the Bike Boom. Their lawyers told them not to sell BMX bicycles because of liability considerations. When the second economic slump hit in 1982, TI-Raleigh was losing money in America, and they finally walked away from the U.S. rights to the Raleigh name and from the U.S. network of Raleigh-franchised dealers.

Huffy reached an agreement with TI-Raleigh in 1982 to use the Raleigh name. Huffy called the new American company Raleigh Cycle Company of America. TI-Raleigh's assembly plant in Enid, Oklahoma closed after the Huffy transfer. For two years, Raleigh Cycle Company of America imported bicycles from Taiwan and Japan. In 1985, they opened a new bicycle factory in Seattle, Washington.

### U.S. Bicycle Importers

The American importers were the most successful. They went to Japan, and later to Taiwan, to have bicycles made to order with their nameplates. West Coast Cycle imported Azuki, Nishiki, and Cycle Pro. Velocipede imported Sekai. Ben Lawee imported Univega. Western States imported Centurion and Diamond Back. The importers had no investment in manufacturing plants or engineering and research departments. They moved quickly with the trends and the boom-and-bust cycles. By and large, their bicycles were good value, and they all used Japanese components.

The Centurion story is a good yarn: Mitchell Weiner was a California bicycle sales agent. In 1969, he got involved in a project to help Raleigh Industries of America import a Japanese copy of the Raleigh Grand Prix. The corporate parent, TI-Raleigh, nixed the plan, and Weiner was stuck with 2,000 bicycles painted in Raleigh Grand Prix colors, including the pin stripes. So he stuck Centurion labels on the bicycles, and sold them at a profit. This led to the formation of Western States Imports, which imported Japanese bicycles and sold them to dealers at lower wholesale prices than comparable Raleighs or Schwinns. The 1975 Centurion LeMans had a

Left: Fig. 11.8. The 1975 Centurion LeMans had a SunTour gear-train and a Takagi Tourney alloy cotterless crankset.

Right: Fig. 11.9. The 1983 Specialized Sequoia was equipped with a Sugino AT triple crankset and a SunTour Cyclone gear-train.

SunTour gear-train, an alloy Tourney cotterless crankset, and weighed 28 lbs. (12.7 kg).

In 1975, Mike Sinyard started Specialized Bicycle Imports (SBI) to import Italian components into California. He developed a fine feel for the U.S. bicycle market, and imported products to fill niches. The SBI Turbo was the first high-performance clincher tire to catch on in the U.S. In 1982, SBI imported the Allez racing bicycle, the Sequoia sport-touring bicycle, and the Expedition loaded-touring bicycle from Japan. The Sequoia was a sports-touring bike with a Sugino AT triple crankset, a SunTour Cyclone gear-train, and SunTour Superbe brakes. Many bike stores added SBI to their line to satisfy the growing market for $700 bicycles.

Many American companies were attracted to unusual bicycle gear-trains. In 1982, Excel, a division of Beatrice Foods, started making low-priced pedals, hubs, and derailleurs in Illinois. In 1983, they imported the Cambio Rino gruppo from Italy with Excel nameplates, and they offered the Excel Cambiogear, which was a 16-speed expanding chainwheel with a 3 to 1 range. It was made of a graphite-fiber composite. Excel managed to convince a few makers of low-priced bikes, but the concept did not succeed, and a year later it was gone.

The 1983 Houdaille Exo Powercam was an American drive-train innovation that was marketed under a succession of owners. The 1977 version was called Facet Biocam; the 1978 version was called Selectocam. This device was well built and did what it claimed to do: The pedals rotated in circles, while the cam altered the mechanical advantage so that the feet went slower through the top and bottom of the stroke and faster through the power stroke. The pedaling style took about 300 miles to get used to: It felt like pedaling standing up, and it was uncomfortable to experienced cyclists who had learned to spin. Although it was expensive, it attracted a small devoted following. A Selectocam-equipped bike finished second in the 1979 Paris–Brest–Paris *randonnée*, and Lon Haldeman won the 1983 Race Across America using an Exo Powercam. Dating back to the Simpson Lever Chain, promoters had learned to hire top athletes to win races with their inventions. Sometime in the mid-1980s, Houdaille stopped underwriting the losses, and production ceased.

Above: Fig. 11.10. The1983 Houdaille Exo Powercam allowed a higher gear at a lower cadence.

Right: Fig. 11.11. The 1983 Excel Cambiogear was a 16-speed expanding chainwheel with a 3-to-1 range.

During much of this period, I was the Technical Editor for *Bicycling* magazine. In that capacity, I performed tests and wrote about 100 technical articles for them. In 1977, I tested five different touring derailleurs and rated them in order:

1: SunTour Cyclone GT

2: SunTour VGT

3: Shimano Crane GS

4: Shimano 600 GS

5: Campagnolo Rally.

In 1984, I tested and rated 37 narrow-range, 36 mid-range, and 39 wide-range rear derailleurs and wrote up the results in *Bicycling*. I had two different test machines—one for racing and the other for touring derailleurs (see Chapter 14 for details of these machines and the tests I performed). It took three hours per derailleur to conduct the capacity and shifting performance tests. I measured how the shift followed the lever movement. I down-rated derailleurs that required pushing the lever beyond the shift point and then re-centering. By the end of this program, I could in-

spect a rear derailleur and predict how it would shift.

Using my standard for precise, predictable shifting, SunTour's slant-parallelogram derailleurs shifted best. Simplex and Shimano double-pivot derailleurs shifted well. Single-pivot Campagnolos and their many clones shifted poorly. The old books said to flick the lever past the shift point and then re-center, which is what you had to do for a poorly shifting rear derailleur. Serious cyclists had learned this technique to shift with Campagnolo derailleurs, and they did not realize that this "overshifting" was actually a defect.

I established that the Huret Duopar, the Shimano SuperPlate, and the SunTour Tech series were the best-shifting rear derailleurs for touring and for mountain bikes.

### Bikecentennial

This decade was a great time for long-distance bicycle touring in the U.S. The magazines were full of stories of tours and details of how to set up your bike and other equipment for touring. Dozens of different touring bicycles were available.

Left: Fig. 11.12. Contributing author Ron Shepherd on Bikecentennial in 1976. His gearing setup, with a 22-tooth "granny" chainwheel, allowed him to ride where others had to walk.

Right: Fig. 11.13. The $820 Trek 720, introduced in 1984, was author Frank Berto's favorite touring bike. It had long chainstays, braze-on mounts for front and rear racks, and a wide-range geartrain with bar-end shifters.

Bikecentennial organized tours across the U.S. in 1976, giving a tremendous boost to bicycle touring. The idea started in 1973, when Don and Lys Burden and Greg and June Siple arrived in San Francisco, about a third of the way through their Hemistour, an 18,000-mile (29,000-km) bike tour from Anchorage, Alaska, to Tierra Del Fuego, on the southern tip of Chile, in South America.

Greg and Don conceived the idea of organizing a 3,000-mile (4,800-km), 76-day, tour across America in the summer of 1976 to celebrate the U.S. bicentennial. The Siples eventually finished Hemistour in 1975, while the Burdens spent the next three years making Bikecentennial a reality.

Before it was over, 100 or so underpaid volunteers had established headquarters in Missoula, Montana. They laid out a 4,250-mile (6,830 km) route from Astoria, Oregon, to Yorktown, Virginia; prepared maps and guide books; recruited and trained 600 ride leaders; and arranged accommodation at all of the stops en route.

In addition to the full 76-day tour, there were 12-, 21-, and 35-day partial tours. Four thousand riders took part, and just over two thousand completed the entire route. Four hundred came from overseas, including contributing author Ron Shepherd.

Ron led a group of 33 Australian cyclists. Most of them used TA triple cranksets with a 26-tooth inner chainwheel. Ron combined a TA 44/32 pair with a 22-tooth granny chainwheel and an Atom 11–32 freewheel. With 26-inch wheels, this gave him a range of 18 to 104 in.

American riders were puzzled by the Australians' low gears, which were rarely used in the well-graded roads over the Rockies. But when they came to the steeper sections of the Ozarks and the Appalachians, they astonished everyone with their hill-climbing ability. "Look at those Aussies go!" was a standard comment.

The Bikecentennial tour was so successful that when it was over, the organization refused to disband. It became Bikecentennial Inc., which later changed its name to Adventure Cycling Association. It now has 30,000 members and organizes tours and prepares the best bicycle touring maps in the U.S.

Bikecentennial also served as a large-scale road test for touring bicycles and components. The standard "alpine-geared" 10-speed was clearly inadequate, and most of the low-end derailleurs were not cut out for continuous duty.

At about the same time, two long-time cross-state touring events attracted thousands of recreational riders. TOSRV (Tour of the Scioto River Valley), which began in 1962, is a two-day, 208-mile (335 km) round trip in Ohio. RAGBRAI (Des Moines Register Annual Great Bicycle Ride Across Iowa) is a 500-mile (800 km) ride that takes a week and attracts about 10,000 riders. By 1981, there were similar rides in Minnesota, Wisconsin, Oklahoma, and Florida.

## Touring Bikes

As a result of these touring events, American cyclists quickly learned the difference between genuine loaded-touring bikes and pseudo touring bikes. By the late 1970s, they knew what French cycle tourists had known for decades: A loaded-touring bike needs a long wheelbase, a shallow head angle with lots of fork rake, heavy-duty tires, good brakes, and front and rear panniers.

It also needs a low enough low gear. I like to think that Ron Shepherd and I had something to do with that part. We preached the gospel of genuine "granny" inner chainrings. For a long time, the smallest inner chainwheel available was the 26-tooth one on the venerable TA Cyclotouriste. The TA crankset had problems with front derailleur interference, so many people used the Stronglight 100, which allowed only a 28-tooth granny. In 1981, Sugino introduced the Aero Tour triple crankset, which had an inner 74 mm diameter bolt-circle that took a 24-tooth granny chainring. The Aero Tour was a much better touring crankset than the TA Cyclo-

Right: Fig. 11.14. The 1984 Specialized Expedition was a $700 loaded-touring bike. It's dressed up here with generator lights, Blackburn Lowrider racks, and Tailwind panniers.

Left: Fig. 11.15. The 1984 Cannondale ST 500 had a Sakae triple crankset, SunTour Superbe Tech derailleurs, and Gran Compe sidepull brakes.

touriste, and soon there were many copies.

In 1981, Blackburn came out with Low-Rider racks and panniers, which made touring with a full load a lot more comfortable due to increased stability of the loaded bike. Interestingly, Jim Blackburn's 1980 stability tests agreed with Daniel Rebour's 1946 recommendations.

The market for loaded-touring bicycles peaked in 1984, with more than 40 different models. Most of the major manufacturers had at least one model, and many offered two or three.

In 1984, I road-tested the top-of-the-line loaded-touring bikes from Cannondale, Trek, and Specialized. The Trek 720 was my favorite. It had long chainstays, braze-on mounts for front and rear racks, and a wide-range geartrain with a Huret Duopar rear derailleur, Simplex front derailleur, and SunTour Barcon shifters Afterward, I had a near copy of the Trek 720 built to order by Columbine. I rode that bike on several long tours, including Land's End to John O'Groats.

The 1984 Specialized Expedition had a Sugino Aero Tour triple crankset, SunTour MounTech derailleurs, and 700 x 32C tires. It's perfectly capable of a ride across the country today.

The 1984 Cannondale ST 500 had a Sakae triple crankset, SunTour Superbe Tech derailleurs, and Gran Compe sidepull brakes. It did not handle as well as the Trek or the Specialized.

The 1984 Fuji Touring Series IV was the bargain. At only $440, it had a Sugino Aero Tour triple crankset, SunTour MounTech derailleurs, and Dia-Compe cantilever brakes. Fuji still makes a couple of affordable loaded-touring bikes as of 2005.

The loaded-touring bicycle market peaked out because of mountain bikes. With low gears, wide tires, and good brakes, they fulfilled many of the requirements for touring. A second reason was that a reliable loaded-touring bike does not wear out, so most of them are still around.

Left: Fig. 11.16. The 1984 Fuji Touring Series IV had a Sugino Aero Tour triple crankset, SunTour MounTech derailleurs, and Dia-Compe cantilever brakes. At $440, it was the bargain amongst touring bikes.

## Mountain Bikes

The development of the mountain bike was the most significant change during this decade. A few hundred mountain bikes were sold in 1979 and 1980. For the next four years, annual sales increased by an order of magnitude each year. About a million mountain bikes were sold in 1984, and by then, this trend had affected the type of bicycle sold in America. Mountain bikes soon outsold road bikes. A few years later, the mountain bike invaded Europe.

Road bikes had become light, fussy, and delicate, which intimidated new riders. You did not have to be a bike nut to ride a mountain bike, and most of them were sold to new cyclists—and not necessarily for off-road use. They were user-friendly, easy to ride, and they rarely got flat tires. They were much closer to the old French touring bikes of the 1940s.

This development was a similar sea change to the switch from the high ordinary bicycle to the safety bicycle in the 1890s. The mountain bike drastically altered the bicycle market in the next decade. To survive, both bicycle and component makers had to adapt to the switch from road bikes to mountain bikes.

The mountain bike originated and matured on the slopes of Mount Tamalpais, just a few miles from where I live in Marin County in northern California. It comprised a combination of features and components that had been invented earlier. The six key characteristics were the following:

- fat tires;
- upright riding position;
- flat handlebars;
- wide-range derailleur gearing;
- powerful brakes;
- heavy-duty frame designed for off-road use.

The mountain bike was developed by people experimenting with conventional bicycles racing down the slopes of "Mount Tam." They quickly learned that the frames and forks of the Schwinn Motorbike and the nearly identical Schwinn Excelsior held up better than other bikes in strenuous off-road riding. Until about 1985, almost all mountain bikes used Schwinn's frame geometry.

Numerous one-of-a-kind fat-tired, derailleur-geared roadsters with good brakes had been built elsewhere before 1975, but nothing came of them. In Marin, the conditions were right for the chain reaction.

Having researched and written a book about the subject, I am convinced that no single person invented the mountain bike. They just evolved by one development after another. It started in the early 1970s, when a small number of counter-culture young men started riding old coaster brake newsboy bikes, endearingly referred to as "clunkers," on the dirt trails of Mount Tamalpais just for kicks. The first timed and recorded clunker races were held on a two-mile long downhill dirt fire road that became known as Repack. Most of the early clunkers used coaster brakes, and the name Repack came from the need to repack the burned out grease in the coaster brakes after a fast downhill run.

Gary Fisher, Charlie Kelly, Joe Breeze, Tom Ritchey, and Mike Sinyard were all deeply involved in the initial development of the mountain bike.

Gary Fisher was born in 1952. He rode and built fat-tired clunkers from 1973 onward. Fisher and Kelly started a cottage industry, taking old newsboy bicycle frames and converting them into derailleur clunkers. The Schwinn Excelsior frame was preferred because it could take abuse. They used an eclectic mixture of components. The shortage of old Schwinn frames led to Kelly's request of Joe Breeze to build the first custom mountain bike, the Breezer. Fisher and Kelly founded MountainBikes (spelled as one word but with a capital B in the middle), the first company to sell mountain bikes in volume. In 1979, they ordered ten custom-built frames from Tom Ritchey patterned after the Breezer. Today, Fisher is president of the Gary Fisher Bicycles, a division of Trek.

Left: Fig. 11.17. The 1935 balloon-tire Schwinn Motorbike and the nearly identical Schwinn Excelsior held up better than other bikes in strenuous off-road riding.

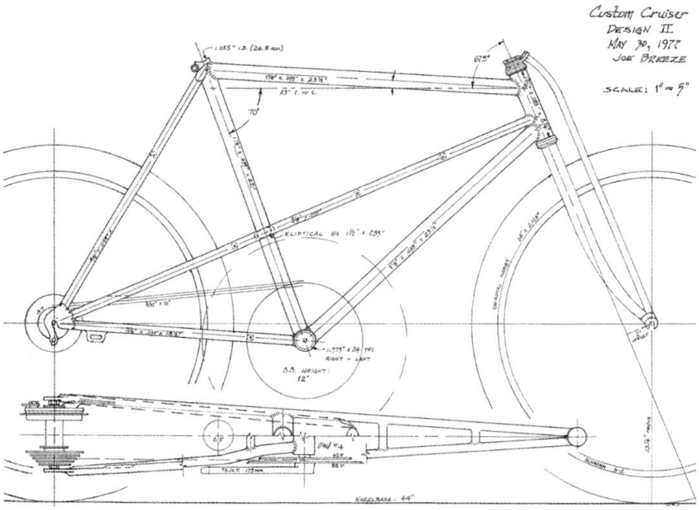


Right: Fig. 11.18. Joe Breeze's 1978 Breezer was the first purpose-built mountain bike with all new components. He based the frame geometry on the dimensions and angles of the Schwinn Excelsior.

Charlie Kelly was born in 1945. With Joe Breeze, Kelly organized the first Repack race in 1976. It was also he who first had the idea of getting a purpose-built mountain bike frame made instead of continuing to use heavy old frames as the basis for an off-road bike. He was Gary Fisher's partner in MountainBikes from 1979 to 1982. Today, he operates a piano moving business and occasionally writes mountain biking articles.

Joe Breeze was born in 1954. In 1977, he built the first batch of ten Breezer mountain bikes. Today, he lives in Fairfax, California, and heads his own company, Breezer Bikes, which designs and imports modern lightweight transportation bikes. The original Breezer was the first purpose-built mountain bike with all new components. The first model had fork braces. The next nine had BMX forks.

Above: Fig. 11.19. Joe Breeze with the first Breezer. In 1977, he made the first batch of ten Breezer mountain bikes.

Right: Fig. 11.20. Gary Fisher, shown here in a 1977 photo, rode and assembled fat-tired clunkers from 1973 onward.

Tom Ritchey was born in 1956. He was a prolific frame builder. In 1979, he built the first dozen frames for MountainBikes, which sold for $1,300 fully built up. He made about 40 Ritchey MountainBikes in 1979. Gary Fisher and Charlie Kelly sold them to buyers in Marin County, California. These sales established that there was a significant market for mountain bikes. From 1979 to 1982, Ritchey probably built half of the frames for the first 2,000 mountain bikes. Today, he lives in La Honda, California, and heads his own company, Ritchey Designs. He still brazes frames for the top-of-the-line Ritchey bicycles.

Mike Sinyard was born in 1949. In 1982 Sinyard shipped a Ritchey MountainBike to Japan and got Toyo to make carbon copies—the Stumpjumper, the first mass-produced mountain bike that was sold in regular bike stores. It used SunTour AR derailleurs, SunTour thumb shifters, Mafac cantilevers, and a TA Cyclotouriste triple crankset. A year later, it was all Shimano Deore. Specialized advertised very little in 1982, since the Toyo factory could not make them fast enough, and the $900 Stumpjumper was never in stock.

Joe Breeze had based the frame geometry of the Breezer on the dimensions of the Schwinn Excelsior. Tom Ritchey used the dimensions of the Breezer to build frames for Ritchey MountainBikes. Mike Sinyard used the dimensions of the Ritchey Mountain-Bike. Finally, the rest of the industry copied the dimensions of the Specialized Stumpjumper. Thus was born what was referred to as "Marin" geometry.

In 1980, the going price for a Ritchey MountainBike was $1,300. You paid in advance, and a few weeks later (if you were lucky), you picked up your bike. It was called a Ritchey Mountain-Bike because Tom Ritchey built the frame. This was standard practice for custom-built bicycles. However, Tom did not own any part of MountainBikes, the company started and owned by Gary Fisher and Charlie Kelly. This led to problems, and eventually to an unhappy breakup at the end of 1983.

Half a dozen entrepreneurs in the San Francisco Bay area made mountain bikes, but

MountainBikes was the market leader. About 200 mountain bikes were sold in 1980. Fisher and Kelly sold three quarters of them from their storefront in Fairfax. Tom Ritchey turned out frames in batches of 100. He charged Fisher and Kelly $150 for a bare frame without forks.

The early Marin builders used a mixture of components that had survived the Repack testing laboratory: Mafac cantilever brakes, Magura moped brake levers, TA Cyclotouriste triple cranksets with long cranks, Huret Duopar rear derailleurs, Simplex front derailleurs, and SunTour Power Ratchet thumb shifters.

The Duopar shifted well on wide-range gearing. The Simplex front derailleur had enough travel to clear the fat rear tire and shifted (sort of) OK on the TA Cyclotouriste triple crankset. The TA was available with long (180 mm or 185 mm) crank arms, and it was cheap. Mafac cantilever calipers were cheap and readily available.

Production of mountain bikes grew each year by an order of magnitude, from tens to hundreds to thousands. By 1981, component sales were in the

Above: Fig. 11.21. Charlie Kelly, shown here in a 1978 photo, was Gary Fisher's partner in MountainBikes from 1979 to 1982. He and Joe Breeze organized the first Repack race in 1976. He was the first writer to cover mountain biking, and he was first with the idea of having a mountain bike frame custom built.

Below: Fig. 11.22. Tom Ritchey, shown here in a 1988 photo, built the first dozen mountain bike frames for MountainBikes.

Right: Fig. 11.23. The Repack race course was the proving ground for the development of the mountain bike.

Below: Fig. 11.24. The hallmark of all the early mountain bikes was the SunTour handlebar-mounted thumb-shifter.

thousands and growing. None of the French companies bothered to ask why they were not selling complete gruppos. The TA Cyclotouriste crankset was typical of French marketing in the 1980s. Every user hated the narrow gap between the crank arm and the outer chainring, because the front derailleur had to be slanted inward to keep the outside of the front derailleur cage from hitting the inside of the crank. Alain Breuil, who owned TA, was a good friend of Fred DeLong. Fred and I told him that if he did not change the design of the Cyclotouriste, the Japanese would eat his lunch. He didn't, and they did. However, in a triumph of tradition over technology, Specialités TA is still in business as of 2008, and the TA Cyclotouriste crankset and matching chainrings are still available.

The 1982 *Bicycling Buyer's Guide* listed 280 bicycles. There were just three mountain bikes: the $1,095 Proteus, the $880 Specialized Stumpjumper, and the $250 Schwinn Sidewinder. The Sidewinder was a 10-speed with sidepull brakes, so it hardly qualified as a mountain bike.

The Marin County entrepreneurs, including Tom Ritchey, Gary Fisher, and Charlie Kelly, did not appear in the *Buyer's Guide,* although they sold 500 MountainBikes in 1982. They were probably too busy to fill out the questionnaire.

Some 3,000 Univega Alpina Sports were sold in 1982. It used SunTour Cyclone GT derailleurs and Mighty shifters, a Sugino Pro-Max double crankset, and DiaCompe cantilevers.

The success of the 1982 Specialized and Univega mountain bikes got attention. In 1983, Cycle Pro, Diamond Back, Fuji, KHS, Lotus, Marukin, Miyata, Murray, Raleigh-USA, Ross, Schwinn, Specialized, Trek, and Univega were all selling mountain bikes to the mass market at prices from $250 to $750.

Mountain bikes took about 10% of the U.S. market in 1983. Specialized's $500 Stumpjumper Sport was the market leader, and if you wanted to spend more than $1,000 for a mountain bike, there were a dozen or so custom makers.

By 1984, almost all of the major bike makers had at least one mountain bike model, and many makers had two or three. Mountain bikes took about 20% of the U.S. bicycle market. The market broadened to include hybrids, or city bikes, which were cheaper, lighter mountain bikes with medium-width tires and medium-range gearing.

The Japanese component makers climbed aboard the mountain bike bandwagon. Aluminum alloy rims for 26-inch wheels and skin-wall tires lowered weight and improved braking and handling. First-class triple-chainring cranksets with 24-tooth inner chainwheels were available from Sugino, Sakae Ringyo, Specialized, Takagi, and Shimano.

SunTour's MounTech gruppo was on about 60% of the bikes sold between 1983 and 1985. It

Left: Fig. 11.25. This late model Repack Racer, now on display at the Shimano museum, in Sakai, Japan, is claimed to be a replica of Gary Fisher's 1976 derailleur clunker. It has a Schwinn Excelsior frame.

Above: Fig. 11.26. Mike Sinyard, shown here in a 1983 photo, introduced the Stumpjumper, the first mass-produced mountain bike that was sold in regular bike stores.

was SunTour's version of the Duopar., and was the first purpose-built mountain bike rear derailleur on the scene. It shifted beautifully, but it wore out quickly in dirty off-road service.

Shimano Deore XT derailleurs were on the rest. The Deore XT mountain-bike gruppo came out half a year later than SunTour's MounTech, and it was a more rugged package. The rear derailleur had much better seals, and Shimano took over most of the market.

Once SunTour and Shimano made full mountain-bike gruppos, there was no market for French components. The window of opportunity for Huret, Mafac, Simplex, and TA had been open from 1979 to 1983, and now it slammed shut.

The whole fat-tire phenomena took place almost in my back yard. I watched first with derision, then with disbelief, and finally with awe. I remember telling Gary Fisher that his first ten custom-built MountainBikes had saturated the market. I missed the point: By 1980, lightweight road bikes had become fussy and unreliable, and the much more user-friendly mountain bike opened up a whole new market.

The market was ripe for change, and the San Francisco Bay area had the appropriate culture for the mountain bike to develop and thrive. Mountain bikes attracted millions of new bicycle riders. Most mountain bikes were ridden on pavement rather than off-road, but that was not the point.

The mountain bike did not evolve from the 10-speed road bike. Instead, its sire was the prewar Schwinn coaster-brake roadster. Many of the buyers had not ridden a bicycle for years. These inexperienced but affluent adult riders did not want to fiddle with gear shifting. This created a de-

Left: Fig. 11.29. The 1982 Specialized Stumpjumper was the first mass-produced mountain bike to be sold nationwide. It was equipped with SunTour AR derailleurs, SunTour thumb-shifters, Mafac cantilever brakes, and a TA Cyclotouriste triple crankset.

Below: Fig. 11.27. The TA Cyclotouriste crankset was typical of French marketing in the 1980s: flawed but unchanged. Being the only triple crankset readily available at the time, it was standard on most early mountain bikes.

Right: Fig. 11.28. The $500 Univega Alpina Sport, introduced in 1982, was the best-selling mountain bike. It was equipped with SunTour Cyclone GT derailleurs and Mighty shifters, a Sugino Pro-Max double crankset, and DiaCompe cantilever brakes.

mand for more reliable, easier-shifting components.

## Bicycle Racing

During this decade, bicycle racing began to get serious attention in the U.S. Most American racers equipped their bikes with a Campagnolo Nuovo Record gruppo if they could afford it.

At the 1976 Montreal Olympic Games, George Mount finished 6th in the Individual road race, the highest American placing ever.

Time trialing became popular, and many bicycle clubs organized evening or weekend time-trials. The United States Cycling Federation published annual records for the 25-mile (40 km) distance.

The 1982 Race Across America (RAAM) lasted for 10 days and got great publicity. The 1978 Hawaii Ironman Triathlon race was a major news event. It combined a 4 km (2.4-mile) swim, a 180 km (112-mile) bike race, and a 42 km (26.2-mile) run.

Greg LeMond won the 1983 World Championship road race and finished third in the 1984 Tour de France.

The 1984 Olympics were held in Los Angeles, and the home team cyclists did very well: Connie Carpenter-Phinney and Rebecca Twigg finished first and second in the woman's road race. Alexi Grewal took the gold medal in the men's road race, and Steve Hegg took the gold medal in the 4,000-meter pursuit. The American 4,000-meter pursuit team finished second to Australia. Mark Gorski and Nelson Vails took gold and silver respectively in the 200-meter sprint.

All of this from a country that had not won a single Olympic cycling medal since 1912. The success of American bicycle racers was great publicity for the sport.

### Tour de France

In 1981, Jonathan Boyer was the first American to ride in the Tour de France. He finished 32nd. He finished 12th in 1983. In 1983, Greg LeMond was on the Renault Gitane team, helping Bernard Hinault. LeMond won the 1983 World Championship race. In 1984, he finished third in the Tour de France to teammate Laurent Fignon.

### Speed and Distance Records

In 1984, Francesco Moser raised the unpaced one-hour record to 51.15 km (31.71 miles). Moser's bike had a large-diameter disc rear wheel, which was an aerody-

*Table 11.1. Tour de France winners, 1975–1984*

| Year | Rider (Country) | Bicycle | Derailleur |
|---|---|---|---|
| 1975 | Bernard Thévénet (France) | Peugeot | Simplex |
| 1976 | Lucien Van Impe (Belgium) | Gitane | Campagnolo |
| 1977 | Bernard Thévenet (France) | Peugeot | Simplex |
| 1978 | Bernard Hinault (France) | Gitane | Campagnolo |
| 1979 | Bernard Hinault (France) | Gitane | Campagnolo |
| 1980 | Joop Zoetemelk (Holland) | Raleigh | Campagnolo |
| 1981 | Bernard Hinault (France) | Gitane | Campagnolo |
| 1982 | Bernard Hinault (France) | Gitane | Campagnolo |
| 1983 | Laurent Fignon (France) | Gitane | Simplex |
| 1984 | Laurent Fignon (France) | Gitane | Campagnolo |

Above: Fig. 11.30. Bernard Thévénet's 1975 Tour de France bicycle was a stock Peugeot PY1 with Simplex derailleurs and a Stronglight crankset.

namic feature that contributed to his record.

The International Human Powered Vehicle Association (IHPVA) began holding speed contests in 1974. The IHPVA placed no restrictions on the design of the vehicle, except that it had to be human-powered and there could be no stored energy. Setting an IHPVA record required a combination of a very strong athlete and superlative aerodynamics. The 200-meter IHPVA record allowed 5 miles to get up to speed before entering the measured distance.

In 1974, Ron Skarin set the first IHPVA 200-meter record at 69.2 km/h (43.0 mph). This record was raised almost every year since, until Dave Grylls went 94.8 km/h (58.9 mph) in 1980.

In 1979, Ron Skarin set the first one-hour IHPVA record at 51.3 km (31.8 miles). In 1984, Freddie Markham raised the record to 60.35 km (37.5 miles).

The motor-paced speed record is a test of both courage and strength. Few people relish the idea of pedaling at 150 mph, even behind a large hood on a race car. The technology is largely based on motorcycles. To keep the cadence within reason requires a huge chainwheel so the bicycle is usually towed up to about 60 mph.

## France

French bicycle makers fell on hard times after the Great American Bike Boom. French domestic sales were down, and imports of inexpensive bicycles into France increased from 350,000 in 1974 to 680,000 in 1984.

From 1974 on, the American market was saturated with 10-speeds, and Peugeot and Motobécane did not even bother to exhibit at the 1976 New York bike show. Part of the problem was that the U.S. Consumer Products Safety Commission (CPSC) mandated a simulated laboratory crash test for bicycle frames in 1976. This test contributed nothing to consumer safety, but it eliminated the popular French and Italian club racers from the U.S. market.

Some French companies lost their identity during those years: Motobécane equipped their "definitely French" medium-priced Mirage with SunTour derailleurs and freewheels, and Gitane equipped their 600 with Shimano 600 components. Only Peugeot stayed loyal to Simplex.

To compete with component gruppos from Campagnolo, Shimano, and SunTour, the French companies formed marketing combines. Spidel was a combination of Mafac, Maillard, Stronglight, and Simplex. FREXA was a combination of Huret, Maillard, Peyrard, Rigida, Poutrait-Morin, and Jeunehomme. Spidel and FREXA were gruppos in name only: The cosmetics did not

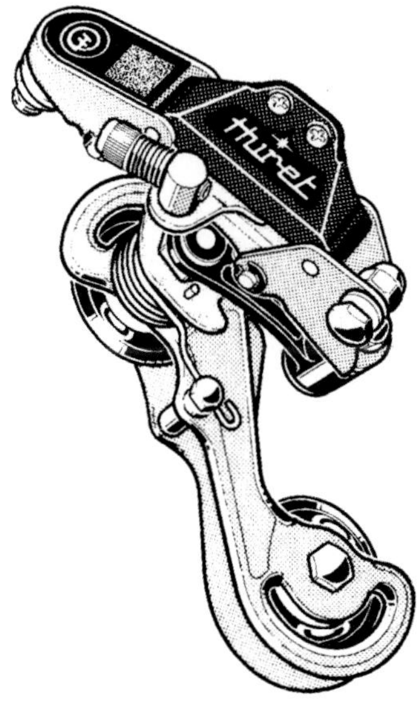


Left: Fig. 11.31. In August 1973, Allan Abbot reached 149 mph (240 km/h) paced by a race car with a windscreen in the back.

Right: Fig. 11.32. The Huret Duopar was the best-shifting wide-range rear derailleur available during this period. The steel-cage version would be called Eco-Duopar. For the techno-snobs there was also a fancied-up version with some titanium bits

match, and the bike makers had to place separate orders with each company.

French bicycle and component makers missed the switch to mountain bikes. By 1984, no French bicycle or component company had a significant share of the U.S. market for either road bikes or mountain bikes. Only Sedis chains retained a significant share of the world market.

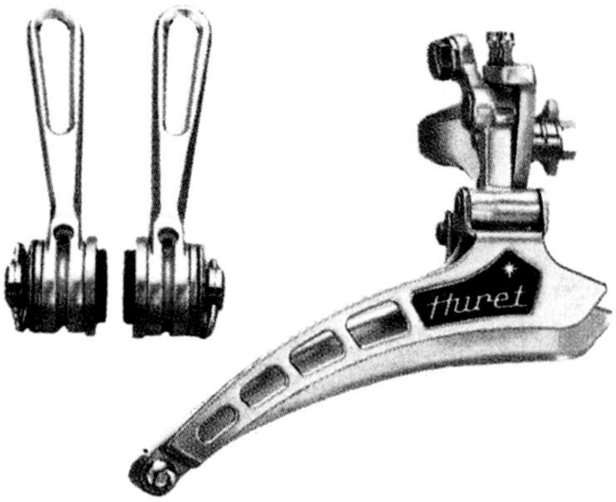


Above: Fig. 11.33. The 80-gram Huret Jubilee front derailleur was the lightest available, but it was too flexible for positive shifting.

Below: Fig. 11.34. Huret's 1976 selection of rear derailleurs. The Allvits were obsolete by this time. The shroud on the Allvit Safety was to meet a CPSC regulation.

## Huret

Huret continued to make the Allvit right up to 1983. The Allvit Safety was given a shroud to meet the CPSC regulation.

Huret made the Jubilee front and rear derailleurs throughout this decade. They were the lightest derailleurs available. One of Huret's problems through most of this decade was that their front derailleurs and shift levers were worse than their rear derailleurs. The Jubilee front derailleur was too flexible to shift positively.

In mid-1975, Huret introduced the Success, a lightweight titanium version of the Challenger. In mid-1976, they introduced the Eco, a low-priced steel version of the Challenger. The Eco sold well in Europe, but it could not compete with SunTour or Shimano in the U.S. market.

Huret introduced the Duopar at the Paris bike show at the end of 1975. It would remain the standout touring rear derailleur for 10 years. The Duopar had a free-floating second parallelogram that allowed the cage to move up or down by more than an inch. It had more capacity than any other touring derailleur. It shifted smoothly and reliably over 13 to 34 wide-range freewheels combined with wide-range triple cranksets. The Eco Duopar was introduced in 1981. It used steel instead of titanium and cost half as much.

I used Duopars on most of my bicycles. It was my reference for evaluating other touring derailleurs. SunTour's derailleurs came close, but they would not shift onto the small sprockets as positively as the Duopar.

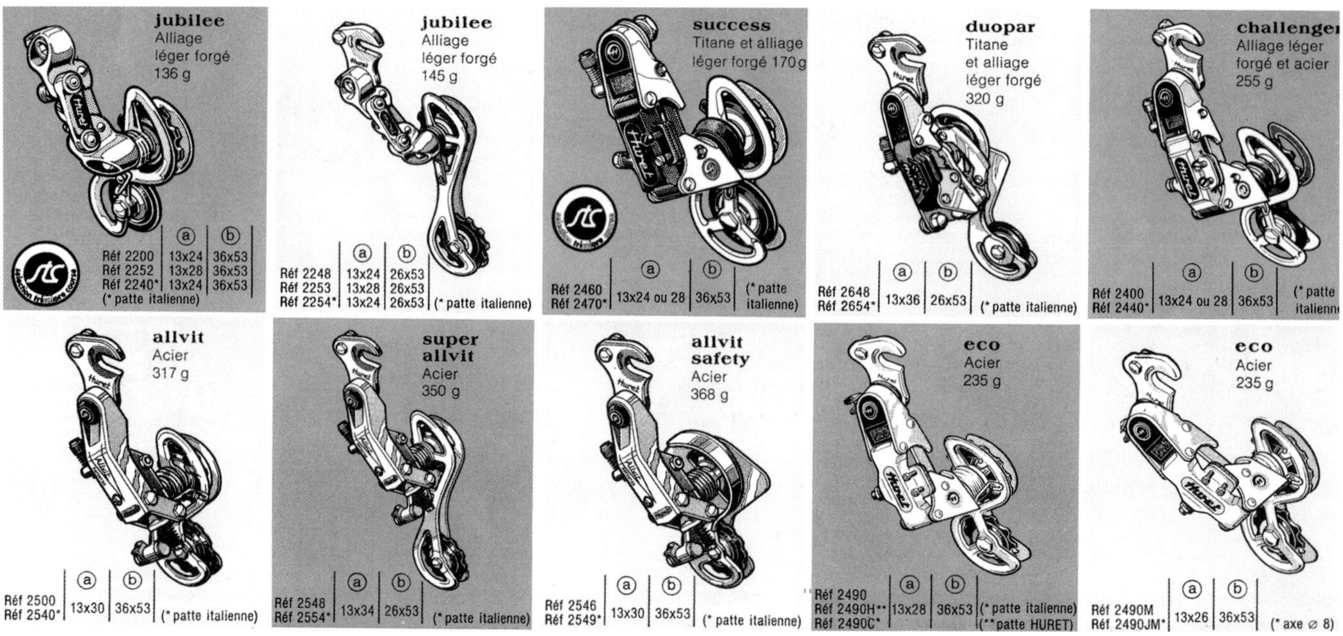

Jan Heine recently tested Duopars on several 1980s bicycles. He found that it shifted easily and silently. He also found that it tended to jump out of gear under hard acceleration. I suspect this was due to some kind of chain/sprocket incompatibility. These derailleurs were used on many of the early mountain bikes.

The Duopar shifted so well that both Shimano and SunTour established "Beat Duopar" design teams. Huret's patent of the second parallelogram prevented immediate copies at the time.

In 1979, Huret introduced the 3-speed Sensomatic 3 with an indexed shift lever. It was designed for the juvenile market, competing with Shimano's Positron indexed derailleur.

In 1981, Fichtel & Sachs bought majority ownership of Huret. Fichtel & Sachs was a division of the German Mannesman conglomerate. They made F&S coaster brakes and hub gears, as well as derailleurs for the German market. The cash cow was their low-horsepower 2-stroke Sachs engine for mopeds and scooters. Sachs subsequently purchased Maillard (hubs and pedals), Sedis (chains), and CLB (brakes), allowing them to offer an almost complete Sachs-Huret gruppo. The problem was that Sachs was saddled with those French companies' old factories, with outdated tooling and equally outdated management practices.

By 1983, Sachs engineer Jürgen Lannoch had taken over the new designs division at Huret. The company became Sachs-Huret, and they introduced the Rival and the Commander derailleurs. The Rival was a well-designed conventional two-pivot rear derailleur that replaced the venerable Allvit. The Commander was an indexed rear derailleur with the indexing cam on the derailleur, suitable for 5- and 6-speed gearing. It was never sold in the U.S.

There was also a Commander-Orbit, a hybrid system that combined a 6-speed freewheel and derailleur with a 2-speed Sachs Orbit geared hub. In 1984, Sachs introduced the Pilot, which was a rear derailleur with an indexed shift lever. Sachs did not aggressively market indexed shifting, and it was not specified on bicycles sold in the U.S.

Above: Fig. 11.35 A and B. The 1983 Sachs-Huret Commander was an indexed 5- or 6-speed rear derailleur with an indexing cam on the derailleur (instead of on the shifter, as is customary these days).

Above: Fig. 11.36. The 1975 Simplex LJ 2000T rear derailleur had a "dropped" parallelogram, like those made by the Japanese manufactuerrs.

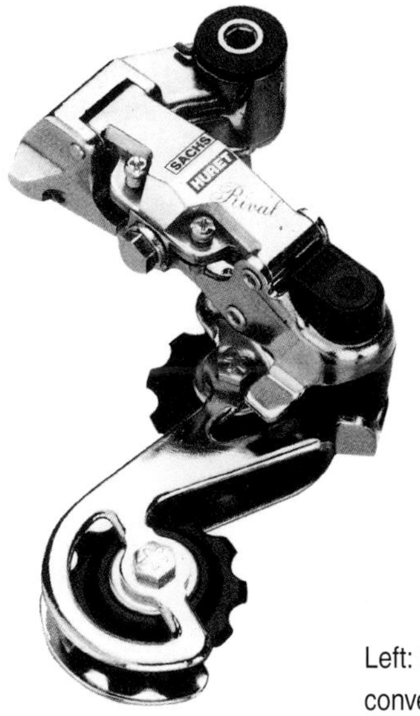


Left: Fig. 11.37. The 1984 Sachs-Huret Rival was a well-made conventional two-pivot rear derailleur.

**Simplex**

Henry Juy, who had taken over Simplex management from his father Lucien Juy, realized from the beginning that the all-Delrin Prestige was too flexible. Over the years, they added more aluminum alloy to stiffen up the Prestige. The top-of-the-line SLJ 5000 was entirely made of alloy. This derailleur was on the stock Peugeot PY10 that Bernard Thévenet used to win the 1975 Tour de France, proving again

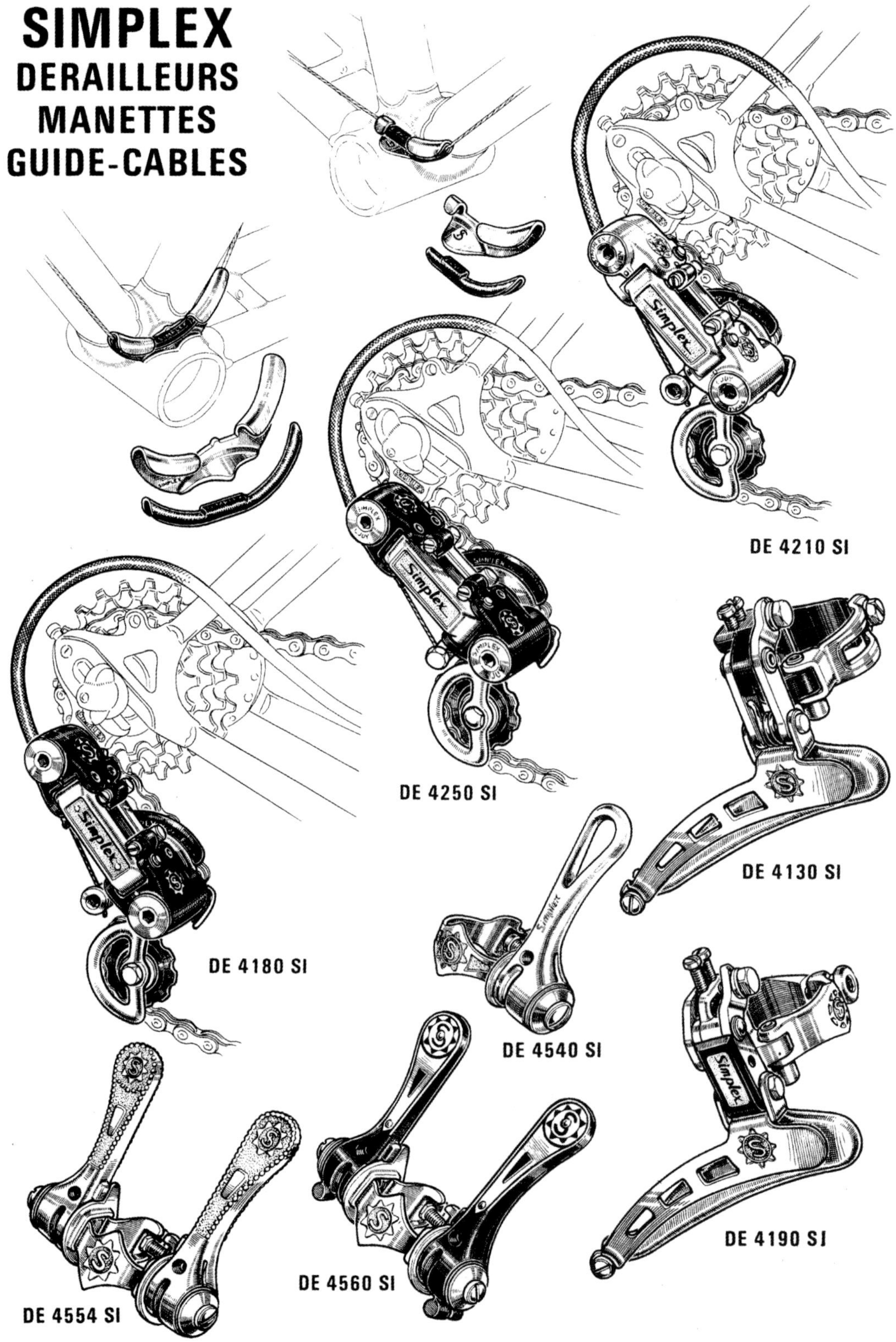


Left: Fig. 11.38. Simplex' top-quality rear derailleurs with regular parallelograms, 1981–1984, as shown in the Bertin-Milremo catalog. The code for the Milremo numbers is: 4210 = SLJ 5500T. 4250 = LJ 4000T. 4180 = LJ 1000T. This page also shows Simplex' best front derailleurs and shift levers. The rear derailleurs shown here did not shift as smoothly as those shown in Fig. 11.39, on the facing page, yet they cost the same to manufacture. Simplex let the buyers decide which type they preferred.

that the rider and not the equipment wins the race.

By 1975, Simplex realized that rear derailleurs with "dropped" bodies and nearly horizontal parallelograms shifted better. Simplex made the LJ 2000T rear derailleur with a dropped parallelogram in 1975. Dropped parallelograms were the hallmark of Shimano derailleurs. My tests indicated that dropped parallelograms shifted more precisely, but Simplex did not put this style into regular production until 1983.

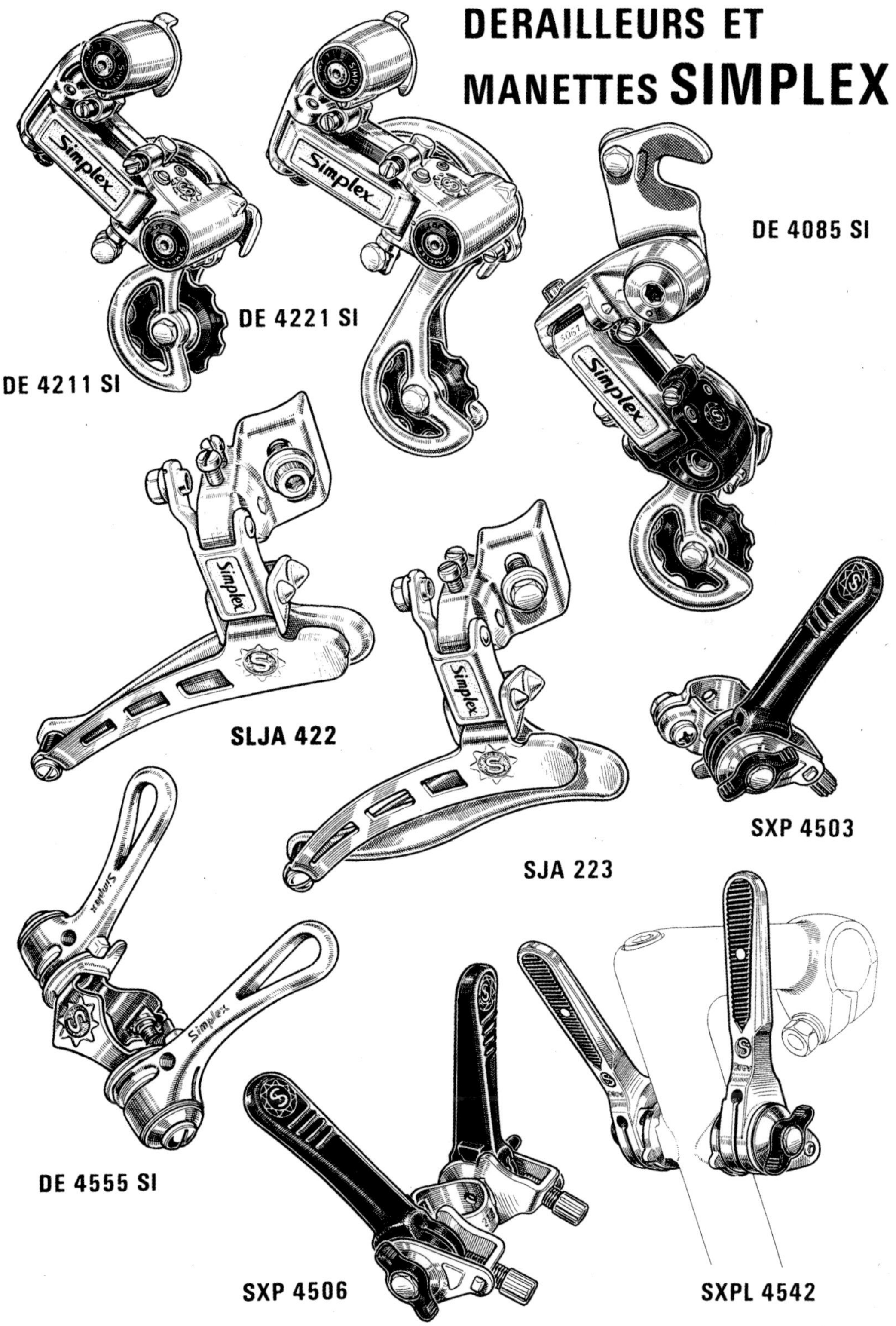


Right: Fig. 11.39. Simplex' 1983 top-quality rear derailleurs with dropped parallelograms. The code for the Milremo numbers is: 4211 = SLJ 6600T/SP. 4221 = SLJ 6600GT/SP. 4085 = S061T/P. Note the racing front derailleur (left) and the touring version (right) with a deeper inner cage. Simplex offered both regular parallelogram and dropped parallelograms during this period. Author Frank Berto tested both varieties and concluded that teh drop-parallelogram derailleurs shifted better than the conventional ones.

Simplex made rugged front derailleurs. They had narrow cages, so they had to be "trimmed" when you shifted on the freewheel, but they could be used with TA Cyclotouriste triple cranksets. They also made racing and touring front derailleurs. The touring versions had deeper inner cages.

The Simplex catalogs listed 18 different rear derailleurs. Simplex supplied OEM bicycle makers with a broad range of very similar derailleurs. In the early 1980s, Simplex made 6000, 5000, 4000, 2000, 1000, 800, 400, 100, and S00 series. The higher numbers had more alloy and less Delrin plastic, and cost more. The 6000 and 5000 models were all-alloy, while the lower-numbered series were alloy and Delrin or steel and Delrin. As for the letter codes, AR (avant) meant front derailleur; CP was a racing rear derailleur with the cage pivot centered between the pulleys. T was a touring derailleur with the cage pivot offset; GT was a grand touring derailleur with a long cage; P included a mounting bracket; and SP was for mounting on a dropout.

By 1984, Simplex was making first-class derailleurs. Greg LeMond won his first Tour de France on a Simplex-equipped bike. To the European market, they introduced the Selematic indexed rear derailleur in 1982.

Simplex made too many different derailleurs, and they had too many employees for their sales. Few bicycle makers specified Simplex. Simplex replacement derailleurs could not be sold in the U.S. because buyers remembered their "cheap plastic" Prestiges and Criteriums. Most French bicycle builders used French, Italian, or Japanese components, depending on the price or assumed customer's preference. Peugeot was loyal to French components and Simplex derailleurs as long as they exported bicycles to the U.S.

### Mavic

Mavic was a long-established maker of top-quality rims. In 1979, they introduced a nearly complete component gruppo, including derailleurs.

The 1982 Mavic 800 rear derailleur was completely rebuildable. In spite of its chunky appearance, it copied Campagnolo Nuovo Record dimensions, and it shifted accordingly. Mavic hubs and bottom bracket used sealed ball-bearings. Mavic made top-quality gruppos but they were rarely seen in the U.S.

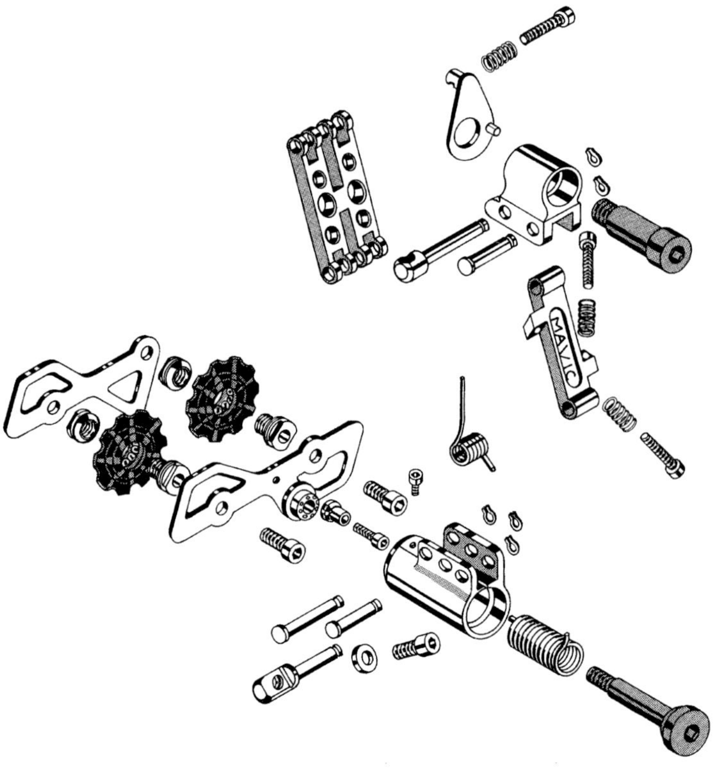

Left: Fig. 11.40. Exploded view of the 1982 Mavic 800 rear derailleur. It was part of a full Mavic gruppo.

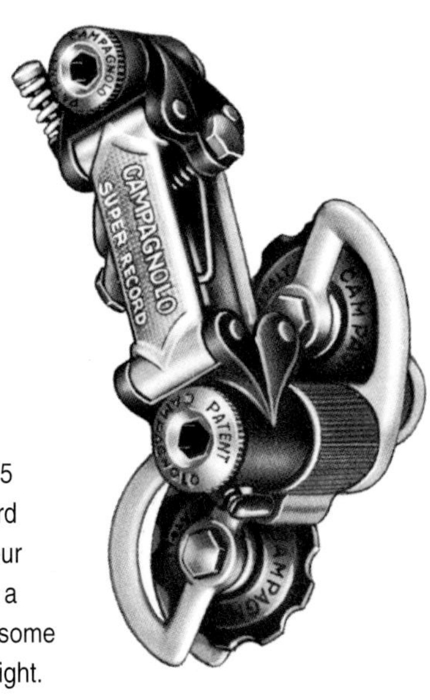


Right: Fig. 11.41. The 1975 Campagnolo Super Record (Model 4001) rear derailleur was a Nuovo Record with a black anodized body and some titanium bits to reduce weight.

## Italy

The Italian domestic market consisted of a large number of transportation bicycles and a small number of elegant racing bicycles. Bicycle production was 2 to 2.5 million through this decade. Exporters concentrated on medium- and high-priced racing models. This small niche market kept growing after the bike boom. Companies like Atala, which had catered to the lower-priced export market, soon disappeared.

In 1975, small Italian component makers, like Galli and Gian Robert, sold low-priced Campagnolo copies. By 1984, this price level belonged to the Japanese. The small Italian makers could not convince the OEM buyers that Campagnolo-copy derailleurs helped to sell bicycles. These companies simply disappeared.

### Campagnolo

Campagnolo came through this decade comfortably, selling complete gruppos and concentrating on the top-of-the-line. Campagnolo did not make stems, handlebars or freewheels because Tullio did not want to compete with his friends at Cinelli and Regina. The Campagnolo Record gruppo with the Nuovo Record rear derailleur changed little from 1970 to 1985, and it sold well. The Campagnolo mystique was well and thriving. The basic design of the Nuovo Record derailleur dated back to the 1963 Record. Campagnolo kept making cosmetic changes instead of a complete redesign.

In 1975, Campagnolo introduced the lightweight Super Record gruppo. The greatest weight-saving came from eliminating the inner supporting ring on the chainrings. Their much-touted "miracle metal" Ergal turned out to be black anodized aluminum alloy. The 1975 Campagnolo Super Record (Model 4001) rear derailleur was a Nuovo Record with a black anodized body and titanium bits to reduce weight. The changes were only

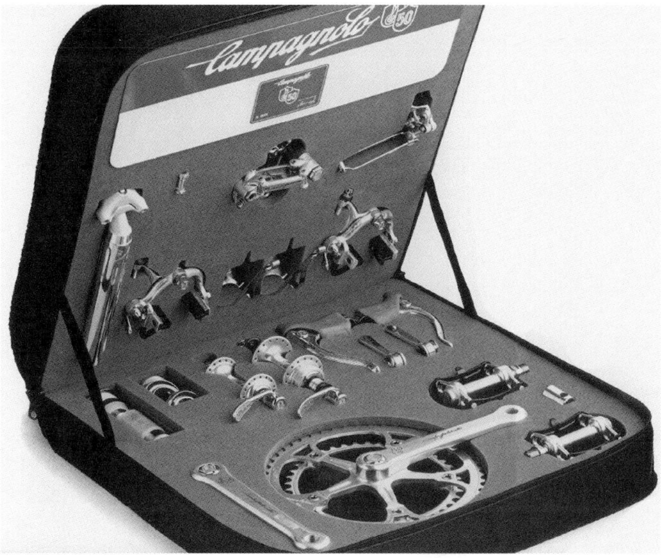

Right: Fig. 11.42. The Campagnolo limited edition 50th Anniversary Record Gruppo came out in 1983, shortly before Tullio Campagnolo's death.

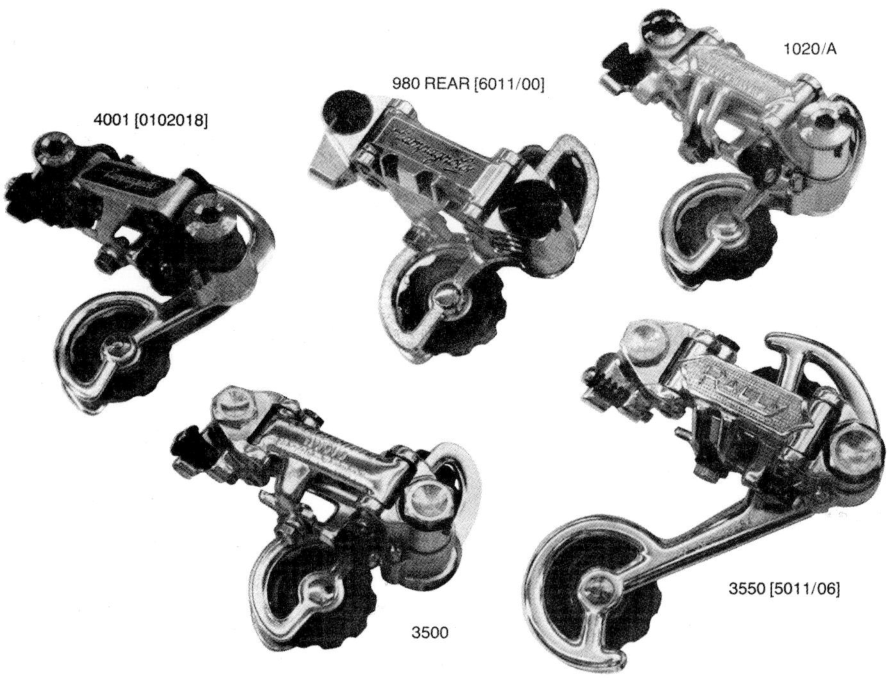


Left: Fig. 11.43. Campagnolo's rear derailleurs as of 1982. Top left to right: Super Record (Model 102018); 980 (Model 6011); Nuovo Record (Model 1020A); bottom: Nuovo Gran Sport (Model 3500); and Rally (Model 3550).

cosmetic, but Super Record sold for a significant premium.

The Rally 3550 touring derailleur came out in 1974, using the cage and the parallelogram of the old Rally and the upper body of the Nuovo Gran Sport. It shifted and sold poorly. A new Super Record rear derailleur (Model 102018) with a different logo on the outer parallelogram came out in 1979.

There were problems with the mid-priced lines. In 1982, Campagnolo came out with new 980

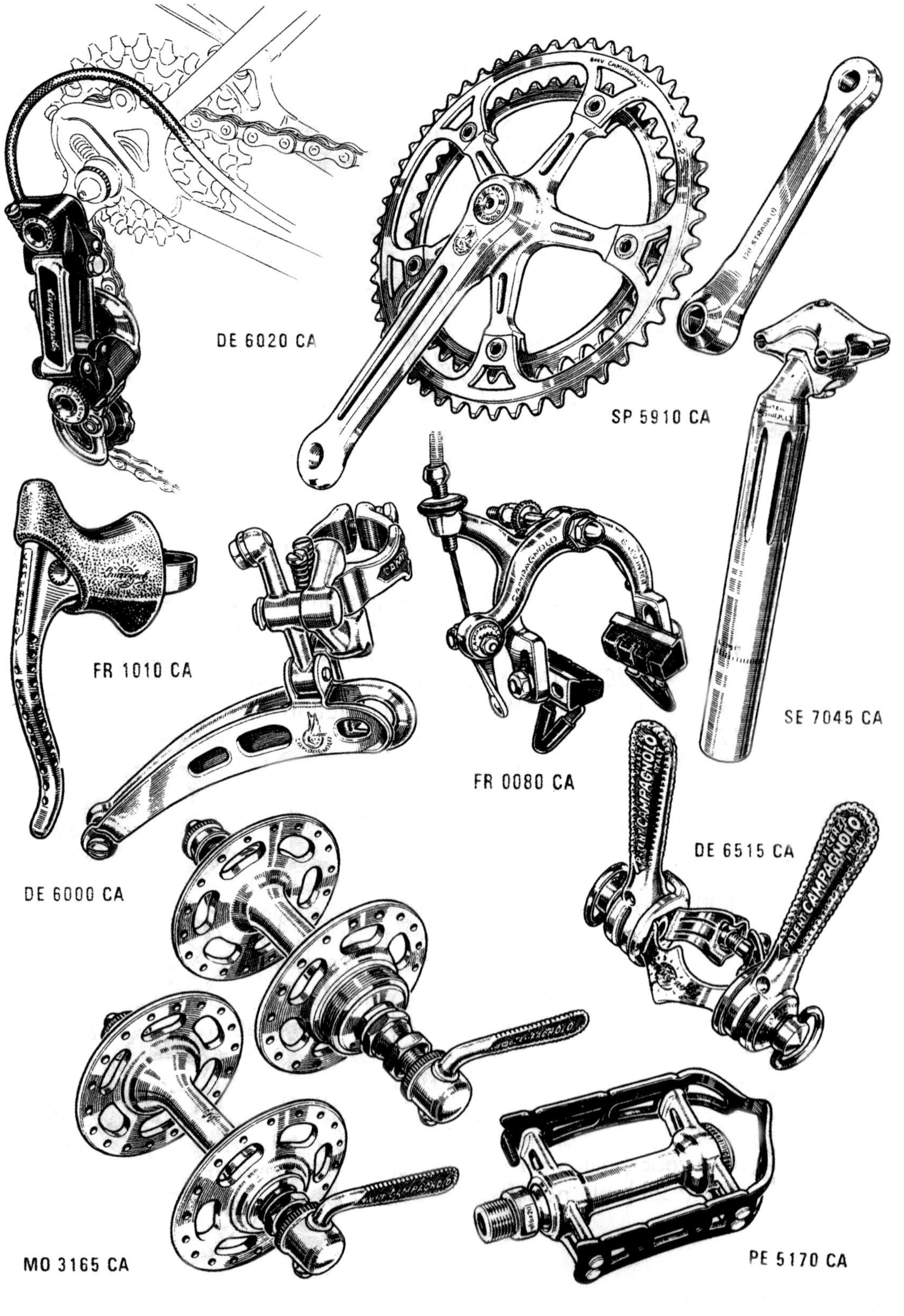


Left: Fig. 11.44. The 1984 Campagnolo Nuovo Record gruppo was a package of top-quality components with matching cosmetics. This final version used the Model 102018 rear derailleur. Compare this gruppo with the 1971 Record gruppo shown on page 207: Campagnolo made these components for 15 years, and most of the parts remained interchangeable.

and Gran Sport gruppos for medium-priced bicycles. These gruppos shared the dimensions and shifting limitations of the Nuovo Record, and they were ugly to boot. About this time, Campagnolo lost control of their distribution system, and Nuovo Record gruppos supposedly meant for OEM makers were being sold by the wholesalers at very low prices. This made it difficult to sell the 980 and Gran Sport gruppos.

The limited edition 50th Anniversary Record Gruppo came out in 1983, just before Tullio Campagnolo's death. The details on the parts were gold-plated, and it came in an elegant fitted wood case.

Tullio Campagnolo died in February 1983, at the age of 81. He had established the standard of quality for all competitors. The name Campagnolo meant quality.

Tullio's son Valentino took over the reins, and one of his first actions was to make a serious run at the mid-price market. Two new gruppos were introduced, Victory and Triomphe. Both gruppos came in two versions—with Corsa (racing) and Leisure (touring) cranksets and derailleurs.

The new gruppos were good-looking, and the prices were right. Unfortunately, their dimensions came from the venerable 1966 Nuovo Record, there were no slant parallelograms, and only one spring-loaded pivot. They did not shift as well as the competition from Japan. Besides, mountain bikes were taking over the world market, and Campagnolo had nothing to offer in that arena.

Campagnolo still competed strongly in the racing arena. In 1983, they came out with a new top-of-the-line gruppo, which they called Record. Never mind that a Record rear derailleur had

Right: Fig. 11.47. The 1976 Galli rear derailleur was a reasonably priced high-quality Nuovo Record copy with different cosmetics.

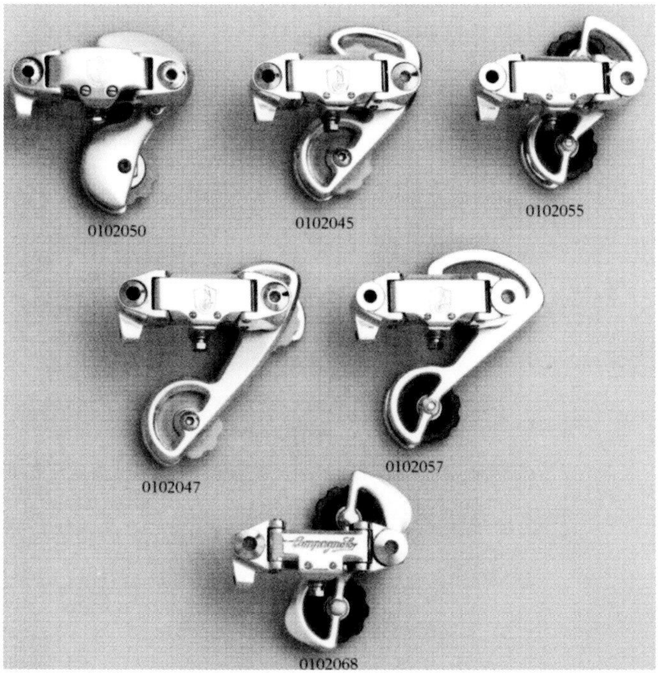


Above: Fig. 11.45. Campagnolo's rear derailleurs, 1983–1987. Top row: C-Record (Model 1020050), Victory Corsa (Model 1020045), Triomphe Corsa (Model 1020055); middle row: Victory Leisure (Model 1020047), Triomphe Leisure, Model 1020057; bottom: 990 (Model 1020068).

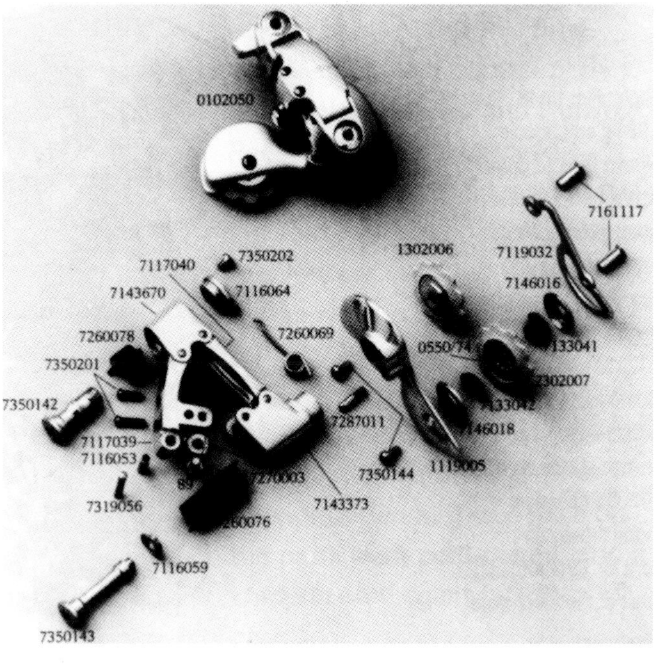


Right: Fig. 11.46. Exploded view of the 1983 Campagnolo C-Record rear derailleur, showing the two spring-loaded pivots.

replaced Gran Sport in 1964, and had been replaced by Nuovo Record in 1966, and supplemented by Super Record in 1976, which was replaced by a revised Super Record in 1982. The new line would be Record. *Basta!* Eventually, everybody, including Campagnolo, called the new gruppo C-Record. The C-Record rear derailleur had spring-loaded top and bottom pivots, so it shifted better, like a Simplex.

Above: Fig. 11.48. The 1984 Ofmega rear derailleur had similar dimensions to Shimano and Simplex 6500 series.

Below: Fig. 11.49. The 1982 Gian Robert System rear derailleur was an inexpensive Simplex copy. It could handle a 13–25 freewheel and a 10-tooth difference on the chainwheels.

#### Other Italian Derailleur Makers

Galli, Gian Robert, Gipiemme, Ofmega, and Rota all made derailleurs in Italy during this decade.

Galli entered the derailleur market in 1976 at two price levels, called Sport and Criterium. Galli's derailleurs were reasonably high quality Campagnolo copies with different cosmetics. They survived until about 1987. Galli cooperated with Maillard and Stronglight at the end to provide full Galli gruppos.

Ofmega derailleurs were beautifully polished and expensive. Ofmega's rear derailleur had similar dimensions to Shimano Dura-Ace and the Simplex 6500 series. They made high-quality gruppos from the middle 1970s until 1987 or so, when indexed shifting took over.

Gian Robert continued to make inexpensive derailleurs at least until 1982. The Gian Robert System rear derailleur was an inexpensive Simplex copy. It could handle a 13 to 25 freewheel and a 10-tooth difference on the chainwheels. Ron Kitching carried them, so they had a following in Britain.

Gipiemme and Roto made Campagnolo-copy derailleurs for a few years in the late 70s. They were rarely seen in the U.S.

## Great Britain

British bicycle production fell from about 2 million in 1975 to about 1.5 million in 1984. British bicycle exports fell from 1.1 million in 1975 to 260,000 in 1984,

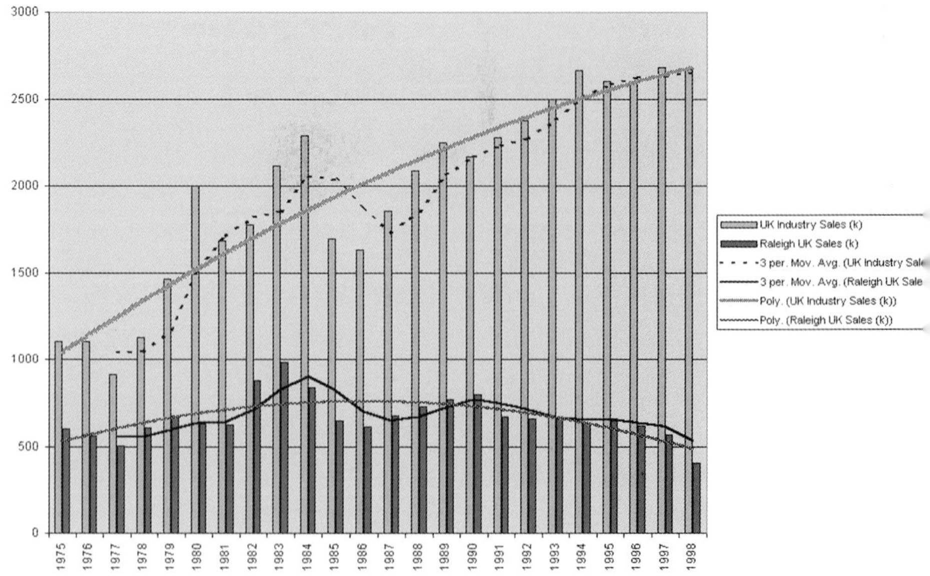


Above: Fig. 11.50. Sales figures for UK total and Raleigh for the period 1975 through 1999. Graph generated by Tony Hadland, based on figures supplied by Raleigh-Diamondback.

reflecting TI-Raleigh's pullout from the U.S. market.

Domestic bicycle sales grew from 1.1 million in 1975 to 2.1 million in 1984. British bicycle production no longer satisfied the home market, and bicycle imports grew from 240,000 in 1975 to 840,000 in 1984. Many of these imports were children's bikes. From the peak in 1954, the mileage traveled by bicycle had declined, until it bottomed in 1974 at a fifth of the 1954 mileage.

The tenfold increase in the price of crude oil in 1974 and the accompanying oil embargo caused a slump in the British economy. Bicycle sales increased, and total bicycle mileage increased again.

In 1975, 42% of sales were single-speed or 3-speed adult bicycles, 26% were drop-handlebar sports bicycles, and 11% were small-wheel adult bicycles. The remainder were children's bicycles. Roughly half of the bicycles were designed for commuting or local transportation.

Right: Fig. 11.51. Raleigh often used a mix of components. This 1977 Raleigh Competition 12 had a Reynolds 531 frame, Campagnolo Gran Sport derailleurs, a Sakae crankset, a 6-speed SunTour freewheel, and Weinmann brakes.

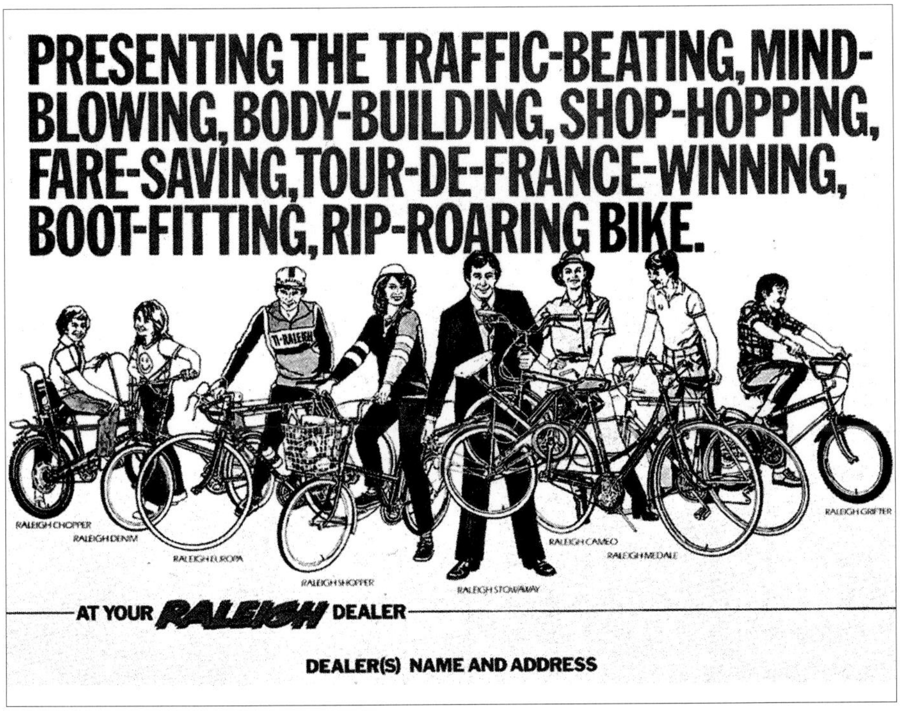


Left: Fig. 11.52. The range of Raleigh bicycles sold in Britain in 1977.

### TI-Raleigh

In 1974, TI-Raleigh made about two million bicycles and exported 70% of them. Most of them went to TI-Raleigh's other worldwide customers, not to the U.S.

The 1977 Raleigh Competition 12 was a typical medium-priced Raleigh. It had a Reynolds 531 frame, Campagnolo Gran Sport derailleurs, a Sakae crankset, a 6-speed SunTour freewheel and Weinmann brakes.

The Raleigh Carlton was a good-quality racing bike that was made in a separate factory. Carlton became the lightweight department at the Nottingham factory in the early 1980s. Raleigh actively supported bicycle racing, and Raleigh-sponsored teams won the Tour de France in 1977 and 1983.

The Raleigh Chopper and the follow-on Grifter high-riser bikes were very successful in Britain. Raleigh got into BMX very late, but their Bomber and Burner models achieved good sales. Raleigh's mountain bikes did not achieve a significant market share.

Sturmey-Archer, which was owned by TI-Raleigh, fell on hard times, both because of competition from Shimano and Sachs, and because of the preference for derailleurs. Four hundred workers were laid off in 1982.

### Moulton

After Raleigh stopped making Moultons in 1974, Alex Moulton began designing the next generation small-wheel bicycles in 1976, referred to as the AM (Alex Moulton) series. He designed and made molds for a new 17x1¼ in. high-pressure tire, which Wolber made exclusively for the AM Series. The frame was made of a triangulated structure of small-diameter tubes, hence the name Spaceframe.

Two Spaceframe Moulton models were launched in 1983: The AM 2 and the AM 7. The AM 2 had a 2-speed Sachs Duomatic hub and was intended for around-town use, while the AM7, equipped with a SunTour Cyclone derailleur and a Regina 7-speed freewheel with sprockets from 9 to 28 teeth, was the sports model. It sold for $500. The factory on the grounds of Moulton's estate in Bradford-on-Avon produced about 1,000 AMs a year.

### Other British Bicycle and Component Makers

Allegro, Comrade, Dawes, Elswick Hopper, Falcon, and Pashley were small to medium size companies that made bicycles for the British domestic market in this decade. Dawes successfully exported touring bicycles.

By the end of this decade, the main surviving British component makers were Sturmey-Archer (hub gears), TDC (cranksets and freewheels), and Cyclo Gear Company (bicycle tools and 2- and 3-speed hybrid conversion kits).

### Deal Drive

Dr. Michael Deal invented the Deal Drive in the mid-1970s. Financing was obtained from British sources, and Deal Drive International was formed to produce and market the invention.

The Deal Drive was a well-built fully automatic expanding-chainwheel with 16 distinct gears and a range of about two to one. It shifted when the pedal force increased. There were ten set-

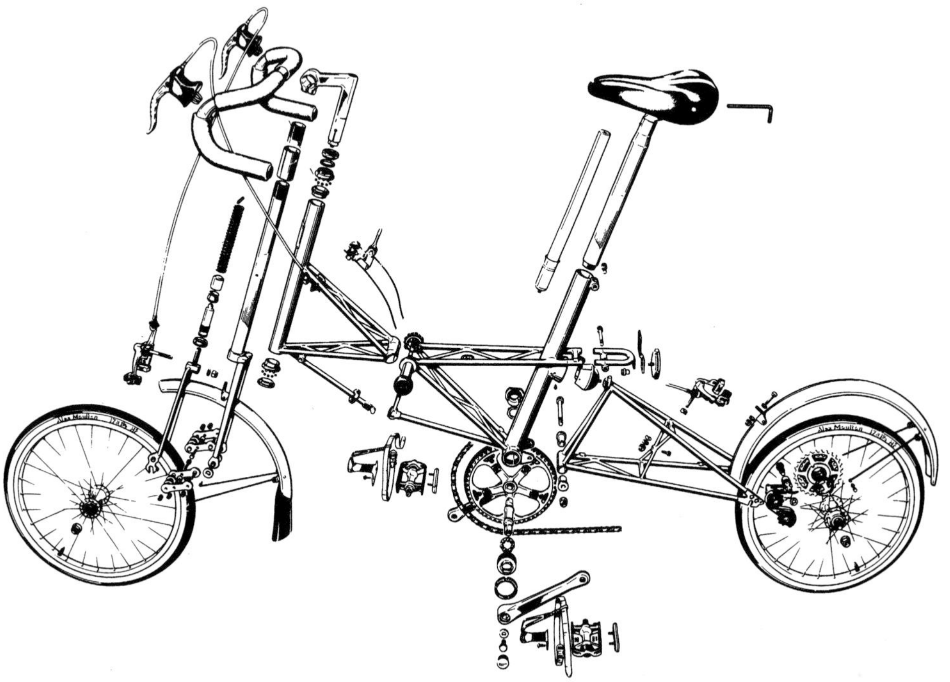

Above: Fig. 11 53. The 1983 "spaceframe" Moulton AM7 was a 7-speed.

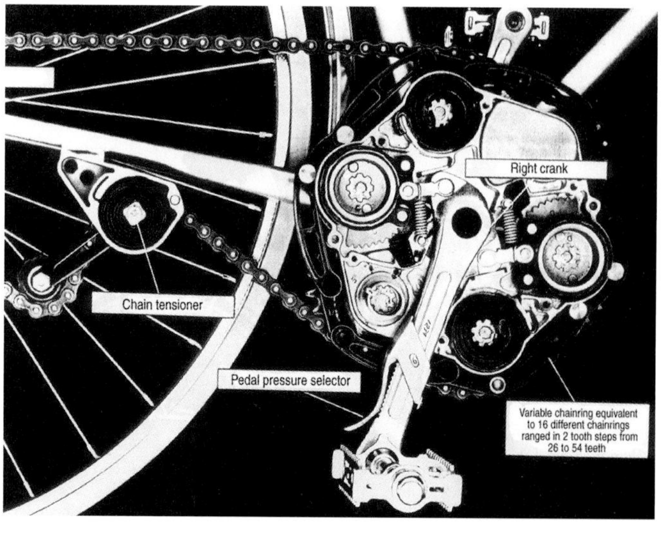


Left: Fig. 11.54. Introduced in 1983, the Deal Drive was an auto-shifting expanding chainwheel with a 2-to-1 range.

Above: Fig. 11.55. The 1984 Zeus Supercronos rear derailleur was a pretty Campagnolo Nuovo Record copy.

Above: Fig. 11.56. The Zeus 2000 rear derailleur was made from 1975 to 1983.

tings of pedal force from 30 to 200 lbs.

It was in production in 1983, and by 1985 it was installed on prototype bicycles made by three different British companies. It was also available as a retrofit crankset. The market for a expensive, heavy narrow-range automatic bicycle transmission proved quite small, and the Deal Drive disappeared.

## Other European Countries

By the mid-1970s, Spain and Germany were the only other European countries with active derailleur production.

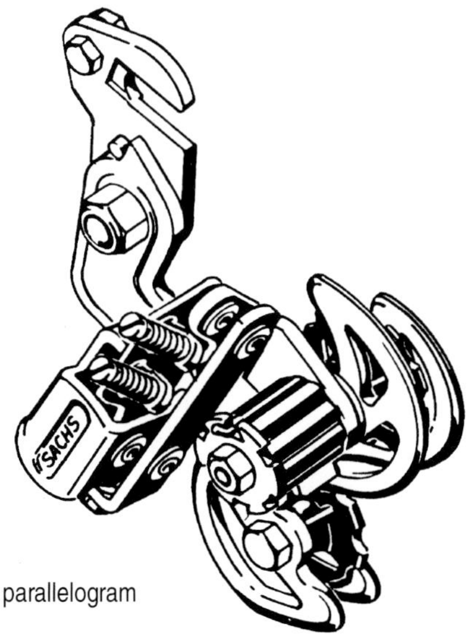


Right: Fig. 11.57. The 1976 Sachs Super Sport had a parallelogram similar to that of the Huret Allvit.

### Spain

In 1975, Zeus introduced the 2000 gruppo, which still copied Campagnolo. Zeus 2000 components were highly polished and drilled for lightness. The Zeus 2000 rear derailleur had a drilled cage and a black anodized titanium body. From 1978 to 1983, Zeus had three gruppos: 2000, Criterium, and Gran Sport. Gran Sport copied Campagnolo Gran Sport, while Criterium and 2000 copied Nuovo Record.

In 1984, Zeus abandoned the low-price market and introduced new higher-price components. The new component gruppos were called Supercronos, New Racer, and Amateur. The derailleurs had Campagnolo Nuovo Record dimensions and performance, with Zeus cosmetics.

The Zeus 2000 freewheel had an alloy body and alloy sprockets. It was the lightest freewheel

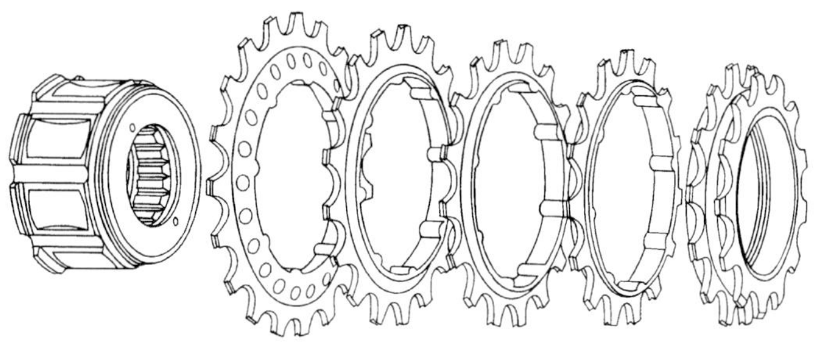

Left: Fig. 11.58. The Zeus 2000 freewheel was made from 1972 to 1986.

Fig. 11.59. The 1982 Sachs 2 x 3 combined a Sachs 3-speed hub with 2 sprockets and a relabeled Huret rear derailleur.

made. It had problems with narrow chains because the alloy sprockets were too wide.

My last Zeus catalog is dated 1986. At the time, Zeus had joined with Maillard to make freewheels. The New Racer and Supercronos derailleurs were still shown. Some time after 1986, Zeus ceased production.

### Germany

Bicycle usage picked up in Germany after the oil crunch of the early 1970s. West Germany made 4.5 million bicycles in 1982. In 1975, Altenberger was still making their chainstay-mounted parallelogram derailleur. The 1976 Sachs Super Sport rear derailleur had a parallelogram similar to the Huret Allvit. Neither was exported outside of the West German market.

Sachs made two hybrid geartrains. Their 2x3 combined a Sachs three-speed hub with a 2-sprocket freewheel and a relabeled Huret rear derailleur. Sachs also made the Orbit, which was a two-speed hub combined with a 6-speed freewheel and derailleur.

Sachs bought Huret in 1981, and the Sachs-Huret products are covered under France, elsewhere in this chapter.

## Japan

In 1977, Japan produced more than 6 million bicycles and exported more than a million. By the late 1970s, labor costs in Japan were as high as they were in Europe, which meant that the Japanese had to compete on quality, performance, and innovation rather than merely on price. At that time, a dollar bought about 250 yen.

In 1975, Azuki, Bridgestone (Kabuki), Centurion, Fuji, Miyata, Nishiki, Panasonic, Sekai, and Takara all exported medium-priced bicycles to the U.S. By 1984, only Bridgestone still exported a full line of Japanese bicycles, and their low-price models were made in Taiwan.

In just five years, from 1973 to 1978, SunTour and Shimano took over the U.S. derailleur market for low- and medium-priced bicycles. Their combined market share went from about 25% in 1973, at the height of the Bike Boom, to 90% by 1978. I estimated market share by counting all of the low- and medium-priced bicycles in *Bicycling* magazine's *Buyer's Guides* from 1973 to 1997. The upper price limit excluded low-volume custom-made models. I counted bicycle models, not total numbers of bicycles, but I do not think this affected the conclusion.

It was similar in Europe. Ron Kitching in Britain and André Bertin in France were major component importers and distributors. They selected components that offered good value and good performance, and sold them widely. André Bertin became the European agent for Shimano in 1975, and Ron Kitching became the European agent for SunTour in 1977. This was the beginning of real competition for Simplex, Huret, and Campagnolo.

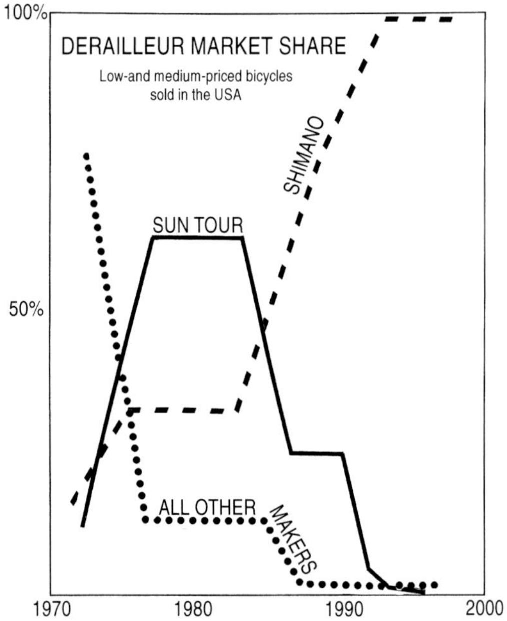


Left: Fig. 11.60. U.S. market share for Shimano, SunTour, and all other derailleur makers. (Source of data: Author's compilation.)

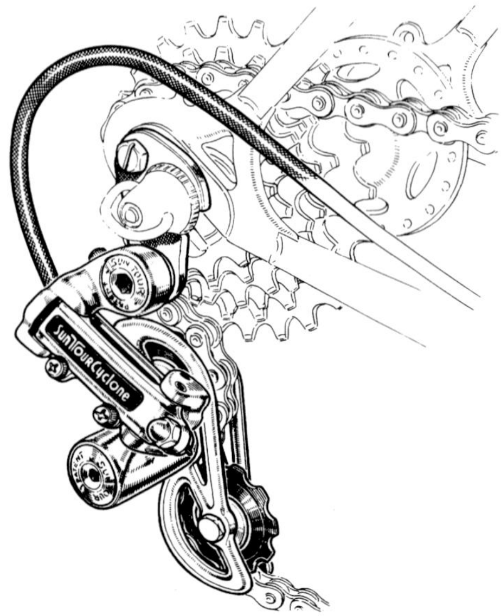


Right: Fig. 11.61. The 1975 SunTour Cyclone was available in short- and long-cage models.

## SunTour

In 1975, SunTour had 291 employees. Their new factory in Shiga made the steel and aluminum alloy forgings. The Mihara and Sakai factories did the machining and assembly.

In 1975, SunTour introduced Cyclone derailleurs, which were lighter and more polished versions of the SunTour V. Cyclone derailleurs shifted splendidly, and the price made them a bargain. Cyclone was available in short-cage (racing) and long-cage (touring) models.

Jan Heine tested a 1975 SunTour Cyclone GT rear derailleur on a Kogswell bicycle with 44/40/24 chainwheels and a14–24 five-speed freewheel. It shifted reliably and predictably with some overshifting.

SunTour did not price their products at what the traffic would bear: Their policy was to set a "fair" price, by adding a modest markup to the production cost.

Below: Fig. 11.62. SunTour's 1976 range of derailleurs included narrow-, medium-, and wide-range models that blanketed the lower and middle price range.

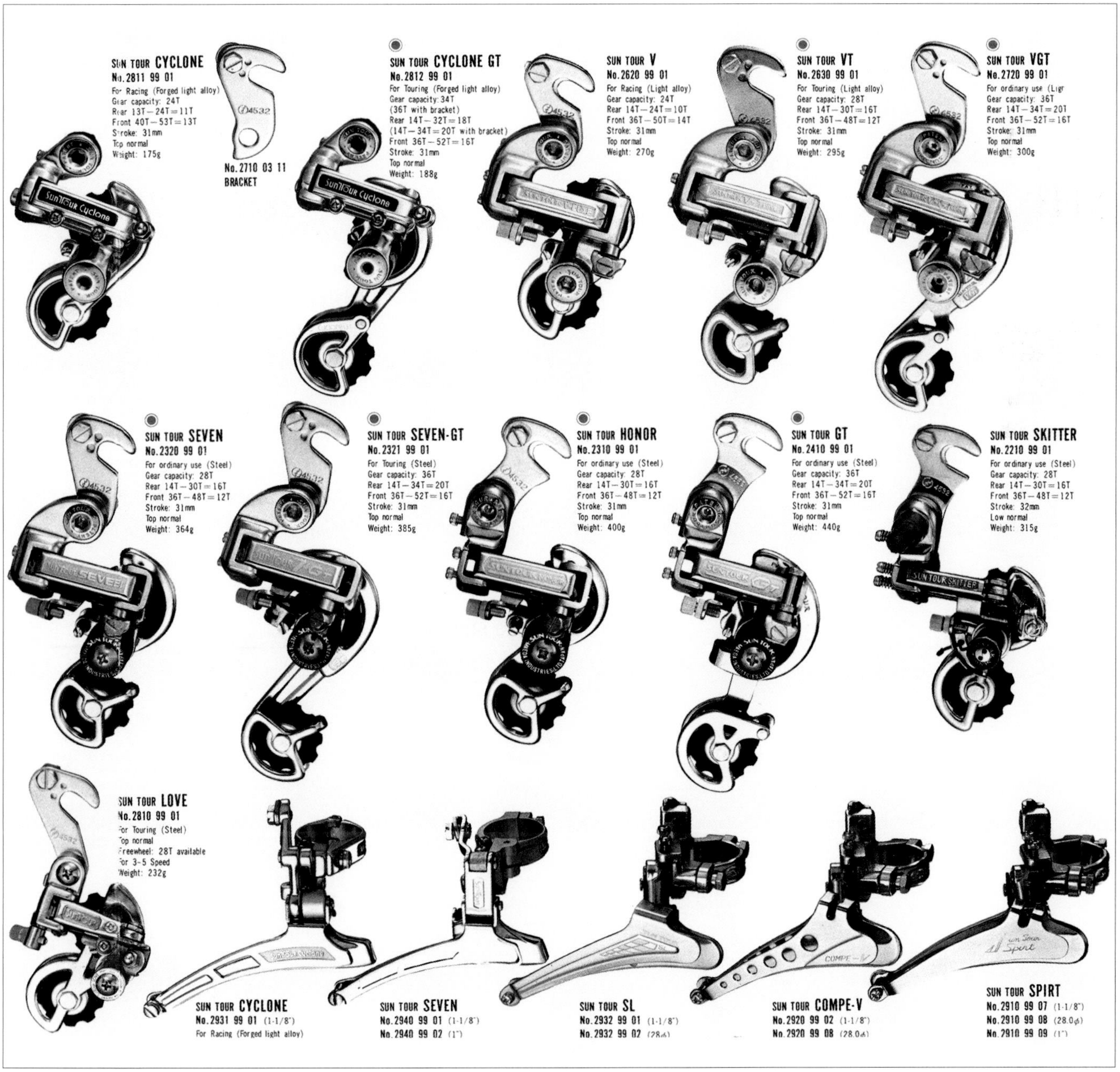

When a Campagnolo Nuovo Record or a Huret Jubilee cost $40, and a Shimano Crane cost $20, the Cyclone cost $16. The Nuovo Record and the Crane both weighed about 200 g (7 oz.), while the Jubilee weighed 140 g (5 oz.). The 175 g (6 oz.) Cyclone was the best-shifting derailleur on the market. However, the result of SunTour's pricing restraint was that their top derailleurs were found on medium-priced bicycles, while their competitor's top derailleurs were found on top-of-the-line bicycles.

SunTour's 1976 range of 11 derailleurs included narrow-, medium, and wide-range models. They blanketed the lower and medium price range.

In 1977, SunTour introduced Vx derailleurs, which were a cosmetic upgrade of the V models. The 1977 Superbe was SunTour's first full gruppo. SunTour made the derailleurs, shift levers, and freewheels. Their partners in the JEX combine provided the crankset, brakes, hubs, pedals, headset, and seatpost. The cosmetics matched, and the price was a bargain. Interestingly, the Superbe front and rear derailleurs were a bit heavier than their Cyclone equivalents, because Superbe was designed to be as reliable and durable as Campagnolo.

In 1978, SunTour introduced the Ultra-6 freewheel and chain. By narrowing the sprockets, spacers, and chain width, the 6-speed Ultra-6 freewheel was the same width as a normal 5-speed, so it fit on a standard 120 mm wide hub. A year later, SunTour introduced the Ultra-7 for 126 mm hubs. SunTour provided zig-zag teeth on the larger freewheel sprockets to help shifting. The Ultra freewheels and the necessary narrow chain did not shift as crisply as the normal width combinations. SunTour started the trend to narrower sprockets, narrower spacers, and narrower chains, which has led us to 10-sprocket freewheels.

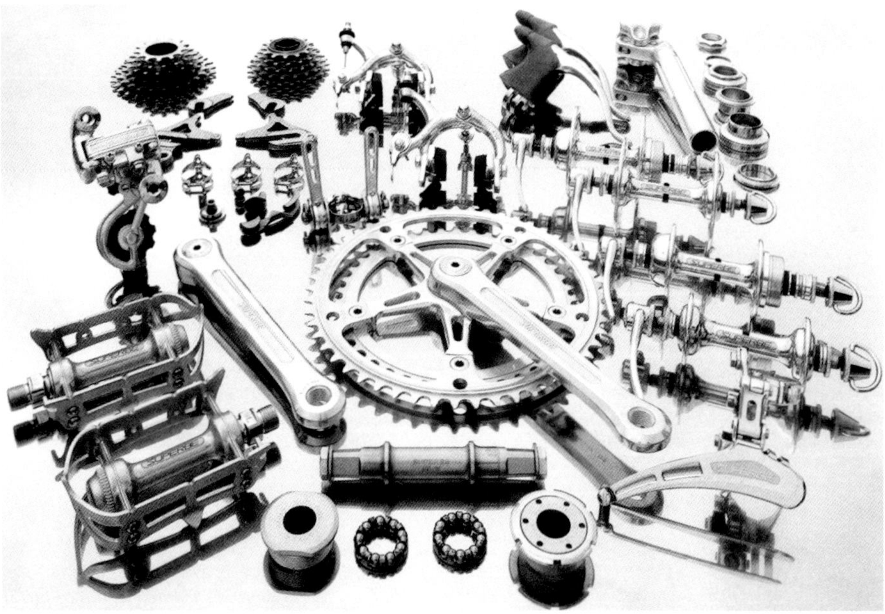

Above: Fig. 11.63. Superbe, SunTour's first full gruppo, was introduced in 1977. It combined SunTour's own components with Dia-Compe brakes, MKS pedals, and a Sugino crankset.

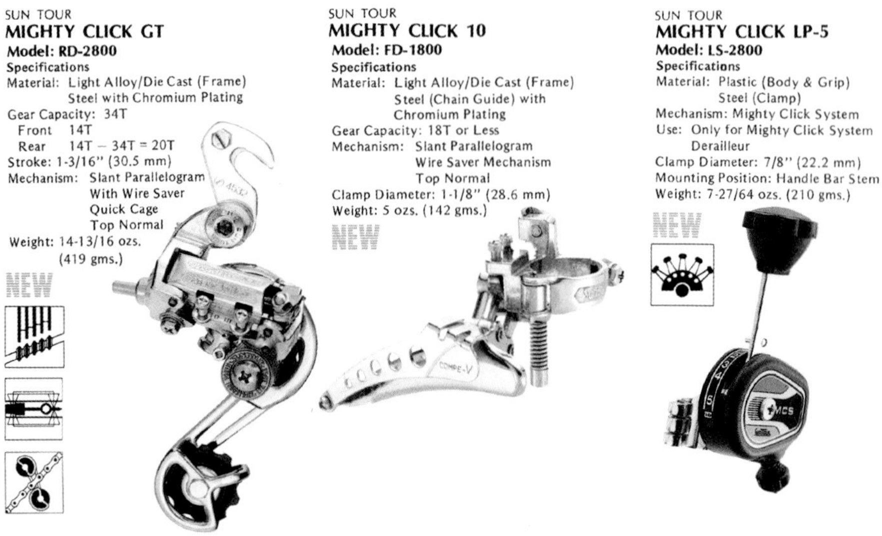


Right: Fig. 11.64. The 1980 SunTour Mighty Click indexed front and rear derailleurs. Also shown here is their indexed stick shifter for use with these derailleurs.

The 1980 SunTour Mighty Click front and rear derailleurs and matching indexed shift levers were designed for low-priced juvenile riders and were not aggressively promoted. By 1980, SunTour had 330 employees and was making 6 million freewheels and 4 million rear derailleurs a year.

The SunTour 1981 GTL rear derailleur could handle the 38-tooth AG (Alpine Gear) freewheel sprocket. I used a number of these on Schwinn Varsity type 10-speeds to drop the low gear from 38 in. to 28 in.

By 1981, the bicycle industry was using the annual model changes in derailleurs and gruppos to provide excitement in each year's new bicycles. SunTour responded with annual cosmetic changes. Superbe became Superbe Pro. Cyclone became Cyclone II. Newly introduced

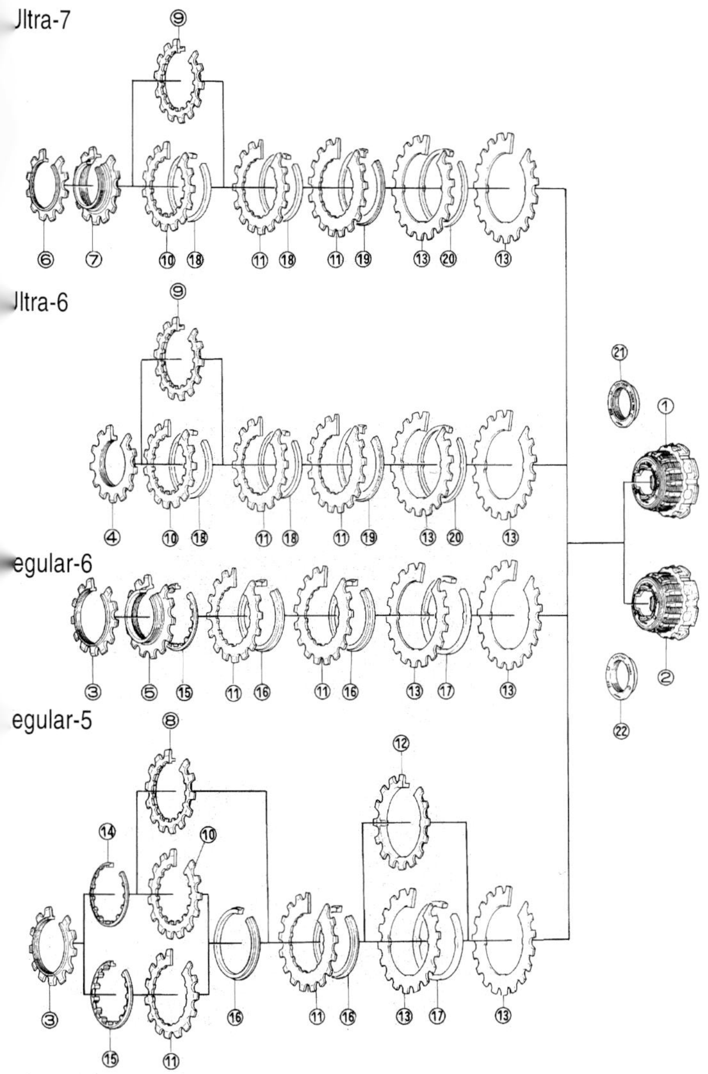

Left: Fig. 11.65. The SunTour New Winner, 1983–1987, was the most versatile freewheel system made. One body handled 12- to 34-tooth sprockets, 5-, 6-, or 7-speeds, and narrow or regular spacing.

Above: Fig. 11.66. The SunTour GTL rear derailleur of 1981 could handle the 38-tooth AG (Alpine Gear) freewheel sprocket.

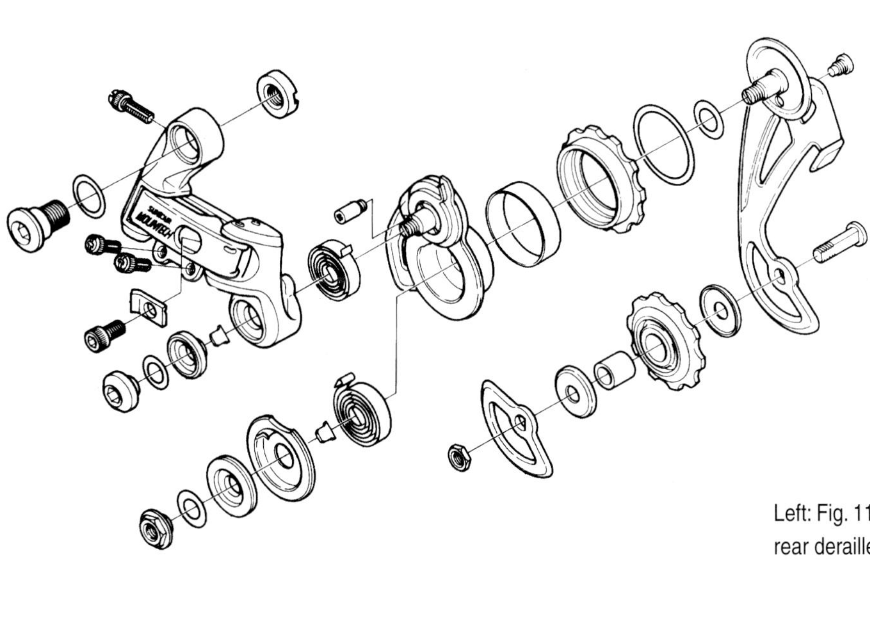

Left: Fig. 11.67. Exploded view of the 1983 SunTour MounTech rear derailleur, showing the problem jockey pulley.

models were aRX, aR, and BL (Blue Line).

1982 was SunTour's peak year. They equipped about 60% of U.S. low- and medium-priced new bicycles, compared to Shimano's 30%. They had 13 different lines of derailleurs, ranging from Superbe Pro to Superbe, Cyclone Mk II, Cyclone, BL, aRX, Vx, aR, Seven, Volante, Honor, GT, and Mighty Click. Most lines had a racing and a touring version. The Vx line came in four different capacities.

SunTour derailleurs, shift levers, and freewheels were the favorites in the replacement market. If you came into a bike shop with a poorly shifting Simplex or Huret derailleur, the standard fix was to replace it with a SunTour.

The 1983 SunTour New Winner was the most flexible freewheel system made. One body handled 12- to 34-tooth sprockets, 5-, 6-, or 7-speeds, and narrow or regular spacing. There were 42 different sprockets available.

SunTour was on top of mountain bike developments, and in 1983 they introduced the SunTour Dirt gruppo. Sugino and Dia-Compe supplied the triple crankset and the cantilever brakes. Everything was on target, except the MounTech rear derailleur. It shifted beautifully and the fatal flaw was not obvious. The large-diameter jockey pulley housed a spring to move the second parallelogram, it was protected by a 1-inch diameter seal that was inadequate. After about a year of road service or a few months of off-road service, the pulley failed, and the MounTech was not rebuildable.

The 1983 SunTour Superbe Tech rear derailleur looked gorgeous. The outer case formed one arm of the parallelogram. The other arm was inside. The Tech family of derailleurs was flawed because of inadequate seals. After a year or so, the pivots would be worn out and the shifting got sloppy, but they could not be rebuilt.

The common element of these failures was the company's small size: In the early 1980s, SunTour had about 20 people working on research and devel-

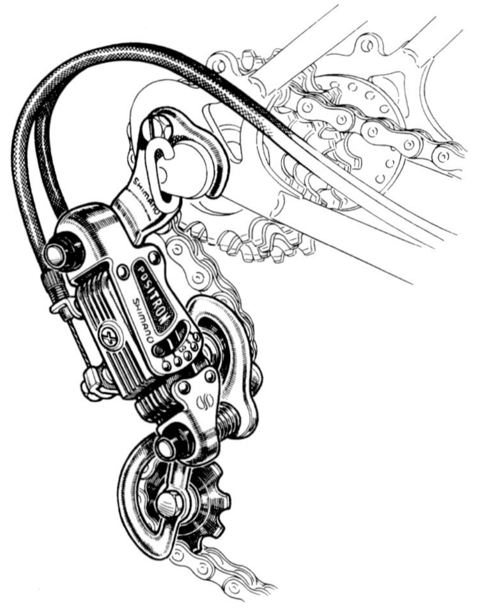

Above: Fig. 11.70. The 1975 Shimano Positron indexed system used two cables and a rear derailleur with built-in indexing.

Above: Fig. 11.68. The 1983 SunTour MounTech was SunTour's answer to the Duopar. It was the first purpose-built mountain bike rear derailleur.

Right: Fig. 11.69. The1983 SunTour Superbe Tech rear derailleur, on which the outer case formed one arm of the parallelogram. The other arm was inside.

Below: Fig. 11.71. This 1978 Shimano Freehub cutaway drawing compares a Freehub on top with a conventional freewheel and hub below. Note the location of the right-hand hub bearing.

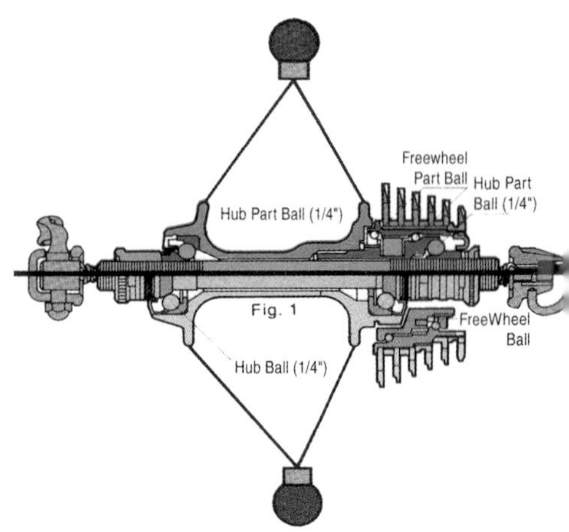

opment, while Shimano had more than 200. MounTech and Superbe Tech were rushed to market with inadequate testing.

### Shimano

In 1975, Shimano products were marketed through the JBM group of companies. JBM consisted of Araya (rims), KKT (pedals), Sakae (handlebars, stems, and cranksets), Takagi (seatposts and cranksets), and Tange (tubing and headsets). As the decade progressed, Shimano expanded to include components that competed with those made by the other JBM members, and they dropped out of JBM.

In 1975, Shimano introduced the Positron indexed system. The shift lever had two cables going to the rear derailleur, which had built-in indexing. Positron was the first indexed system to obtain a significant market share. Positron was designed for inexpensive beginner's bicycles.

In 1976, PPS (Positron Pre-Select) was added. It used a single push-pull cable. The pre-select feature allowed gear shifting without pedaling. The shift took place as soon as you started pedaling. PPS used a crankset with the freewheel in the bottom bracket

Above: Fig. 11.72. The 1978 Dura-Ace EX and 600 EX rear derailleurs had no inner cage plate.

Below: Fig. 11.73. The 1975 Shimano Dura-Ace gruppo. The rear derailleur was called Crane until 1977, and then it became Dura-Ace EX. This illustration, from the Bertin-Milremo catalog, also shows the early U.S.-made Speedwell titanium frame.

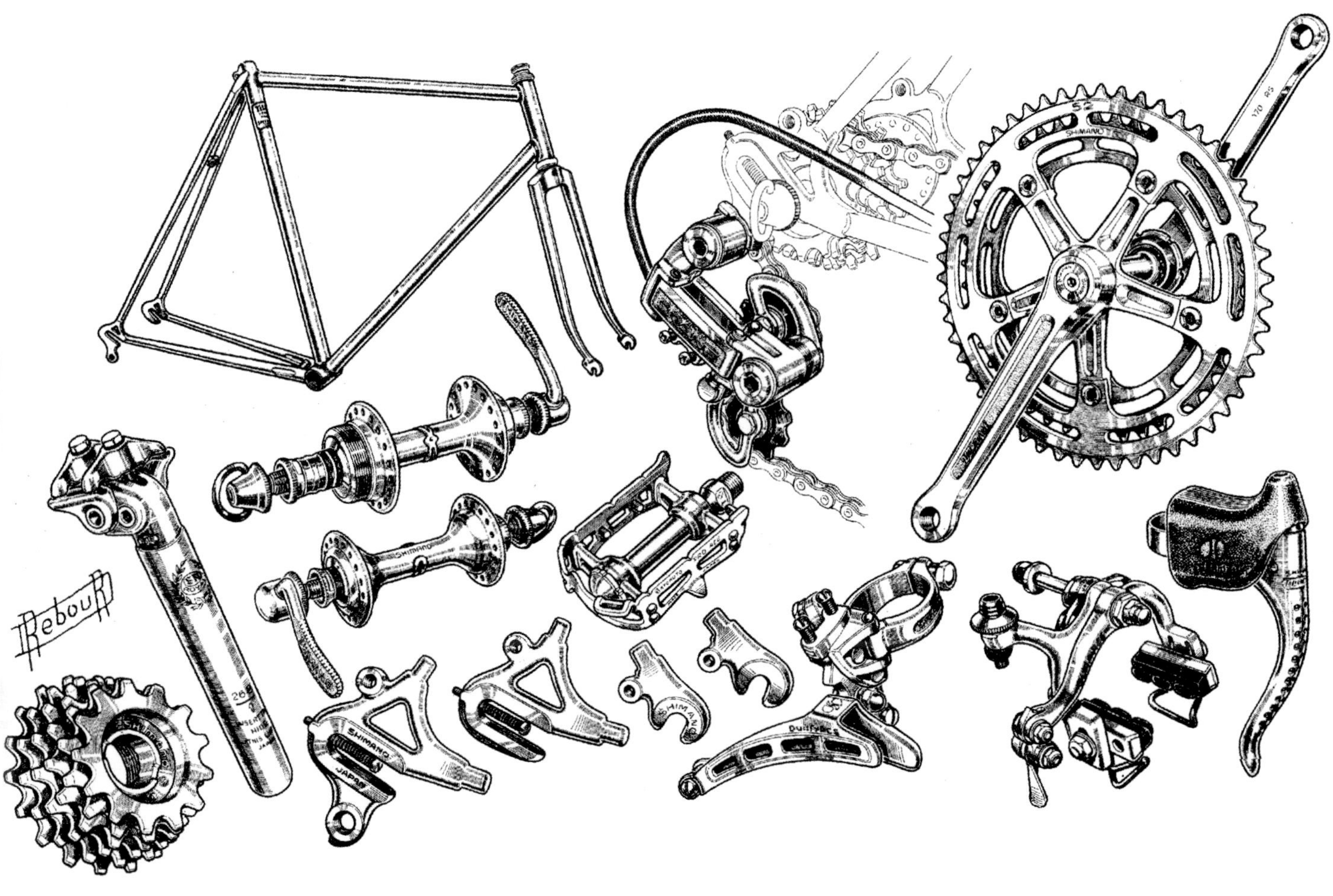

and a special friction-freewheel in the rear hub.

The 1977 version was called Positron II, and Positron III came out in 1980. Positron expired in 1982. The final versions used a 6-sprocket freewheel hub. Positron did not catch on because it cost more, and low-end bicycle buyers would not pay extra for the indexed shifting. Shimano learned from this experience: From this time on, they introduced innovations into the top lines first, and then let them

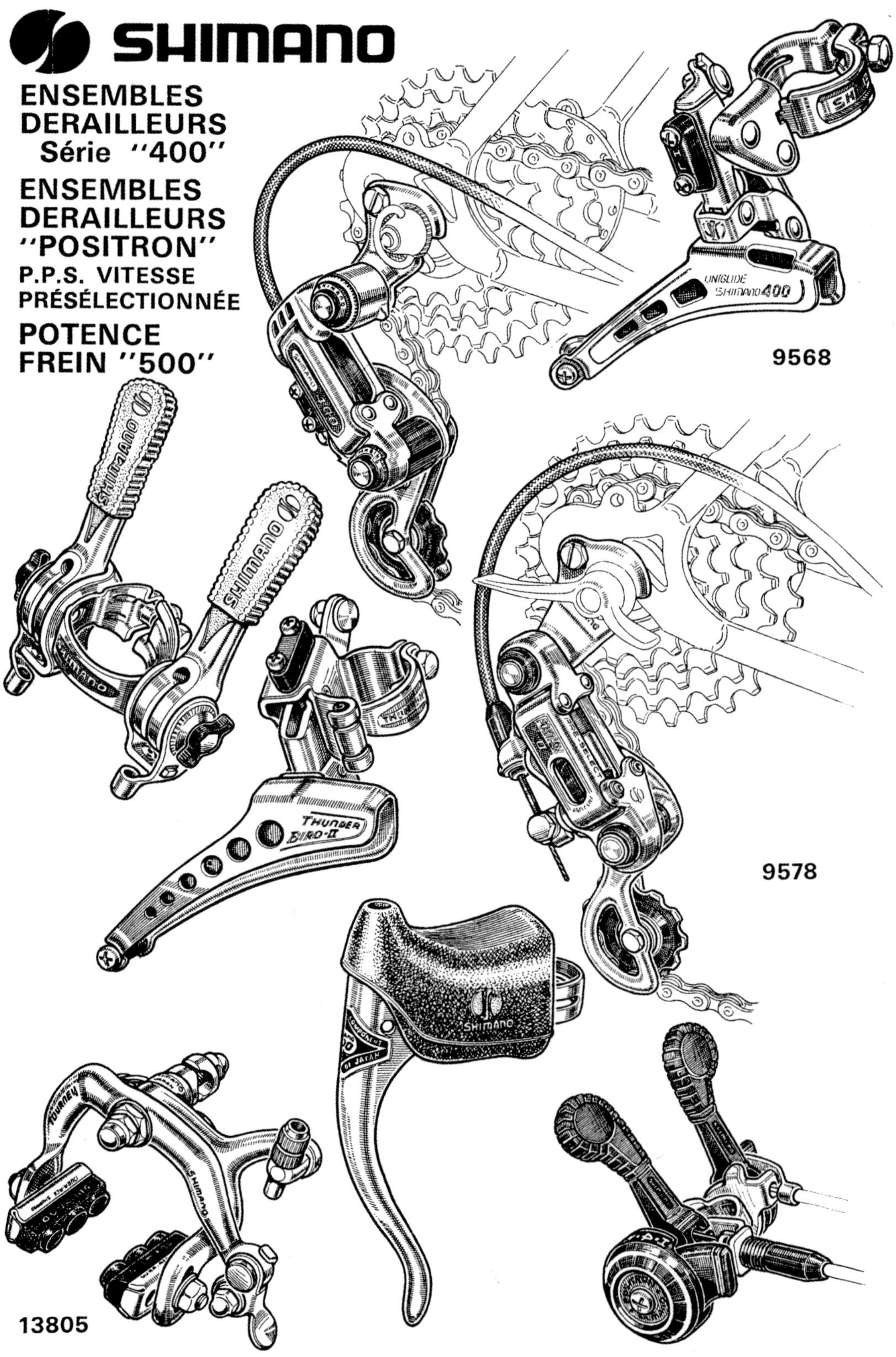


Left: Fig. 11.74. The 1976 Positron II rear derailleur used a single stiff push-pull cable and a front freewheel to allow gear shifting without pedaling. This page also shows the 1979 Shimano 400 derailleurs. Bertin-Milremo was the Shimano distributor for France, and these rare Rebour illustrations come from their catalogs.

"trickle down" to the lower-priced lines.

In 1975, Shimano introduced two complete gruppos, Dura-Ace and 600. The Dura-Ace gruppo competed with Campagnolo Record. The rear derailleur was called Crane until 1977. In 1978, it became Dura-Ace EX.

The Shimano 600 gruppo was not as highly polished as Dura-Ace, but it performed almost as well. Shimano 600 had no significant competition until SunTour in-

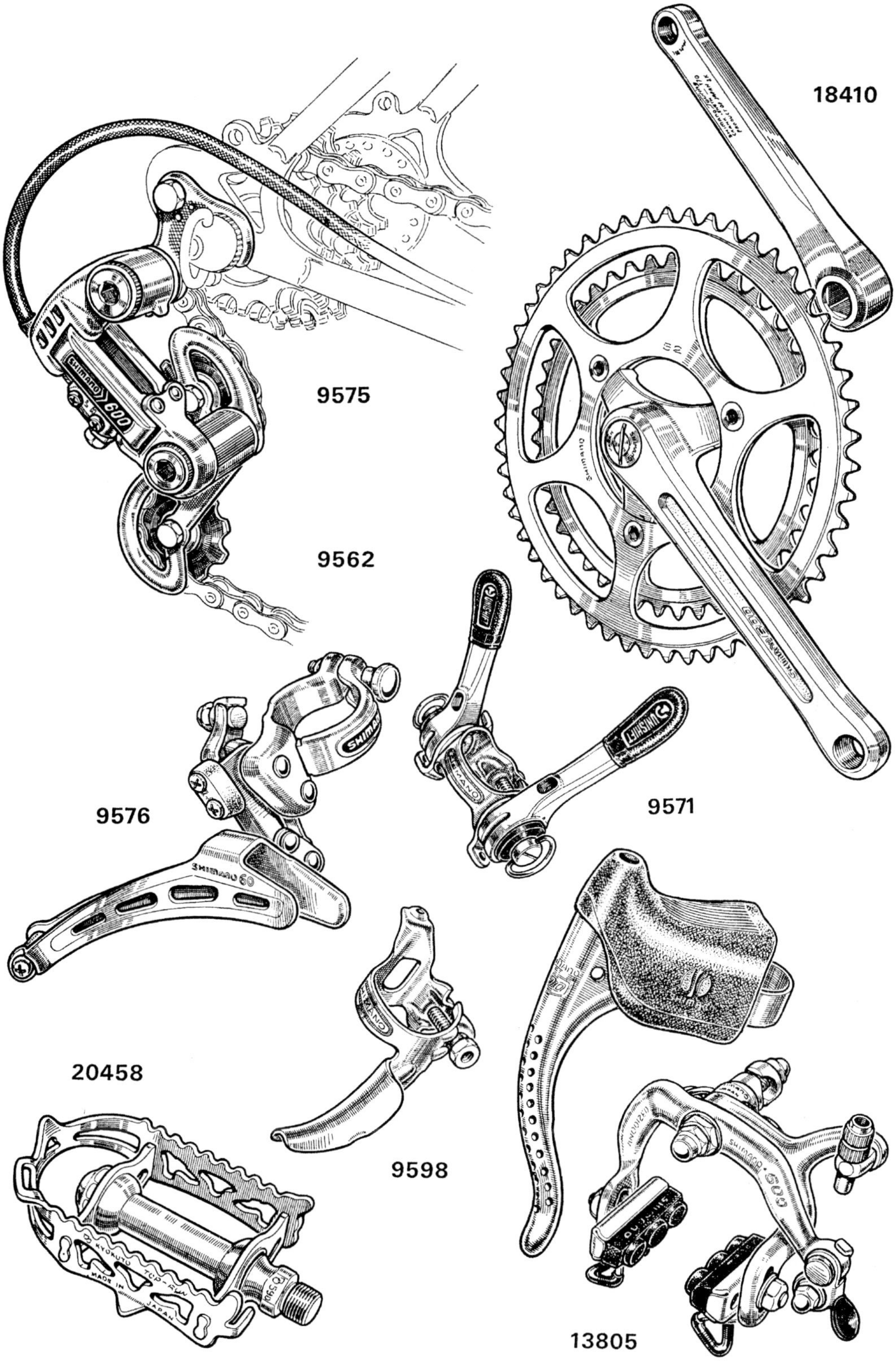


Right and next page: Figs. 11.75 and 11.76. The 1975 Shimano 600 gruppo was not as highly polished as Dura-Ace, but it performed almost as well. The 600 gruppo was a major competitive breakthrough for Shimano. In 1978, it became 600 EX. The 3-pin crankset had a unique bolt-circle, which only stayed in production for three years.

troduced Superbe. The components in the 600 gruppo were for medium-priced bicycles. To many buyers, 600 was Dura-Ace with a bit less polish and a much lower price. The introduction of this medium-priced gruppo was a very effective marketing technique, and the OEMs (Original Equipment Makers) loved it. When Bertin featured the Shimano 600 gruppo in the Milremo catalog, the handwrit-

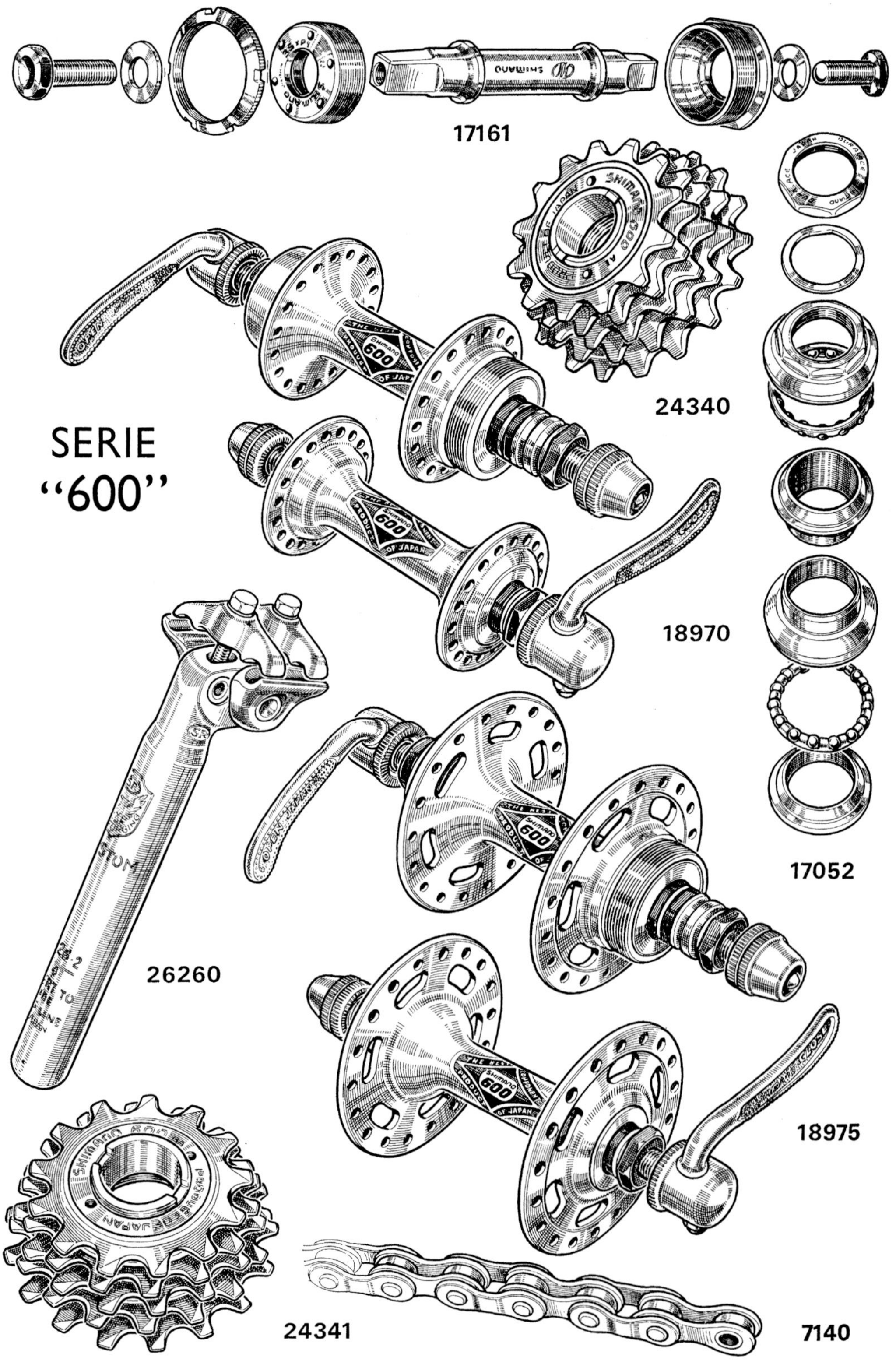


Left: Fig. 11.76. See preceding page for caption (Shimano 600).

Facing page: Fig. 11.77. With a bit more ornamentation, and a cassette instead of a screwed-on freewheel, the 600 EX gruppo remained in production from 1978 until 1982.

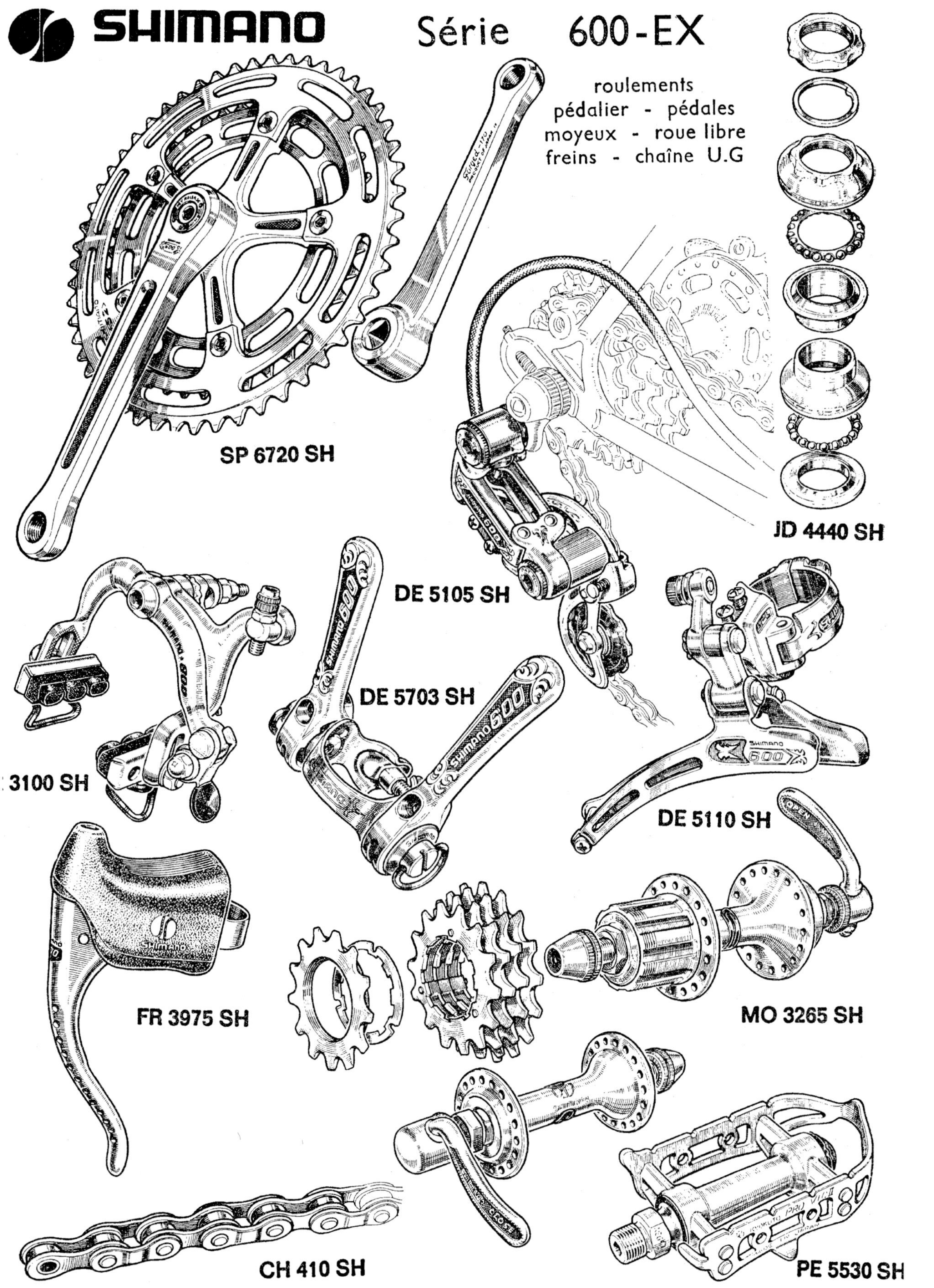
SHIMANO
Série 600-EX
roulements
pédalier - pédales
moyeux - roue libre
freins - chaîne U.G
SP 6720 SH
JD 4440 SH
DE 5105 SH
DE 5703 SH
3100 SH
DE 5110 SH
FR 3975 SH
MO 3265 SH
CH 410 SH
PE 5530 SH

ing was on the wall for Huret and Simplex.

In 1978, Dura-Ace became Dura-Ace EX, and 600 became 600 EX. The 600 EX gruppo made only cosmetic changes to the derailleurs but the 600 EX crankset got five arms, just like Campagnolo and Dura-Ace.

By the mid-1970s, Shimano probably had more people working on research and development than all of their competitors combined. A steady stream of innovations appeared.

Left: Fig. 11.82. Shimano's 1981 Deore touring gruppo.

Top left: Fig. 11.78. In 1976, Shimano introduced the Dura-Ace 10-mm-pitch chainrings, with matching sprocket and chain, for track racing.

Right: Fig. 11.81. Shimano's huge booth at the 1981 New York and Cologne bicycle trade shows included a display of a rider in a wind tunnel.

Above: Fig. 11.79. The 1981 Shimano Dura-Ace AX aerodynamic rear derailleur.

Right: Fig. 11.80. Shimano's non-round BioPace chainrings were heavily marketed from 1983 to 1992—and then quietly disappeared from the catalog.

Dura-Ace 10 was introduced in 1976 as a gear-train designed around a chain with a 10 mm pitch, instead of the conventional ½-inch (12.7 mm) pitch. It was an interesting idea, but no one followed Shimano's lead, and it became a short-lived specialty for fixed-wheel track bikes. Shimano never made derailleurs to match.

Shimano was the first company to understand that the sprocket,

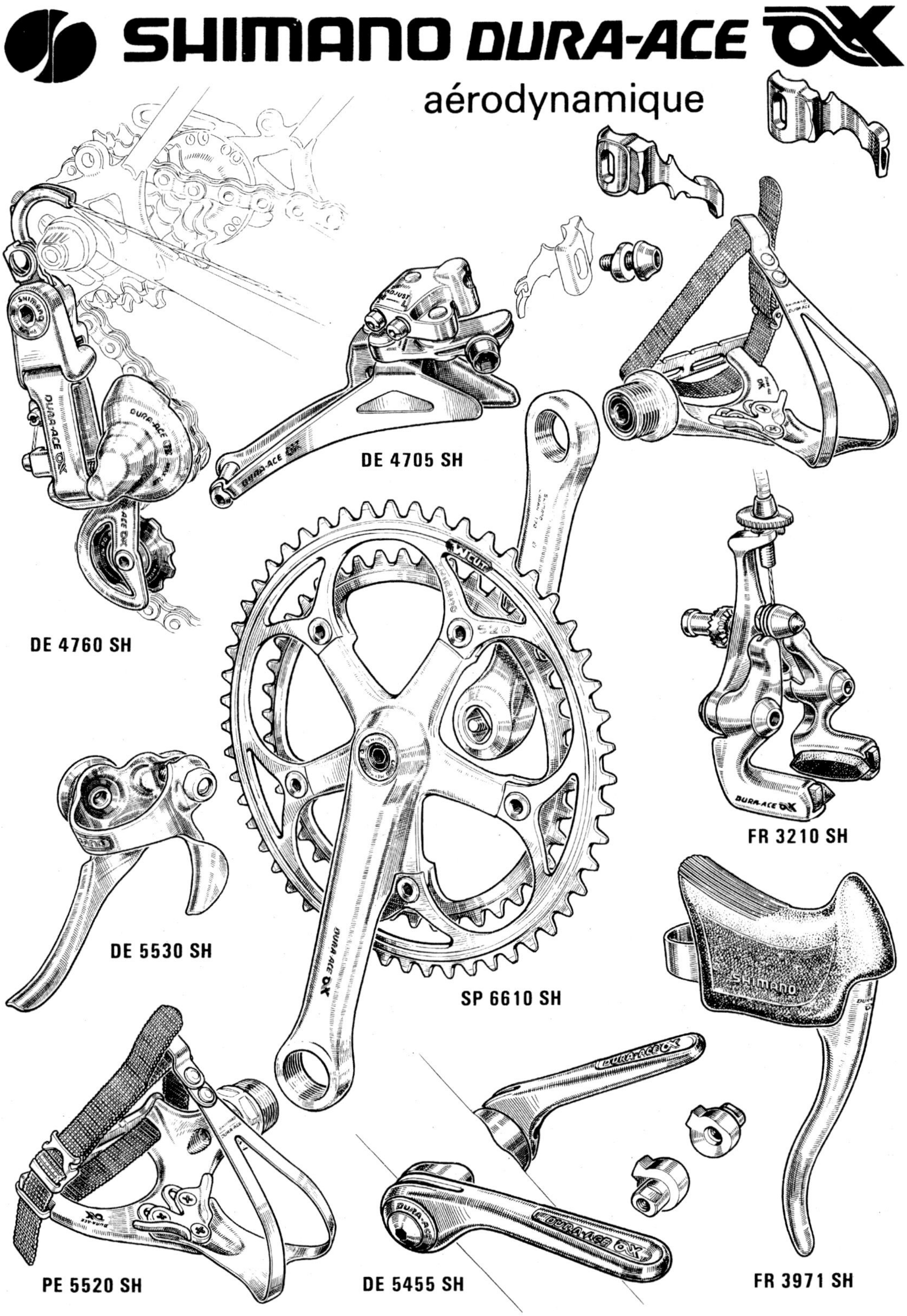


Right: Fig. 11.83. Shimano's Dura-Ace AX was their flagship aero gruppo. The indexed rear derailleur was the weakest item: it was fragile and fussy to adjust. The professional racers rejected it. The dropped pedals and the rear-entry brake levers were good ideas, and at least the latter one has survived: Today, practically all road bike brake levers are of this general "aero" design.

chain, and rear derailleur were an interactive system. In 1977, they introduced the Uniglide chain and chamfered Uniglide sprocket teeth to ease the strain of shifting from sprocket to sprocket. This was followed by the Centeron jockey pulley, which had a slight amount of sideways play, so that the rear derailleur did not have to be completely centered for the chain to run quietly.

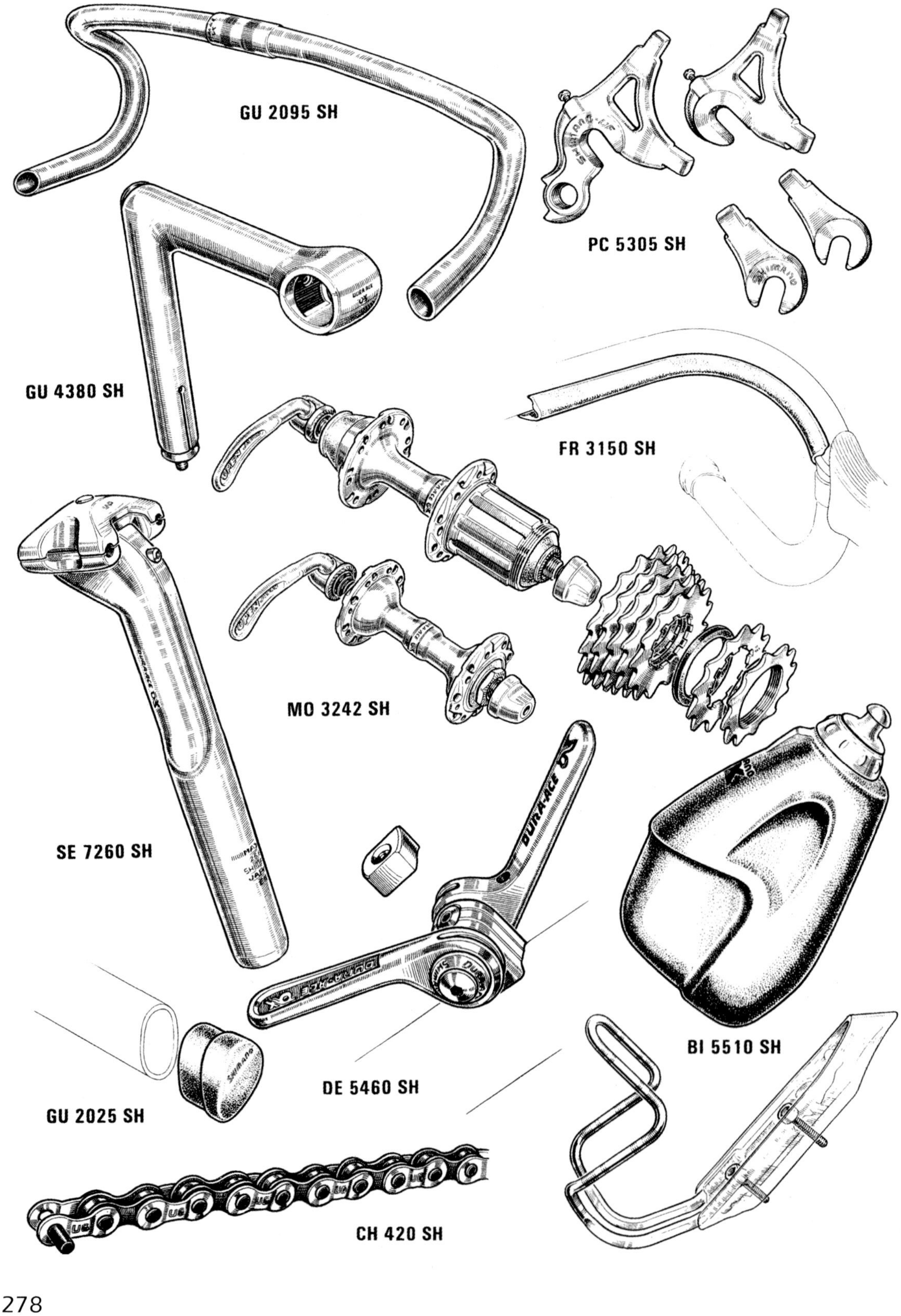


Left: Fig. 11.84. Some concepts that survived from the 1981 Dura-Ace AX gruppo included handlebars with grooves for the rear-entry ("aero") brake cables, a seatpost with a narrow-profile pillar, aerodynamic spoke entry on the hubs, and a narrow water bottle.

In 1978, Shimano introduced the Freehub, which combined the freewheel with the rear hub. This major improvement was available in both the Dura-Ace and 600 gruppos. Shimano did not invent the Freehub; they perfected it. The idea had been around for years. The outboard bearing supported the rear axle close to the dropout, avoiding broken axles. The small-diameter freewheel body allowed 11-tooth sprockets. All of the inner sprockets were splined, giving complete freedom

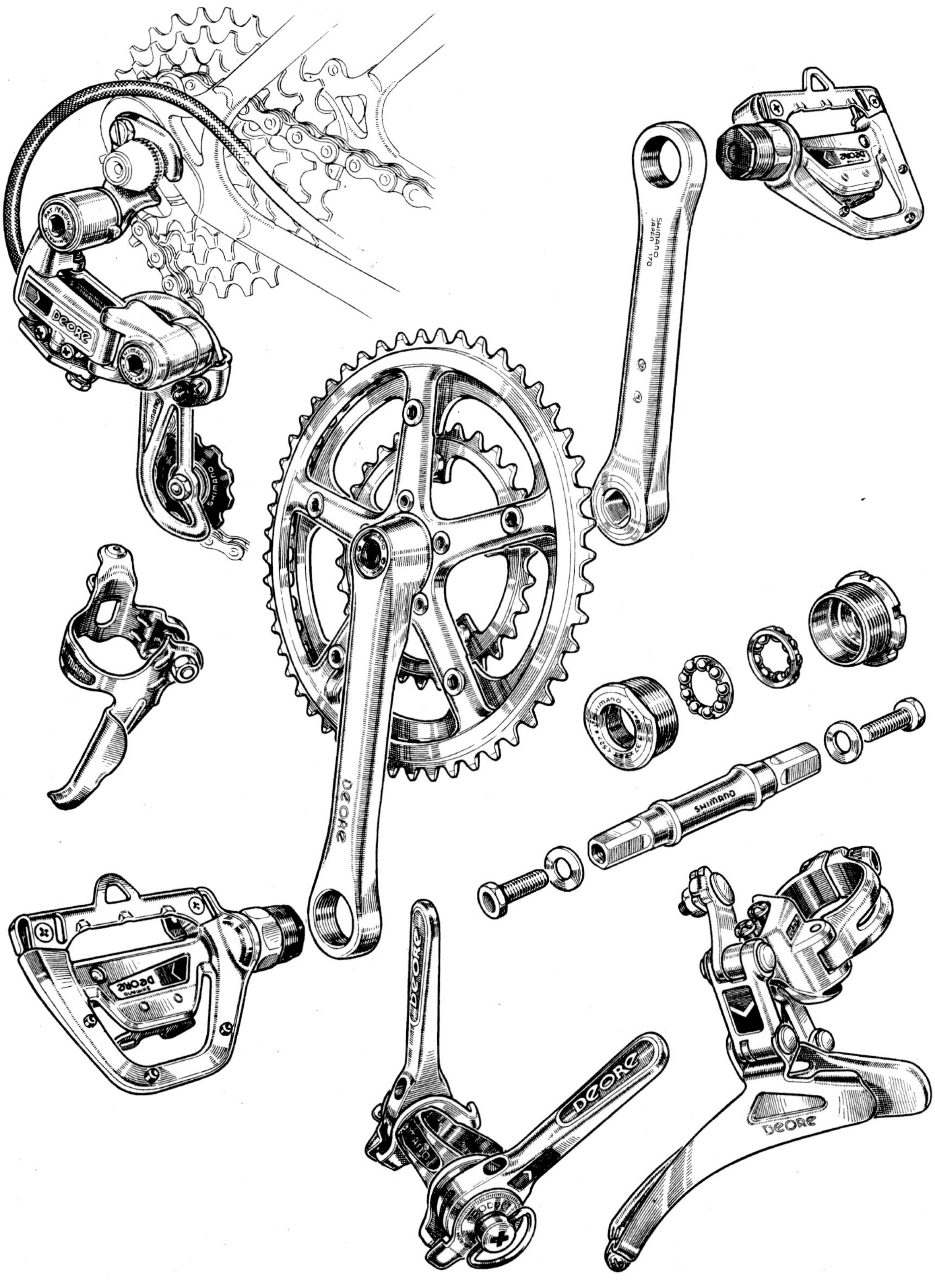


Right: Fig. 11.85. The 1981 Deore touring gruppo worked nicely, and it set up Shimano for the growing mountain bike market. The two-piece crankset could be used with TA chainrings. When Shimano adopted Sugino's 110/74 mm bolt circles in 1985, that became the industry standard. The dropped platform pedals were very comfortable.

of selection. Over the next fifteen years, Shimano's Freehub concept replaced conventional screw-on freewheels and hubs.

The 1978 Dura-Ace EX and 600 EX rear derailleurs saved weight by replacing the inner cage plate with a tab. This allowed the derailleur to snuggle closer to the spokes, which allowed narrower 6-speed freewheels or Freehubs to be used. A

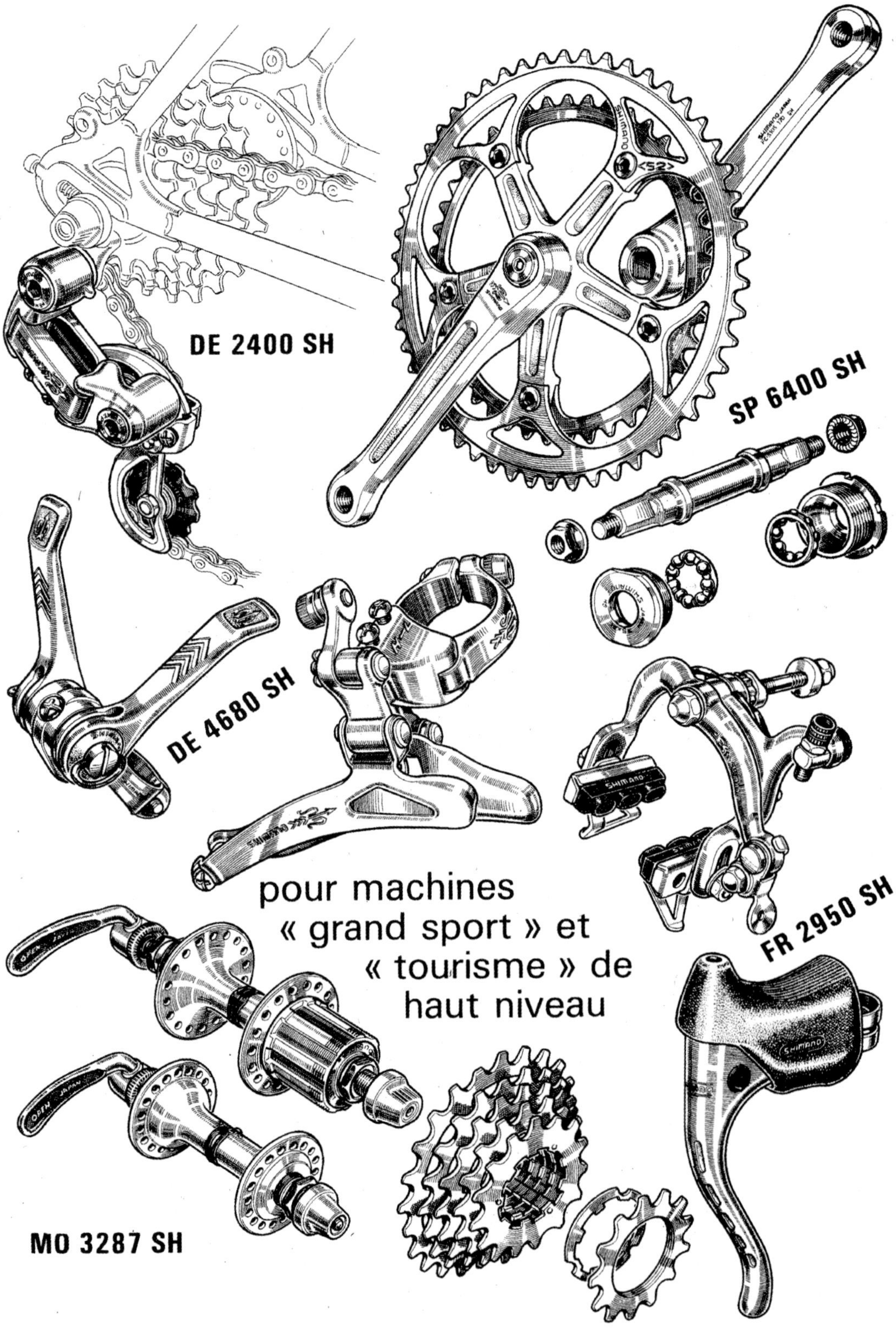


Left: Fig. 11.86. The 1983 Shimano 105 gruppo simply destroyed the competition. It looked expensive, and it performed well. 105-equipped bikes sold well. The French and Italian makers had nothing comparable. Note the expensive-looking 5-arm crankset and the freehub.

few years later, Shimano went back to an inner cage plate.

In 1979, Shimano moved the gruppo concept to a lower price level with Altus/Selecta. Shimano's annual model changes provided bicycle makers with novelty for their own annual model changes.

In 1981, Shimano's marketing experts decided that the 1980s

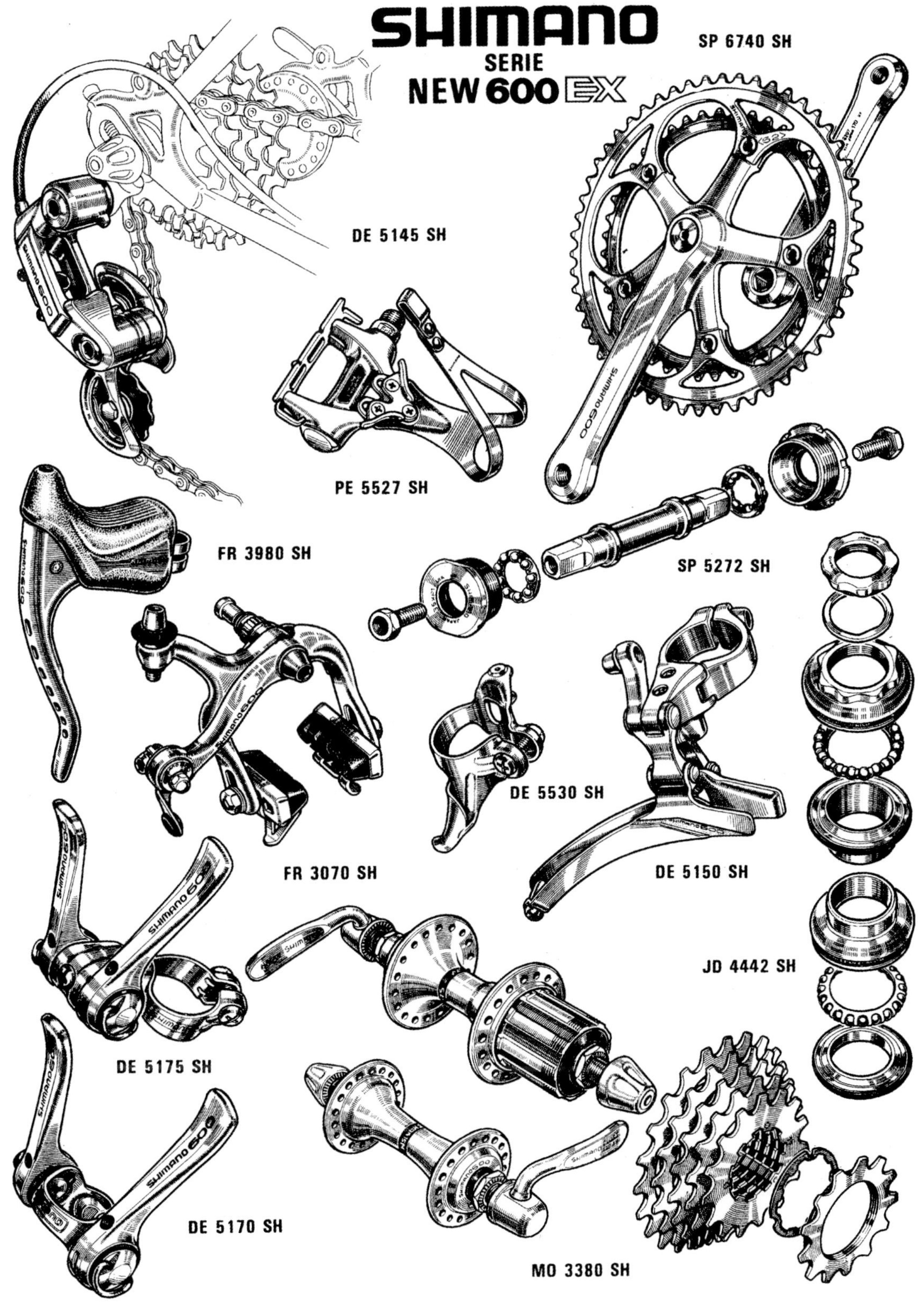


Right: Fig. 11.87. The 1984 New 600 EX gruppo was a cosmetically changed version with more polish. Note the drilled brake levers and the platform pedals. With full gruppos at three price levels, Shimano had no real competition from Campagnolo, Simplex, or Huret. This page from the September 1984 Bertin-Milremo catalog is one of Daniel Rebour's last line drawings.

would be the Decade of Aerodynamics. Shimano had a huge display at the 1981 New York and Cologne Bicycle Shows, including a rider in a wind tunnel.

Shimano had four AX gruppos: Dura-Ace AX, 600 AX, Adamas AX, and Positron AX, from 1982 until 1984. Dura-Ace AX was Shimano's flagship aerodynamic gruppo. The Dura-Ace AX aerodynamic rear derailleur had a little staircase which provided the indexing and maintained a constant chain gap. It shifted beautifully but it was fussy to adjust and fragile. The professional racers rejected it. The dropped pedals and the rear entry brake levers were neat ideas.

Most of the bike makers displayed an aerodynamic bike in case the idea caught on. There was even a Huffy Aero with a full Dura-Ace AX gruppo and flattened "aero" tubing. It sold for $250 when the AX gruppo retailed for $300 and was in short supply. Smart buyers bought the Huffy, cut out the AX gruppo, and threw the rest in the dumpster. The Aero was a department store best seller until Shimano caught on and stopped selling AX to Huffy at the extra-low introductory OEM price.

Shimano pushed aerodynamics aggressively in 1982 and 1983, but it was a losing battle. Everyone knew that the rider, not the bike (let alone the components) was the main source of air resistance. Only Dura-Ace AX survived to 1984. The competitors probably hoped to see Shimano humbled, but it wasn't to be: When it was all over, Shimano just shrugged off Aero and carried on.

The Shimano 1981 Deore touring gruppo worked reliably and it set up Shimano for the growing mountain bike market. The two-piece crankset used TA chainrings at first. When Shimano adopted Sugino's 110/74 mm bolt circles in 1985, that became the industry standard.

In 1983, Shimano's 105 gruppo brought the gruppo concept down to $250 bicycles. The Shimano 105 gruppo looked expensive and performed well. The cosmetics were good, and 105-equipped bikes sold well. By this time, the competition realized that gruppos were essential for survival, but it was too late. This was the time to sell off your Huret and Simplex stock.

Shimano BioPace chainrings were a huge marketing success story from 1983 to 1992. You could hardly sell a bicycle without BioPace. They looked like oval chainrings, but the design was quite different. It was supposed to reduce knee stress.

The superior shifting of the Huret Duopar was hard to compete with, so Shimano set out to design a derailleur that worked as well. They came up with the

Left: Fig. 11.88. The 1984 Shimano Deore SuperPlate rear derailleur had a second parallelogram, like the Duopar.

Right: Fig. 11.89. The 1983 Shimano Deore XT was a complete mountain-bike gruppo. In addition to these derailleurs and thumb shifters, it also included a triple-chainring crankset, cantilever brakes, brake levers, hubs, freewheel, pedals, and a chain.

1984 Shimano Deore SuperPlate rear derailleur, which had a second parallelogram like the Duopar. They got around Huret's patent by spring-loading the second parallelogram. The SuperPlate was expensive to make, and when SunTour's slant parallelogram patent ran out, Shimano wasted no time before incorporatign slant parallelograms in all of their new derailleurs and quietly dropped the SuperPlate.

Both Shimano and SunTour recognized the growth potential of mountain bikes. Shimano spent an extra 6 months designing and testing improved seals for their mountain bike components. SunTour rushed to market and got most of the 1983 mountain bike business, but Shimano was ready with a complete mountain-bike gruppo when SunTour MounTech had problems. Shimano assumed that there must be something magic about the TA Cyclotouriste cranksets, so they copied TA. But all that the mountain bikers wanted was TA's 26-tooth granny chainwheel and long crank arms.

The Shimano and SunTour mountain-bike gruppos spelled the end for U.S. sales of TA Cyclotouriste triple cranksets, Simplex front derailleurs, Huret Duopars, and Mafac cantilever brakes.

Shimano and SunTour offered the bicycle makers a constantly changing array of derailleurs in this decade. Even if many of the changes were purely cosmetic, the European competitors faced a moving target. Each year brought new models with new features. The revised derailleurs were parts of entire gruppos with matching cosmetics.

Campagnolo survived because of the loyalty of the professional racers and buyers of high-priced racing bicycles. The Taiwanese makers survived by offering cut-rate prices to the makers of cut-rate bicycles. There was no market for the smaller European makers, and they disappeared from the U.S. marketplace. A few survived in the much smaller domestic markets in Europe.

### Sugino

Sugino made cranksets, stems, and seatposts. They made cranks with SunTour nameplates for the SunTour gruppos. In 1982, they made the Aero Tour TAT triple cranket with 110 and 74 mm bolt circles. This became the ideal triple crankset for loaded touring and mountain bikes. The 110 mm bolt circle could handle chainwheels as small as 34 teeth, and the 74 mm could handle 24-tooth inner chainwheels. With a 34-tooth freewheel sprocket you had a 19-inch low gear. These bolt circles became the *de facto* standard for mountain bikes when Shimano adopted them. It was the end for the unique TA and Stronglight bolt circles.

## Taiwan

Taiwan had 150 factories making complete bicycles in 1974. Although most of them were quite small, between them, they produced 2.1 million bicycles annually and exported 1.6 million.

1975 was a disastrous year for the Taiwanese makers, as the U.S. market for cheap bicycles dried up. Production decreased

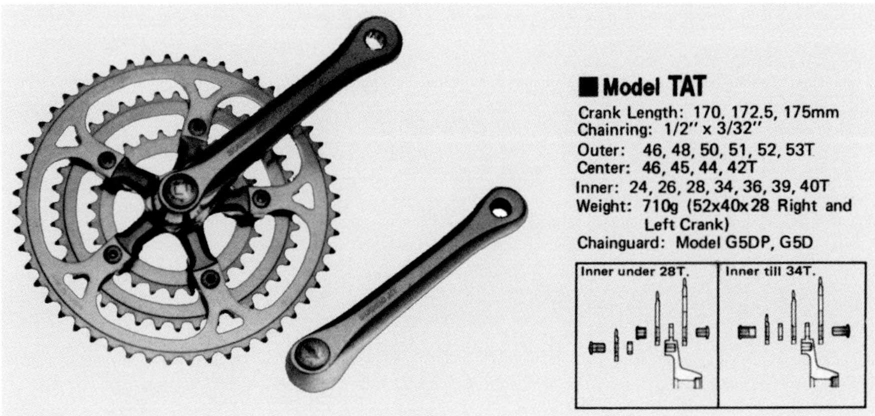


Right: Fig. 11.90. The 1981 Sugino Aero Tour TAT triple crankset introduced the 74 mm bolt circle that allowed 24-tooth inner chainrings. Combined with a 34-tooth freewheel sprocket, that gave a 19-in. low gear.

to 800,000 bicycles. Quality was poor because of cut-throat competition. The government stepped in with an export inspection program, and unqualified plants were shut down. By 1976, only 37 companies were making bicycles in Taiwan.

By the late 1970s, Taiwan was the most significant maker of low-priced bicycles and components. As the Japanese yen appreciated against the dollar, it became harder and harder for the Japanese bicycle makers to compete in the export market, and Taiwan picked took their market share.

Taiwan produced 2.2 million bicycles in 1978 and exported 2.1 million of them. Half of the bicycles imported into the U.S. that year came from Taiwan.

Schwinn's 1977 decision to import bicycles from the Taiwanese company Giant taught the Taiwanese how to make quality low- and medium-priced bicycles. It was also a great boost for Taiwan's reputation. Taiwan slowly outgrew their reputation for low-quality gas-pipe bicycles. Merida, the other major Taiwan bicycle maker, made most of Raleigh's low-priced bicycles.

Four Taiwan derailleur makers—Falcon, Joy Tech, Long Yih, and SunRace—made unabashed copies of Shimano derailleurs. They did not copy SunTour derailleurs, because those were still protected by the slant parallelogram patent.

## Cycling Magazines

In France, *L'Officiel du Cycle* ceased publication in March 1975. However, Daniel Rebour continued to produce line drawings (many of which are reproduced in this book) for commercial accounts, until 1983.

*Bicycling* had a significant technical content in this decade. In 1978, the magazine was purchased by Rodale Press and moved from California to Pennsylvania. Rodale rode the surge of interest in adult cycling that began in the early 1970s.

*Bike Tech*, also published by Rodale, started in 1982 and provided a wonderful forum for technical writers and their tech-savvy readers.

*Bike World* ceased publication in 1979. The *Fat Tire Flyer* was started in 1980 by Charlie Kelly and reported on the mountain bike scene until it ceased publication in 1987.

FAT TIRE FLYER 1

PO BOX 757 FAIRFAX CA.

issue one August/September 1980

Here it is Fat Tire Fans, your first issue of the Fat Tire Flyer. In this initial issue we will fill you in on the local cruiser/bomber/ ballooner/clunker/mountain bike and otherwise fat tire bike scene, racing, riding, touring, and so on.

The most famous bomber racecourse, Repack, has been retired temporarily due to the fact that the road is in very dangerous condition from the winter's rains. We are looking for another downhill course, if possible one without political complications as far as using it goes. Any ideas?

We at the Fat Tire Flyer are seeking input with regard to a number of ballooner related subjects. Anyone who has anything to contribute is welcome to send it in. We can't pay for contributions because the Fat Tire Flyer is a non-profit newsletter produced on a voluntary basis.

The Flyer is an off-shoot of an organization called The Marin County Wilderness Wheelers. The purpose of the Wheelers is to deal with various problems which affect us all, most importantly potential restrictions on bicycling in local parks and water districts. Without an organization to protect our interests we will have no interests to protect. We are encouraging all ballooner riders to get in touch.

Above: Fig. 11.92. The first issue of *The Fat Tire Flyer* was published August 1982 in Fairfax, California, by mountain bike pioneer Charlie Kelly, with Denise Caramagno as editor.

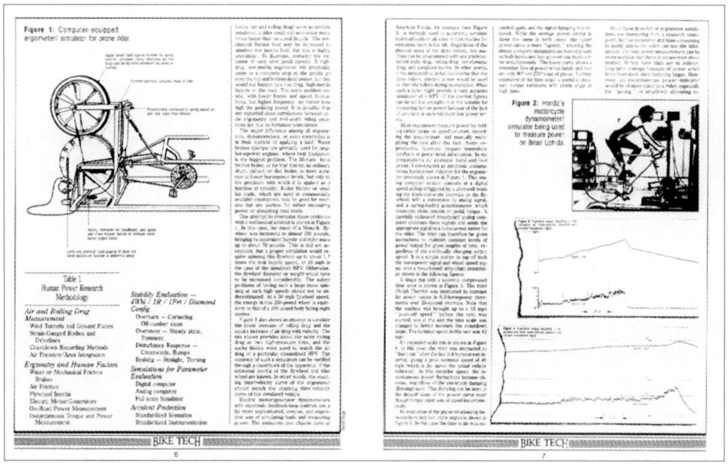

Left: Fig. 11.91. A page-spread from an early issue of *Bike-Tech.* Perhaps a bit more technical than what most people were waiting for, but packed with well-researched technical materials.

# 12. The Rise of Shimano: 1985–1994

MOUNTAIN BIKES and indexed shifting were the two major changes in this decade. These two changes took over the market so quickly that the companies that adapted to rapid change thrived and the companies that moved slowly did not survive.

Shimano not only adapted to the changing market: they instituted most of the changes themselves. By the end of the decade, Shimano dominated the component market. They succeeded so well, that bicycles were classified by their Shimano component gruppos. It was an Ultegra bike, a 105 bike, or a Deore bike. The bike maker's name became secondary.

SunTour put up a good fight, but they were always a year behind Shimano. By 1994, SunTour had become the brand name for Sakae Ringyo, a small Taiwanese component manufacturer (which itself also had its origins in a Japanese company).

For most of this decade, Campagnolo's indexed shifting derailleurs and gruppos for mountain bikes could not be sold on the U.S. market. Campagnolo survived because of the loyalty of the racers.

Sachs-Huret made quality components, but they were always a year or two behind Shimano, and the company could not gain a foothold in the crucial U.S. market.

The small European derailleur companies,, like Galli, Mavic, Ofmega, Simplex, and Zeus, disappeared during this decade, leaving Shimano with about 80%

Right: Fig. 12.1. John Howard, cycling at 152 mph (245 km/h) on the Bonneville Salt Flats in 1986.

of the market, while Campagnolo, Sachs-Huret, and the Taiwanese companies (Falcon, Joytech, Long Yih, SunRace, and SunTour) took care of the remainder.

At the end of this decade, there was renewed interest in something besides mountain bikes or road bikes. Trekking bikes, hybrid bikes, and commuting bikes all appeared.

In 1994, Shimano, Sachs, and Sturmey-Archer all introduced 7-speed hub gears to expand the developing market for alternatives to derailleurs.

Look, a French ski binding company, introduced clipless pedals in 1985. Clipless pedals had clear advantages over the conventional toeclips and straps, and they soon became the standard for racing bikes and mountain bike alike.

## Bicycle Racing

During this decade there was an increase in interest in bicycle racing in the U.S. Much of this was due to Greg LeMond's success in the Tour de France. He won the Tour 3 times, before being "dethroned" by Miguel Indurain, of Spain, who won the next 5 Tours. Campagnolo was still the brand of choice for professional and amateur racers alike, but Shimano began to make inroads.

In 1986, John Howard set a new bicycle land speed record by riding a measured mile in 23.64 seconds, at an average speed of 152.28 mph (245 km/h) behind a Bonneville streamliner with a hood as a wind shield. His single-speed bicycle used an intermediate jackshaft to obtain a 390-inch gear. It covered 111 ft. (34 m) for every turn of the crank. Howard was first towed up to a speed of 60 mph (96 km/h).

In 1993, two British amateurs, Chris Boardman and Graeme

Right: Fig. 12.2. Bernard Hinault, of France, (right) won the Tour de France five times between 1978 and 1985. Greg LeMond, of the U.S., (left) won the race three times, in 1986, 1989, and 1990. After that, the Tour was dominated by Miguel Indurain, of Spain, who won five times in a row from 1991 until 1995.

*Table 12.1. Significant bicycle drivetrain developments, 1978–2009**

| Feature | Shimano | SunTour | Sachs-Huret | Campagnolo | Other |
|---|---|---|---|---|---|
| Indexed Shifting Road Bikes | 1985 | 1987 | 1987 | 1992 (1987) | |
| Indexed Shifting Mountain Bikes | 1987 | 1987 | 1989 | 1990 | |
| Modern Rear derailleurs | 1985 | 1987 | 1988 | 1991 | |
| Freehubs | 1978 | 1989 (69) | 1994 | 1991 | |
| Bushingless Chains | 1986 | 1990 | Sedis 1978 | 1990 | Sedis, 1978 |
| Brake-Shift Levers | 1990 | N/A | SRAM, 2008 | 1992 | |
| HyperGlide Sprockets | 1989 | 1990 | 1994 | 1994 | |
| SuperGlide Chainwheels | 1990 | 1990 | 1994 | 1994 | |
| Compact Chainwheels | 1994 | 1992 | 1994 | 1992 | |
| 7-Sprocket Freewheel | 1981 | 1980 | 1988 | 1987 | |
| 8-Sprocket Freewheel | 1989 | 1992 | 1991 | 1991 | |
| 9-Sprocket Freewheel | 1997 | 2000 | 2001 | 1997 | |
| 10-Sprocket Freewheel | 2004 | N/A | SRAM, 2008 | 2000 | |
| 11-Sprocket Freewheel | N/A | N/A | N/A | 2009 | |
| Clipless Pedals | 1987 | 1990 | N/A | 1988 | Look, 1985 |

* To complete the table, we have also included 9-, 10-, and 11-sprocket freewheels, which were introduced in the next decades.

Obree, broke Francesco Moser's 1984 one-hour record of 51.15 km (31.71 miles). First, Obree went 51.6 km (32.7 miles). Six days later, Boardman went 52.3 km (32.5 miles). Their partially aerodynamic bikes and aero riding positions certainly contributed to their speed.

In 1986, Freddie Markham raised the 200-meter IHPVA record to 65.5 mph (105.4 km/h). In 1992, Chris Huber raised the 200-meter record to 68.7 mph (110.5 km/h). In 1985, Pat Kinch raised the one-hour IHPVA record to 47 miles (75.58 km).

## Major Developments

Table 12.1 shows twelve genuine improvements to bicycle components that happened in this decade. By 1995, these improvements were part of almost every quality derailleur bicycle. To complete the table, I show 9-sprocket and 10-sprocket freehubs, which did not came along until the next decade.

Shimano was not always the inventor. Sometimes they fine-tuned the inventions of others. Items like the freehub had appeared decades earlier. Shimano's improved versions were copied by the competition and became part of the modern bicycle.

*Table 12.2. Tour de France winners, 1985–1994*

| Year | Rider (Country) | Bicycle | Derailleur |
|---|---|---|---|
| 1985 | Bernard Hinault (France) | Hinault | Campagnolo |
| 1986 | Greg Lemond (USA) | Look-Hinault | Campagnolo |
| 1987 | Stephen Roche (Ireland) | Battaglin | Campagnolo |
| 1988 | Pedro Delgado (Spain) | Pinarello | Campagnolo |
| 1989 | Greg Lemond (USA) | Bottechia | Mavic |
| 1990 | Greg Lemond (USA) | Lemond | Campagnolo |
| 1991 | Miguel Indurain (Spain) | TVT | Campagnolo |
| 1992 | Miguel Indurain (Spain) | Pinarello | Campagnolo |
| 1993 | Miguel Indurain (Spain) | Pinarello | Campagnolo |
| 1994 | Miguel Indurain (Spain) | Pinarello | Campagnolo |

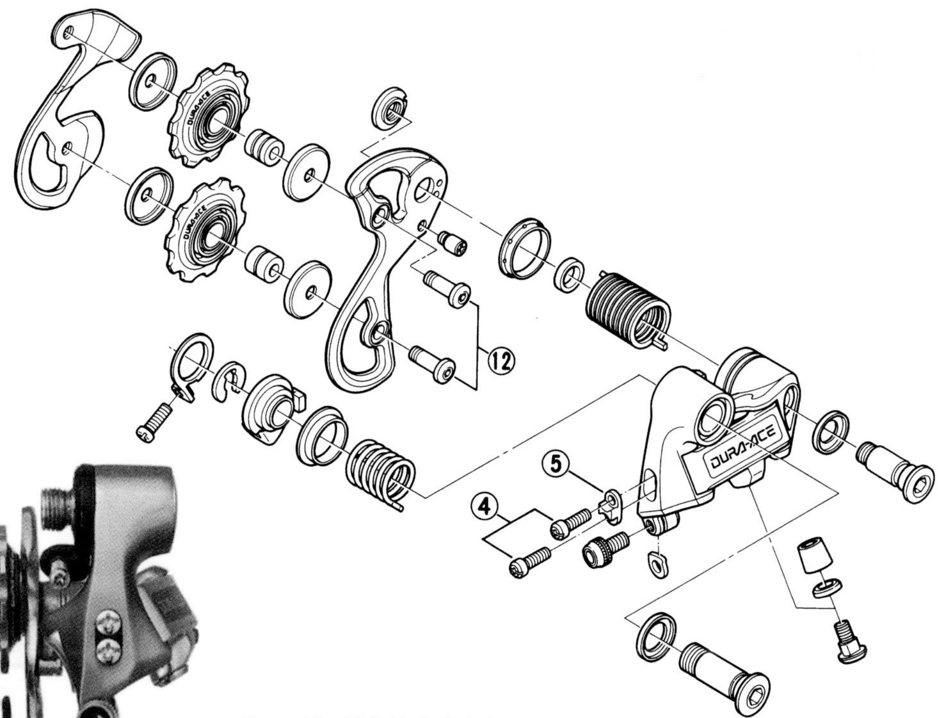


Above: Fig. 12.3. Exploded view of the 1985 Shimano Dura-Ace rear derailleur.

Left: Fig. 12.4. The 1985 Shimano Dura-Ace rear derailleur.

## Rear Derailleurs and Rear Shifting

In 1985, the revised Shimano Dura-Ace rear derailleur, combined two spring-loaded pivots (developed by Simplex) with a dropped horizontal slant parallelogram (developed by SunTour). Shimano used computer programs to optimize the key dimensions and the tensions of the two spring-loaded pivots. This design provided a constant chain-gap and "early" shifting, which were the keys to reliable indexed shifting. By 1988, all of Shimano's rear derailleurs used the Dura-Ace design. Over the next five years, it became the standard, as it was copied by the competition. It is possible to provide a constant chain-gap with a different design but none of the major makers tried very hard.

The 1987 Shimano Deore XT rear derailleur used the same combination of a slant parallelogram, two spring-loaded pivots, and optimum geometry and spring tensions to provide reliable indexed shifting for mountain bikes. It was a Dura-Ace with a long cage. Soon everyone copied the Deore XT and it became the standard mountain bike rear derailleur.

After 1990, the important improvements in rear shifting did not involve rear derailleurs or shifters. They were changes in the shape of the sprocket teeth and positioning of adjacent sprockets. This made it easier and smoother for the chain to move from sprocket to sprocket. Shimano called this *HyperGlide*.

Above: Fig. 12.5. The 1983 Shimano Deore half-step front derailleur. Note the shallow cage.

Above: Fig. 12.7. This 1987 Shimano Deore XT long-cage rear derailleur was for mountain bikes.

## Gruppos, Front Derailleurs, and Front Shifting

A full gruppo consists of front and rear derailleurs, shift levers, hubs, crankset, bottom bracket,

Below: Fig. 12.6. The 1987 Shimano Deore XT crossover front derailleur. Note the deep inner cage plate.

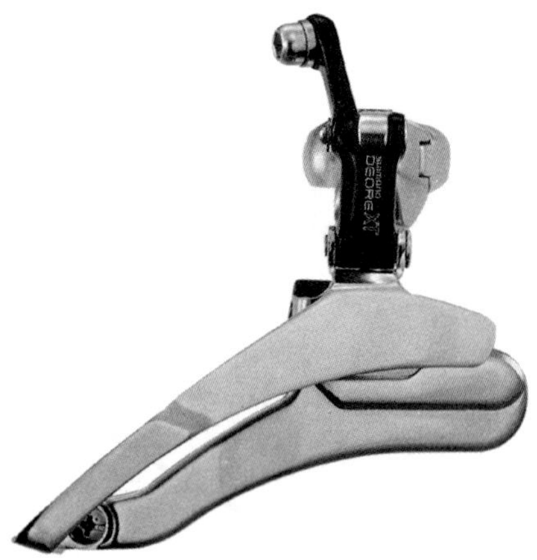


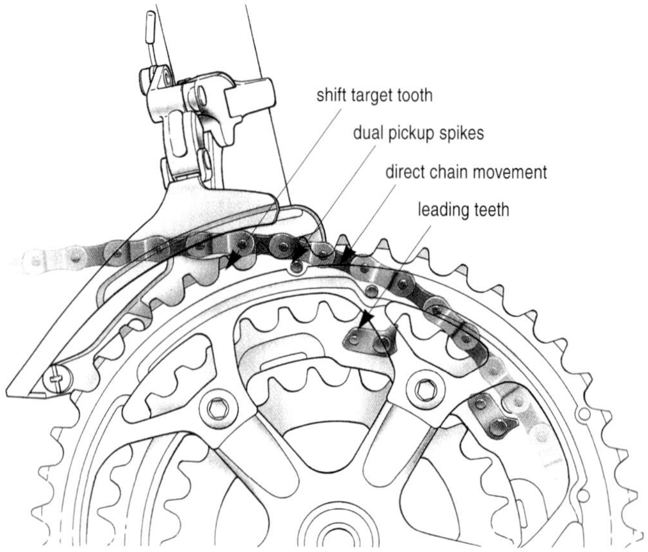


Right: Fig. 12.8. The 1993 Shimano HyperDrive chainwheel, showing the pins and hooks that lead the chain from one chainring to the next.

freewheel (or cassette), brake calipers, brake levers, cables, and casings. It may also include pedals, headset, chain, stem, and seatpost. Campagnolo sold the first gruppos in the 1960s. (If course, the correct Italian plural of gruppo is *gruppi,* but only a diehard bike snob would use that term.)

More than two hundred gruppos were introduced in this decade. Most new gruppos included a "new" front derailleur that was cosmetically different from the previous front derailleur and from the competitor's front derailleurs. There were few innovations in front derailleurs. In fact, you can substitute a 1960 Campagnolo Record front derailleur for almost any modern front derailleur and it will friction shift adequately. The inner cages of modern front derailleurs have various ridges and bumps in the right positions to help shifting a bit.

In a search for uniqueness, some front derailleurs moved forward and some moved upward, as they swung outward during the shift from the small to the big chainwheel. None of these actions made much difference. When testing derailleurs for *Bicycling* in the 1980s, I concluded that the key characteristic was rigidity. Rigid front derailleurs shifted better than flexible front derailleurs.

For a few years, there were two kinds of touring derailleurs. Half-step models had shallower inner cages to clear the middle chainrings of half-step triple cranksets, like 50/45/26. Half-step gearing and half-step front derailleurs disappeared around 1985, when mountain bikes and 7-speed cassettes took over. The crossover front derailleur had a deeper inner cage to do battle with the mountain bike crossover cranksets, like 50/38/26.

The key improvements in front shifting took place on the chainwheels. First, certain chainwheel teeth were cut away so that the shifts took place on the dead centers when chain tension was lowest. Then, the sides of the chainwheels were bulged and pins were added to guide the chain from the middle to the large chainwheel. Finally, little hooks were added to grab the chain during the shifts from the little chainwheel. Shimano led the way with these improvements. As they evolved, Shimano called the various changes HyperGlide C, SuperGlide, HyperDrive, and Front SIS.

SunTour called their revised chainwheels PoweRing, and Campagnolo called them Exa-Drive. These changes, combined

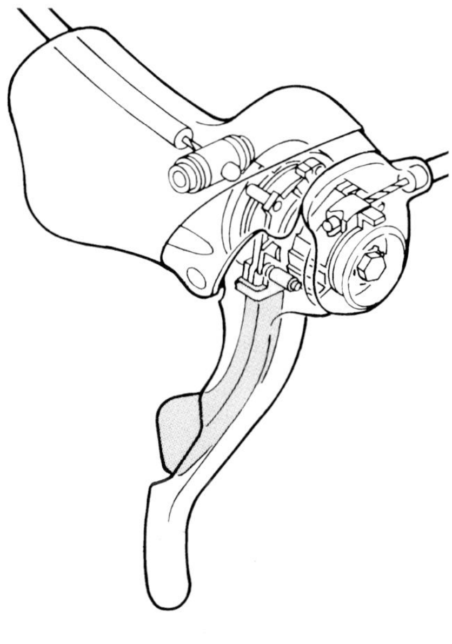

Above: Fig. 12.9. Shimano's 1991 STI (Shimano Total Integration) dual-control levers for operation of brakes and gearing.

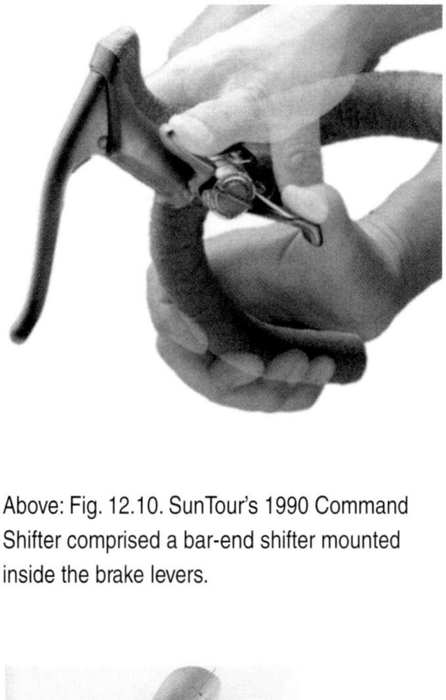

Above: Fig. 12.10. SunTour's 1990 Command Shifter comprised a bar-end shifter mounted inside the brake levers.

Left: Fig. 12.11. The 1987 Shimano SIS thumb shifter.

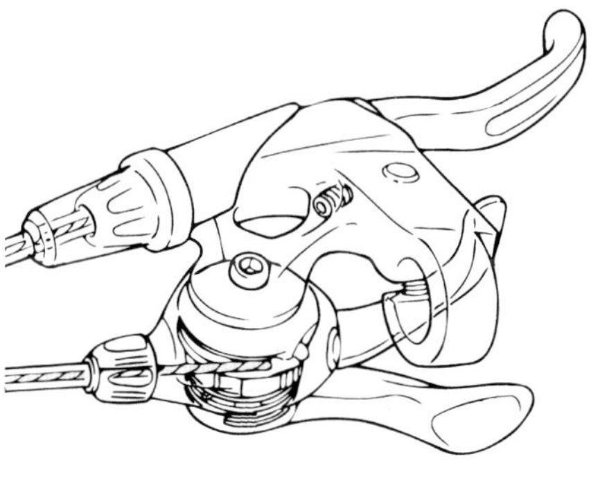

Above: Fig. 12.12. The 1990 Shimano Rapid-Fire Plus shifter had two levers, one for upshifts, the other for downshifts.

with the improved front derailleurs, made indexed front shifts almost as reliable as rear shifts.

### Shift Levers

There was more drama in shifters, as the advertisers inflated minor differences into presumably major advances. In 1985, road bikers could chose between down-tube shifters, bar-end shifters, and stem-mounted shifters. The latter were for beginners. Indexed versions of the three styles were available, usually with a manual shifting option.

The manual shifting option was interesting. If the indexed shifting worked properly, it was not needed. If the indexed shifting did not work properly, the user was unhappy, regardless of the manual option. Shimano's patent on the dual indexed/manual mode caused problems for their competitors.

In 1991, Shimano's STI (Shimano Total Integration) dual control levers combined the brake levers with the indexed shift levers. They dropped the manual option. Campagnolo's Ergopower dual control levers followed in 1992. Sachs used Campagnolo's levers.

SunTour's Command Shifter was a bar-end lever mounted next to the brake levers. It never had a chance because by 1990, almost no one was specifying SunTour.

There was more action with mountain bikes. Until 1986, everyone used either SunTour or Shimano thumb shifters mounted on top of the handlebars. SunTour's power ratchet had a lighter action. In 1987, Shimano and SunTour both offered thumb shifters with indexed or friction modes.

In 1990, the levers became push buttons under the handlebar. One button shifted up and the other shifted down. Shimano called it Rapid Fire and SunTour called it X-Press. In 1990, Shimano dropped the friction mode since it was never used. The 1990 Shimano Rapid Fire Plus shifter had two levers using the thumb for up shifts and the index finger for down shifts. SunTour still offered top-mounted thumb shifters with a friction option.

### SRAM Grip-Shift

SRAM introduced the indexed *Grip-Shift* for road bikes in 1988. It worked like a motorcycle throttle and it was a hard sell because it was heavier and clumsier to use than a bar-end shift lever. It also required drilling a hole in the handlebar.

SRAM's Grip-Shift became popular on low-priced mountain bikes after SRAM revised the design to handle mountain bikes and hybrids. The 1990 CX version came with full-width slippery green grips that required 360 degrees of rotation (something Shimano had figured out on its

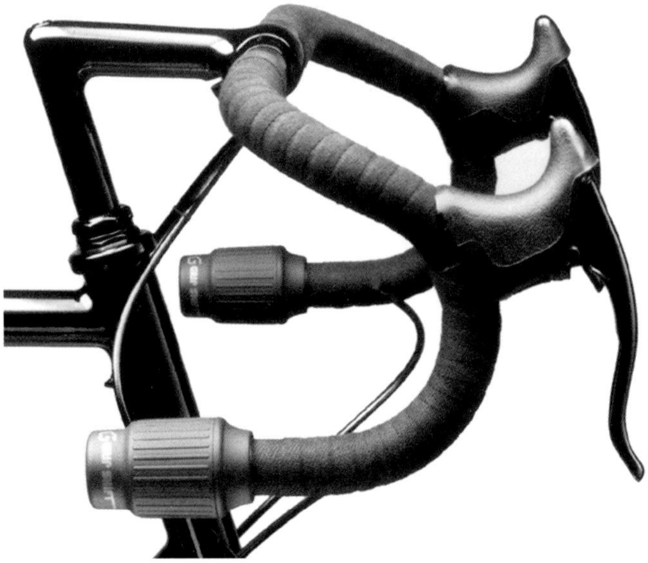

Left: Fig. 12.13. The 1988 SRAM indexed Grip-Shift for road bikes.

Below: Fig. 12.14. The 1992 Grip-Shift for mountain bike use was a twist ring, mounted inboard of the fixed part of the handlebar grip.

Lark shifter 10 years before). Market acceptance was underwhelming. The problem with the early Grip-Shifts, and with Campagnolo's and Sachs' similar twist shifters, was that full-width shifters auto-shifted going over bumps. In 1991, everyone got the message, and by 1992 twist shifters were mounted inboard of the handgrip. This was a superior design.

SRAM obtained a number of patents on twist shifters. A patent is a license to sue, and in 1989, SRAM sued Shimano for patent infringement and for violating U.S. anti-trust laws by monopolizing the component market and requiring tie-in sales. Shimano settled out of court. The exact settlement was not revealed, but Shimano changed their pricing structure to allow bike makers to mix and match gear train components. Shimano did not offer twist shifters for modern derailleur bicycles until 1999 (they had them for the Lark 20 years before).

In the early 1990s, the high-priced yen gave SRAM a price advantage, and their Grip-Shift took a significant share of the low-price market. By 1995, SRAM Grip-Shifts were on more mountain bikes than Shimano shifters. SRAM is a litigious company, and they filed a series of lawsuits against Shimano, SunTour, and SunRace.

Sachs and Campagnolo showed thumb shifters and twist shifters in the catalogs for their mountain-bike gruppos, but neither of them ever became a factor in the U.S. mountain bike market.

## United States

The decade began with a sense of uncertainty. Road bike sales were declining and mountain bike sales were rising. Was the mountain bike a fad or was it the real thing? When would mountain bike sales decline? Few recognized that the mountain bike would take over the market. It would become a "bicycle" and drop-handlebar, skinny-tired bicycles would become "road bikes."

Annual U.S. sales moved in a fairly narrow range between 10 and 13 million bicycles. About half were juvenile bikes with wheels 24 in. (610 mm) or smaller in diameter, and the other half were adult bikes. Quality bikes were about a quarter of the total volume and were mostly sold by neighborhood bike stores. Three quarters were gas-pipe bikes that were sold by mass marketers, auto supply stores, toy stores, department stores, and discount stores. A new type of merchandiser, the bicycle mass marketer, muddled this neat division. Stores like Nashbar, Performance, and REI became large enough to have quality bicycles made with their brand name labels.

As mainstream bicycles became more like commodities, custom-built bicycles caught on. By 1984, 70 different U.S. com-

Above: Fig. 12.16. The 1985 Trek 2000 road bike had an epoxy-bonded aluminum frame.

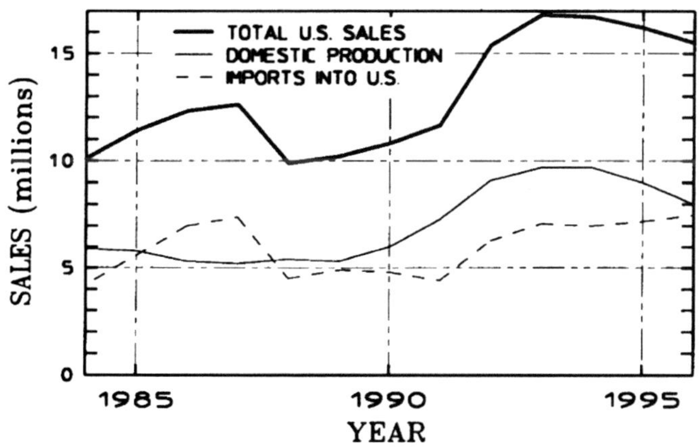


Left: Fig. 12.15. U.S. total bicycle sales, domestic bicycle production, and bicycle imports, 1984—1995. (Source of data: Bicycle Manufacturers' Association of America.)

panies were either building custom frames or selling imported frames.

By 1985, Trek, Raleigh-USA, and Cannondale were selling mainstream bikes with non-steel frames. Buyers could choose from steel, aluminum, titanium, or carbon-fiber/epoxy. By 1988, eight different makers made bicycles with aluminum frames for less than $500. None of the alternative materials replaced steel. More than half of the quality bicycles and all of the low-priced bicycles were made of steel.

The U.S. became the technological leader in both mountain bikes and road bikes. Trek, Cannondale, and Raleigh-USA were the major U.S. makers of quality bikes. They constantly innovated and followed the trends. Sales of loaded-touring bikes declined in 1985 but they continued as a niche market.

### Raleigh-USA

In 1982, Huffy signed a franchise agreement with TI-Raleigh, which created a new company, Raleigh Cycle Company of America. Huffy hoped to use the Raleigh name and the U.S. Raleigh dealers to move up into the high-end bicycle market. They sponsored Raleigh Team USA, which won important victories and helped name recognition. Huffy imported a complete line of bicycles from Japan for two years but it was not very profitable.

Huffy built a Raleigh of America factory in Kent, Washington, to make their Technium bicycles. The Technium frame used oversized aluminum tubing for the three main tubes and alloy steel for the stays and forks. The frame was joined by epoxy-bonded lugs. Initially, there were problems with joint failures and poor paint quality. In 1985, Huffy let the Raleigh franchise agreement expire, although they sold Raleigh bicycles for two more years.

In 1987, Derby International bought Raleigh from Tube Investments. Derby was a management investment company that already owned Nishiki and Kalkhoff. When Derby bought Raleigh, Raleigh-USA and Raleigh-UK once again had a common owner.

### Schwinn

Schwinn made a turnaround in the late 1980s, but by then 80% of their bicycles were made by Giant in Taiwan. When Schwinn entered into a joint venture with China Bicycle Company in 1987, Giant started selling Giant-labeled

Left: Fig. 12.17. Introduced in 1983—after long delays—the $35 Avocet 20 was the first electronic cyclometer. It could display speed, trip and total distance, and elapsed time.

Above: Fig. 12.18. The 1985 Raleigh 480 was made in Kent, WA. It sold for $300, equipped with an SR 52/40 crankset, SunTour 14–28 freewheel, and SunTour derailleurs.

Left: Fig. 12.19. The 1985 Specialized Rockhopper set the standard for moderate-price mountain bikes.

bicycles in the U.S. The Giant-Giants were identical to the Schwinn-Giants, but cheaper. Schwinn bought 51% of Csepel, a Hungarian bicycle company, and this venture was a disaster.

Schwinn's domestic production ceased in 1991 with the closure of the Greenville, Mississippi factory, and Schwinn became an importer. Schwinn was 60 million dollars in debt, and the banks refused to extend more credit. By 1993, Schwinn was bankrupt, and the Schwinn name was sold to Scott Sports and Zell Chilmark, an investment group. Interested readers are referred to Judith Crown and Glenn Coleman's excellent book *No Hands* for the sad story of how Ed Schwinn ran his company into the ground.

### Other Makers

The market share for imported bikes peaked in 1987. This coincided with a period of high inflation and devaluation of the dollar. Huffy, Murray, and Roadmaster were the major makers of juvenile bikes and adult gas-pipe bikes. Their highly automated factories could compete with imports from low-wage countries. Ross bicycles covered a wider price range.

In 1988, more than half of the bicycles sold in the U.S. were made in the U.S.

Trek made a full line of middle and high-priced road bikes and mountain bikes with steel and aluminum frames. The 1985 Trek 2000 racing bike used an epoxy-bonded aluminum frame and a Shimano Dura-Ace indexed gear train. It cost $1,200. In 1986, Trek followed with the 1500 and the 1000 using the same frame and less expensive components. By 1994, Trek was importing a lower-priced line of adult and children's bikes. In 1994, Trek bought Fisher Mountain Bikes, Klein, and Bontrager. As Schwinn declined, Trek became the largest U.S. bike maker.

Cannondale continued with large diameter aluminum frame tubing and they concentrated on the middle- and high-priced markets.

Specialized was one of the largest importers of quality bicycles, designed by Specialized. The 1985 Specialized Rockhopper brought down the prices of quality mountain bikes with a Shimano Deore XT gruppo down to $400. In 1985, Specialized tried to compete in the component market. They imported a nearly complete component gruppo, but it did not thrive.

Rock Shox introduced suspension forks for mountain bikes in 1989. Mountain bikes with full suspension (front and rear) appeared in 1990. By 1991, there were a dozen different full suspension models. The Fisher RS-1 was an early dual-suspension mountain bike. Motorcycle designer Mert Lawill had designed the rear suspension.

By 1992, dual-suspension bikes were available for less than $600. By 1993, 40% of the mountain bikes had some form of suspension.

Above: Fig. 12.20. The 1991 Fisher RS-1 dual-suspension mountain bike.

Right: Fig. 12.21. The 1991 Cannondale SM 1000 mountain bike had a SunTour XC Pro indexed gear-train.

Much of the design for full suspension was based on motorcycles. Motorcycle engines produce a steady torque at high RPM. Bicycle engines (people) produce variable torque at low RPM. Many of the early rear suspensions bobbed up and down as the rider pedaled. This defect was called "biopacing," a put down of Shimano's oval chainwheels. As front suspension and full suspension took over, mountain bike prices went up.

Hybrids took over the mid-priced ($200 to $400) market. Hybrids went from 6% of the 1990 market to 20% of the 1991 market and stayed at that level. In 1994, Specialized and Cannondale tried to interest the Americans in city bikes. The $500 Specialized Globe 7 city bike had a Shimano Nexus 7-speed hub gear with coaster brake. It was fully equipped with lights, fenders, rack, chain guard, and kick stand. However, U.S. buyers were not quite ready and city bikes met with little success.

Road bike sales went down year by year. After bottoming in 1994, at about 2% of the market, sales turned up a bit. There were no cheap road bikes. The inexpensive bike market belonged to hybrids.

1992, 1993, and 1994,were hard years for U.S. bicycle dealers. Sales were down and prices were up with the devaluation of the dollar and the appreciation of the yen. $200 to $250 was the hot price range. In 1993, many makers combined Shimano derailleurs and hubs with SRAM Grip-Shifts, SR or Sugino cranksets, Sedis chains, and Dia-Compe brakes to save money and to establish a unique image. This lasted for about four years and then the buyers decided they wanted all-Shimano gruppos.

BMX, (Bicycle Moto-Cross), was born again in 1993, and BMX bikes took a significant share of the U.S. market. By 1994, shock absorbing front forks were standard on mountain bikes selling for more than $250.

## Europe

The mountain bike caught on in Europe in 1990, about five years after the U.S. When the European makers were slow to satisfy the demand, U.S. companies like Trek, Cannondale, and Specialized exported bicycles to Europe. Scott was especially successful and became a major European supplier.

By 1993, trekking bikes were becoming popular in Europe. This was a marriage between the old city bike with fenders, lights, and luggage racks and the mountain bike with derailleur gearing and high-quality components.

In 1993, the European Economic Community (EEC) slapped a 30.4% anti-dumping tariff on bicycles imported from China. This was on top of the normal 17% import duty.

Germany was the largest single bicycle market in Europe. Sales averaged about 6 million bicycles per year in this decade. Imported bikes went from less

Left: Fig. 12.22. The 1994 Specialized Globe 7 city bike, inspired by the design of the kind of bikes popular in Europe, had a Shimano 7-speed hub gear.

Right: Fig. 12.23. The 1985 Sachs-Huret New Success long-cage rear derailleur.

than 5% of the market to 36% in 1992. Hower, no derailleurs were made in Germany. The derailleurs for Sachs were made in France by Huret (see below) bined with an indexed 6-sprocket freewheel and a derailleur. The Sachs Orbit 2×3 was a three-speed hub combined with two sprockets and a derailleur. Sachs-Huret must have smiled at Shimano's Indexed Shifting in 1985. They had been offering their indexed Commander and Orbit

## France

France was the second-largest bicycle market in Europe. French bicycle makers had difficulty competing with Taiwan and China and their share of the French domestic market fell steadily. By 1993, mountain bikes had taken two-thirds of the market, and mass-marketers had 70% of bicycle sales.

In 1985, Gitane, Bertin, and Peugeot were still exporting medium-priced road bicycles to U.S. dealers for the tiny Francophile market. They were normally equipped with Sachs-Huret Rival derailleurs. Peugeot started making mountain bikes in 1985. Gitane was the last completely French bicycle to be exported to the U.S.

Right: Fig. 12.24. The 1987 Sachs-Huret ARIS New Success rear derailleur had a dropped slant parallelogram. It was part of a gruppo that included ratchet shifters that allowed the rider to select either indexed or non-indexed shifting.

Below: Fig. 12.25. The 1989 Sachs-Huret ARIS Rival ATB (i.e., mountain bike) rear derailleur indexed well, but it did not succeed in the U.S. market.

### Sachs-Huret

Sachs-Huret made well-designed, well-built components, but they were always a year or two behind Shimano.

In 1985, the Sachs-Huret line still included the Jubilee with its drilled-out cage, the Duopar, and the Eco. These were all carryovers from Huret's better days.

The Sachs Commander system was a two-speed hub com-

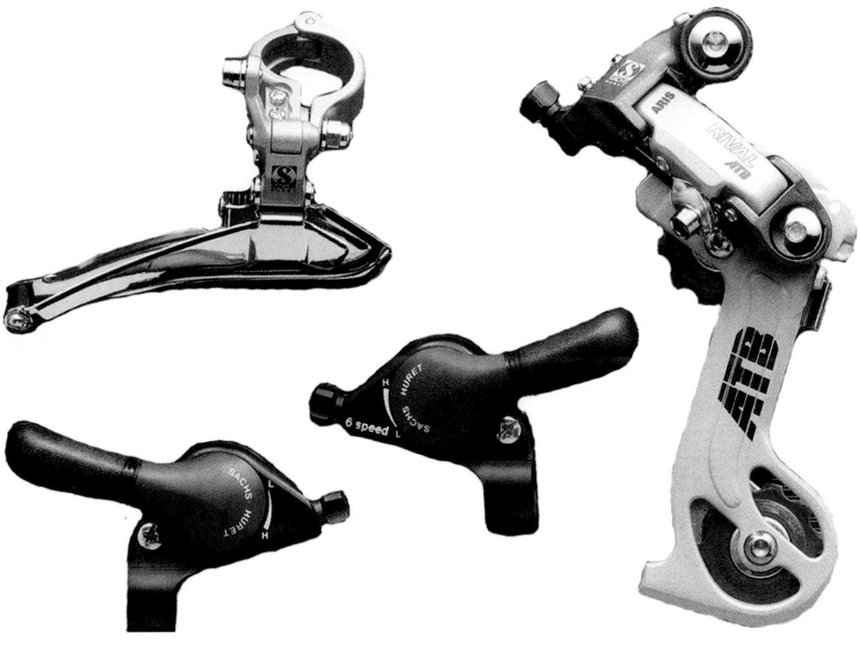

2×3 for years, and U.S. sales were non-existent.

The first Rival and New Success gruppos appeared in 1986 with hubs, cranksets, and freewheels from Sachs-Maillard. The Sachs-Huret New Success rear derailleur had dropped horizontal parallelograms and two spring-loaded pivots.

In 1989, the gruppos included brakes from Modolo. Sachs-Huret had three derailleur price levels: New Success, Rival, and Rider. All three had optional long cage rear derailleur and ratchet thumb shifters. Sachs-Huret was at least thinking about mountain bikes.

In 1990, Sachs-Huret introduced ARIS (Advanced Rider Index System). The New Success rear derailleur was revised to include a slant parallelogram. The ARIS gruppos included a Sachs-Sedis chain and a Sachs Maillard freewheel with a special tooth profile. ARIS provided good indexed shifting because Sachs recognized the requirements for a constant chain gap, early shifting, and a matching gear train. However, there were no U.S. sales.

In 1988, the ARIS New Success and ARIS Rival derailleurs got revised cosmetics. In 1989, there were touring and mountain bike versions of ARIS Rival. Sachs also introduced the Elysée, which was a pre-select indexed shifting group for city bikes.

The Huret name disappeared from the catalogs and derailleurs in 1991 and Huret's Incheville factory became Sachs Industries S.A. Both Gérard and Alain Huret (the third generation) are still alive, but they rarely answer letters, so it is hard fill in the blanks on the last chapter of the Huret history.

In 1991, Sachs made a serious run at Shimano with six road bike and five mountain-bike gruppos, covering the whole price spectrum. ARIS became ARIS Plus. The ATB gruppos offered cantilever or drum brakes, sealed bearing hubs, oval chainrings, and an optional Power Grip twist shifter. 8-speed freewheels were available, but it was all in vain for the U.S. market. Only Gitane used Sachs.

There were no major introductions in 1992 or 1993. Sachs concentrated on the European market, and looked for chinks in Shimano's armor. Sachs hydraulic brakes and the 3×7 (a heavy-duty three-speed hub with an indexed 7-speed rear derailleur) were sold to individual mountain bikers through retail bike stores. Sachs never cracked the OEM (Original Equipment Maker) market in the U.S.

In 1994, Sachs made another run at the U.S. market with seven road gruppos and five mountain-bike gruppos covering a broad price range. Sachs introduced a sealed-bearing freehub. Sachs road gruppos used Campagnolo's *Ergo* combined brake-shift levers with a different indexing cam to match Sachs derailleurs. The 7-speed Sachs Torpedo Super 7 hub gear had a range of 284%. It went with the Sachs 5-speed and three-speed hubs for city and trekking bikes.

### Simplex

Simplex' 1985 bankruptcy was the only way to comply with the restrictive French labor laws. Henry Juy reopened with 40 employees, producing a small range of products. They called their three derailleur lines Gold, Silver, and Bronze.

Left: Fig. 12.26. The 7-speed Sachs Torpedo Super 7 hub gear was sold from 1994 to 1999, and then became a SRAM component.

Right: Fig. 12.27. The 1992 Simplex Vallée rear derailleur was the end of the line for Simplex.

In 1987, Simplex merged with Ofmega. No Simplex-equipped bikes were displayed at the U.S. shows after 1987. The 1992 medium-priced Simplex Vallée rear derailleurs had dropped slant parallelograms and two spring-loaded pivots. They probably would have index shifted quite well.

By 1995, Simplex was owned by Sratec, a company that made locks for Peugeot cars. They made a small range of Simplex derailleurs and the Retrofriction downtube shift levers.

### Mavic

Mavic made the 801/851 single-pivot rear derailleur from 1986 to 1992. The cage pivot fit into a slot and could be adjusted to match the sprocket selection. It was completely rebuildable. Mavic used Simplex Retrofriction downtube shift levers. As Campagnolo had meanwhile learned, single-pivot derailleurs cannot provide a uniform chain gap, which is required for indexed shifting.

In 1993, Mavic introduced the 840 rear derailleur for road bikes and the 845 for mountain bike. These were Shimano copies with slant parallelograms and two spring-loaded pivots. An indexed shift lever was available for it.

In 1994, Mavic made the *ZMS* (ZAP Mavic System) 8000 microprocessor-controlled electric rear derailleur. It was elegant and expensive ($1,700 for the complete gruppo). Push buttons triggered a solenoid that allowed the jockey pulley to shift to pre-selected positions. The shifting force came from the rotation of the jockey

Above: Fig. 12.28. The 1993 Mavic 840 rear derailleur for road bikes was a copy of the Shimano Dura-Ace.

Above: Fig. 12.31. The 1986 Mavic 801 rear derailleur was a Campagnolo Nuovo Record copy with different cosmetics.

Left and right: Figs. 12.29 and 12.30. The 1994 Mavic ZAP 8000 was the first microprocessor-controlled electronically shifted rear derailleur. It did not hold up well under wet conditions.

pulley. The microprocessor and the batteries were stored in the handlebar. Mavic introduced ZAP to a limited set of professional racers in 1994 and 1995. It had startup problems, especially in wet weather. Mavic quietly dropped out of the component business in 1996 to concentrate on rims and hubs.

## Italy

Italy makes more bicycles than any other country in Europe. Most of these are heavy-duty utility machines, but there is active support for bicycle racing and there are numerous racing clubs.

The U.S. bike makers churned out medium- and high-priced road bikes with frames made of exotic fibers and metals. Each year, the latest model made last year's model obsolete.

Italy was different. Italy had dozens of small and medium-sized custom builders of bicycles with steel frames, and they thrived in this decade. If you wanted a traditional road bike with built-in prestige, you bought it from a small Italian company like Basso, Bottechia, Casati, Cinelli, Ciocc, De Rosa, Giordana, Gios, Guerciotti, Masi, Mondonico, Pinarello, Pogliaghi, Simoncini, or Tommasini. The classic Italian racing bike used the most expensive Campagnolo gruppo that the buyer could afford. This was the market that saved Campagnolo in this decade as they battled with mountain bikes and Shimano indexed shifting.

### Campagnolo

Campagnolo assumed that mountain bikes and indexed shifting would be fads. The decade was an ordeal for Campagnolo, because the OEM buyers who specified components for new bikes, and the people who bought new bikes showed an overwhelming preference for mountain bikes and for indexed shifting. If anybody in Campagnolo USA reported on the state of the U.S. market, nobody in Vicenza was listening.

Campagnolo tried to compete in the low- to middle-priced road bike market, and later in the mountain bike market. However, their lower-priced products failed in the marketplace, because they were ugly and their poorly shifting friction-shifting derailleurs were not competitive.

In addition, components were now part of the annual bicycle model change. Campagnolo could no longer offer a derailleur like the Nuovo Record, that would remain unchanged for 15 years.

Campagnolo lost control of their wholesale market. They sold large numbers of Nuovo Record gruppos to pseudo bike makers, and these gruppos worked back into the cut-rate retail market to compete with Campagnolo's mid-priced gruppos. Nuovo Record gruppos were still showing up on new bicycles in 1988, three years after production ceased.

Campagnolo survived because of the loyalty of the racers. By 1993, they were concentrating on medium- and high-priced racing bikes, and flourishing.

The 1985 catalog still included the old Super Record, Nuovo Record, and 980 rear derailleurs, but it did not show the C-Record, or the Victory and Triomphe gruppos for racing (Corsa) and touring (Leisure), even though they had been on the market for a year. The new derailleurs had different cosmetics but they still had the single-pivot and the basic dimensions of the Nuovo Record.

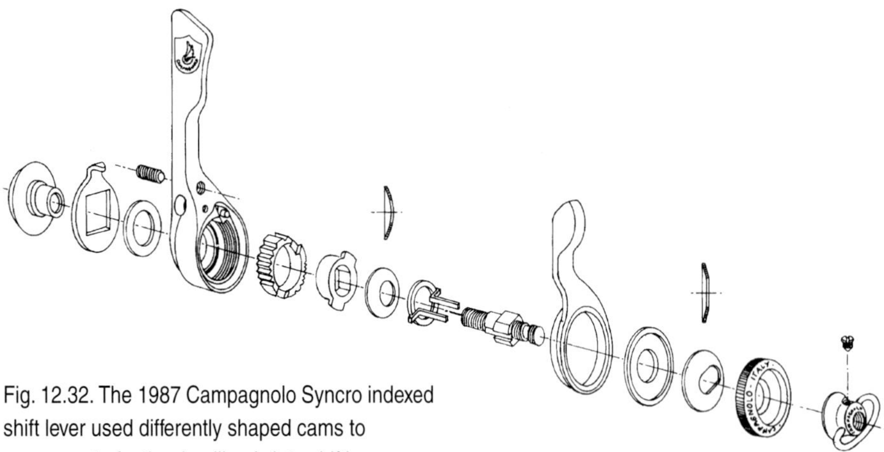

Fig. 12.32. The 1987 Campagnolo Syncro indexed shift lever used differently shaped cams to compensate for the derailleur's late shifting.

The lower-priced Triomphe and Victory gruppos sold poorly.

Campagnolo started making rims and disc wheels in 1986, and this market grew and prospered.

In 1987, Campagnolo reacted to indexed shifting with the Syncro downtube shift levers for "non-competitive cyclists." The Syncro indexed shift lever used 4 different cams in an attempt to compensate for different late-shifting rear derailleurs. The 1988 Syncro shift lever had 7 different cams.

Syncro was doomed because Campagnolo did not understand the laws of indexed shifting. First, indexed shifting requires a complete system of indexed shift lever, rear derailleur, freewheel sprockets, chain, cable and casing. Second, an indexed shifting rear derailleur has to provide a constant chain-gap and shift early. None of Campagnolo's 1987 rear derailleurs were suitable for indexed shifting.

Campagnolo still hoped that indexed shifting would go away. Bianchi understood the requirements of the U.S. market and they equipped their 1987 Campione D'Italia with SunTour AccuShift.

Nuovo Victory and Nuovo Triomphe gruppos were introduced in 1987 for racing and touring. The cage design was revised, moving the pivot point closer to the jockey pulley, but there was neither a slant parallelogram nor a spring-loaded top pivot. I became unpopular with Campagnolo by describing the LX as a SunTour VGT without the slant parallelogram.

In 1988, Campagnolo introduced the Croce D'Aune road gruppo, which was slightly less expensive than the C-Record. It was named after the legendary mountain pass where Tullio Campagnolo conceived the quick-release in a freezing race in 1927. The Croce D'Aune rear derailleur had a tie rod that moved the jockey pulley backward as it shifted onto the bigger sprockets.

They also introduced the Chorus gruppo. The Chorus rear derailleur had a single pivot and a unique adjustable slant parallelogram with two slants, 5° and 30°. There were short-cage and long-cage versions. Even with a slant parallelogram, Chorus rear derailleurs and Syncro levers did not make beautiful music.

Above: Fig. 12.33. The 1988 Campagnolo Chorus rear derailleur with adjustable slant parallelogram.

Right: Fig. 12.34. The 1988 Campagnolo Croce D'Aune rear derailleur with a tie rod.

Campagnolo hated to copy anyone. Perhaps the Croce D'Aune and Chorus derailleurs resulted from an edict to design better-performing derailleurs without copying Shimano. They did not succeed, and both models died an early death.

In 1989, Valentino Campagnolo tried to make his company competitive with Shimano across the whole component spectrum, with very high-, high-, middle-, and low-priced gruppos for both mountain bikes and road bikes. These sold so poorly that they are now collector's items. By the end of the decade, Campagnolo had abandoned mountain bikes and low-priced painted gruppos, and concentrated on expensive road bikes.

The mountain bike was clearly the wave of the future and not a fad. Between 1989 to 1995, Campagnolo introduced Centaur, Xenon, Euclid, Icarus, Olympus, Themis, and Record OR gruppos. Campagnolo tried thumb shifters, twist shifters, centerpull brakes, cantilever brakes, powder-coated finish, and polished aluminum finish. Nothing succeeded. None of the gruppos made the slightest impression on the U.S. market for mountain bikes. Today, these names only interest derailleur collectors. In 1996, Campagnolo abandoned the mountain bike market.

The Athena and Xenon road gruppos were also introduced in 1989. The Athena rear derailleur was less expensive than the Chorus and did not have the slant parallelogram. Xenon was Athena with a gray powder-coated finish instead of polished alloy. The brake lever housings and the pedal bodies were made of resin polymer. The powder-coated gruppos were just plain ugly, and this did not help Campagnolo's image. There were no new introductions in 1990, when Campagnolo hit the bottom.

The year 1991 was the catch-up year. Campagnolo decided to copy Shimano, just as long as it worked. Campagnolo's huge display at the Anaheim show included five road-bike gruppos: C-Record, Croce D'Aune, Chorus, Athena, and Xenon; four mountain-bike gruppos: Euclid, Centaur, Olympus, and Themis; a road tandem gruppo, and even a tandem mountain-bike gruppo. The 1989 Centaur was Campagnolo's first mountain-bike gruppo. The 1991 Centaur rear derailleur was a Deore copy, and it probably indexed shifted quite well.

All of the 1991 rear derailleurs had slant parallelograms and two spring-loaded upper pivots. Campagnolo made their first 8-speed cassette freehubs. The sprockets had HyperGlide-like teeth to improve indexed shifting. Campagnolo strongly recommended using a Rohloff chain to match the new sprockets. Syncro was dropped. With the new, properly designed rear derailleurs, sprockets, and chains, Campagnolo did not need Syncro levers with their built-in overshift.

1992 saw Ergopower integrated shift brake levers for the C-Record, Chorus and Athena gruppos. The downtube shift levers became index only. Croce D'Aune and Xenon were dropped. 1992 sales were up significantly, helped by a favorable ratio between the lire and the yen.

Above: Fig. 12.35. The 1994 Campagnolo Record road gruppo was a beautifully polished set of parts, but by now, Shimano had made big inroads in the top road bike sector.

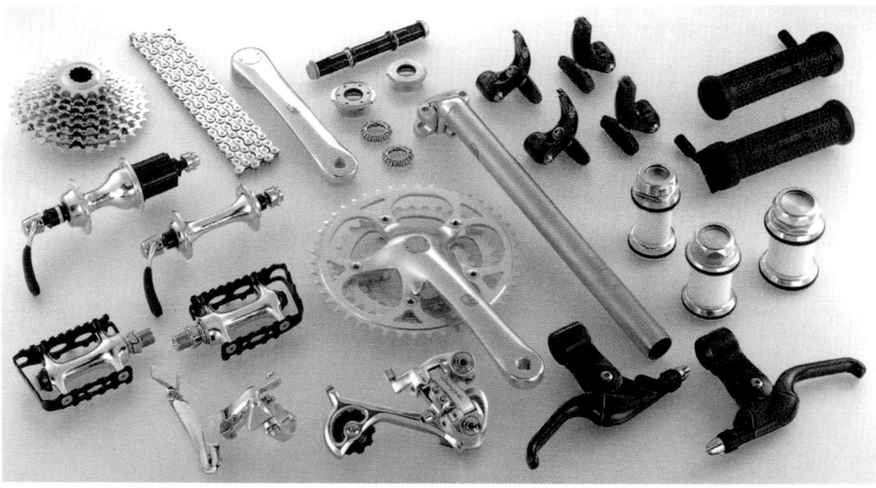

Right: Fig. 12.36. The 1989 Centaur was Campagnolo's first mountain-bike gruppo. It was not well received in a market already "owned" by Shimano and SunTour.

1993 was a pause year. The inner cage of the road rear derailleurs were changed. A new Veloce road gruppo competed with Shimano 105. It was Athena with a gray powder-coat finish. The worst was over, and Campagnolo's 1993 sales of gruppos for medium-priced ($800 and higher) road bikes set a record.

1994 was Campagnolo's 60th anniversary. The new lower-priced Stratos road gruppo in polished alloy competed with Shimano's RX 100. The cassette sprockets became more HyperGlide-like and the chainwheel got SuperGlide-like hooks and ramps. Campagnolo called these changes ExaDrive. Record Ergopower integrated brake-shift levers got a carbon-fiber body. The Record road gruppo got double-fulcrum sidepull brakes to replace the Delta brakes, which had never caught on. The 1994 Record road gruppo combined Shimano's basic designs with Campagnolo's mystique.

## Great Britain

In this decade, British bike sales ran between 2 and 3 million per year. Domestic production ran between 1.1 and 1.2 million, and the rest were imported. Two out of three bike sales were mountain bikes. Bicycle ownership went from 11 million in 1985 to 15 million in 1990, and 20 million in 1995. This compared to 22 million private cars. There were about 3,000 independent bicycle dealers, but one mass marketer, Halfords, had about 40% of the bicycle retail market.

Tube Investments (TI) sold Raleigh to Derby, a management company in 1988. Derby completely revamped Raleigh's management and hired a new CEO, Alan Finden-Crofts. Mountain bikes finally took hold in Britain and Raleigh was a major maker. This was followed by hybrids and suspension mountain bikes. The Nottingham factory remained open and Raleigh remained the largest British bike brand with about one third of the market. Almost all of Raleigh's derailleur bicycles used Japanese components, although a few used Sachs-Huret. No British derailleur maker had survived.

The Sturmey-Archer AW 3-speed was unchanged. The Sturmey-Archer Sprinter 5-speed hub provided gears of –33%, –21%, 1:1, +26.5%, and +50%. The first models required two cables, which were actuated by a single shift lever. The 1993 Sprinter 5-speed hub needed only a single cable. It was available with either a coaster brake or a drum brake.

### Moulton

Dr. Alex Moulton introduced the second generation, AM series, "space-frame" Moulton small-wheel bicycle in 1983. Moulton had pioneered small wheels, high-pressure tires, full suspension, and a one-size-fits-all unisex

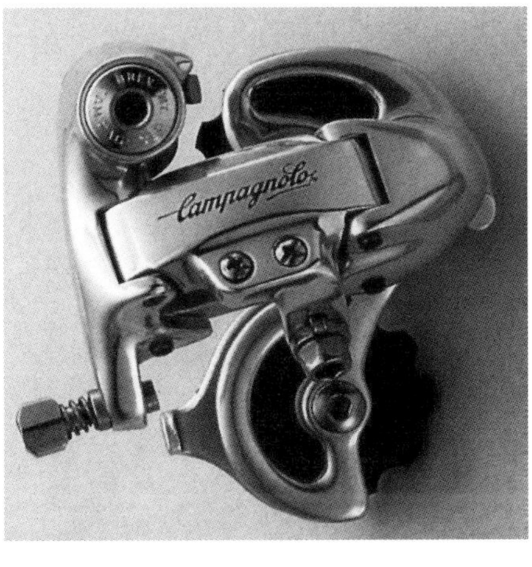


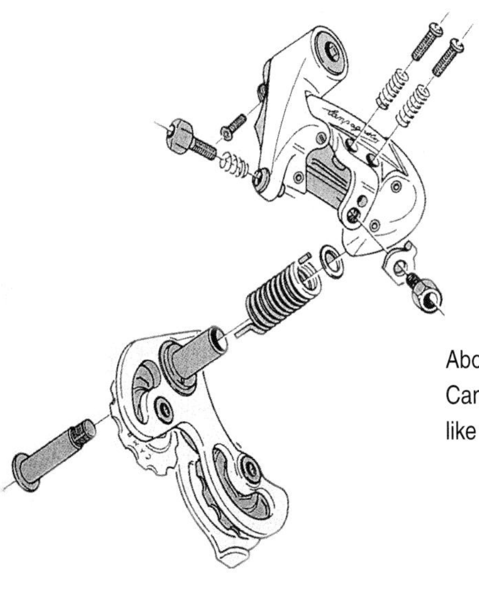

Above and left: Figs. 12.37 and 12.38. The 1994 Campagnolo Record rear derailleur looked remarkably much like the Shimano Dura-Ace.

Right: Fig. 12.39. The 1992 Sturmey-Archer Sprinter 5-speed hub gear with drum brake.

frame. These concepts were continued in the AM series. The new frame was made from small diameter tubes and could be separated into two halves for portability or storage. It was available with a wide range of gear trains. In 1986, Jim Glover set an IHPV record of 51.3 mph (82.5 km/h) on a faired Moulton.

In 1988, Moulton made a mountain bike called the ATB. Moulton licensed Pashley to build and market a cheaper version, the APB. It was available with a 14-speed Shimano gruppo and STI brake-shift levers. Other models used a Sturmey-Archer three-speed hub or Sachs or SunTour derailleur gear trains. The first 9-speed freewheel was installed on the 1989 Moulton Speed. The special Hügi freehub was a 9 to 16 straight block. The derailleur was a SunTour GPX.

## Japan

In 1985, the U.S., Canada, Germany, and Britain were suffering massive trade deficits with Japan. The five countries met in mid-1985 and reached the G-5 agreement. Over the next six months, the value of the yen went from 240 per dollar to 125 per dollar, and by 1994 it was 100 yen per dollar.

The "Yen Shock" had a major effect on the bicycle export market. Japanese low- and medium-priced bicycles could not compete profitably and that market shifted to Taiwan and later to China.

Many Japanese component companies set up branch factories in Taiwan, but the Taiwan construction industry could not build the factories fast enough. Shimano had built their offshore factory in Singapore in 1973 and they quickly shifted production of the lower-priced lines to Singapore.

At the start of this decade, Japanese makers were still exporting full lines of bicycles to U.S. dealers. After the 1985 yen shock, Japanese wages rose and the yen appreciated against the dollar. Soon, only higher-priced bicycles were exported from Japan while the lower and mid-priced bicycles were made in Taiwan. In 1993, Miyata stopped exporting Japanese bikes to the U.S. By 1994, Bridgestone, Japan's largest bicycle maker, producing 1.5 million units annually was the last Japanese company to export to the U.S.

About 8 million bikes a year are sold in the Japanese home

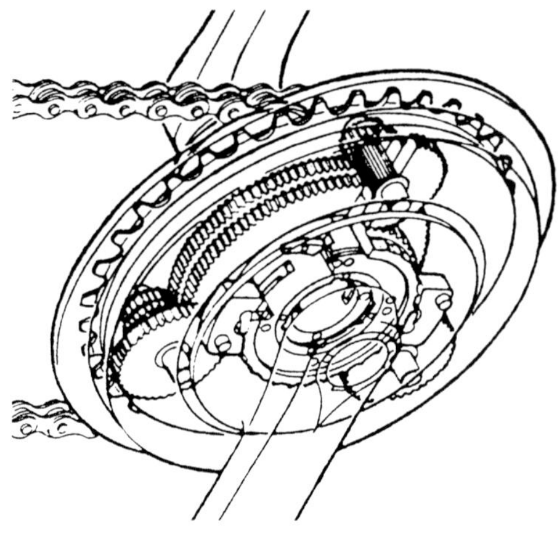

Above: Fig. 12.40. The 1994 Shimano FG-5S10 three-speed bottom-bracket gear.

Left: Fig. 12.41. The 14-speed Alex Moulton APB 14 (All Purpose Bicycle) was built by Pashley, in Stratford-on-Avon, England. It sold for $2,000 at the time of its introduction in 1994.

market, where bicycles are a vital part of the transportation system. They are used for shopping and to get to the train stations, where they wait by the thousands for the return trip. The standard Japanese commuter bike, called a "light cycle," is an inexpensive single-speed with a band brake. The up-scale model is called a "fashion cycle." Maintenance is minimal, and when a bike gives problems, it is replaced with an identical model. The 1994 Shimano FG-5S10 3-speed bottom-bracket gear was an effort to upgrade the Japanese domestic bicycle. It failed because in Japan, housewives handle the family finances, and they resisted the extra cost.

### SunTour

SunTour peaked in 1985, when they offered the best combination of performance and value in conventional friction-shifting gear trains. By 1987, the market demanded indexed shifting. For the next three years, the performance of SunTour's AccuShift was about a year behind Shimano. SunTour's market share plummeted, and in the expensive yen environment, they ran out of time and resources to compete with Shimano.

SunTour's 1985 catalog showed 26 different rear derailleurs and three complete road-bike gruppos: Superbe Pro, Superbe, and Cyclone. The top-of-the-line Superbe Pro racing rear derailleur was handsomely styled and attractively priced. The maximum sprocket was 23 tooth and the wrap capacity was 26 teeth. The SunTour XC mountain-bike gruppo had a new rear derailleur which eliminated the troublesome MounTech jockey pulley and new roller cam brakes licensed from Wilderness Trail Bikes.

Nevertheless, SunTour's share of the U.S. market fell from 60% to 50%. SunTour's biggest problem was the revaluation of the yen, which ate up their profit because most of the orders had been written in foreign currencies, rather than yen. Their response to Shimano Dura-Ace

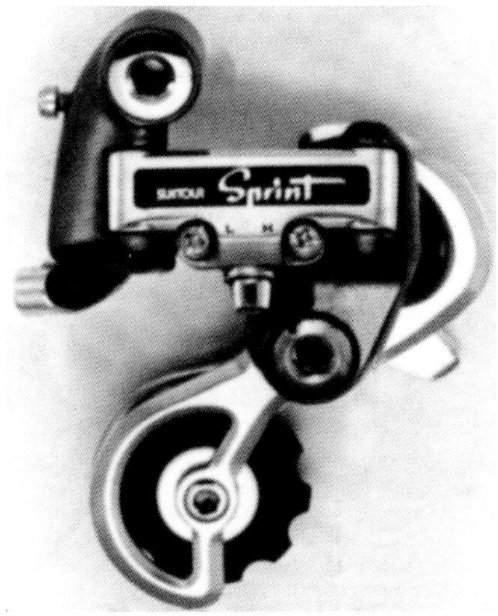


Above: Fig. 12.44. The 1986 SunTour Sprint rear derailleur had a capacity of 24 teeth.

Above: Fig. 12.42. The 1985 SunTour Superbe Pro was a splendid friction-shifting rear derailleur.

Right: Fig. 12.43. The 1985 SunTour XC mountain-bike gruppo featured a new XC rear derailleur, which eliminated the troublesome MounTech's troublesome jockey pulley.

with SIS was postponed by a year.

1986 was a retrenchment year. SVx replaced the old Vx. The 1985 Vx GT was the last survivor of the VGT series from the early 1970s. SunTour introduced the Sprint gruppo, priced between Cyclone and Superbe, to compete with Shimano's 600 Ultegra. Three jockey-pulley diameters allowed precise tuning to the freewheel. Sprint was splendid value and everything looked expensive, but it was not indexed. When Shimano added SIS indexed shifting to the 600 Ultegra and L gruppos in 1986, SunTour had nothing to compete and their U.S. market share fell to about 40% and Shimano passed SunTour.

SunTour went all out to catch up with indexed shifting in 1987. The SunTour AccuShift family provided indexed shifting across the board for 5 road-bike and 4 mountain-bike gruppos. Ten indexed rear derailleurs combined with eleven indexed shift levers and nine freewheels with wide- or narrow-spacing and 5, 6, or 7 sprockets.

However, SunTour lacked the resources to replace all of their rear derailleurs in one year. The road bike rear derailleurs were new and they had a spring-loaded top pivot. SunTour called it Twin Tension.

The mountain bike rear derailleurs did not have spring-loaded top pivots. They had longer chain gaps and the indexed shifters included built-in overshift. They also needed a "stiffer" chain. SunTour did not make chains. They recommended five different narrow chains for the different derailleurs. You could not mix and match chains.

Above: Fig. 12.45. This 1986 SunTour Cyclone with Tri-Pulley was a special version made exclusively for Nishiki.

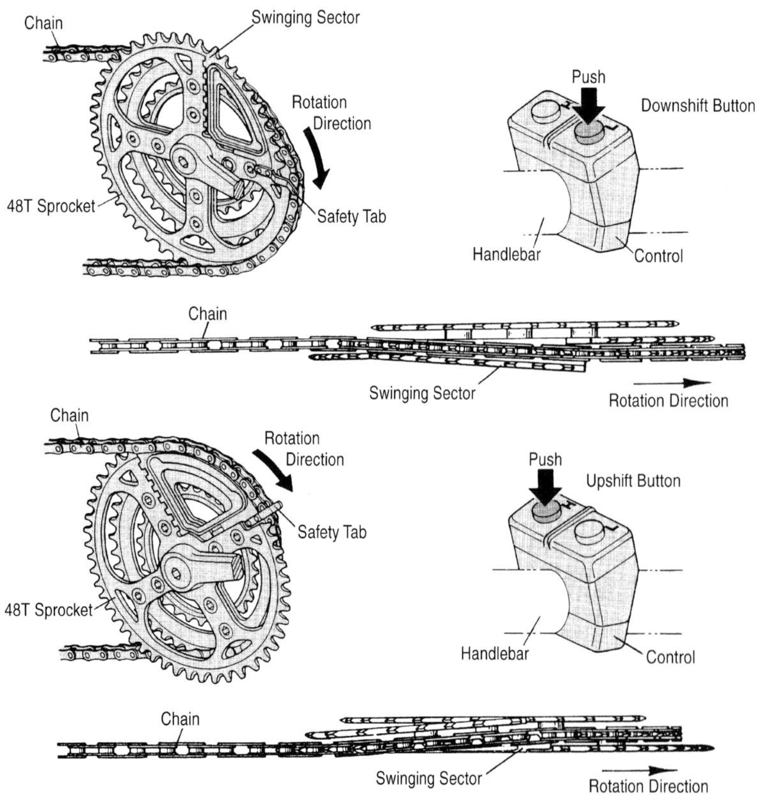


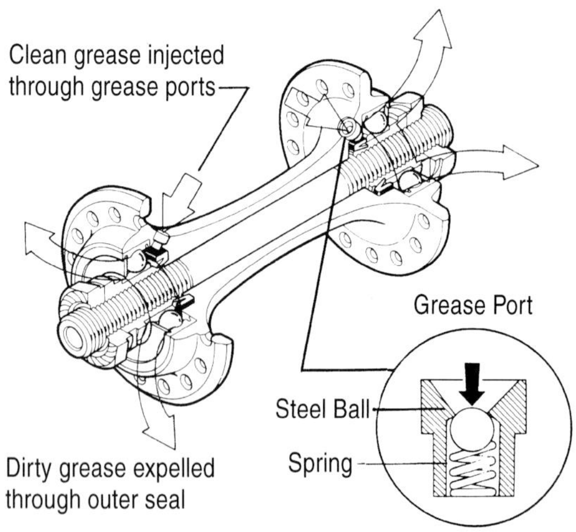


Above: Fig. 12.46. SunTour's 1991 GreaseGuard hubs, bottom bracket, headsets, and pedals—licensed from California mountain bike pioneers Wilderness Trail Bikes—were ideal for sloppy conditions.

Left: Fig. 12.47 A and B. The 1991 SunTour BEAST was licensed from Browning. It had to be recalled because of shifting problems.

Facing page: Fig. 12.48. The 1987 SunTour AccuShift family included 5 road-bike and 4 mountain-bike gruppos.

SunTour tried to do too much in one year. AccuShift bikes did not index shift well, and they did not stay properly adjusted very long, especially in cases where the makers of lower-priced bikes tried to combine their inventories of cheap freewheels, hubs, and chains with SunTour's indexed derailleurs and shift levers. The bike stores soon learned to prefer Shimano SIS. SunTour's share of the 1987 U.S. market dropped to about 30%.

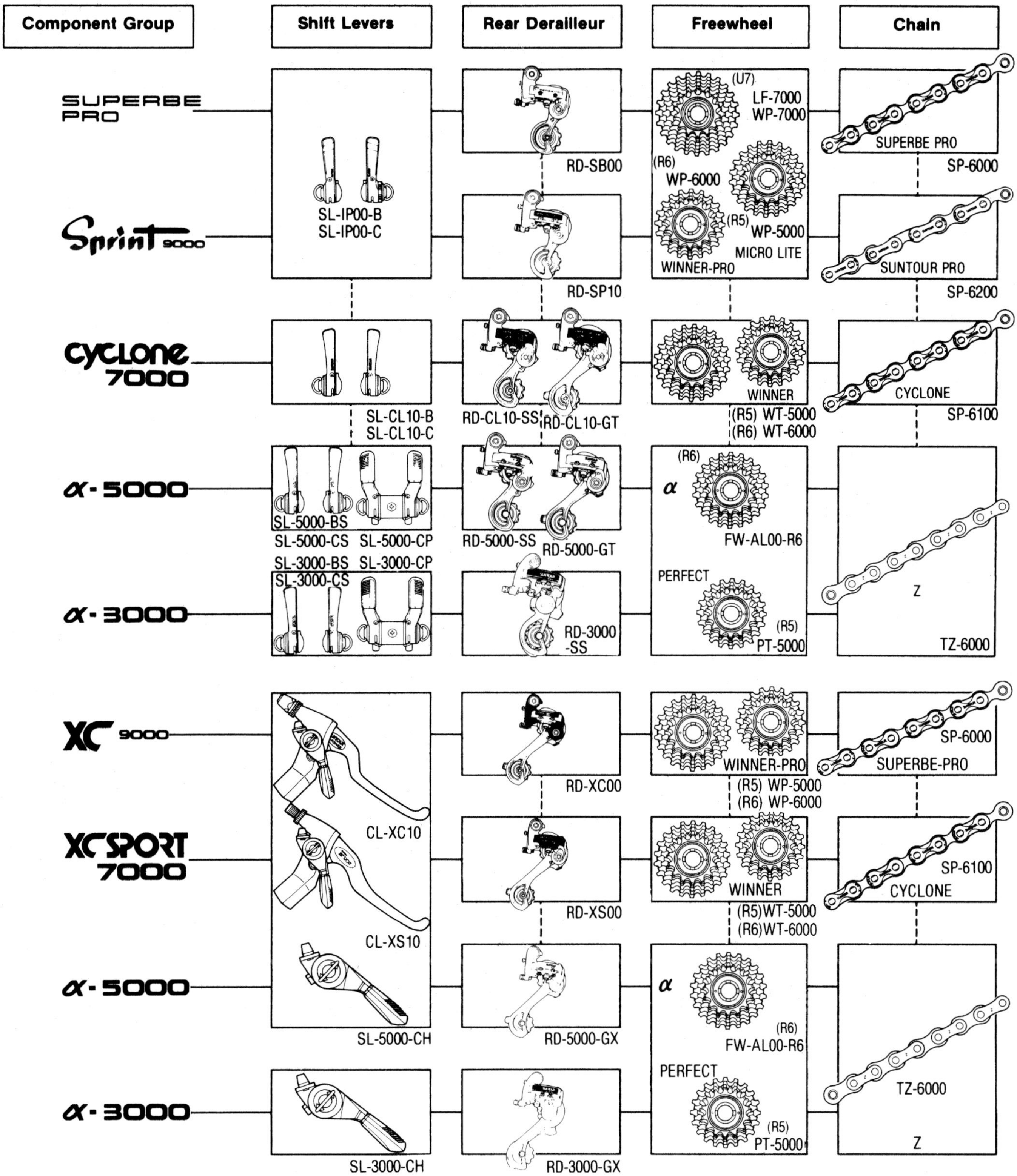

SunTour had their Taiwan factory up and running by 1988. Taiwan produced the low-end (non-indexed) components. The 1988 mountain bike rear derailleurs got spring-loaded top pivots which improved their indexing. The Alpha 2000 derailleurs brought indexed shifting down to the $200 price range. SunTour's market share dropped to 25%. The Shimano Deore gruppo dominated the profitable high-end mountain bike market.

For 1989, there were four new AccuShift road gruppos, GPX, Olé, Edge, and Blaze, for mid-priced ($350 to $600) road bikes. There were two new mid-priced mountain-bike gruppos, XCE 4050 and XCM 3040, and four new low-priced mountain bike rear derailleurs. The top of the line XC9000 gruppo got a 7-speed freehub.

Almost all of SunTour's rear derailleurs now had two spring-loaded pivots. There was a shortage of indexed shifting components and both SunTour and Shimano were sold out. SunTour did well in the low- and medium-priced markets, and their market share rose to about 30%.

SunTour USA had been headquartered in New Jersey since the early 1970s. In 1990, they moved to Marin County in Northern California to be closer to the mountain bike action.

SunTour's shifting performance caught up with Shimano in 1990. AccuShift Plus shaped sprockets competed with HyperGlide. PoweRing chainwheels competed with SuperGlide. There was a new flexible bushingless chain and X-Press push button shifters for mountain bikes. All of the mountain rear derailleurs were redesigned to work with the revised sprockets and chains.

To avoid a "me too" image, SunTour took licenses on the Browning swinging gate crankset and on GreaseGuard hubs and headsets. The SunTour BEAST (Browning Electronic AccuShift Transmission) eliminated the front derailleur and provided almost instantaneous shifts. The name was prophetic. The prototypes worked perfectly, but the Taiwan production line could not maintain the critical clearances and the production BEASTs would not shift reliably. They had to be recalled and the bicycles retrofitted with triple chainrings and front derailleurs. Fortunately, few had been delivered.

SunTour Grease-Guard hubs, bottom bracket, headsets, and pedals were ideal for sloppy conditions. They were licensed from Wilderness Trail Bikes. A small fitting allowed grease to be injected to flush out water. The public preferred Shimano's "sealed" bearings, even though they were less effective for really wet service, such as riding through streams.

By 1990, SunTour and Shimano were close in performance, and SunTour had some unique features. However, a "self-fulfilling prophecy" set in: SunTour offered 12 road and 15 mountain derailleurs over the whole price range, but the people writing OEM specifications chose Shimano, especially for high-priced bicycles. They assumed that Shimano would sell better. There were fewer SunTour-equipped bikes. Often there was just a "token" SunTour bike, while all the other

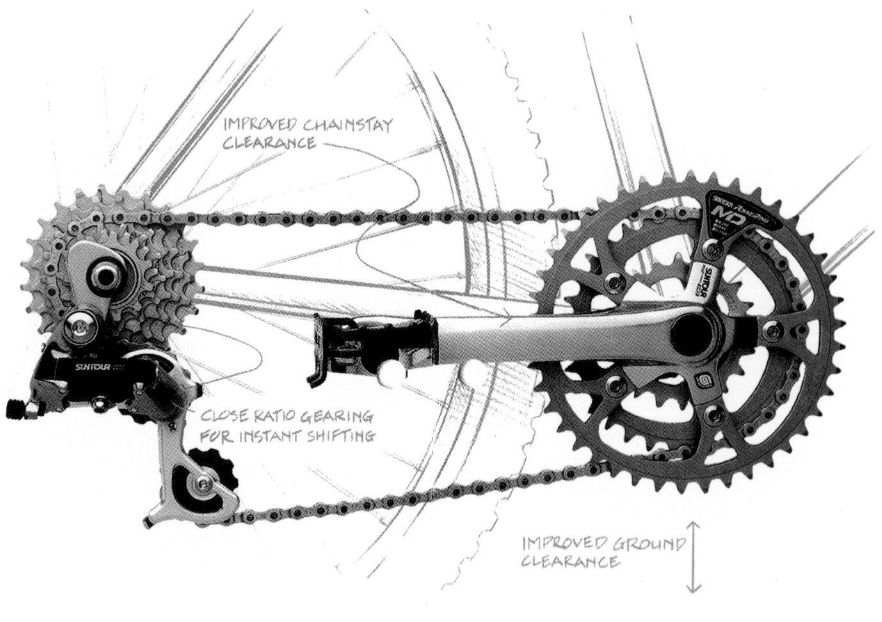


Left: Fig. 12.49. The 1992 SunTour MicroDrive was designed around small chainrings, to save weight and improve ground clearance.

bikes were Shimano-equipped. As a result, the bike shops and the buying public assumed that there must be something wrong with SunTour. Their market share slipped back to 25%.

There were no major introductions in 1991. There was still a large demand both in the U.S. and Europe. The SL road gruppo replaced Sprint. There were cosmetic changes (and new model numbers) on many of the derailleurs. Four low-priced gruppos were dropped. SunTour and SR (Sakae Ringyo) merged their marketing. SunTour's market share remained at about 25%.

By 1992, both SR and SunTour were in serious financial trouble. Mori Industries Inc. purchased SR, which then bought Maeda Industries (SunTour). SR made handlebars, chainwheels, pedals, and seatposts. Mori was a major Japanese steel company. The new company was named SR-SunTour. SunTour closed their California sales office and moved to Sakae's office in Kent, Washington.

In 1992, SunTour introduced PowerFlo freewheel sprockets with "gates" for easier shifting. SunTour's 1992 MicroDrive was a significant innovation. Smaller bolt circles allowed chainrings like 42/32/20, which saved weight and improved ground clearance. New short-cage rear derailleurs were provided to go with the MicroDrive cassettes. It was a good idea, and Shimano picked it up two years later.

*Bicycling* Magazine's 1992 *Buyer's Guide* listed 70 road bikes selling for less than $1,400. All 70 were Shimano-equipped. At higher prices, there were also Campagnolo, Mavic, or SunTour gruppos. SunTour's market share was down to 10%. There were fewer gruppos in 1993. SunTour was selling front suspension forks. The gear trains got a twistgrip shifter and 8-speed freehubs.

The 1993 SunTour S-1 rear derailleurs were designed for hybrid and city bikes. They mounted on a special boss on the chainstay. The horizontal slant parallelogram harkened back to the 1930s Nivex or the 1960s Altenberger. SunTour S-1 had good features. You could dump the bike on the right side with no damage to the derailleur. The indexing was in the rear derailleur rather than the shift lever. It was a good design and it might have succeeded ten years earlier when SunTour was a dominant supplier. In 1993, with SunTour on the ropes, no one would specify S-1.

SunTour shut down one of their factories and sold the land. They were down to 150 employ-

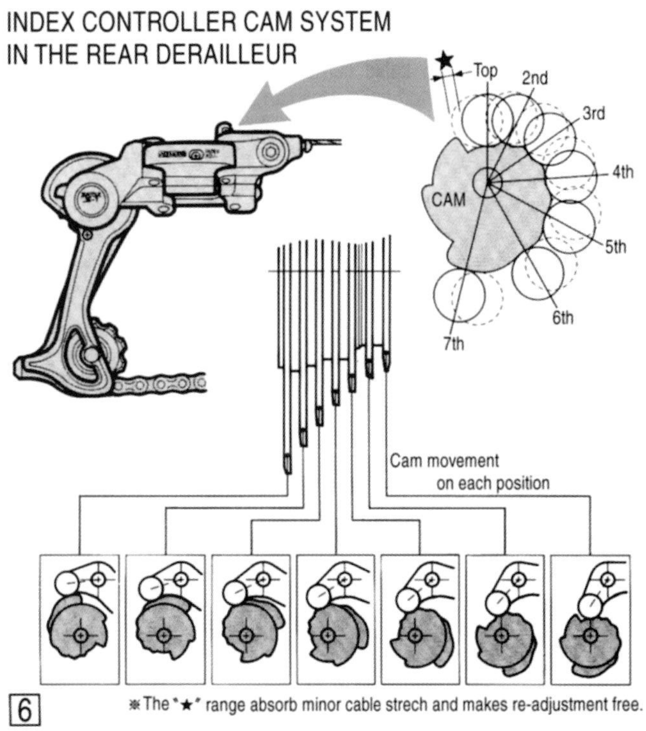


Above and right: Figs. 12.50 and 12.51. The 1993 SunTour S-1 rear derailleur had built-in indexing, but it required a chainstay braze-on.

ees. They had only five percent of the U.S. market, and to add to their misery, 100,000 or so SunTour coaster brakes with defective springs had to be recalled.

In 1994, SunTour cut back to the products that they hoped they could sell. The mountain-bike gruppos offered twistgrip shifters or thumb shifters. The three road gruppos, Superbe Pro, SL and Blaze were unchanged. SunTour provided 6- and 7-speed screw-on freewheels with Power-flo teeth and ramps. S-1 rear derailleurs were featured for city bikes. By the end of the year, SunTour was offering closeout prices on their medium- and high-priced components. The SunTour factory in Osaka closed its doors.

In 1985, SunTour had about 50% of the U.S. derailleur market for quality bicycles. Over the decade, SunTour introduced more than 120 new rear derailleurs and they still fell victim to the fast changing market. SunTour was too small to maintain the research and development needed to survive. The 1985 devaluation of the yen was the overriding disaster that ate up SunTour's financial reserves. The decade saw the "Sunset of SunTour."

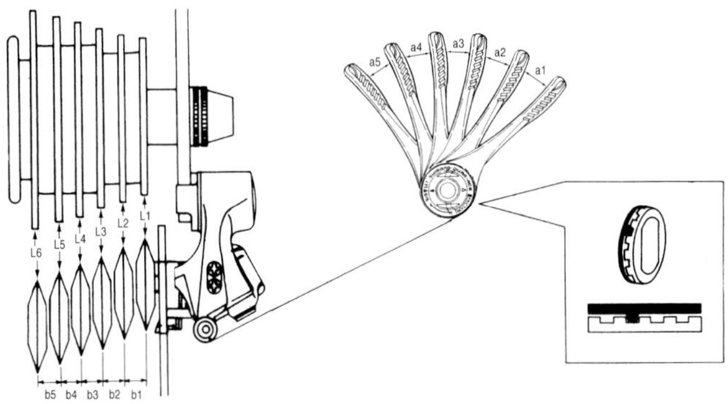

Right: Fig. 12.52. Shimano's 1985 SIS Dura-Ace was a complete indexed shifting system.

Below: Fig. 12.53. The 1987 New 105 gruppo had SIS shifters and SLR Light Action brakes.

### Shimano

Shimano recognized that the yen shock was permanent. To survive, Shimano added value to their mass-produced products so that the lower-priced models looked like and performed like the higher-priced models.

Shimano introduced New Dura-Ace with SIS (Shimano's acronym for Shimano Index System) at the 1984 fall trade shows. The indexed shift lever, cable, casing, rear derailleur, chain, and freewheel sprockets all worked together to provide precise positive shifts each time the lever moved to the next click position.

However, it was a low-key introduction. The New Dura-Ace gruppo was available with or without SIS shift levers. The SIS downtube levers could be switched from index to friction mode. Indexed shifting did not seem like a big deal at the time. Shimano had promoted indexed Positron derailleurs in 1977 and indexed Aero AX derailleurs in 1981, and neither had gone anywhere at the time. SunTour had offered indexed Mighty Click derailleurs since 1980. In fact, the 1969 SunTour catalog showed a 5-speed Click Shifter. Sachs-Huret had offered the indexed Commander since 1983.

Shimano focused the marketing of Dura-Ace SIS on high-priced racing bikes. Eddy Merckx, who had used Campagnolo equipment for his entire racing career, endorsed SIS. The test writers for the magazines all spoke favorably about how SIS worked like push-button tuning on a radio.

Shimano pushed SIS Dura-Ace as a complete gruppo, so that they could control the interactions between the gear train components. They prepared a comprehensive installation manual explaining precisely how the derailleur had to be installed so that the indexing would work properly.

Looking at the 1985 New Dura-Ace rear derailleur, it was obvious that SunTour's slant parallelogram patent had expired and that Shimano had quietly added a slant parallelogram to their basic derailleur with two spring-loaded pivots. It was not obvious that the SIS Dura-Ace was the first computer-optimized rear derailleur. The dimensions of the parallelogram pivots and the cage geometry had been optimized along with the tensions of the two springs. The goal was to maintain a nearly constant chain gap in all gears. SIS Dura-Ace shifted early on all gears, and that was the key to crisp, positive indexed shifting.

Shimano did not retrench after the yen shock. They kept hiring engineers and expanding their research and development. They moved their U.S. headquarters to a new building in Irvine, California in 1986.

Shimano's advertising heavily promoted BioPace non-round chainrings. BioPace was first available in the 600 gruppo in 1983 and was optional in all gruppos by 1987. The public liked the idea, even though few could feel the difference. Most of Shimano's competitors were forced to offer optional oval chainwheels or something similar. BioPace geometry was quite different from oval chainrings, but the main result was the same: a lower pedaling cadence. For the next few years, BioPace went

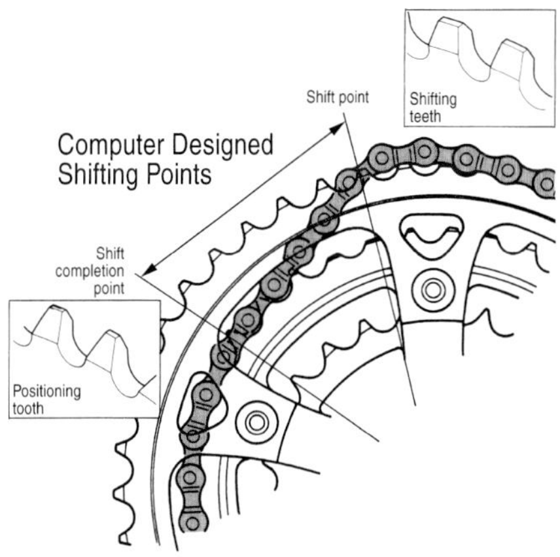


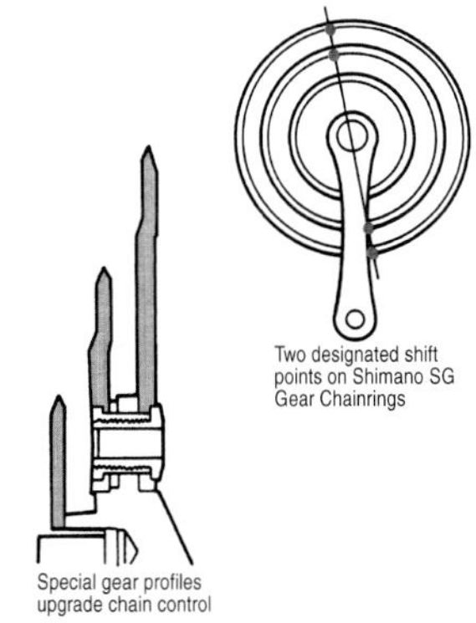


Left: Fig. 12.55. Shimano's 1991 SuperGlide chainwheels featured cut-down teeth, so that shifts took place at the dead centers.

Right: Fig. 12.56. The 1987 Santé gruppo was positioned between 600 and Dura-Ace.

Below: Fig. 12.54. Shimano's 1989 HyperGlide sprockets made rear indexed shifting significantly easier.

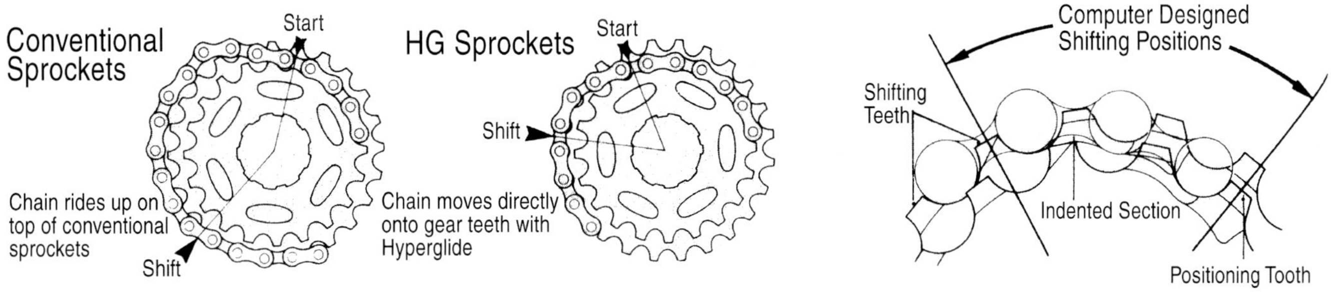

through some modifications, gradually becoming ever rounder, until it was phased out in 1993.

In 1986, SIS trickled down to the 600 and L Series road-bike gruppos. Shimano also introduced their first narrow bushingless chain, copied from the Sedisport. SIS worked well and it was a huge marketing success. Indexed shifting was hot. It was hard to sell a medium-priced bicycle without SIS. All of the road tests of SIS bicycles were favorable. Even the conservative racers were responding to the "never miss a shift" message. Shimano began to dominate the component market.

Santé and New 105 were the major introductions in 1987. The Santé gruppo was priced between 600 and Dura-Ace. It was aimed at sport cyclists rather than genuine racers, and it never found a niche. There were also new Deore XT, Deore, and MS mountain-bike gruppos with indexed shifting. SIS became standard across almost the whole Shimano line. Dura-Ace and Santé went from six to seven speeds.

In 1988, the Singapore factory made their first SIS components, making SIS available in all Shimano gruppos. Even the low-priced Exage gruppos had front and rear indexed shifting. In 1988, 600 became 600 Ultegra and got 7 speeds.

In 1989, Dura-Ace got 8 speeds. Deore XT II, Deore II, and 105 got 7-speeds. The Bio-Pace chainrings got rounder, so you could barely feel the "bio."

The 1989 introduction of HyperGlide sprockets provided a major improvement in shifting ease. HyperGlide sprockets had "release" teeth and narrowed teeth so that the chain could slide smoothly from one sprocket to the next cog without climbing on to the top of the teeth and then dropping down. I likened them to freeway on-ramps and off-ramps. HyperGlide made rear indexed shifting significantly easier and riders could feel the difference. HyperGlide sprockets had to be keyed to the freehub so they could only be assembled in one alignment. Gear freaking possibilities were very limited because only a limited number of combinations were available.

In 1990, Shimano provided STI integrated brake and shift levers for road bikes. The object was to allow steering, braking, and shifting with your hands on the handlebars. On mountain bikes, the

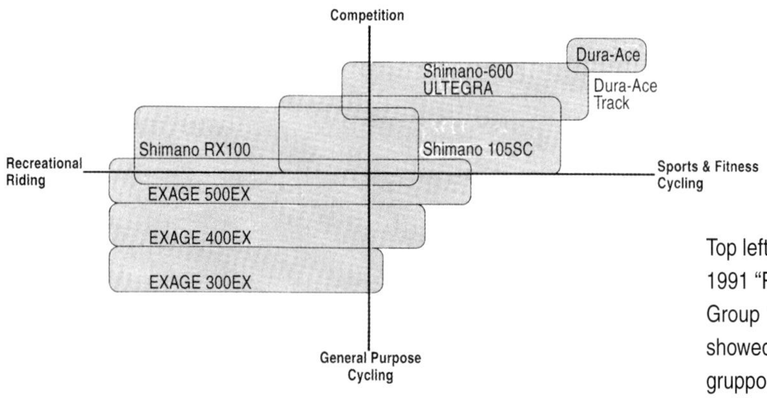

Top left: Fig. 12.57. Shimano's 1991 "Road Component Group Positioning" graph showed how Shimano's gruppos blanketed the market.

Bottom left: Fig. 12.58. Shimano HyperDrive front shifting was upgraded in 1993 with more pins and hooks.

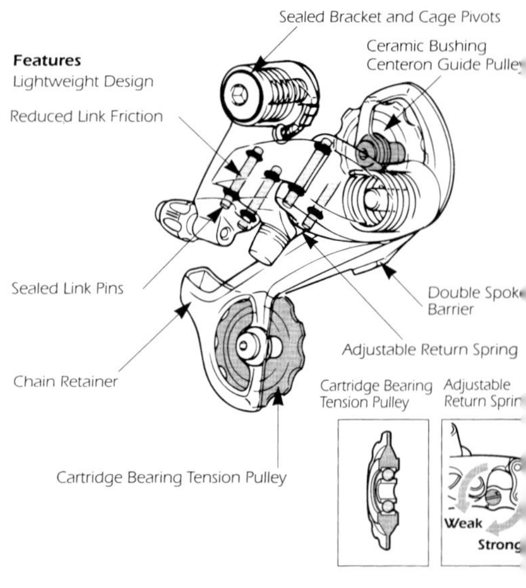


Right: Fig. 12.59. The 1992 XTR RD-M900 rear derailleur.

push buttons were mounted under the handlebars to allow shifting with your hands on the grips.

Santé was quietly dropped in 1990. It never found a niche. It was one of Shimano's few marketing failures in this decade. 600 Ultegra and 105 SC both moved up the price/quality ladder to fill the gap left by Santé. The new RX-100 road gruppo slotted in under 105 SC. The Exage gruppos got cosmetic improvements and new numbers.

In 1991, *SuperGlide* chainwheels cut down certain teeth so that the shifts always took place at the dead centers when the chain was under low stress.

Shimano recognized that with the high-priced yen, the Taiwanese competitors could always beat them on price. They developed value-added innovations to make Shimano equipped bicycles more attractive to buyers and introduced these innovations faster than they could be copied and brought to market.

Shimano started using robots for component assembly in 1982. By 1984, they were designing and building their own SARA (Shimano Automated Robot Assemblers). By 1991, they had 700 robots in service.

From the earliest days of the company, the three Shimano brothers committed the company to technical development. Ten percent of the employees were engineers. Ten percent of the profits were spent on research and development.

Shimano celebrated their 70th anniversary in 1991. The old Deore gruppo split into Deore DX and Deore LX, improving coverage of the medium-priced mountain bike market. HyperGlide sprockets and SuperGlide chainwheels were used in all Shimano gruppos.

In 1992, 600 Ultegra got 8 speeds and STI integrated brake-shift levers. The 1992 XTR gruppo was a super-premium mountain-bike gruppo provided for cost-is-no-object mountain bikers. The XTR RD-M900 rear derailleur could handle a 32-tooth sprocket and the total wrap-up capacity was 33 teeth. The 8-speed XTR cassette had the sprockets riveted to an aluminum spider to save weight. The 1992 70GS gruppo was made for $200 mountain bikes.

In 1993, 105 SC got 8 speeds and STI integrated brake-shift levers. There were two new CX gruppos specifically for city bikes with tiny 42/32/20 chainrings copied from SunTour MicroDrive.

Four low-priced Altus gruppos replaced the three GS gruppos for hybrids and inexpensive mountain bikes. Front shifting was upgraded again, this time with hooks on the middle chainring and pins on the outer chainring. It was called either HyperDrive, Dual SIS, or SG-X (Shimano's Madison Avenue types could not decide).

BioPace was quietly dropped in 1993. It may or may not have had an ergonomic advantage, but it had been a great sales booster for nine years.

"Ride Light" was Shimano's buzzword for 1994. The rear derailleurs were revamped with lighter springs and less friction. Compact chainrings, like 42/32/22, were used in the mountain-bike gruppos. This was called HyperDrive C. There were new Alivio and STX mountain gruppos and an MJ gruppo for kid's bikes. The RX100 gruppo got 8-speeds and integrated brake-shift levers.

In 1985, Shimano had about 40% of the U.S. market for quality bicycles with three road-bike

Right: Fig. 13.60. The three Shimano brothers, photographed in 1977. From left to right: Shozo, Yoshizo, and Keizo. Together, they guided Shimano through this decade of remarkable development for the company.

gruppos, one mountain-bike gruppo, and two low-priced sets of derailleurs and shift levers. Over the next ten years, Shimano introduced more than 100 new rear derailleurs. By 1994, Shimano had more than 90% of the U.S. market. They had twelve mountain-bike gruppos, four road-bike gruppos, two city-bike gruppos and one youth-bike gruppo. This period had truly been the "Shimano Decade."

## Cycling Magazines

*Bicycling* did not have a significant technical content after 1991. *Bicycle Guide* changed ownership in 1989, and gradually lost its technical and historical content. The *Fat Tire Flyer* ceased publication in 1987. *Bike Tech* was the main source of technical articles until it ceased publication in 1988. Dr. Chester Kyle started *Cycling Science* in 1989 to provide a forum for technical writers.

*Mountain Bike* started in 1985 and was acquired by Rodale in 1989. Its main competitor was *Mountain Bike Action*. A number of other American mountain bike magazines, like *Bike* and *Dirt Rag*, sprang up with lots of pictures to cater to readers with short attention spans.

The Veteran Cycle-Club in Britain publishes two quarterly magazines. *News & Views* is their newsletter, and *The Boneshaker* is their journal. Both contain some excellent historical material. The Wheelmen is the comparable American veteran bicycle club organization. Their quarterly magazine, *The Wheelmen*, also has considerable 19th-century historical content.

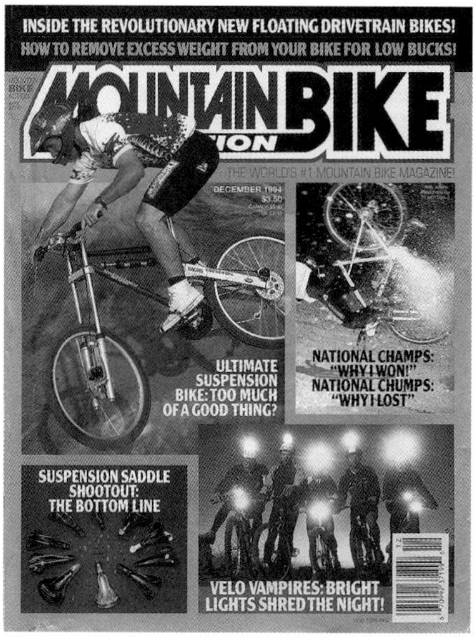


Above and right: Figs. 12.61 and 12.62. From "Funk" to "Hype": This decade witnessed the mountain bike's coming of age (or losing its values, depending on how you look at it). The funky early publication *Fat Tire Flyer* (above), gave way to financially more secure competitors like *Mountain Bike Action* (right).

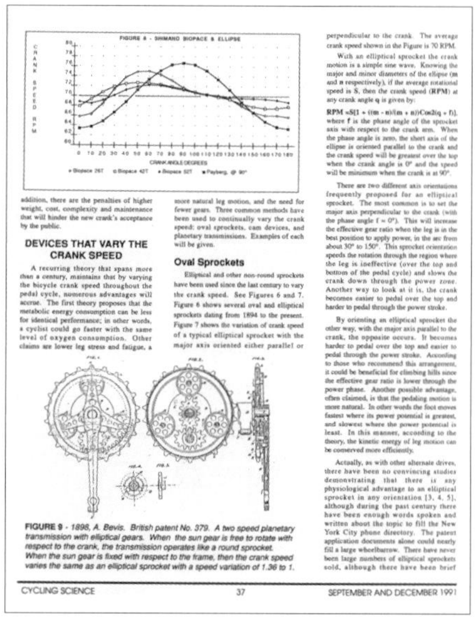


perpendicular to the crank. The average crank speed shown in the Figure is 70 RPM.

With an elliptical sprocket the crank motion is a simple sine wave. Knowing the major and minor diameters of the ellipse (m and n respectively), if the average rotational speed is S, then the crank speed (RPM) at any crank angle q is given by:

RPM = S[1 + ((m - n)/(m + n))Cos2(q + f)],

where f is the phase angle of the sprocket axis with respect to the crank arm. When the phase angle is zero, the short axis of the ellipse is oriented parallel to the crank and the crank speed will be greatest over the top when the crank angle is 0° and the speed will be minimum when the crank is at 90°.

There are two different axis orientations frequently proposed for an elliptical sprocket. The most common is to set the major axis perpendicular to the crank (with the phase angle f = 0°). This will increase the effective gear ratio when the leg is in the best position to apply power, in the arc from about 30° to 150°. This sprocket orientation speeds the rotation through the region where the leg is ineffective (over the top and bottom of the pedal cycle) and slows the crank down through the power zone. Another way to look at it is, the crank becomes easier to pedal over the top and harder to pedal through the power stroke.

By orienting an elliptical sprocket the other way, with the major axis parallel to the crank, the opposite occurs. It becomes harder to pedal over the top and easier to pedal through the power stroke. According to those who recommend this arrangement, it could be beneficial for climbing hills since the effective gear ratio is lower through the power phase. Another possible advantage, often claimed, is that the pedaling motion is more natural. In other words the foot moves fastest where its power potential is greatest, and slowest where the power potential is least. In this manner, according to the theory, the kinetic energy of leg motion can be conserved more efficiently.

Actually, as with other alternate drives, there have been no convincing studies demonstrating that there is any physiological advantage to an elliptical sprocket in any orientation [3, 4, 5], although during the past century there have been enough words spoken and written about the topic to fill the New York City phone directory. The patent application documents alone could nearly fill a large wheelbarrow. There have never been large numbers of elliptical sprockets sold, although there have been brief

addition, there are the penalties of higher weight, cost, complexity and maintenance that will hinder the new crank's acceptance by the public.

DEVICES THAT VARY THE CRANK SPEED

A recurring theory that spans more than a century, maintains that by varying the bicycle crank speed throughout the pedal cycle, numerous advantages will accrue. The first theory proposes that the metabolic energy consumption can be less for identical performance; in other words, a cyclist could go faster with the same level of oxygen consumption. Other claims are lower leg stress and fatigue, a more natural leg motion, and the need for fewer gears. Three common methods have been used to continually vary the crank speed: oval sprockets, cam devices, and planetary transmissions. Examples of each will be given.

Oval Sprockets

Elliptical and other non-round sprockets have been used since the last century to vary the crank speed. See Figures 6 and 7. Figure 6 shows several oval and elliptical sprockets dating from 1894 to the present. Figure 7 shows the variation of crank speed of a typical elliptical sprocket with the major axis oriented either parallel or

FIGURE 9 - *1898, A. Bevis. British patent No. 379. A two speed planetary transmission with elliptical gears. When the sun gear is free to rotate with respect to the crank, the transmission operates like a round sprocket. When the sun gear is fixed with respect to the frame, then the crank speed varies the same as an elliptical sprocket with a speed variation of 1.36 to 1.*

CYCLING SCIENCE 37 SEPTEMBER AND DECEMBER 1991

Above: Fig. 12.63. Typical page from *Cycling Science*. It was a good technical "zine," published by Chester Kyle to fill the gap left when *Bike Tech* ceased publication.

# 13. Into the 21st Century: 1995–2010 Part I. Shimano

THIS WAS a very competitive period. Shimano continued to dominate the component market for quality multi-speed bicycles, but they lost market share to SRAM and to lower-priced components from Taiwan and China. I have divided the wealth of materials for this most recent period into two parts. This chapter deals with Shimano, the dominant player. Campagnolo, SRAM, and everyone else will be covered in chapter 14.

In this chapter, I will first cover the individual developments over this period. Then I will list Shimano's product introductions year by year.

In 1995, Shimano had about 85% of the U.S. market for components for quality multi-speed bicycles. Although there were no reliable statistics, my guess is that in 2008, Shimano had about half of multi-speed component business. They were the world's largest bicycle component maker, with 6,500 employees worldwide.

In 1995, Shimano had fifteen gruppos: five road, four mountain, and six comfort. By 2008, Shimano had 23 gruppos, seven road, nine mountain, and seven comfort.

Shimano SPD (Shimano Pedaling Dynamics) pedals had appeared in 1991, and Shimano became a major supplier of shoes and pedals. In 2000, wheels became wheel systems and numerous competitors offered complete wheel systems with non-Shimano hubs. Shima-

Right: Fig. 13.1. Lance Armstrong, shown here after his first Tour de France victory in 1999, won all seven of his Tours using Shimano-equipped Trek bicycles.

no entered the market for complete wheels to protect their market for front and rear hubs.

Shimano Inc. is a publicly traded company in Japan. Shimano American is a privately owned company. Shozaburo Shimano, the founder, died in 1958. Management passed to the second generation, his sons, Shozo, Keizo, and Yoshizo.

Keizo Shimano, the middle brother, died in January 1995 at age 62. He had led Shimano's research and development, and established the long-range policies and the corporate environment that served Shimano so well.

John Neugent, who headed Shimano's competitor Sachs Bicycle USA until the SRAM takeover, wrote about Keizo's death, saying:

> In my opinion, the demise of Huret, Simplex, SunTour and others was caused by the inability of the companies to react to Shimano's successful product innovation. In order, these successful innovations were; first good quality index shifting, next good quality mountain bike groups, and finally HyperGlide. Each of these introductions left the competition trying to play catch up. None of the companies either could react or did react in time.
>
> I think Keizo's death is a landmark occasion to mourn the most powerful man in the bicycle industry ever. If one were to look to one individual (although many say that all three brothers were equally important), I think Keizo was the man responsible for the components that remade the bicycle. Keizo Shimano had more influence than any other person in the history of the bicycle. A truly great and remarkable man.

During this period, Shimano transferred management to the third generation. In 2002, Yoshizo Shimano stepped down as President of Shimano Inc., and Shozo's son, Yozo, became the new President. Yoshizo's son, Kozo, became President of Shimano American, and Keizo's son, Taizo, took on R&D and Planning in Japan.

## Bicycle Racing

Shimano (and Trek) had the good fortune to sponsor the U.S. Postal racing team, later Discovery Channel team, headed by Lance Armstrong. Armstrong's record seven consecutive Tour de France victories from 1999 to 2005 captured the U.S. public's imagination and interest in cycling.

Left: Fig. 13.2. Keizo Shimano, shown here in a 1977 photograph, headed Shimano's Research and Development until his untimely death in 1996.

*Table 13.1. Tour de France winners, 1995–2009*

| Year | Rider (Country) | Bicycle | Derailleur |
|---|---|---|---|
| 1995 | Miguel Indurain (Spain) | Pinarello | Campagnolo |
| 1996 | Bjarne Riis (Denmark) | Pinarello | Campagnolo |
| 1997 | Jan Ullrich (Germany) | Pinarello | Campagnolo |
| 1998 | Marco Pantani (Italy) | Bianchi | Campagnolo |
| 1999 | Lance Armstrong (USA) | Trek | Shimano |
| 2000 | Lance Armstrong (USA) | Trek | Shimano |
| 2001 | Lance Armstrong (USA) | Trek | Shimano |
| 2002 | Lance Armstrong (USA) | Trek | Shimano |
| 2003 | Lance Armstrong (USA) | Trek | Shimano |
| 2004 | Lance Armstrong (USA) | Trek | Shimano |
| 2005 | Lance Armstrong (USA) | Trek | Shimano |
| 2006 | Oscar Pereiro (Spain) | Pinarello | Campagnolo |
| 2007 | Alberto Contador (Spain) | Trek | Shimano |
| 2008 | Carlos Sastre (Spain) | Cervélo | Shimano |
| 2009 | Alberto Contador (Spain) | Trek | SRAM |
| 2010 | | | |
| 2011 | | | |

### Speed and Distance Records

In 1996, Chris Boardman raised the one-hour record to 56.4 km, (35.0 miles) riding a partially streamlined bicycle in the "superman" position. In 2000, Boardman rode a classic track bike 49.44 km (30.7 miles).

I liked his "classic" record because it provided a comparison with Merckx's 1974 classic hour record of 49.43 km (30.7 miles). It also provided a direct comparison of modern rider positions versus the old UCI rules.

In 2002, Sam Whittingham raised the International Human Powered Vehicle Assn. (IHPVA) 200-meter record to 130.4 km/h (81.05 mph). In 2002, Guido Martens rode a streamlined recumbent 82.6 km (51.3 miles) in one hour.

## The Comfort & Transportation Trend

In this period, Shimano expanded their program to widen the U.S. adult bicycle market beyond sport and recreation. Shimano wanted to get Americans out of their automobiles for short trips. Each year, Shimano sponsored design contests for alternatives to road bikes and mountain bikes. By 2002, the emphasis was on "Transportation" bikes. Shimano stayed with their commitment to expand the adult market and to make cycling an attractive alternative to driving.

Shimano quietly shifted their development and promotion emphasis to practical bicycles with Nexus hubs or Nexave derailleur geartrains. From 2005 onward, new derailleurs for road and mountain bikes came at a slower pace.

Hub gears and hybrid gear-trains (derailleur plus hub gears) multiplied, especially combined with electronic controls. By 2008, computer-controlled automatic shifting of both derailleurs and hub gears was practical and reliable. Unfortunately, it was also expensive.

By 2008, with the world experiencing "Peak Oil" production and constantly rising gasoline costs, Shimano's early decisions appear prescient.

## Off-Road Components

By 2006, Shimano was making mountain bike gruppos for four

Above: Fig. 13.3. The 2008 Raleigh Coasting, at $450, shows the concept of the transportation bicycle with Coasting and Cyber controls—but why no lights and fenders?

Left: Fig. 13.4. The 2008 K2 Coasting was a $700 high-end model with internal wiring and a suspension seat post. This one without even as much as a luggage rack.

kinds of mountain bike racing and two types of casual riding:

- XTR: for serious cross-country racing on lightweight hardtail bikes or short travel (less than 4-inches) full-suspension bikes.
- XT, LX, and Deore: for cross-country riding on full-suspension bikes with five inches of suspension travel.
- Hone: for all-mountain riding on terrain with small jumps and drops on full-suspension bikes with 6 inches of travel.
- Saint: for freeride racers on downhill courses with large jumps and drops on heavy-duty full-suspension bikes with more than 6 inches of suspension travel.
- Alivio, Acera and Altus: for casual riders on smooth off-roads on short-travel full-suspension or hardtail bikes.
- Tourney: for city bike riders on paved roads on mountain bike look-alikes.

## Technical Developments

Every year, I attend the Interbike trade show to look at the new developments, and to add to my collection of catalogs. Shimano's catalog grew from 86 pages in 1995 to 268 pages in 2008. My stack of fourteen catalogs is six inches (15 cm) thick.

Derailleur gearing remained Shimano's core product, and the majority of bicycles sold in the U.S. are equipped with derailleur gearing,

### Road Bike Derailleurs

The 1985 Dura-Ace 7400, the first SIS (Shimano Indexed System) rear derailleur, was right the first time, and only minor changes were made over the next 15 years. Ultra-premium buyers are a conservative bunch, and they don't want annual "gee-whiz" changes to destroy product identity

### Front Derailleurs

In the 1980s, I once called front derailleur shifting "mechanical sadism." Shimano ended the cruelty by revising front derailleurs a bit and chainwheels a great deal. Anyone comparing 1980s front shifting with today's gear-trains can appreciate the dramatic improvement.

Shimano's continuing design program, called HyperDrive, reduced the force needed to derail the chain from one chainwheel to another. The main features were spikes on the outer chainring to help lift the chain, and contoured teeth to provide a gap for the chain to enter. Aligned chainwheels provided a special path so that the chain was continuously

Left: Fig. 13.5. The 1997 Klein Mantra with XTR gruppo.

Right: Fig. 13.6 A and B. In 1995, Shimano introduced Top-Swing front derailleurs (bottom) as an alternative to their conventional front derailleurs (top).

engaged during the transfer. HyperDrive allowed lighter front derailleur springs, which allowed lighter-action shift levers. In 1995, Shimano offered Top-Swing front derailleurs as an alternative to the normal mounting. Most full-suspension mountain bikes needed a mounting stub on the seat post for the front derailleur. Top-swing front derailleur reduced the length of the stub.

### Number of "Speeds"

From 1996 to 1999, Shimano went from 7- to 8- to 9-sprocket freehubs. Each change began with the highest-priced gruppos, and then trickled down in the next few years.

By 1999, Shimano had 9-speed cassettes with matching shifters, chains, and chainwheels in the top three road gruppos (Dura-Ace, Ultegra, and 105) and the top three mountain-bike gruppos (XTR, Deore XT, and Deore LX). The 9-speed cassette was the same width as the old 8-speed. The extra sprocket was provided by thinning the sprockets and the spacing between them. This required a very narrow chain that required thinner chainwheels. The 9-speed components were interchangeable between gruppos.

In 2004, four years after Campagnolo had introduced 10 sprockets, Shimano provided 10 sprockets on Dura-Ace. They converted Ultegra to 10 sprockets in 2005, and 105 to 10 sprockets in 2006.

Shimano reluctantly joined the move to 10-sprockets. In my opinion, this was because the Shimano engineers knew that 7 or 8 sprockets were the optimum. The ninth and tenth sprockets provided no advantage —just more duplicate gears. The narrow chains required by nine and ten sprockets were weaker, less reliable, and wore out faster. However, in a market driven by numbers, expensive racing bicycles *had* to have at least 10 speeds.

### Revised Sprocket Shapes

Reliable rear shifting was possible with 9-speeds and then 10-speeds because of the reshaped sprockets. This has been an ongoing process for Shimano. Basically, it was three steps forward and then two steps backwards. Each time they make a significant shifting improvement, like Interactive Glide, they add a "Speed" and a narrower chain and shifting gets worse.

### STI Shift Levers

In 1991, Shimano had introduced indexed shifting and STI (Shimano Total Integration) combined brake-and-shift levers for road bikes. In short order, a road bike that did not have indexed shifting and something like STI levers could not be sold.

This had an unforeseen effect on the road bike market. STI levers were complicated and expensive and protected by patents. It was infeasible to make cheap copies. Campagnolo's similar Ergo Power combined brake-shift levers were equally expensive. The cost of the STI levers and the matching indexed derailleurs set a floor price of road bikes. Low-priced road bikes, the old drop-handlebar "ten-speeds" of the 1970s, simply disappeared. With higher prices, the market share of road bikes fell dramatically.

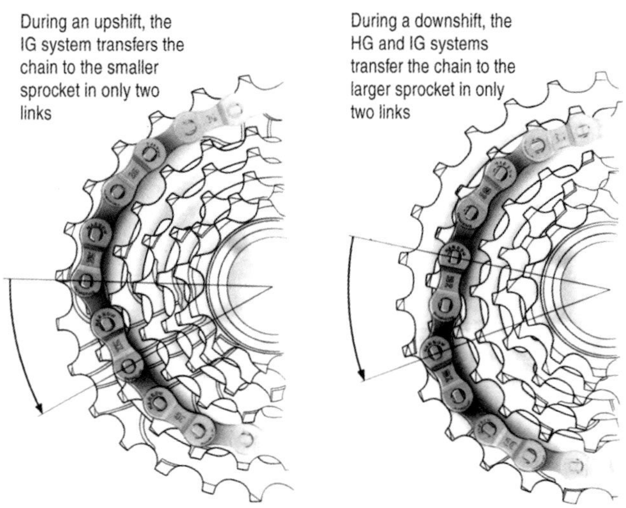


Right: Fig. 13.7. In 1995, Shimano's IG sprockets were reshaped so that shifts took place over only two links of chain. The sprockets were narrowed to provide downshift and upshift release ramps.

### Hub Gearing

Shimano hub gears date from the rebirth of the company after WW-II. Hub gears fit in with Shimano's program to make bicycles more user-friendly.

In 1995, Shimano, Sachs, and Sturmey-Archer all introduced 7-speed hub gears. All three hubs had a gear spread of about 260%. Since then, both Shimano and its competitors have introduced versions with even more gears—all the way up to 14 speeds in the case of the Rohloff Speedhub. The Shimano 3-speed Inter-3 had a 184% range. The 4-speed Inter-4 also had a 184% range. The 7-speed Inter-7 had a 244% range. The 2004 8-speed Inter-8 had a range of 304%. All came with optional coaster (back-pedal) brakes or roller brakes.

### Shimano's Market Share

From 1995 onward, most U.S. bike makers stopped specifying complete Shimano gruppos to cut costs. On middle-priced bikes, they substituted SRAM or Taiwanese hubs, cranksets, headsets, pedals, and brakes, often with the bike maker's logo. Shimano derailleurs and shifters were usually specified. As a result, Shimano's share of the cost of a new bicycle fell from about half to about a quarter.

### Ultra-Premium Products

Shimano, Campagnolo and SRAM all understood the rules of the ultra-premium boutique market: The higher the price, the higher the perceived value. This has always been true for $1,000 bottles of Lafitte Rothchild wine, or $15,000 Rolex watches, or $300,000 Rolls Royces.

An ultra-premium product is perceived as the best, even though it may perform only slightly better than the rest of the field. Don't charge 10% more for the extra performance, or 50% more for the extra production cost: charge whatever the traffic will bear. Raise the price until lower sales balance the higher profits. That's what happened not only to Shimano Dura-Ace and XTR, but also to Campagnolo Record and SRAM Red.

Left: Fig. 13.8. The Cannondale Street, equipped with Alivio gear-train. During this period, Shimano shifted some of its product development toward the developing market for this type of bicycle.

Above: Fig. 13.9. The 2004 Nexus Inter-8 had the highest efficiency of any Shimano hub gear.

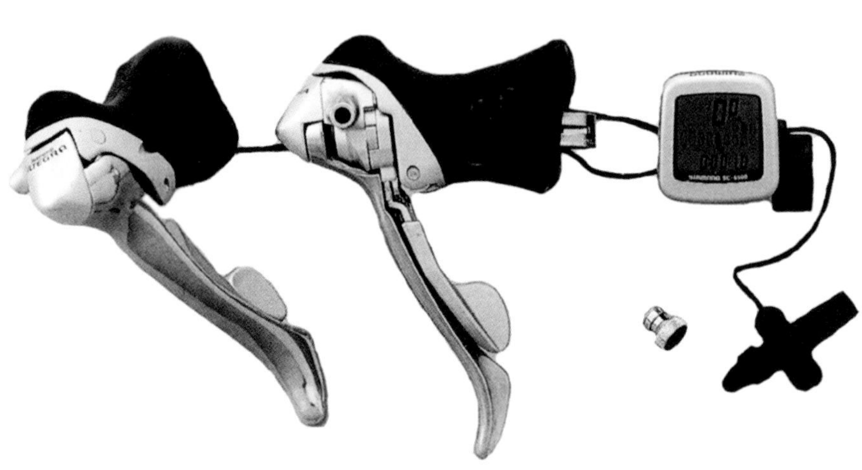

Left: Fig. 13.10. The 1998 Shimano Flight Deck computer had both shift levers wired into the display, so that it could show gear and cadence without requiring a crank sensor.

## Shimano Year by Year

### 1995

This was a significant year, because Shimano switched emphasis from mountain bikes and road bikes to city bikes. They switched from racers and pleasure riders to commuters. "Next Generation" was the 1995 buzzword to cover Nexus hub gears and Nexave wide-range derailleur systems that targeted non-competitive adult cyclists. Inter-7 was the new 7-speed hub gear. STX, Alivio, Acera and Altus were new gruppos for recreational mountain bikes. 700 CX and 400CX were new gruppos for hybrid bikes.

The RSX road gruppo had 7-speed SIS and STI brake/shift levers, bringing the gruppo concept to a lower price point.

Interactive Glide (IG) was the buzz word for easier shifting. IG cassette sprockets provided shifting on ramps and off ramps for both upshifts and down shifts. IG rear derailleurs, chains and chainwheels were part of the system.

HyperDrive-C was the term for a "compact" mountain bike drivetrain using 42-32-22 chainrings and 11–28 cassettes.

### 1996

The new XTR mountain gruppo was the big news. It used titanium for the rear axle, the freewheel body, and the large sprockets. The XTR rear derailleur was lighter and had sealed bearings and a coil return spring. The XTR front derailleur's inner and outer cages moved independently. It was wider in the low-gear position than in the high-gear position. The bottom bracket had needle bearings. A short hollow spindle splined into the cranks. The HollowTech XTR crankset had crank arms that were forged and then TIG welded to form a hollow box. The gruppo cost around $1,600 and weighed a pound (450 g) less than 1995 XTR.

The 105-SC gruppo got a triple crankset, making a 24-speed road gruppo. In my opinion, the 30-tooth inner chainwheel was a bit large for loaded touring. Only the RX-100 gruppo provided a really low gear.

Tourney gruppos were supplied for economy recreational bicycles.

Front IG was the name for spikes and pins on the chainrings to make front shifting more reliable.

Having optimized derailleur gear-trains until they almost owned the market, Shimano increased emphasis on Nexus hub gears. The 1996 Inter-7 had a 244% range. The Inter-4 had a 184% range.

### 1997

Dura-Ace was completely redesigned and got "9 speeds" or more correctly, 9 sprockets on the cassette, resulting in 18 speeds with double chainrings. The weight of the gruppo was reduced by more than a pound (about 500 g). There was a new RD-7700 rear derailleur with 11-tooth pulleys.

The cassette had an 11-tooth minimum. The large sprockets

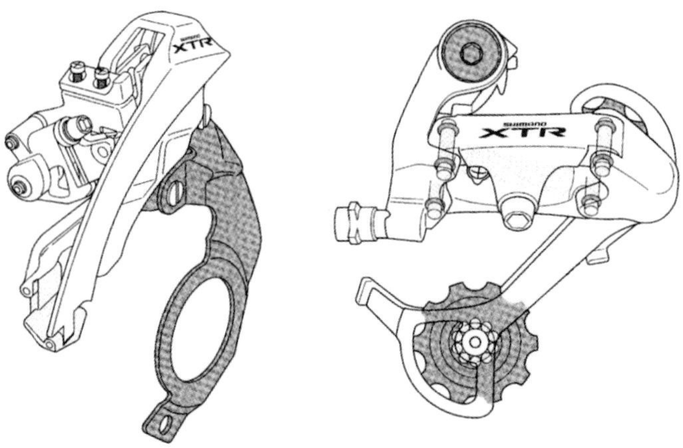


Left: Fig. 13.11. Shimano's 1996 XTR front and rear derailleurs for mountain bike use.

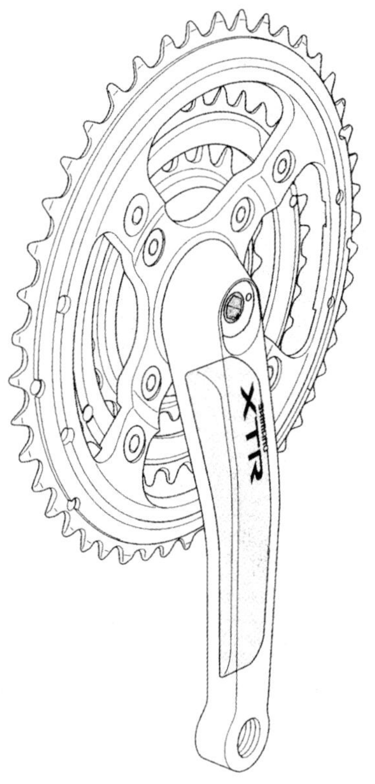


Right: Fig.13.12. The 1996 Shimano XTR crankset was one of the first cranksets with hollow cranks.

were titanium. The ninth speed required a narrower chain and narrower spacing on the sprockets and chainwheels.

### 1998

In 1998, Shimano produced a limited edition 25th Anniversary Dura-Ace gruppo. Only 5,000 were produced and the retail price was $2,200.

Ultegra got the Dura-Ace treatment, including hollow forged crank arms, hollow splined bottom bracket spindle, and a 9-sprocket cassette, shedding ¾ of a pound (340 g) in the process. Ultegra also shed its powder-coated finish and became a gruppo for bikes selling as high as $2,000. It was interchangeable with Dura-Ace in most features and dimensions. Ultegra got an optional 52/42/30 triple crankset.

The 105 gruppo also went to 9 speeds.

Newly introduced was the Nexave (which stands for "Next Generation Avenue") gruppo. It had an extra-wide-range derailleur gear-train for city bikes and recreational cyclists. The 34-tooth sprocket was reborn in the 11–34-tooth, 8-speed cassette. Combined with either 48/38/28 or 42/32/22 cranksets, it provided whopping 600% gear range.

The sprocket selection was unique, with seven medium steps from the 11-tooth to the 26-tooth sprocket and a huge step from the 26-tooth to the 34-tooth to give a "stump-puller" low gear. Nexave offered either hub brakes or cantilever brakes. The Nexave rear derailleur worked "backwards." The spring pushed inward for lower gears, and the cable pulled outward for higher gears. Both levers have the same action, which is easier for beginners. This brought back memo-

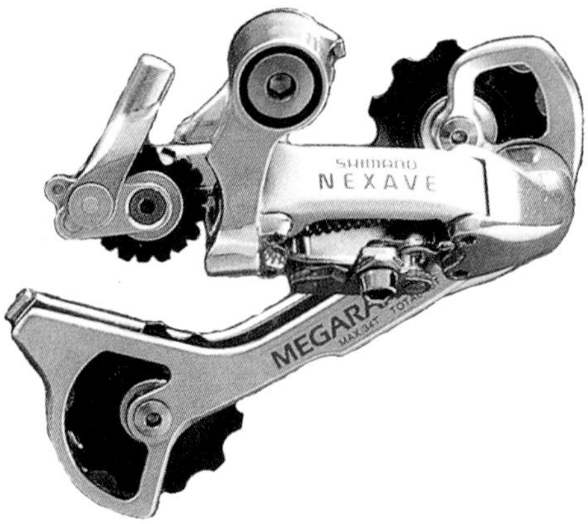


Left: Fig. 13.13. The 1998 Shimano Nexave MegaRange rear derailleur was ideal for wide-range gearing.

Right: Fig. 13.14. This 1996 Nexus demonstration bicycle reflected Shimano's program for Nexus-equipped city/commuter bikes.

Left: Fig. 13.15. The 1997 Shimano Dura-Ace rear derailleur took advantage of easier-shifting sprockets and improved chains.

Right: Fig. 13.16. In 1997, the Shimano Dura-Ace cassette got 9 sprockets.

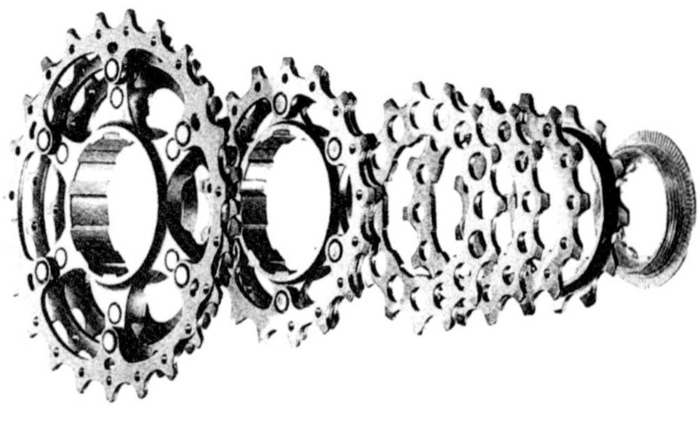

ries of SunTour's old reverse-action front derailleurs.

Also for city bikes and "comfort bikes, Shimano promoted their Nexus 3-speed, 4-speed, and 7-speed gear hubs, along with a newly added a Nexus hub generator and a Nexus anti-lock front brake.

Also introduced this year was the Flight Deck digital control system, with push buttons built into the right-hand STI brake-shift lever. Both levers were wired into the display so that the computer could display the gear (gear inches or development) and compute the cadence.

On the comfort-bike side, the newly introduced Auto-D was a full automatically controlled 3-speed hub gear. The electronic shifter analyzed your cadence and speed and shifted for you. When the battery went dead, you were stuck in low gear.

### 1999

The XTR rear derailleur offered optional reverse action (Low Normal). The coil spring in the parallelogram had lower tension to provide easy downshifting on hills. This was possible because of easy shifting HyperGlide sprockets.

The top mountain-bike gruppos got wide-range Mega-9 (12–34 or 11–32) 9-speed cassettes, and derailleurs to handle them.

The Nexave sports touring gruppos were divided into 8-speed T400 and 7-speed T300 gruppos.

The 1999 Nexus Auto-D control unit was updated to provide automatic shifting for the newer Nexus 4-speed hub for bikes in the $350 to $550 range.

### 2000

Dura-Ace, Ultegra, 105, DX, Tiagra, and Sora were the six road-bike gruppos. Sora was a newly introduced "economy" road-bike gruppo.

XTR, Deore XT, Deore LX, Deore, Alivio, Altus, and Acera were the seven mountain-bike gruppos.

Nexave C 900, Nexave T400, Nexave T300, C201, C101 and C050 were the sport-recreation-comfort-bike gruppos.

After Shimano won a patent lawsuit with SRAM over twistgrip shifters, the Tourney gruppo became available with optional Revoshift twist shifters. Most of Shimano's gruppos had an optical display to show the selected gear.

For downhill racers, the Shimano Skunk Special Project Team developed Shimano Airlines. A small compressed air tank provided power for approximately 400 push-button-activated shifts by a special rear derailleur.

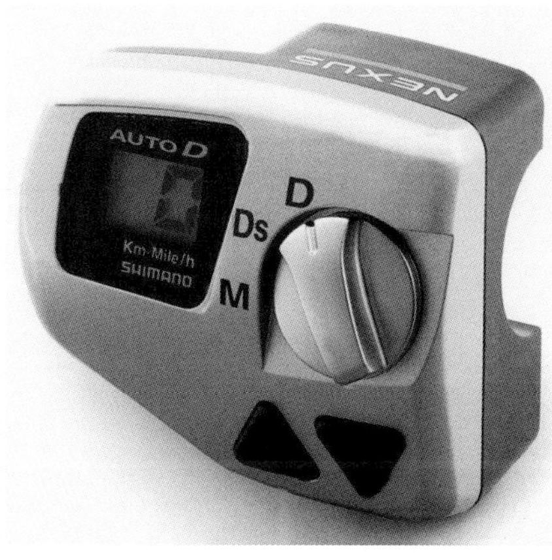


Above and left: Fig. 13.17 and 13.18. The 1999 Nexus Auto-D control unit offered three modes to the Nexus 4-speed hub: automatic (D), accelerated automatic (Ds), and manual shifting (M).

## 2001

Flat-bar controls with optical displays were available for the road gruppos.

C 900 and C 950 were new sport-recreation-comfort derailleur gruppos. C 900 was the top-of the line with new front and rear derailleurs and hydraulic brakes.

The C 900 front derailleur used SPCM (Shimano Power Change Mechanism) to take power from the crank to reduce the shift lever force. This seemed like a lot of complication, but it was the first step towards an electronically controlled, auto-shifting front derailleur.

## 2002

Shimano celebrated their 80th Anniversary.

Di2 (Digital Integrated Intelligence) was incorporated into a limited edition premium comfort gruppo, Nexave C 910. Computer control replaced conventional manual control. It automatically shifted the 9-speed rear derailleur, and controlled the suspension. The control computer shifted the rear derailleur using three pre-set modes, depending on the rider's

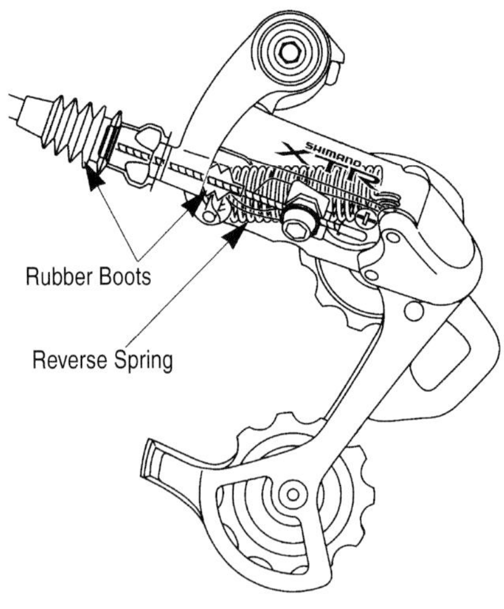


Above: Fig.13.19. The 1999 XTR rear derailleur offered optional reverse action.

Right: Fig. 13.20. The C 101 gruppo was Shimano's least expensive recreational gruppo for 2000.

Below: Fig. 13.21. How "Low Normal" works. Note that both derailleurs now downshifted with the same lever action.

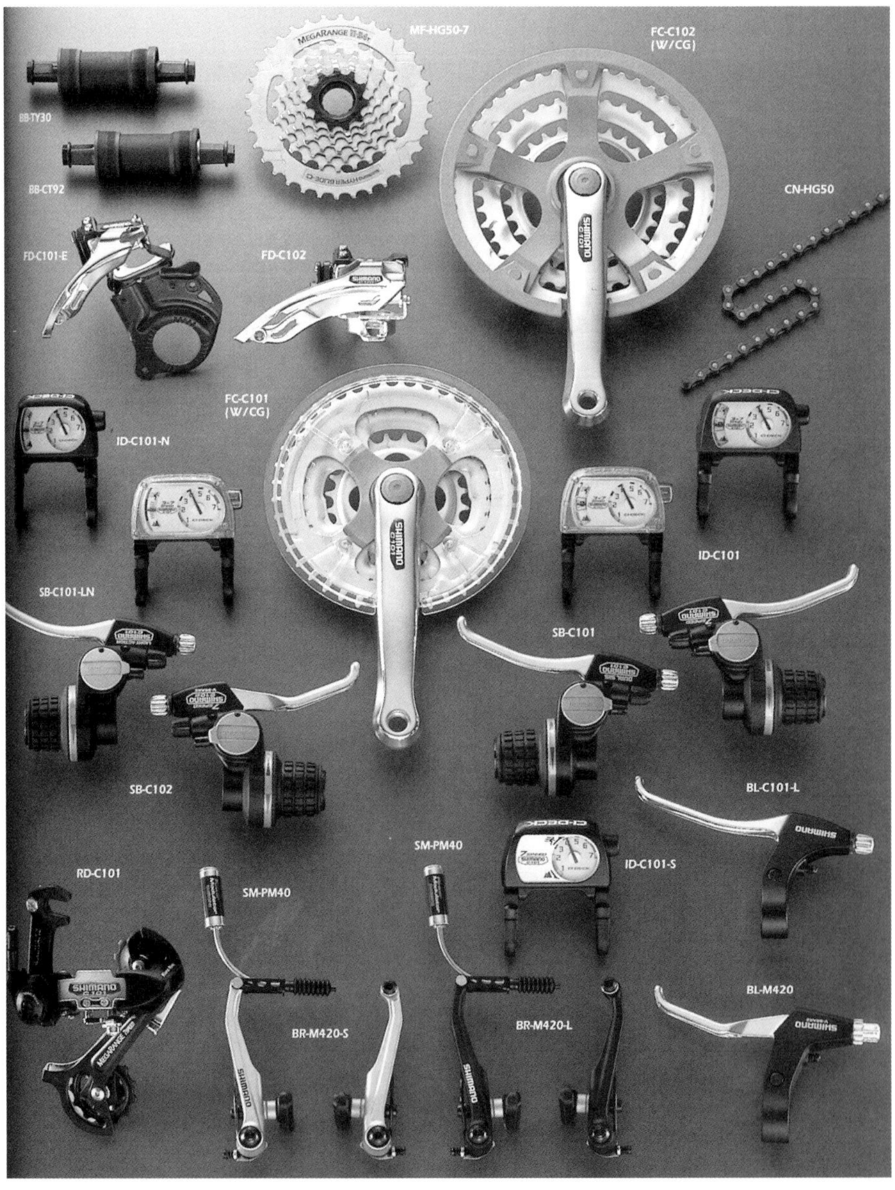


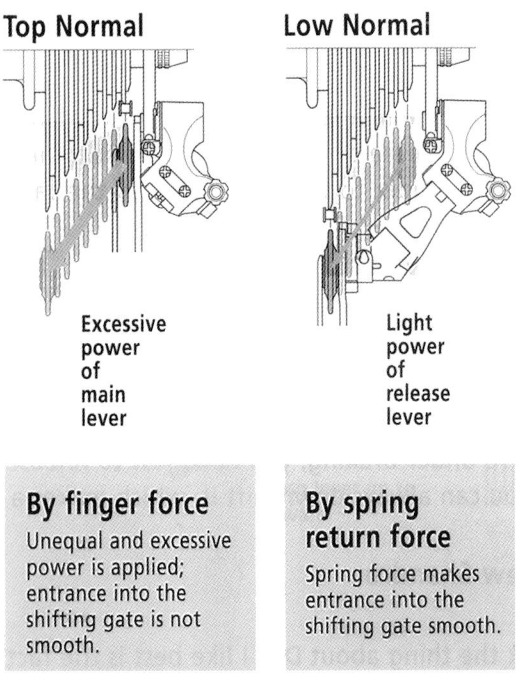

strength. A push button shifted the SPCM-assisted front derailleur. The shock absorbers were adjusted automatically, depending on the gear. This was Shimano's demonstration of the state of the art in 2002. It was either too sophisticated or too expensive, and it did not reappear until 2004. Also in 2002, Dura-Ace got a triple crankset option.

## 2003

The 2003 XTR mountain gruppo was completely revamped. The right hand crank and spider and the hollow spindle became a single unit. The left hand crank clamped to the spindle. The bottom bracket bearings were outboard of the bottom bracket, saving weight and increasing rigidity. The brake and shift levers were combined into a single lever. You pulled back to brake and lifted up or pushed down to shift gears. This was a significant change but it was unpopular because of auto-shifting on rough trails.

Nexave -C 500 and C 050 were new comfort gruppos.

The October 2003 Interbike show, where products for the 2004 season were introduced, was a Shimano tour de force. Lance Armstrong had won his fifth consecutive Tour de France on a bicycle equipped with prototypes of the new 10-speed Shimano Dura-Ace gruppo. The huge Shimano booth was full of new-for-2004 products. Shimano had about 50 sales people to man the booth, and still the visitors were waiting in line.

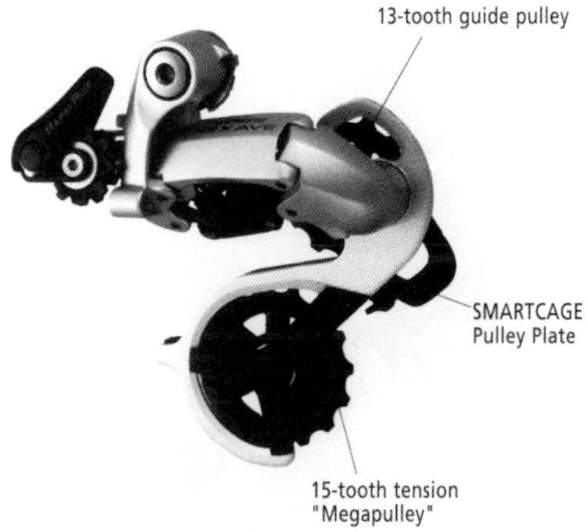


Above: Fig. 13.26. For 2002, Shimano's rear derailleurs got larger pulleys.

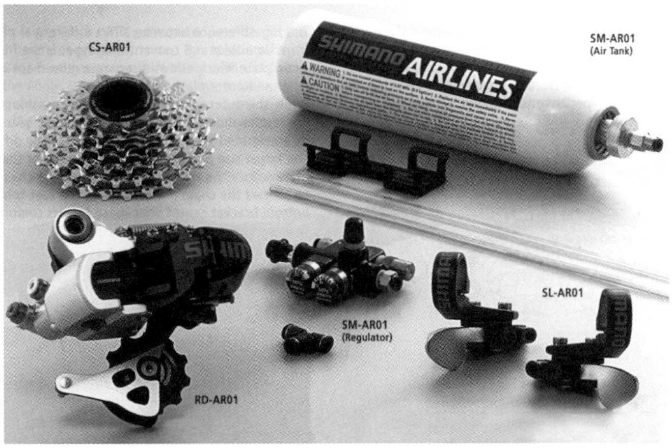


Right: Fig. 13.22. The 2000 Shimano Airlines was a pneumatically actuated rear derailleur for downhill racers.

Above right: Fig. 13.23. In 2001, Shimano introduced flat-bar controls for the 9-speed road gruppo.

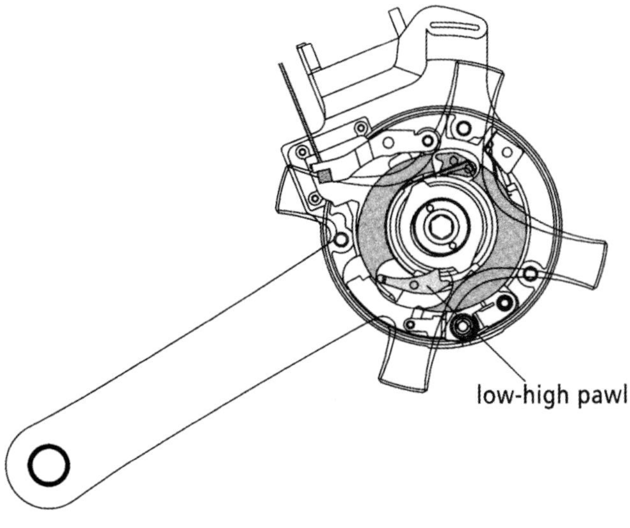


Left: Fig. 13.24. The 2001 Shimano SPCM 2001 self-powered front derailleur.

Bottom right: Fig.13.25. The mechanically controlled 2001 C 900 front derailleur.

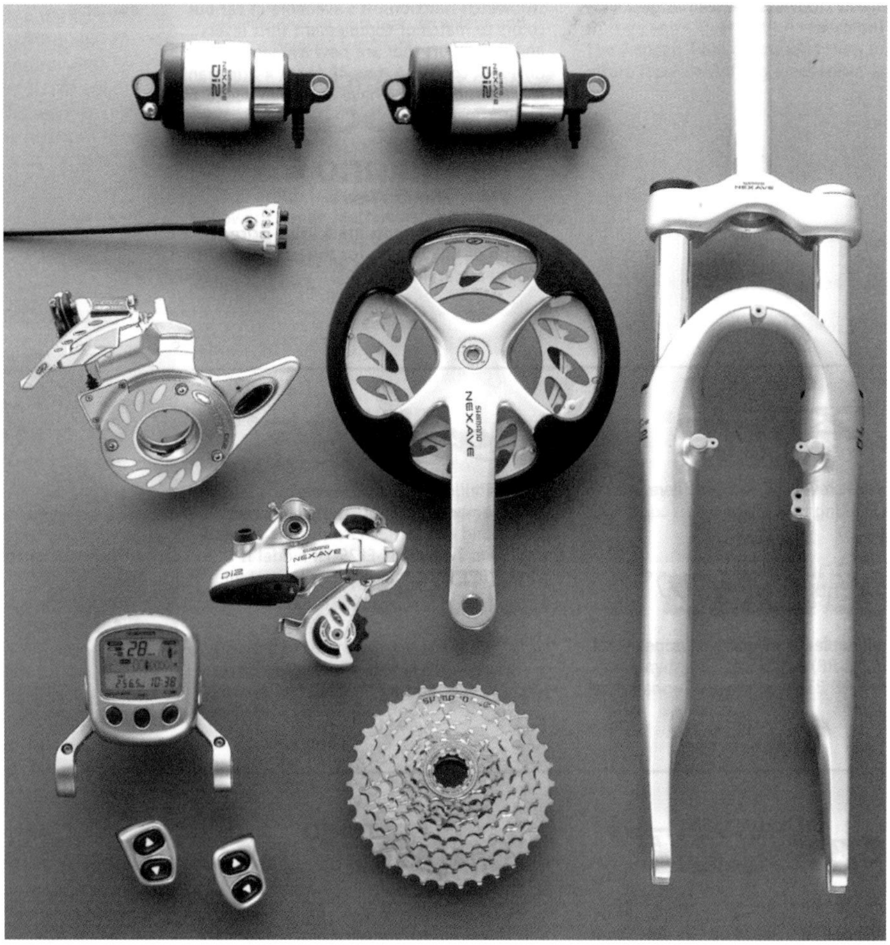


## 2004

Success caught up with Shimano, and they were unable to fill all of their orders in a timely fashion. Dura-Ace 7800 got 10 sprockets. I'm too much of a retro-grouch to shout "Hosanna." I did not like Campagnolo's switch to 10-speeds in 2000, and I did not like Shimano

Left: Fig. 13.27. Introduced in 2002, the Nexave C 910 gruppo had digitally controlled gears and suspension.

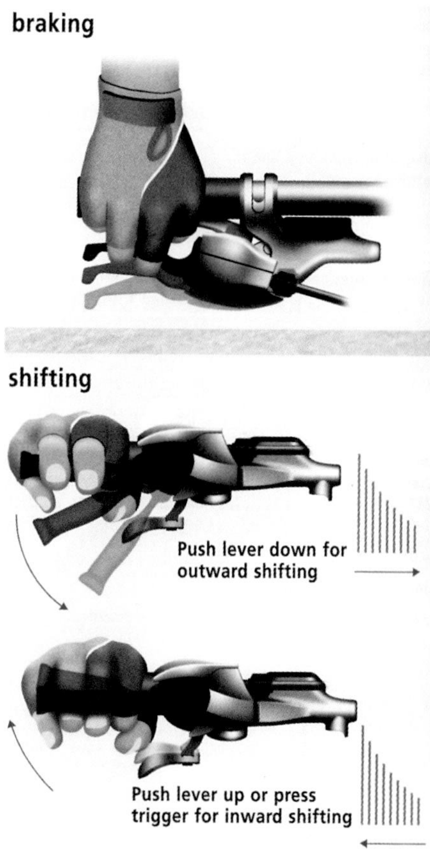


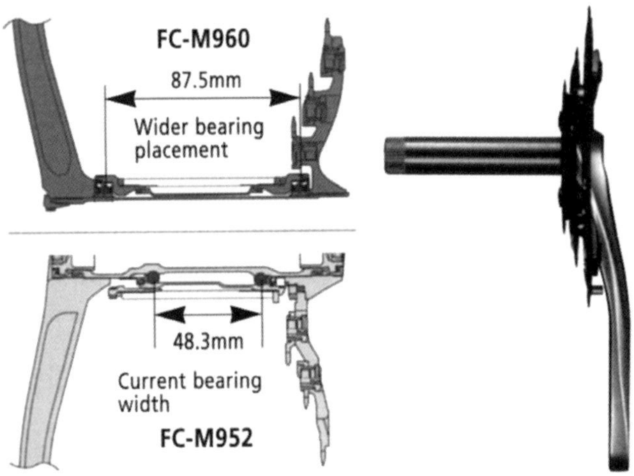


Left: Fig. 13.28. Hollow bottom bracket spindle and outboard–mounted bearings on the 2003 XTR.

Right: Fig. 13.29. The 2003 XTR Dual Control mountain bike shift lever provided single-lever braking and shifting Pulling in applied the brakes; up or down shifted the gears.

Left: Fig. 13.30. Information overload? The 2002. Shimano CI Deck included a separate digital speedometer, trip odometer, clock/stopwatch, plus two dials to show what gears you're in.

10-speeds any better in 2004. The rest of the gruppo had that combination of good looks and good specifications that shout, "Buy me!"

The Nexave C 810 comfort gruppo pulled together the previous three years of development. It included a computer that con-

Above: Fig. 13.32. LinkGlide was Shimano's name for the 2004 tweaked sprocket profile to reduce the required rear-derailleur shifting effort to facilitate electronic shifting.

Right: Fig. 13.31. With hydraulic disk brakes, the 2004 Saint gruppo was designed for extreme mountain biking conditions.

Right: Fig. 13.33. The 2004 Shimano Dura-Ace gruppo with hollow spindle crankset.

trolled the rear derailleur, the front derailleur, the front suspension fork, and the rear suspension. The computer double-shifted the rear derailleur when the front derailleur shifted over the triple crankset. It included a Flight Deck cycle computer, and a Freehub dynamo with front and rear lighting. Shimano realized that for comfort and city bikes, a dynamo was required to avoid unsafe night-riding situations due to dead batteries. Shimano called this tour de force Ecovision.

The Intego hybrid gear-train combined an 8-speed 11–33-tooth freehub with a three-speed hub gear.

The Capreo gruppo for small-wheel bicycles provided a 9-tooth sprocket. Alex Moulton was consulted in its development.

Saint was the new super-heavy-duty gruppo for extreme off-road users. I chuckled at the name, because Interbike included three off-road demo race courses, from which seven test riders had to be evacuated by ambulance, and another three by helicopter.

Hone was a new gruppo for serious off road riders. The rear derailleur mounts on the rear axle, instead of on the dropout, to provide more crash resistance. It was not as bulletproof as the Saint gruppo, but it was more attractively priced.

Shimano introduced the Nexus Inter-8, an 8-speed hub gear with a gear spread of 307%. The Inter-8 hub gear had a very light shifting action and was Shimano's most efficient hub gear unit.

## 2005

Ultegra got the 10-speed Dura-Ace treatment, with the integrated crank and hollow spindle. The triple crankset used two bolt-circles. Although Shimano only offered 53/39/30 chainrings, the inner bolt-circle allowed the use of a 24-tooth "granny" gear.

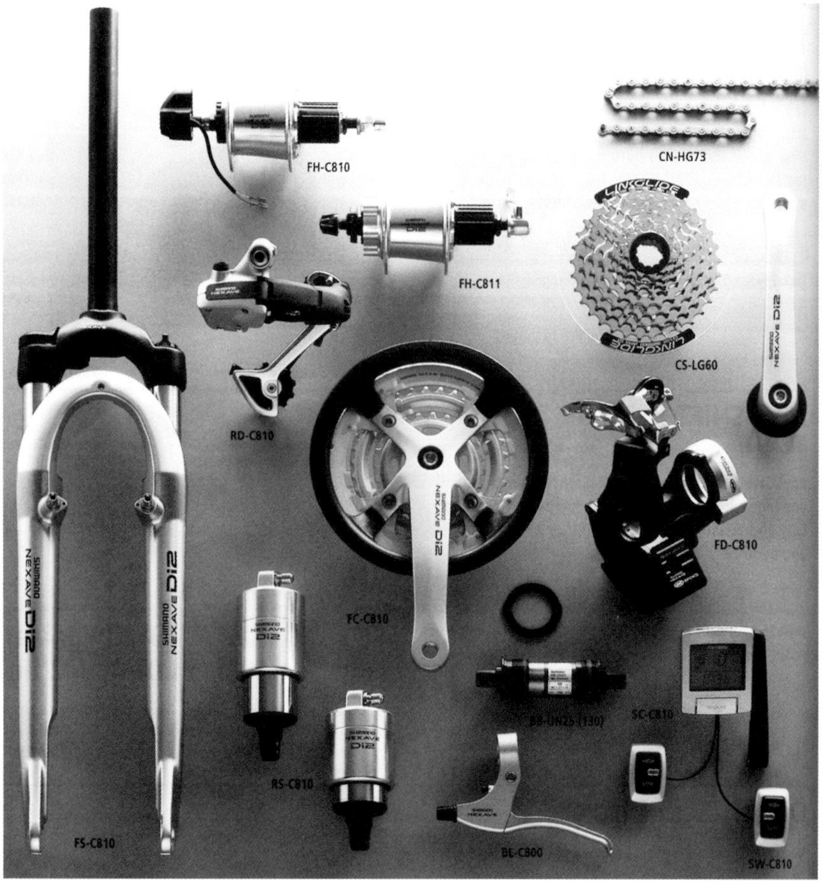


Above: Fig. 13.34. The 2004 Nexave C 810 gruppo showed the auto-shifting front and rear derailleurs and the electronically adjusted front and rear shocks.

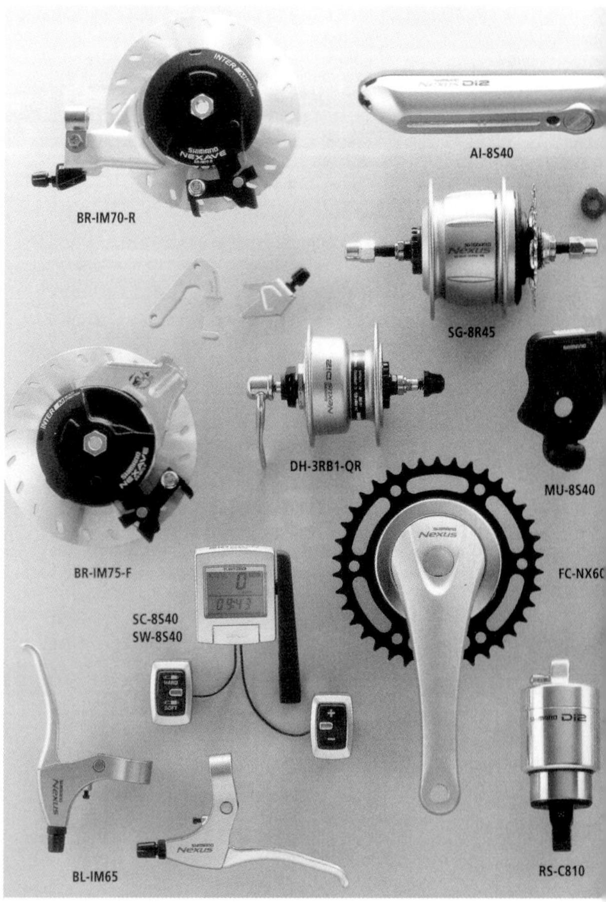


Right: Fig. 13.35. The digitally controlled 2005 Cyber Nexus gruppo.

I installed the 24-tooth chainwheel on my Moulton and found that both derailleurs shifted just fine. However, my speed with the 14-inch Low (24/34 × 20-inch wheels) was too low for balance. (I knew that there had to be a limit.)

## 2006

The 105 gruppo got the full package of 2004 Dura-Ace upgrades: 10-speeds, HollowTech cranks, Compact Drive and integrated bottom bracket. Everything was interchangeable between Dura-Ace, Ultegra, and 105 except the polishing and the price.

Similarly, the new Deore gruppo got the full package of the 2005 Deore XT and Deore LX upgrades.

Electronic rear shifting continued to be a work-in-progress. The Nexave RD-C 810 electronically controlled rear derailleur could handle an 8-sprocket 11 to 33 cassette and a 22-tooth difference on the chainwheels (46-34-24), giving a 5½ to 1 gear range.

The self-powered front derailleur could be shifted by a pushbutton on C 600 Nexave or by the computer on C 810 Nexave

## 2007

There was a new XTR gruppo. The Rapidfire Plus shift levers were separate from the brake levers, which and allowed quicker shifts.

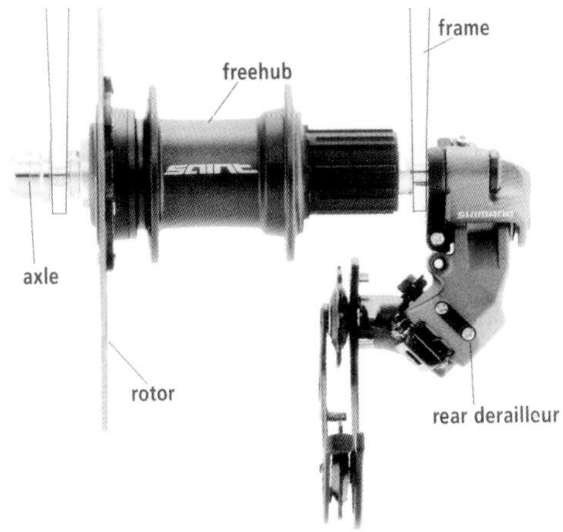


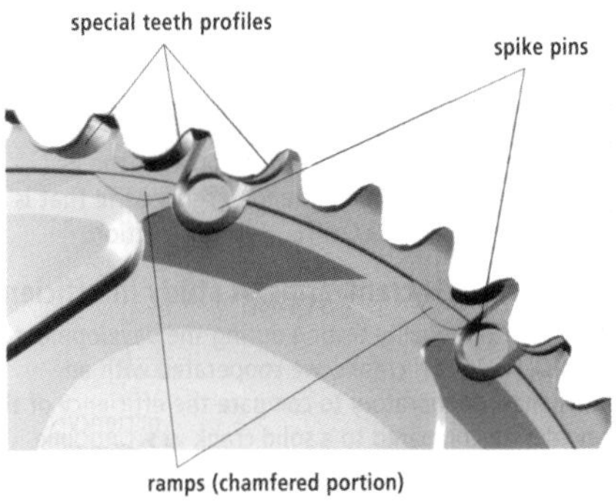


Above: Fig. 13.36. The 2005 Hone rear derailleur. Like the Saint rear derailleur, it was mounted on the rear axle instead of on the rear dropout.

Left: Fig. 13.37. In 2005, SG-X chainrings refined the improvements in front shifting.

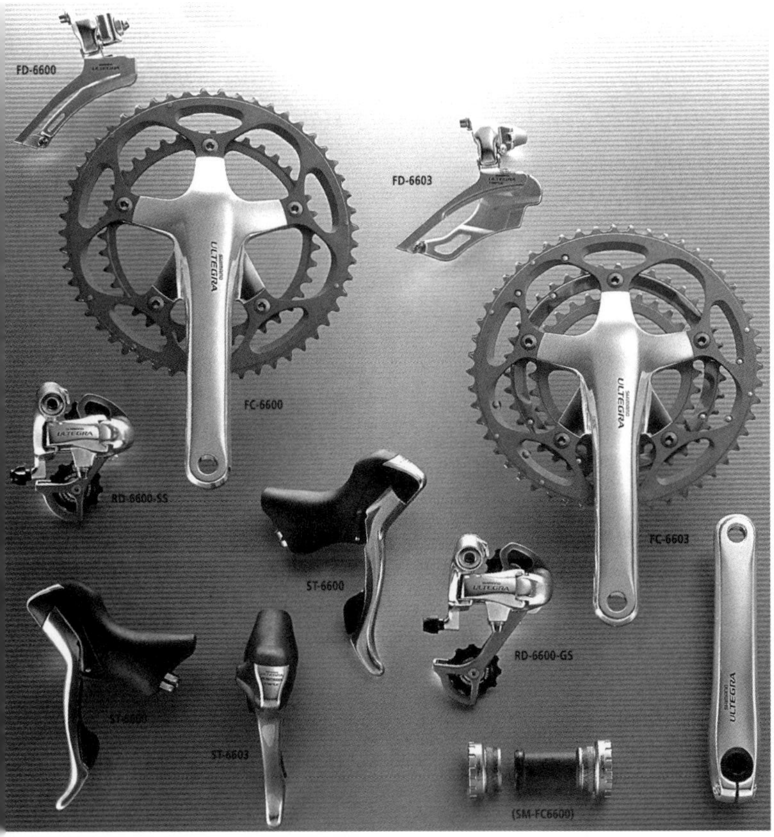


Left: Fig. 13.38. The 2005 Ultegra gruppo. Note the different front derailleurs for double and triple cranksets.

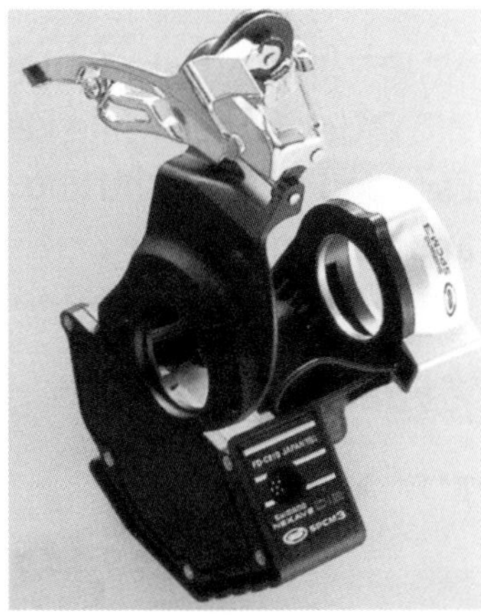

Right and below: Figs. 13.39 and 13.40. The 2006 C 810 Nexave electronically controlled front and rear derailleurs.

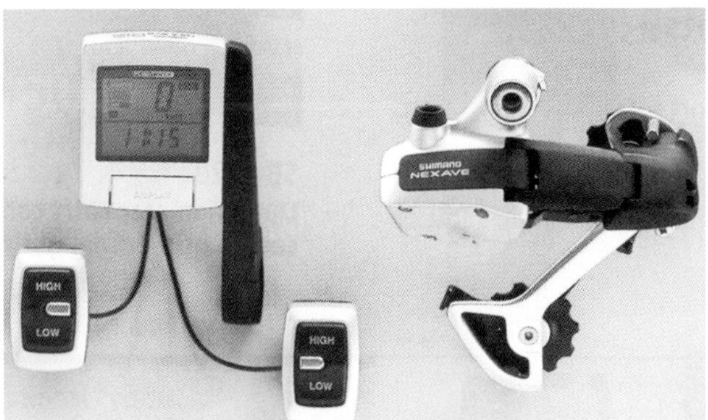

The parallelogram shape of the rear derailleurs emphasized the X. Top-normal and Low-Normal versions were available.

Alfine was a new gruppo for premium-level bikes. It combined the 8-speed hub gear and hydraulic disc brakes.

Nexus Auto-D with Inter-3 provided economical fully automatic shifting. The front hub dynamo powered for the shifting and the lights. The control package on the chainstay provided 8 shifting modes. The 3-speed hub included a back-pedal coaster brake

Shimano's "Coasting" was more of an environmental concept than a gruppo. In 2007, Shimano launched a major advertising campaign directed at the 160 million Americans who don't ride bicycles and who don't understand gear shifting or hand brakes. Coasting bicycles got rid of these hang-ups with a generator hub, which powered a completely auto-

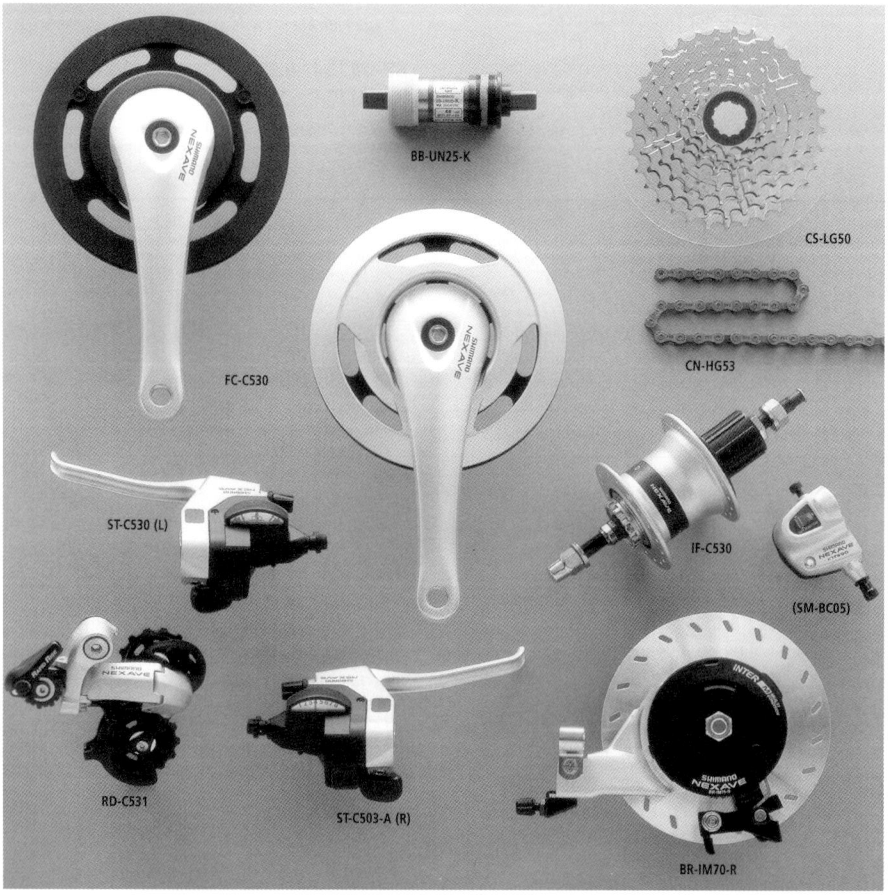


Above: Fig. 13.41. The 2005 Nexave C 530 Intego gruppo was a hybrid, combining a three-speed hub with a derailleur and an 8-speed cassette.

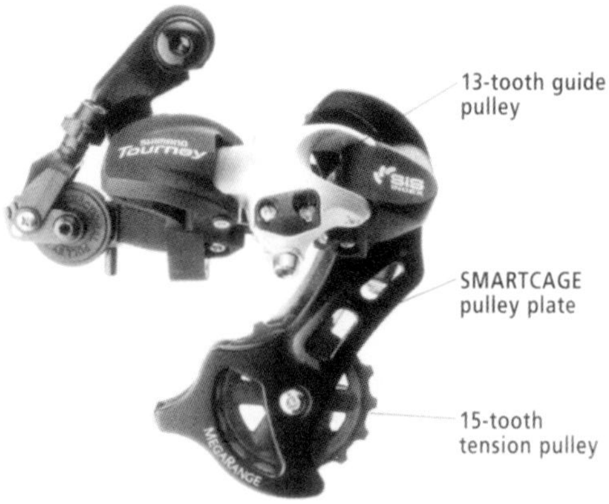


Above: Fig. 13.43. For 2006, Smartcage and larger pulleys increased the capacity of the rear derailleur.

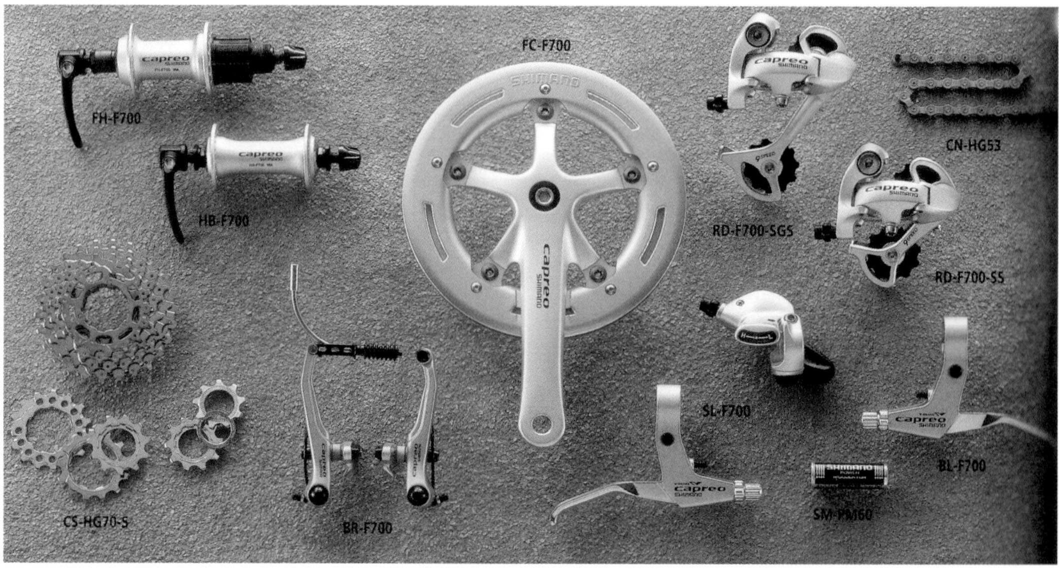


Right: Fig. 13.42. The 2005 Capreo gruppo for " compact" bicycles, i.e. those with small wheels, such as folding bikes.

matic-shifting 3-speed gear hub and a back-pedal rear brake. At $400 to $800, Coasting bicycles were not inexpensive. Coasting used the same technology as the Nexus Auto-D, with different packaging and promotion. Over the next few years, the several manufacturers introduced Coasting-equipped bikes, but their sales were disappointing.

## 2008

As the price of Shimano's top-of-the-line gruppos increased, there was a large gap to the mi-priced level. With the deflated dollar, there were no longer many buyers for bikes with Dura-Ace and XTR, which by now had become $5,000+ bicycles.

Ultegra SL was a new gruppo to fill the price gap beneath Dura-Ace. Deore XT was a new gruppo to fill the price gap beneath XTR. In both cases the polish and the cosmetics were closer to Dura-Ace and XTR.

Shimano's mountain bike derailleurs came in Top-Normal and Low-Normal versions. XTR and Deore XT offer a third version, called Shadow, which had a low profile for riding on narrow trails. All versions offered GS standard cages with 33 teeth capacity and SGS long cages with 45 teeth capacity.

The Sora road gruppo got 9-speeds and new cosmetics.

## 2009

The Dura-Ace road gruppo was redesigned after four years. The STI levers got a shorter throw and adjustable brake lever position. Cranksets with Compact 50-34 chainrings were introduced. The cranks have a new profile and a hollow outer chainring.

The big news was the availability of electronic shifting on Dura-Ace. First introduced on "comfort" bike components, electronic shifting has now found its way to the top of the line. The Flight Deck computer was now wireless and could also interact with a heart-rate monitor.

The electronically shifted Dura-Ace worked like a charm on the roller-mounted bike shown at the Interbike trade show. It was first used by several riders on Shimano-sponsored teams in the 2008 Tour de France.

If you want to try it for yourself, be advised that Shimano considers it a bargain at "only

Right: Fig. 13.44. The 2007 XTR gruppo.

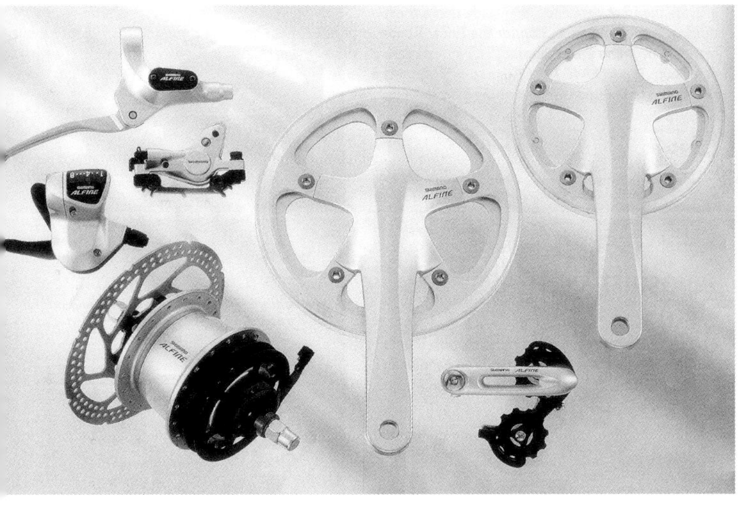

Below: Fig 13.45. The 2007 Alfine gruppo with hub gear and disk brake. The item that looks like a rear derailleur is just a chain tensioner.

$1,500 more" than the regular (mechanical) Dura-Ace gruppo, on which Sastre won that year's Tour de France.

The Saint mountain bike gruppo was lightened some more, and XT was tweaked further.

## 2010–2012

For the first time, Simano's dominance in all sectors of the market was being challenged by SRAM's range of products, bringing to a (temporary) end Shimano's decade of Tour de France victories.

In the 2011 Tour de France, Shimano-equipped bicycles were once again victorious, whereas SRAM could not repeat their 2009 successes, and Campagnolo was used by only a few teams.

Shimano's digital shifting trickled down to the Ultegra gruppo for 2012, at a price point that would make it accessible to a larger segment of riders, beyond those professionals on Shimano-sponsored teams.

Right Fig. 13.46. The 2008 Shadow rear derailleurs were designed for "aggressive" riding and use on narrow "singletrack" trails.

Below: Fig. 13.47. The 2008 Deore XT gear-train, showing three of the rear derailleurs: Low Normal/ standard cage, Top Normal/ long cage, and. Shadow/ long cage.

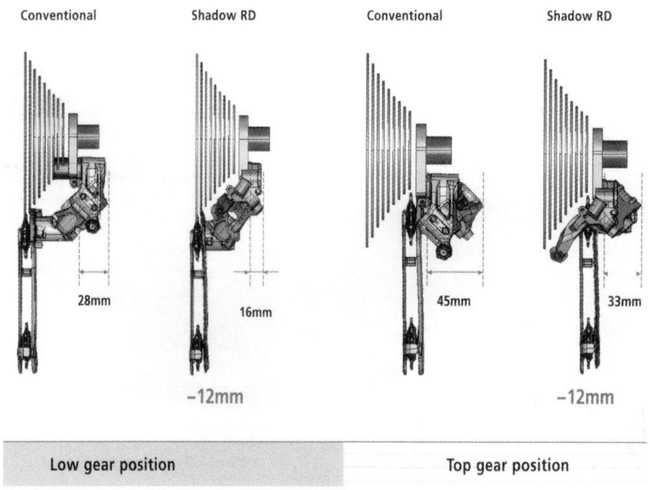

Above: Fig. 13.48. The 2009 Saint rear derailleur.

Below: Fig. 13.49. Electronic shifting is finally here. These were the electronic components of the 2009 Digital Dura-Ace gruppo. The control unit was contained in the front derailleur housing.

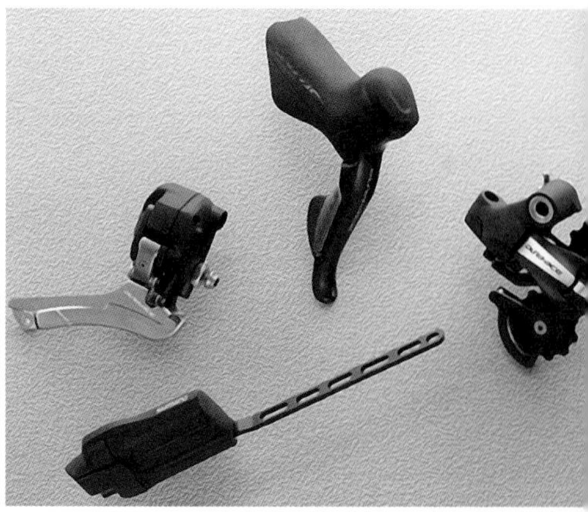

# 14.

# Into the 21st Century: 1995–2012 Part II. SRAM, Campagnolo, Etc.

AT THE TURN of the millennium, the playing field changed. Production shifted from Japan, first to Taiwan, and then to China. At the same time, Shimano lost some ground to other component makers. This chapter will cover those others: SRAM, Campagnolo and the smaller makers of bicycle gear-trains and their individual components.

## Bicycle Production

The bicycle was still a major part of world transportation. In 1995, World Watch Institute reported that 114 million bicycles were built. For 2006, the Japan Bicycle Promotion Institute reported that 117 million bicycles were built that year—about three times world automobile production. The majority were single-speed transportation bicycles; perhaps 20% were multi-speed bikes. China made about 68% of the world's bicycles. Chinese bicycles with Chinese components took over the U.S. low-priced mass market.

The Bicycle Manufacturers' Association (BMA) closed shop in 1994. The accompanying statistics are based on estimates prepared by the Bicycle Council and *Bicycle Retailer & Industry News*.

The graphs of Fig. 14.2 and 14.3 respectively show developments over the most recent 15-year period and the long-term trend of annual bicycle sales. Total sales went from about 2

Right: Fig. 14.1 To celebrate their 100th anniversary in 1995, Schwinn made 5,000 replicas of the 1950 Black Phantom.

million bicycles in the late 1940s to 18 million bicycles in 2007 (the latest year for which figures were available).

Annual U.S. bicycle sales averaged 16 million for the period. This was made up of 11 million adult bicycles (wheel size more than 20-in. diameter), and 5 million children's bicycles (wheel size 20-in. diameter or less). Bicycle imports went from 6 million in 1992 to a peak of 20 million in 2002. Britain, France, and Italy, the major exporters during the 1970s "bike boom," had less than 1% of the U.S. market in this period.

In 1995, dual-suspension bikes dominated the high-end mountain bike market. Prices came down quickly, and by 1996 you could buy a "double squishy" bike for $400. Disc brakes became common. The growth of mountain bike racing led to different categories of racing mountain bikes with suspension, gearing, and brakes designed for the different kinds of races.

Mountain bike market share peaked at 80% in 1992 and decreased to about 50% by 1996. By 2006, sales of mountain bikes suitable for off-road riding were about a quarter of the market.

"Comfort" bikes first appeared in 1999. Those bikes had wider tires, suspension handlebars or forks, and a suspension seatpost. BMX bikes, city bikes, comfort bikes, commuter bikes, hybrid bikes, loaded-touring bikes, mountain bikes, nostalgia bikes, recumbents, road racing bikes, tandems, transportation bikes, and trekking bikes all competed for a share of the market.

China became the major exporter of low-priced bicycles and Chinese bicycles took over the U.S. market for children's bikes and low-priced adult bikes.

Huffy and Brunswick, the last surviving U.S. mass-market manufacturers, tried to compete with Chinese labor costs and failed.

Huffy closed their factory in 1998 and put their labels on Chinese imports. In 2004, Huffy declared bankruptcy.

Brunswick made a serious effort to continue building bikes in the U.S. They made bicycle frames in a highly automated U.S. factory, and shipped the frames to a "maquilador" factory in Mexico for painting and assembly. In 2001, both the U.S. and the Mexico factories closed.

Taiwan continued to make and export medium-priced bicycles. The typical "U.S." bicycle maker imported their medium-price bicycles from Taiwan and their inexpensive bicycles from China.

Nobody made very much money making or selling bicycles and the industry consolidated. Barracuda, Bianchi, Cannondale, Gazelle, GT, Mongoose, Roadmaster, Raleigh, Ross, Schwinn, and Univega all got new owners. Companies went in and out of bankruptcy, sometimes two or three times.

In 1999, Raleigh ended 112 years of bicycle manufacture in Nottingham and sold the factory site. Raleigh-USA closed their Kent, WA factory in 2001.

Trek, Cannondale, and Specialized were the main American makers of quality bicycles. They thrived until about 2000, and they developed significant export markets. Cannondale got into the motorcycle business in 2001 and rode it all the way to bankruptcy. They emerged with a new owner. Trek made their top-quality bicycles in the U.S. and imported the less expensive models. Trek absorbed Gary Fisher and Klein, which became divisions of Trek.

By 2008, U.S. annual domestic bicycle production consisted of about 200,000 quality bicycles made by Trek, Cannondale, Specialized, GT, and numerous smaller custom builders. U.S. bi-

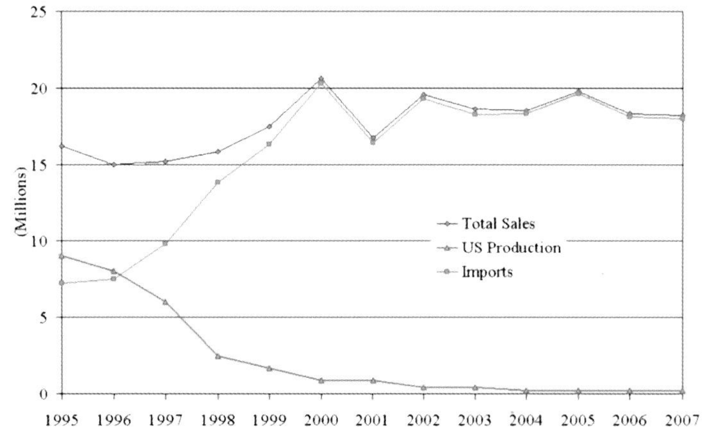


Left: Fig. 14.2. Graph of U.S. domestic bicycle production, bicycle imports, and total bicycle sales, 1995 to 2007. (Source of data: *Bicycle Retailer and Industry News.*)

cycle production ceased to be a significant part of the market afer 2000, so Figs. 14.1 and 14.2 no longer show Domestic Production for those years. There is still a significant number of custom builders, both for regular bikes and for tandems and recumbents.

## U.S. Bicycle Sales

The U.S. population doubled between 1945 and 2004, passing 300 million in 2007. Per capita bicycle sales grew from one bicycle per 70 Americans to one bicycle per 15 Americans. The explosion of bicycle sales during the American 10-speed boom from 1971 to 1974 was an aberration from the long term trend.

The U.S. retail market for adult bicycles was split 50-50 in this period. The mass marketers (ToysЯUs, Costco, Wal-Mart, etc.) had about half of the market, and their bicycles sold for around $100. The 7,000 specialty bike shops and the sporting goods stores supplied the more expensive half of the market. Porsche and Mercedes Benz failed in their effort to sell mountain bikes in automobile showrooms.

In the late 1990s, U.S. sales of quality derailleur adult bikes (the kind sold by bike stores) were about 2 million per year during this period. Seven companies—Trek, Specialized, Giant, Schwinn, Raleigh, Cannondale, Schwinn-GT, and Fuji—had two-thirds of this market. A dozen smaller companies fought for the leftovers. The average price was around $300. Trek and Cannondale made 20% of these quality bikes in the U.S., while the remainder was imported. Taiwan supplied the medium-priced bikes, and China supplied the cheap ones.

The road bike's share of the market increased from 4% in 1995 to 10% in 2004 to 12% in 2006. Many of the new road bikes had triple cranksets. Front shifting had become so reliable that you could carry a small inner chainring for the occasional long steep hill.

The baby boomer's babies became teenagers, and a hot market for BMX bikes developed. Imports of 20-inch-wheel bikes (mostly BMX bikes) went from 1.4 million in 1996 to 5 million in 2003.

As the mountain bike market declined, the bike-makers looked for the next hot trend. Unfortunately, there wasn't just one. Genuine loaded-touring bikes made a bit of a comeback, but many of the touring bicycles were for "credit-card" tourists, for cyclists who don't carry a load but travel light and sleep in motels.

Some nice commuter bikes, such as the Specialized Globe, were available in 1995, but they did not catch on at the time. The Globe reappeared in the 2005 catalog, with a rear rack, fenders and lights. Maybe $4.00 per gallon gasoline will finally get Americans to consider bicycles for short trips.

The difference between commuter bikes, comfort bikes, city bikes, hybrid bikes, and transportation bike became blurred. The combination of features, including wider tires, flat handlebars, a rear rack, fenders, lights, and an easy-shifting gear-train varied between makers.

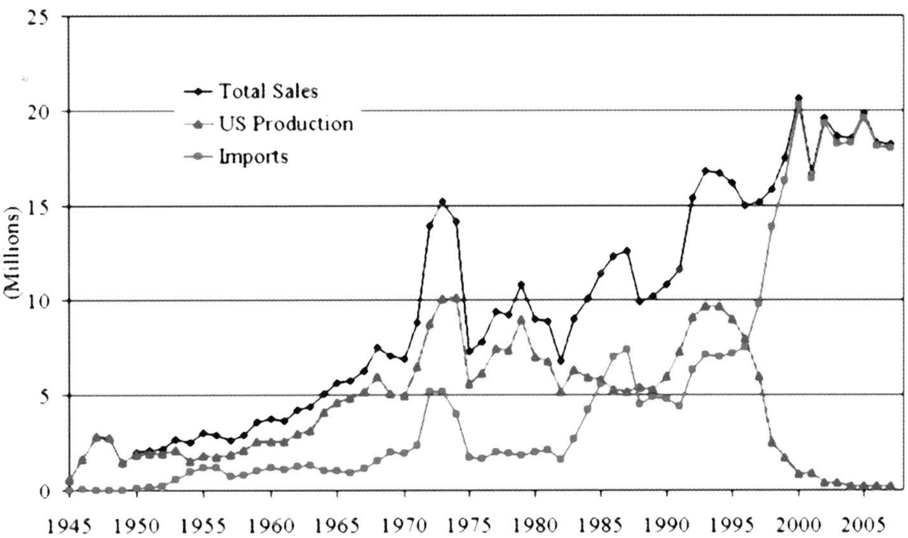


Right: Fig.14.3. Graph of U.S. domestic bicycle production, bicycle imports, and total bicycle sales, 1945 to 2007. (Source of data: The Bicycle Council.)

### Recumbents

There has been a revival of interest in recumbent bicycles. They have a core of enthusiastic user. Most of the recumbents are made by small companies. Recumbents often have interesting drive-trains. They climb hills slowly, so they need an extra-low low gear.

With low frontal area and streamlining, they can go very fast, so they need an extra-high high gear as well. This requires a wide range between high and low gears. Many long-wheelbase recumbents use an intermediate jackshaft to support the long chain. With a second rear derailleur on the jackshaft, I could hypothesize 300 speeds. (3 on the chainwheel, times 10 on the jackshaft, times 10 on the rear wheel).

### Tandems

Tandems can have interesting gear-trains. The DaVinci tandem takes the power from the two sets of pedals to an intermediate jackshaft, which rotates at twice the pedal cadence. The chainwheels may have 12, 18, 24, and 30 teeth (equivalent to 24, 36, 48, and 60). Combined with an 11–28-tooth cassette, this provides gears from an 18-inch low to a 140-inch high gear. The two freewheels allowed either partner to coast while the other one keeps pedaling.

### Comfort Bikes

In 1999, Comfort bikes were the hot category. The idea was to seduce first-time adult buyers with flat handlebars mounted higher than the seat, low bottom brackets for easy mounting, soft-riding tires, soft saddles on suspension seatposts, and suspension handlebars. The concept caught on, and by 2008 most manufacturers offered several comfort models.

### Transportation Bicycles

In 2003, Shimano began a crusade for Transportation bikes, and other component makers also started marketing their components for this kind of machine in the U.S. By 2008, the major makers were offering multiple models, though not everyone used the same terminology and called them Transportation bicycles. I liked the complete line of 11 transportation Breezers from Joe Breeze, the Marin mountain-bike pioneer. The Dutch and Danes would immediately recognize this category. It's the kind of bicycle they rode every day for shopping, touring, and commuting.

Typically, these bikes have a hub gear, a front hub dynamo, lights, fenders, a rear rack, and 26×1.5-inch tires. In 2005, Cannondale introduced its version, called Street, also with a rack, fenders, and chain guard.

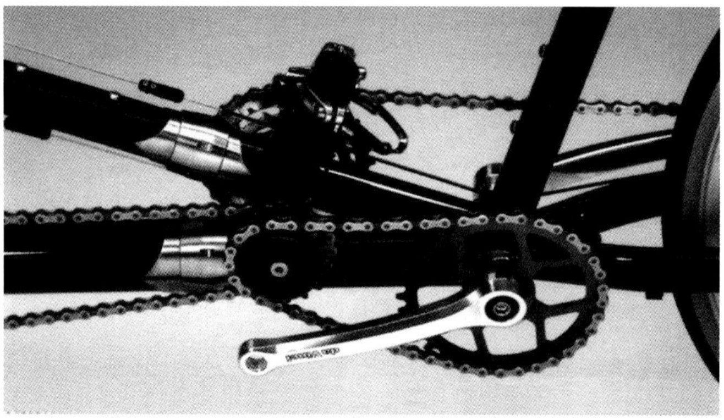

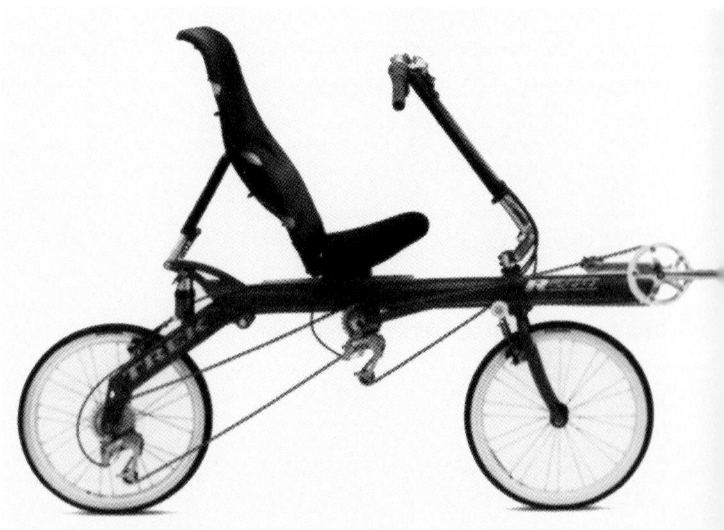

Above: Fig. 14.4. This 2005 DaVinci tandem had an intermediate jackshaft.

Right: Fig. 14.5. The 2001 Trek R 200 recumbent used an intermediate drivetrain with two SRAM rear derailleurs. It had 21-speed gearing.

### Electric Bicycles

By 1997, electric bikes had a significant share of the Japanese market. Japanese laws limited their top speed and required that the motor provide only pedal assistance. By 2007, the Chinese market was in the millions, driven by the availability of minimum cost, minimum performance, electric bicycles.

Ten different electric bicycles were on display at the 1997 Interbike trade show. There were trade-offs between cost (about $1,500), battery weight (about 20 lbs.), performance (about 20 mph), and range (around 20 miles). There were two categories: electric mopeds and electric-assisted pedal bicycles.

The electric moped ran until the battery went dead, and then you pedaled home. Electrical pedal-assistance is a better idea, but it requires a sophisticated control system. You dial in the level of electrical boost. The harder you pedal, the more electric boost, and the faster you go. It feels like riding with a strong tailwind.

More than a dozen electric bikes were exhibited at the 1998 show, including models from major makers like Trek, GT, and Peugeot. They used either lead-acid or nickel-cadmium batteries. There were purpose-built electric bikes and conversion kits. The purpose-built electric bikes were more expensive, but their controls were usually more user-friendly.

The purpose-built bikes either had the motor in the front or rear hub, or at the bottom bracket. The battery was mounted low between the wheels. In 1998 prices ranged between $700 and $1,500. The conversion kits were added to standard fat-tire bikes. The motor was perched above the front or the rear wheel, and drove a friction roller against the tire. The battery was either inside the frame or on a rear rack.

Fewer electric bikes were displayed at subsequent shows. Everyone waited for Lee Iacocca's company, EV Global, to hit the market in 2001. Giant was a partner in EV Global, but in spite of the lots of publicity and marketing, EV Global did not survive.

Trek sold the ElekTrek with a Yamaha power-pack for a few years with little publicity or success. SRAM made the Sparc electric rear hub in 2001, but almost no one in the U.S. specified it.

The ZAP (which stands for Zero Air Pollution) ElectriCruiser was a typical add-on electric power package. Its lead-acid battery was designed for a riding speed of 14 mph (22.5 km/h). The kit sold for $650, but in spite of that modest price, ZAP did not survive.

Above: Fig. 14.6. A typical modern transportation bike: The Breezer Uptown 8, with step-through frame, 8-speed hub gearing, hub-generator lights, lock, kickstand, chainguard, fenders, and a rear luggage rack.

Right: 14.7. The 2008 Schwinn Continental used a removable lithium polymer battery and a lightweight motor in the front hub, cutting the electric drivetrain weight to about 10 pounds. Note the shaft-drive driving a Shimano Nexus Inter-8 hub gear.

Hope springs eternal, and a fresh crop of electric bikes appeared at each Interbike show. Ten electric bikes were shown at the October 2004 Interbike show, most of them for the first time. Giant offered the Revive Spirit electric-assist comfort bike with Panasonic electric assistance and Shimano Inter M gear-train.

Hybrid automobiles and laptop computers have resulted in better batteries. "Better" means smaller, lighter, more powerful, longer lived, more efficient and faster charging. The 2007 Schwinn line set a new standard. The total weight of the motor and battery was 10 pounds.

To date, U.S. buyers have not been attracted to electric bicycles and the bicycle shops have not figured out how to sell electric bikes. Were they motorcycles or pedal bicycles? It depended on the power, the top speed, and the applicable state law. Were they a zero-pollution answer to global warming? Not really. They just relocated the pollution from the vehicle to the electric generating plant. As of 2008, they were still trying to get a foothold in the U.S. market. My sense is that really expensive (say $5.00 per gallon) gasoline will generate a market.

## SRAM

SRAM is an American company owned and run by one man, Stanley R. Day. During the 1995 to 2008 period, SRAM became

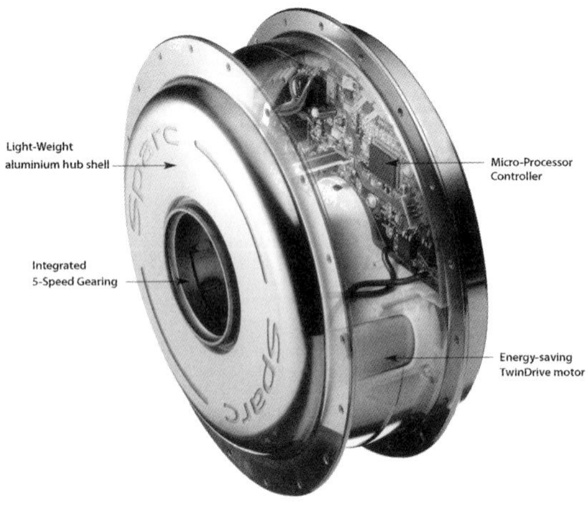


Left: 14.8. The 2001 SRAM Sparc electric-drive rear hub.

Right: Fig. 14.9. The 1998 ZAP ElectriCruiser was a typical add-on electric power package that could be installed on a regular mountain bike.

Below: Fig.14.10. At $1,700, the 1997 EV Charger was a well-designed electric bike. The motor assisted the pedals. With its 20-lb (9 kg) battery, it had a 20-mph (32 km/h) top speed and a 20-mile (32 km) range.

Above: Fig.14.11. The 2005 Giant Revive Spirit was a semi-recumbent bike with electric pedal assistance.

the major competitor to Shimano, the market leader. Shimano's products during this period were covered in the preceding chapter.

SRAM continued its program of buying competitors and incorporating their products. Many of the acquisitions were not particularly successful but SRAM persisted. They developed an in-house engineering capability to produce unique SRAM designs. The 2006 SRAM road bike gruppos, Force and Rival, made SRAM a serious competitor in the entire component market.

In 1995, their main products were inexpensive mountain bike twist shifters, which were compatible with Shimano derailleurs. Because of patent problems, Shimano did not make twistgrip shifters for mountain bikes, and SRAM found a significant market for their low-priced Grip-Shifts. SRAM Grip-Shifts came in three quality levels, and many OEMs used them to save money or to provide uniqueness. SRAM also made front and rear derailleurs, but they were rarely specified.

Above: Fig. 14.12. SRAM's President Stan Day, photographed in 2001. He has successfully expanded his company from a one-trick pony—maker of twistgrip derailleur controls—to a major player in the bicycle component game.

In 1996, SRAM announced ESP (Extra Sensory Perception). The ESP 900 mountain bike rear derailleur had main body parts made of carbon-fiber-reinforced epoxy. It featured a steeply-angled parallelogram, a coaxial pulley, and one-to-one cable movement. It did not have a spring-loaded top pivot (B-spring), so it was an updated SunTour VGT. After making 10,000 ESP 900s, SRAM discovered manufacturing problems that had caused poor shifting or breakage, and they had to recall all of the ESP 900 rear derailleurs.

The 1997 ESP 9.0 rear derailleur had a carbon-composite body. SRAM also introduced the less expensive 7.0 and 5.0 rear derailleurs to compete at lower price levels.

At the end of 1997, SRAM bought Sachs from Mannesman. This was the beginning of SRAM's program to become a genuine worldwide competitor to Shimano and Campagnolo.

SRAM's purchase of Sachs was the big news at the fall 1997 bike show. Everyone speculated on the outcome, because the two companies had such different cultures.

Sachs was a division of the Mannesman Corporation, a major German heavy-industry conglomerate. Starting in 1985, Sachs had methodically acquired the best of the surviving French component makers: Huret, Maillard,

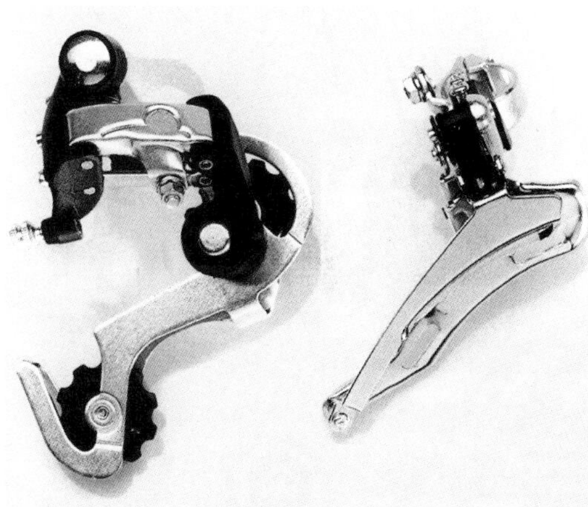

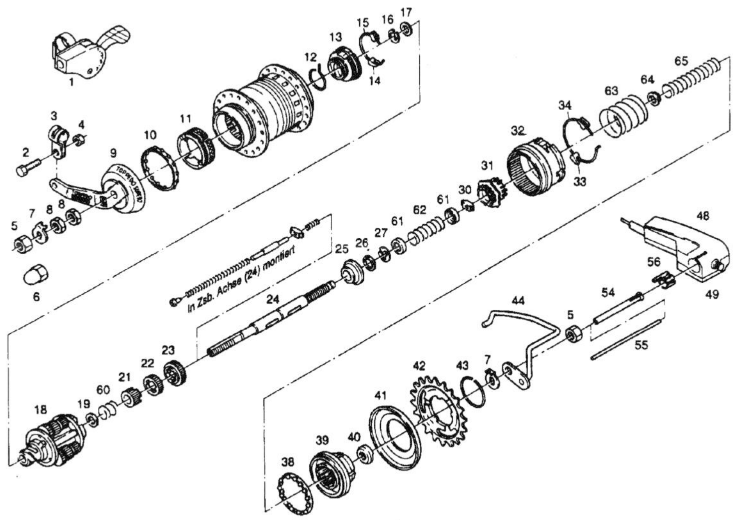

Above and left: Figs. 14.13 and 14.14. When SRAM took over Sachs in 1998, they inherited both the Sachs-Huret line of derailleurs (above) and Sachs' own line of hub gears and brakes, such as the Super 7.

and Sedis. Sachs sold medium- and high-priced gruppos. They were reasonably successful in Europe, but they never became a factor in the U.S. From 1995 to 1997, Sachs promoted hub gears and hybrid hub-derailleur geartrains for the city/trekking/hybrid market, picking up the crumbs that fell from Shimano's table.

SRAM reportedly paid 30 million dollars for Sachs' bicycle component division. They agreed to build a new factory in Schweinfurt and to retain all of Sachs' German employees. This was a significant takeover. It was the end of gear-train research and development in France and Germany. Japan, Italy, and the U.S. took over.

It was too late for changes, so SRAM continued to offer both SRAM and Sachs gear-trains in 1998. SRAM's ESP 9.0SL, 9.0, and 7.0 rear derailleurs were suitable for 8-speed cassettes. The SRAM 5.0 was a 7-speed. In 1998, all of SRAM's rear derailleurs got spring-loaded top pivots, so they became Shimano Deore copies instead of SunTour VGT copies.

SRAM's ESP rear derailleurs used their unique 1:1 actuation ratio. The Sachs Di.R.T. rear derailleurs had shorter cable travel, so they were compatible with either Shimano shifters or SRAM Grip-Shifts.

If a twistgrip shifter wraps the cable around the handlebar and connects to a conventional rear derailleur—which needs about half an inch of cable travel—the clicks will be very close together. The early twistgrip shifters included a mechanism to provide shorter cable travel. SRAM made a virtue of necessity and designed the ESP rear derailleurs for about an inch of cable travel. This was the source of their "1:1 actuation ratio." By 1998, SRAM had sold 34 million Grip-Shifts.

In 1999, SRAM dropped all of the Sachs road-bike gruppos. Sachs had been purchasing combined brake-shift levers from Campagnolo, and SRAM had problems with that relationship. SRAM dropped all references to the Sachs brand. SRAM also dropped the Spectro (née Elan) 12-speed hub gear.

SRAM offered both ESP derailleurs and German Sachs Di.R.T. (Direct Response Technology) derailleurs. The new names were SRAM Plasma, SRAM Quarz, SRAM Neos and SRAM Centera. SRAM also offered hubs, brakes, freewheel, cassettes, and chains. The chains were made by SRAM Portugal. The freewheels and cassettes were made by Maillard.

SRAM offered the complete line of Spectro 3×7 hub-derail-

Above: Fig. 14. 17. The Shorty twist-grip shifter remained SRAM's top mountain bike shifter up to 2003.

Above: Fig. 14.15. The 1996 SRAM ESP 900 rear derailleur was essentially an updated SunTour VGT copy.

Left: Fig. 14.16. The 1999 German-made SRAM Plasma rear derailleurs were quite different from SRAM's own 1999 ESP derailleur models

Right: Fig. 14.18. The 1999 SRAM ESP 9.0 SL was still a SunTour VGT, but now with composite knuckles.

leur hybrid systems, and 3-speed, 5-speed and 7-speed hub gears, which were now called "Spectro by SRAM." Matching Spectro twistgrip shifters were also available.

In 2000, SRAM dropped the last Di.R.T. derailleurs. After that, all SRAM rear derailleurs had 1:1 actuation.

I always liked the SRAM-Sachs Spectro 3×7 hybrid gearing system, which combined a three-speed hub with a 7-speed derailleur. It was not a big seller. In 2001, the 3×7 became the SRAM 27-speed DualDrive.

In 2001, SRAM started promoting components designed for the comfort and city bike markets with 3-speed, 5-speed, and 7-speed hub gears with a wide variety of brakes.

In 2002, SRAM purchased RockShox. RockShox made front and rear suspensions for mountain bikes and was active in mountain bike racing.

In 2004, SRAM purchased Avid, who made cable–operated and hydraulic brakes. They also purchased TruVativ, who made cranksets, seat posts, handlebars, and stems. With a wider range of products and larger engineering staffs, SRAM design teams embarked on a crash program develop a full road bike gear-train, including combined brake-shift levers

The SRAM 2004 range included Impulse trigger shifters and X.O, X.9, and X.7 rear derailleurs for mountain bikes. The company had a good year because of the increased demand caused by Shimano's inability to meet demand.

In 2005, SRAM introduced a new X-9 rear derailleur with the same features as the X-O. Impulse trigger shifters became available for their three top lines of mountain-bike gruppos. They also introduced a top- mounted front derailleur.

Above: Fig. 14.19. The 2001 SRAM-Sachs Dual-Drive combined a 9-speed 11-to-34 cassette with a three-speed hub to get a 578% gear range.

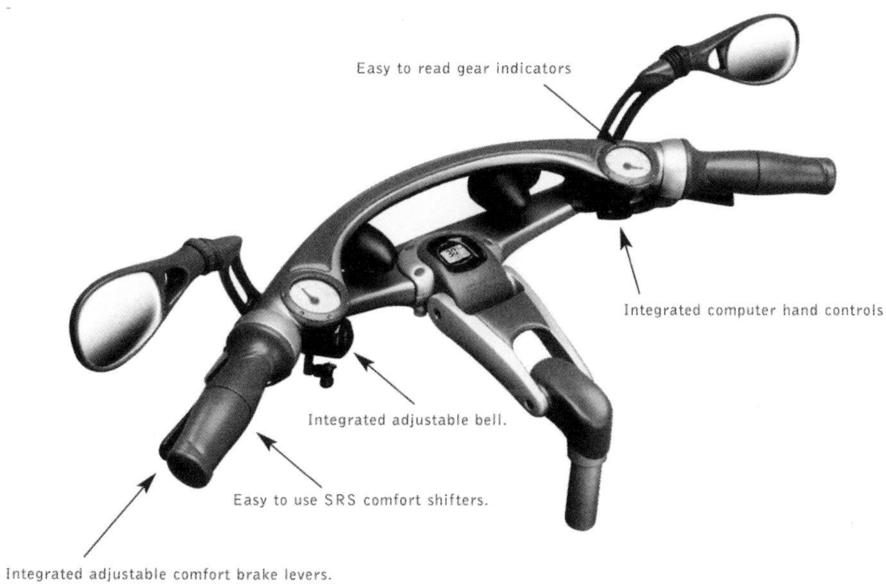


Above: Fig. 14.20. The German-designed 2001 SRAM SmartBar combined a suspension handlebar with a cyclometer, headlights, gear indicators, twist shifters, rear-view mirrors, and a bell.

Above: Fig. 14.21. The 2003 SRAM X-O rear derailleur was the company's top mountain bike derailleur.

Below: Fig. 14.22. With the 2004 X-9 Trigger Shifter, SRAM finally offered an alternative to its GripShift mountain bike shifters.

By 2006, SRAM had become a truly international company, with manufacturing facilities in the U.S., Mexico, Germany, Ireland, France, Portugal, Taiwan, and China. They provided a broad range of gear-trains for mountain bike, city bikes and comfort bikes. SRAM also owned Rock-Shox suspension forks, Avid brakes and Truvativ cranksets.

For three years, SRAM's designers fought their way through the minefield of Campagnolo and Shimano patents. Finally in 2007, they produced their own combined brake-shift levers for road bikes. This allowed SRAM to offer two complete road bike gruppos. SRAM Force was priced between Dura-Ace and Ultegra and SRAM Rival was priced between Ultegra and 105. The buzzword for the shift levers was DoubleTap. A short inward push of the separate shift lever moved the rear derailleur to a smaller sprocket. A long push shifted to a larger sprocket and even longer push shifted 2 or 3 sprockets at a time.

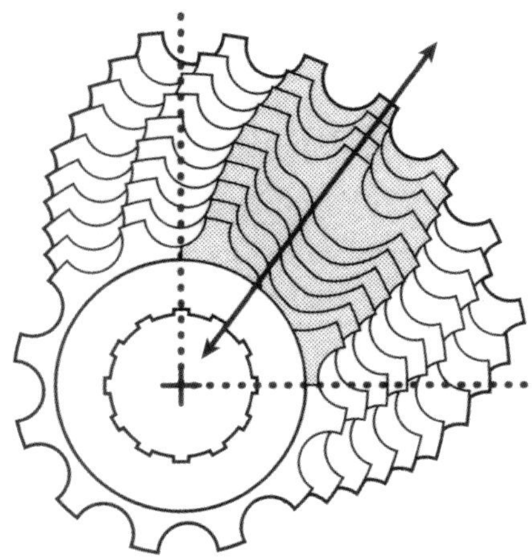

Above and below: Figs. 14.23 and 14.24. SRAM's OpenGlide sprockets and PowerLock chain joint.

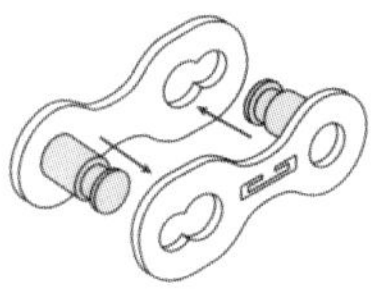

Above, below, and left: Figs. 14.27, 14.28, and 14.29. The 2006 SRAM DualDrive combined a 3-speed hub gear with a 9-sprocket 12–32 cassette and a rear derailleur.

Above: Fig. 14.25. The 2005 SRAM X-5 rear derailleur for mountain bikes.

Right: Fig. 14.26. The 2007 SRAM Force gruppo with Double-Tap brake-shift levers.

SRAM did not copy Shimano's Hyperglide sprocket profiles either. They simply deleted one tooth from the each of the small sprockets to provide a shifting lane. They called this OpenGlide Technology. SRAM also used a simple master link (PowerLock) to join the chain. After 19 years, SRAM could finally compete with Shimano across the board for road bikes, mountain bikes, and comfort/city bikes.

SRAM's comfort bike gruppos were called i-Series. SRAM had inherited a first-rate group of hub gear design talent from Sachs. In 2008 the German engineers produced the first 9-speed hub gear, the i-Motion 9, with a range of 340%. It was available with a coaster brake, a disk brake, or the i-Brake (a sealed drum brake).

The 2008 Red road gruppo was a high point for SRAM. Twenty years after their first Grip-Shift, SRAM's three road gruppos (Red, Force, and Rival) were providing serious competition to Shimano's Dura-Ace, Ultegra, and 105, and Campagnolo's Record, Chorus, and Veloce. So far none of SRAM's road gruppos offer a triple crankset for the tourists, and perhaps the wide-range 20-speed mountain bike XX system, introduced in 2010, may offer all the flexibility tourists need.

Red was designed for cost-is-no-object professional road racers. The main upgrade from Force and Rival was titanium and carbon fiber for light weight. It provides serious competition for Campagnolo Record and Shimano Dura-Ace.

The 2009 Tour was a success for SRAM-equipped bicycles, after a decade-long streak of Shimano successes. For the 2010 model year, SRAM-equipped bicycles were

Left: Fig. 14.31. The 2008 SRAM Red front derailleur had a titanium cage to get the weight down to 58 grams.

Below: Fig. 14.32. The 2008 Red rear derailleur had a titanium parallelogram spring and ceramic bearings. It weighed just 153 grams.

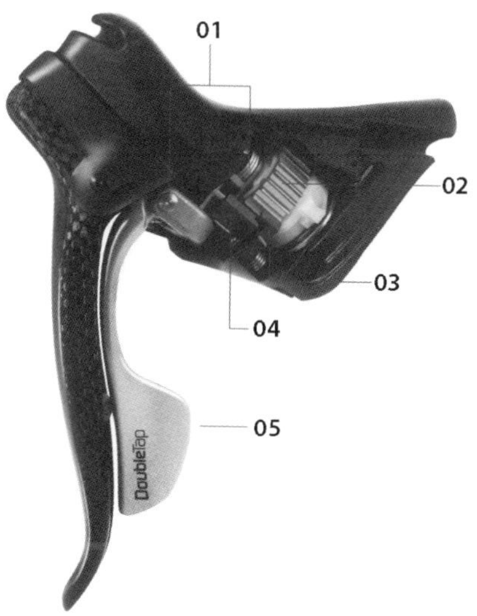


Above: Fig. 14.30. The 2007 SRAM DoubleTap lever uses fewer parts than the competition. 01 is the actuator; 02 is the ratchet wheel; 03 is the holding pawl; 04 is the transport pawl; 05 is the dedicated shift lever.

Above: Fig. 14.33. The 2008 SRAM i-Motion 9 hub gear.

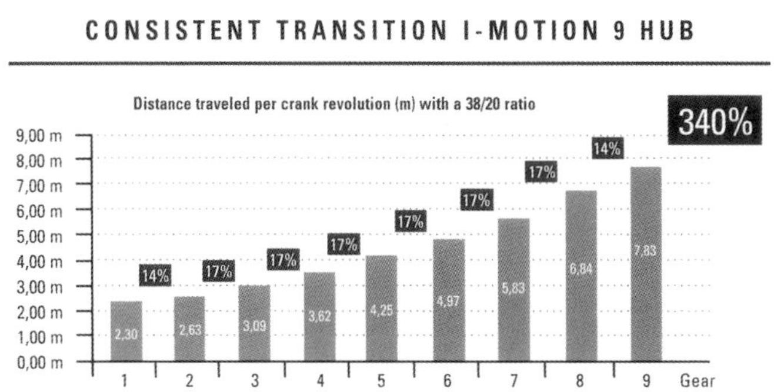


Right: Fig. 14.34. The gear ratios of the i-Motion 9 hub gear are equally distributed over its 1-to-3.4 range.

available in all categories from many bicycle manufacturers.

For 2012, SRAM introduced something called WiFLi, a wide-range 20-speed setup using an 11–32-tooth cassette and small chainrings. Does not seem such a good idea, on account of the resulting higher chain force.

## Campagnolo

Campagnolo waned as SRAM waxed. Campagnolo went downhill from 1996 to 2008. In 1995, they benefitted from the international exchange rates that resulted in cheaper lire and dearer yen. They had about half of the component market for expensive road bikes. "Road" was the definitive word because road bikes were in decline and mountain bikes, city bikes, and all of the other flat-handlebar bikes were the growth part of the market.

Campagnolo's tentative efforts to share the flat handlebar market failed. However, Campagnolo became a major supplier of rims and complete wheels.

Campagnolo's market was road racers and expensive road bikes. Over the period, Campagnolo used carbon-fiber to compete with Shimano in the weight race. By 2007, they were suffering from an expensive Euro.

In 1995, the Record gruppo got a low-profile crankset and a cartridge bottom bracket. Wide-range touring derailleurs were available with Athena, Veloce, and Mirage. The Stratos gruppo was replaced by two low-priced gruppos, Mirage and Avanti, which were Veloce with an unpolished aluminum finish.

The Record OR mountain-bike gruppo was unchanged in 1995, except for steel compact chainrings.

In 1996, the Record OR mountain-bike gruppo was dropped. With a sigh of relief, Campagnolo abandoned mountain bikes. So few Campagnolo mountain bike components were sold from 1989 to 1996 that they are now collector's items.

In 1996, the Record gruppo reduced weight with titanium bits and pieces in the shift levers, rear derailleur, hubs, and sprockets.

Campagnolo continued to make long-cage rear derailleurs and triple cranksets for the tourists. They realized that gray powder-coat paint did not help their image, and the lower-priced gruppos went back to a polished and anodized finish.

Campagnolo continued their weight-loss program. From 1996 to 1999, they cut the weight of all of their gruppos. This often involved use of titanium or reinforced carbon-fiber. Between 1995 and 1997, Campagnolo cut

Above: Fig. 14.36. The 1995 Record rear derailleur showed the traditional Campagnolo elegance.

Right: Fig. 14.35. SRAM's 2008 Red cassette showed the meaning of cost-no-object. The seven larger sprockets were CNC-machined from a hollow forged steel billet.

Left: Fig. 14.37. Despite Campagnolo's best efforts to distinguish their Ergopower brake-shift levers from Shimano's STI, they were eventually judged to infringe on Shimano's patent, forcing them into a licensing agreement with Shimano.

the weight of the Record gruppo by 400 g (11 oz.).

In 1997, Record and Chorus got 9 speeds, and the sprockets, chainwheels, and rear derailleurs were revised for the narrower 9-speed chains.

In 1998, Athena and Veloce got 9 speeds, and light-action dual-pivot brakes became standard on all of the gruppos. In 1999, Ergopower shift levers were made from carbon fiber. In 1999, Mirage also got 9 speeds.

Campagnolo moved into the new century. After unsuccessful forays into mountain bikes and the lower-priced market, they returned to their roots—top-quality components for high-end racing and sport touring bicycles. They progressed by evolution, introducing each improvement with the top-of-the-line Record gruppo, and letting it trickle down to Chorus, Athena, Veloce, and Mirage. Everything was interchangeable between Campagnolo's gruppos. The main difference between gruppos was the name on the derailleur and the amount of polishing. The company became a major supplier of high-end wheels and rims.

In 2000, Record and Chorus got 10 sprockets. Campagnolo was able to squeeze the extra sprocket into the same cassette width by narrowing the spacers from 2.8 mm to 2.4 mm. A new very narrow chain was provided. There were two new rear derailleurs. The racing model had a 26-tooth maximum sprocket and a 27-tooth total capacity. The touring triple model had a 28-tooth maximum sprocket and a 37-tooth total capacity.

I admired Campagnolo's achievement in providing 10 sprockets, but it was a high price to pay for an insignificant advantage. Clock-making tolerances were required, and everything from before 2000 became obsolete. 10 sprockets (often called 10 speeds) was a poor idea. Perhaps if *Bicycling* magazine had been doing engineering tests in 2000, they could have established that 10 speeds shifted worse than 9 speeds, and that the narrow chains were weaker and wore out faster. Nothing like that happened, and 10 speeds slowly took over the high-priced

Above: Fig. 14.38. In 2000, Campagnolo was first off the block with 10-sprocket cassettes, a dubious improvement in the author's expert opinion (not to mention the 11-speed version introduced nine years later).

Right: Fig. 14.39. The 2000 Daytona gruppo slotted between Chorus and Veloce, and has since been renamed Centaur.

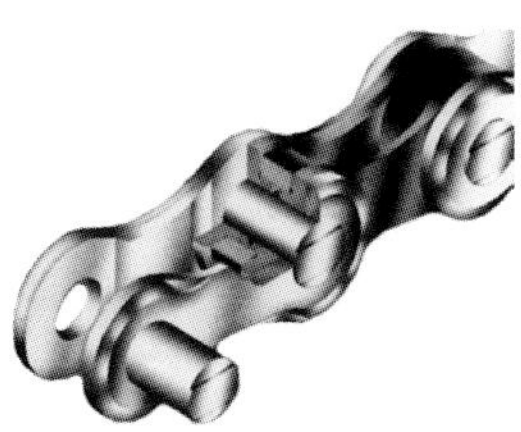

Below: Fig. 14.40. Campagnolo made their own extra-narrow bushingless chains for the 9-speed gruppos in 1997. The chain plates had a nickel-Teflon finish.

market, and eventually trickled down to the Veloce level.

Campagnolo added the Daytona gruppo in 2000. It included a triple crankset with a matching triple front derailleur and long- and short-cage rear derailleurs. The 30-tooth minimum inner chainring and the 28-tooth maximum sprocket were suitable for sports-touring, but not for long distance loaded touring. Also in 2000, they added a carbon-fiber seatpost and ErgoBrain, a cycle computer with cadence indication.

In 2001, Campagnolo added a 29-tooth maximum sprocket. Daytona got 10 sprockets. There were four different Record derailleurs, 10-speed for up to 26-tooth, 10-speed for up to 29-tooth, 9-speed for up to 26-tooth, and 9-speed for up to 29-tooth and a triple crankset.

In 2002, the Record and Chorus 10-speed gruppos got a triple crankset with a long-cage rear derailleur and a triple front derailleur. Centaur replaced Daytona because of a trademark problem.

In 2003, the Record crankset got carbon-fiber cranks and spider (the star-shaped attachment that holds the chainrings). A new low-priced gruppo, Xenon, was added. Campagnolo had six full gruppos: Record, Chorus, Centaur, Veloce, Mirage, and Xenon. The top four gruppos offered 10-speed freewheels. All six gruppos offered a choice of double and triple cranksets.

In 2004, Chorus got carbon-fiber cranks, and the Record rear derailleur got carbon-fiber cage

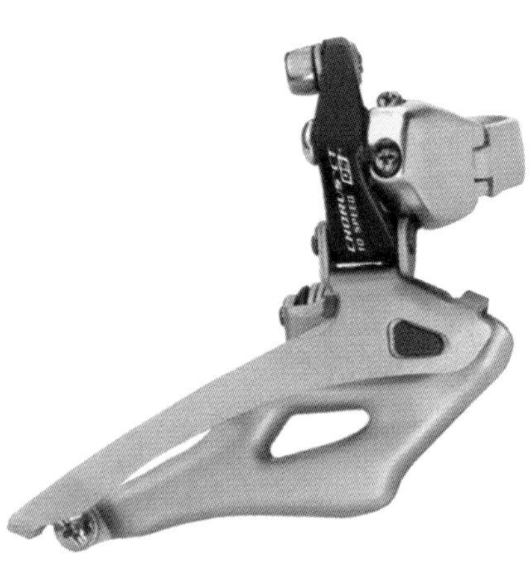

Above: Fig. 14.41. Campagnolo's 2006 front derailleur for use with flat-bar controls.

Above: Fig. 14.43. The three 2004 Record rear derailleurs have slightly different cage lengths. The carbon-fiber cage plates use crossed fibers for maximum stiffness at a minimum weight.

Below: Fig. 14.42. The 2001 Record rear derailleur had carbon-fiber parallelogram arms and a revised cage shape.

Below: Fig. 14.44. For 2005, Campagnolo introduced new flat-handlebar controls with carbon-fiber shift levers. For a while, it looked like Campagnolo might be trying to reenter the mountain bike market.

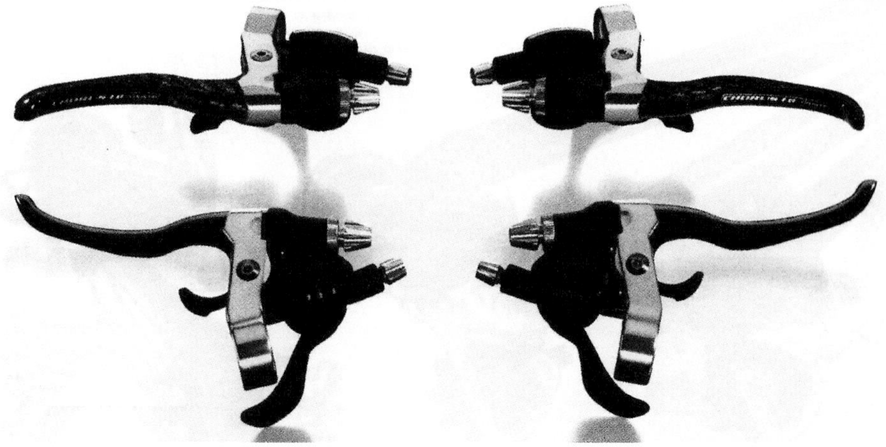

plates. They now offered 24 different rear derailleurs to cover the six gruppos, three capacities, and 9- and 10-speed freewheels.

For 2005, Campagnolo added flat-bar shifters for all gruppos except Record. The front shifter required a special front derailleur. There was a CT compact drivetrain that used 48/34, 50/34, or 50/36 chainrings. This also required another new front derailleur. Each gruppo had six different front derailleurs.

There were no significant changes to Campagnolo's drivetrain in 2006 but new cantilever brakes were available.

For 2007, Campagnolo introduced the Ultra-Torque crankset and bottom bracket. The hollow spindle was split in the middle with each half permanently fixed to cranks. The spindles are connected by a Hirth joint. The design allowed a lighter spindle and wider spacing of the outboard bearings. The Ergopower combined brake/ shift levers were simplified. The left hand lever was redesigned for a lighter pull. Xenon, the lowest–price gruppo, got 10 speeds. As an interesting aside, Campagnolo made 42 different front derailleurs to cover the combinations of 6 gruppos plus the triple gruppo, standard and compact chainrings, braze-on and clip-on mounting, and two different down tube diameters. This also showed the problems in shifting ultra-narrow chains between close-spaced chainwheels.

2008 was Campagnolo's 75th Anniversary, but no new product introductions. For 2009, Campagnolo tried to up the ante in the sprocket count race—with an 11-speed cassette, matched by an even narrower (5.5 mm) chain and very skinny chainwheels. It did not turn out to be a winner, and nobody else followed them

Above and below: Figs. 14.45 and 14.46. The 2007 Campagnolo Ultra-Torque bottom bracket relied on a Hirth joint to connect the two parts of the hollow spindle.

Above and right: Figs. 14.47 and 14.48. For time-trial and triathlon use, Campagnlo still offers bar-end shifters and brake-only levers (as do Shimano and SRAM).

Above and below: Figs. 14.49 and 14.50. Campagnolo's 2009 Record 11-speed rear derailleur and cassette.

down this path. By 2011, most professional teams had abandoned Campagnolo in favor of Shimano's digital shifting.

The rather industrial looking rear derailleur to go with it got ceramic ball bearings—also something of dubious benefit, especially on such a minimal-load bearing application as derailleur pulleys (see Chapter 15 for more information about this particular issue).

## U.S. Derailleurs & More

From 1995 to 1997, it was hip to use something other than Shimano, even if it cost more, and especially if it was made in America. This allowed small U.S. companies with computer numerical control (CNC) machines to make components and sell them at a premium.

A small cottage industry developed to supply the anti-Shimano market. The first products were round items like hubs, bottom brackets, and headsets, which could be churned out by automatic screw machines. Brakes and cranksets came next. The CNC products were usually lighter—often because they didn't include the safety factor built into Shimano's designs.

In 1995, four small U.S. companies sold CNC-machined aluminum rear derailleurs in anodized colors. The Paul Powerglide was a SunTour VGT copy in red or blue anodized aluminum. It used brass bushings and titanium bolts and was completely rebuildable. With HyperGlide sprockets and modern chains, it had an easier task than the 1970s VGT. Paul sold about

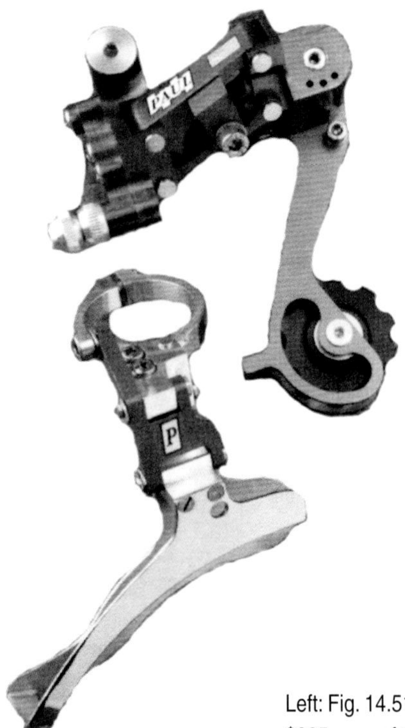

Above: Fig. 14.52. The folksy named Joe's derailleur was another expensive Deore copy.

Left: Fig. 14.51. The 1995 Paul Powerglide rear derailleur was a $325 copy of the SunTour VGT.

Above: Fig. 14.53. The White Industires LMDS derailleur.

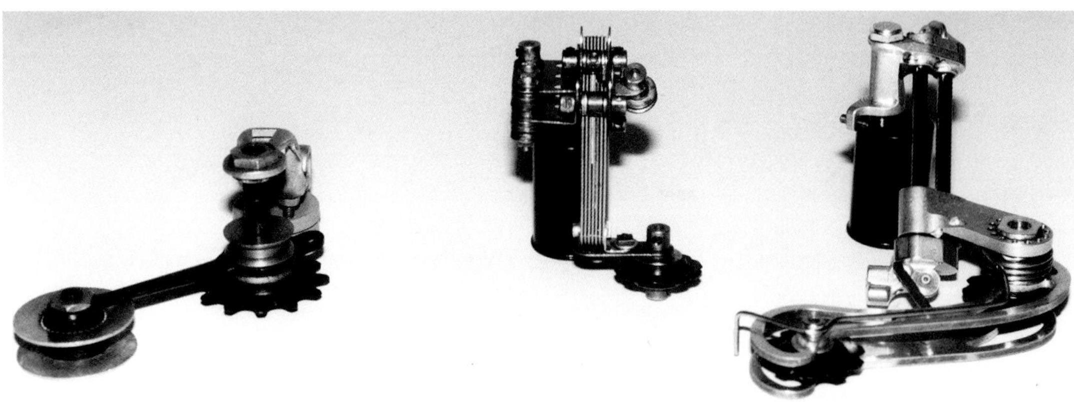

Left: Fig. 14.54. Evolution of a derailleur design, 1930–1994. Left: The 1930 Cyclo Tourist slides in and out horizontally. Middle: Meccano prototype. Right: The LMDS rear derailleur jockey pulley slides on two rails sloped to maintain an optimum chain gap.

4,000 derailleurs a year in 1995 and 1996.

The Precision Billet Gorilla rear derailleur copied the Deore XT's dimensions. It had stainless steel pivots and hardware and sealed ball-bearing pulleys. They sold 4,000 derailleurs in 1994 and 10,000 in 1995.

The Joe's rear derailleur was another Deore copy that lasted for two years. It was made in Denver, Colorado, by Joe Glader's company, Prototype Machining. When Shimano raised the ante in 1997 with XTR, Joe's abandoned the bicycle component market. It turned out to have been a blip in the market. Shimano's 1997 XTR derailleur was lighter, and it shifted better. By 1997, the anti-Shimano buyer paid a premium for a heavier, poorer shifting, derailleur.

White Industries' Linear Motion derailleur System (LMDS) was technically the most interesting, and I had a bit to do with it. Doug White's factory in Ignacio, California, made sealed bearing hubs, bottom brackets, and cranksets. In 1994, I agreed to help Doug design a rear derailleur, provided that it would not just be another Shimano copy. I made two prototypes with my 1930s Meccano set. One was an improved Huret Allvit, and the other was an improved Cyclo Touriste.

Doug liked the improved Cyclo Touriste. Two years, and six prototypes later, it hit the market. The rear derailleur jockey pulley carrier slid on two rails on linear ball bearings. The rails were sloped to maintain an optimum 1½-inch (38 mm) chain gap in all gears. It used two cables, just like its Cyclo grandfather. The cable slack provided self-centering. In spite of splendid shifting performance, LMDS derailleurs never broke into the mass-market.

## Browning

The Browning swinging-gate electronic front shifter first appeared in 1982. In 1990, it became the SunTour BEAST, and it failed along with SunTour. In 1997, it became the Browning Red Shift, which had both front and rear shifters, providing 12-

Left: Fig. 14.55. Author Frank Berto's Columbine loaded-touring bike with White Industries LMDS derailleurs. The twist shifters were mounted on stubs on the stem. Note the S&S couplings to allow the bike to fit in a suitcase.

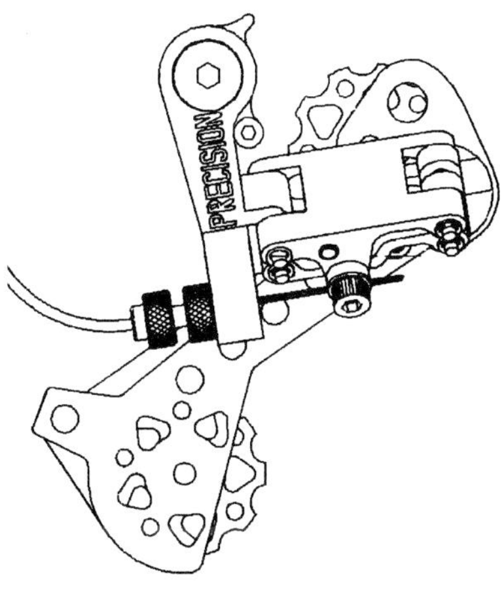


Above: Fig.14.56. The 1994 Precision Billet Gorilla rear derailleur was a $220 Deore XT copy.

Right: Fig. 14.57. The 1997 Browning Red Shift combined a 3-speed swinging-gate chainwheel with a 4-speed swinging-gate freewheel.

speed gearing. This included electronic fully automatic or manual shifting controls. It worked just as advertised, but the gear-train market had changed dramatically from the 1980s.

A 12-speed Browning Red Shift could not be sold for twice as much as a 24-speed Shimano derailleur gear-train with all of the HyperGlide features. No OEM maker would specify Browning. Browning survived as a family business on Bainbridge Island in Washington State. In 2003, they were developing automatic transmissions for electric bikes.

### Mavic

Mavic dropped out of the derailleur business in 1996 to concentrate on rims and hubs. They could not resist the lure of the derailleur, and in 1999 they came back with the Mavic Mektronic wireless electronic rear derailleur and cyclometer. This disappeared from the catalog in 2002. Mavic was the last purely French company to make derailleurs.

### Falbrook/NuVinci

Falbrook Technologies introduced the NuVinci Continuously Variable Planetary (CVT) hub gear in 2007. This was the first time that a completely stepless wide-range transmission has gone beyond the prototype stage. Three different companies specified it for their bicycles. The first generation was expensive, heavy, bulky, and has about the same range as a 7-speed hub gear.

My editor has been using an early production version for around-town cycling in San Francisco and reports that it seems like a winner—but not for a hardcore cyclist who wants to be in control of his gear ratios and enjoys the challenge of shifting. It's easy enough to get used to once you get the hang of if. If price and weight can be further reduced—the current version weighs in at 9 lbs. (4 kg)—this device may have a future in the realm of city- and comfort-bikes. Since this category was the industry's strongest growth sector in 2008–2009, the timing seems to be right.

## Others Manufacturers

The bicycle remained a significant element of the European transportation network. In 1992, 27% of all journeys were made by bicycle in Holland, 18% in Denmark, and 11% in Germany. The European Union countries had a significant tariff on bicycle imports from China, so their domestic factories still made many of the bicycles sold in Europe.

In addition, there are still some smaller Japanese companies, besides Shimano, active in the component field. Finally there are a number of component makers in Taiwan, China, and India.

Above: Fig. 14.58. The1999 Mavic Mektronic electronically controlled rear derailleur had a capacity of 30 teeth and could handle a maximum 26-tooth sprocket. The electronics did not hold up well under wet conditions.

Left and Above: Figs. 14.59 and 14.60. The Falbrook NuVinci infinitely variable hub gear used balls to transmit torque between an input cone and an output cone.

### Pre-SRAM Sachs-Huret

Sachs survived in the bicycle component business until 1998 because they made excellent hub gears and coaster brakes, which were popular in Europe, especially in Germany and neighboring countries. Sachs also made first-class derailleur bicycle geartrain components, but they never managed to break into the U.S. market.

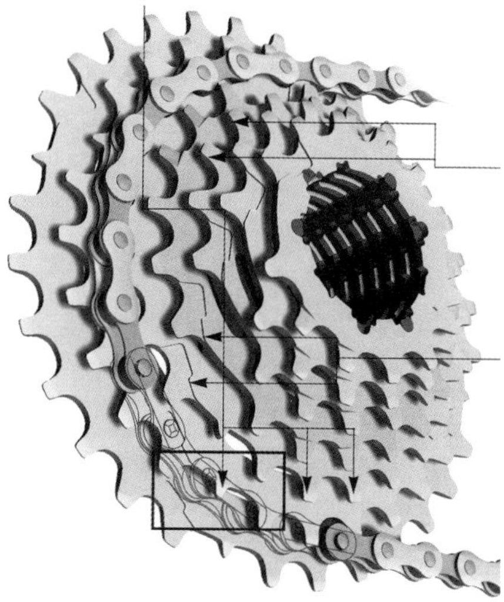

Above: Fig. 14.61. The Sachs 1995 PowerGlide cassette sprockets had been copied from Shimano's HyperGlide design.

The 1994 Sachs New Success rear derailleur was a quality product using the basic design of the 1984 Shimano Dura-Ace. It could handle a 28-tooth maximum sprocket, and the total capacity was 28 teeth. Sachs' major effort to crack the U.S. market in 1994 was a failure.

In 1995, they cut back to four road/trekking gruppos and three mountain-bike gruppos, and competed where they did not have to go head-to-head with Shimano. The 1995 Sachs PowerGlide cassette sprockets had shifting lanes and cut-down teeth to improve shifting ease. Something similar was done on the chainwheels.

In 1995, Sachs introduced the PowerGrip Pro, which was a top-quality twistgrip shifter to operate either Shimano or Sachs rear derailleurs. Sachs ended their long relationship with Sedis and started making chains under the Sachs brand name.

In 1996, Quarz, Neos, Centera, 3000, and Traxx were new names for the Sachs mountain bike gruppos. Bandix was the kid's gruppo. The New Success road rear derailleur got a new cage. The 700 road gruppo was dropped. Sachs front derailleurs got Inverse Action, with the cable pulling the cage onto the smaller chainwheels, and a spring pushing it onto the larger chainwheels. This good idea had first been used by SunTour in the 1970s.

Sachs came up with a Speedtronic electronic shift control and display for the Super 7 hub gear. Sachs emphasized Shimano compatibility.

Above: Fig. 14.62. The 1994 Sachs New Success rear derailleur was based on the 1984 Dura-Ace.

Left: Fig. 14.63. The 1996 Sachs 3×7 hybrid hub was a combination of a 3-speed geared hub and a 7-sprocket cassette. The Sachs 3-speed hub was rugged enough for mountain bike service.

Right: Fig. 14.64. The 1998 Sachs Plasma rear derailleur was the last Sachs product to be introduced prior to the SRAM takeover.

Sachs' 1998 buzzword was "Di.R.T." (Direct Response Technology). Di.R.T. derailleurs came in Plasma, Quarz, Neos, Centera, and Traxx mountain-bike gruppos. The three road-bike gruppos, New Success, 5000, and 3000 were unchanged. The Di.R.T. Plasma rear derailleur had front cable entry to reduce shifting force, especially when the derailleurs were covered with mud. The bodies were fiber-reinforced thermoplastic. There was also an innovative Di.R.T. front derailleur that used a coil spring.

Sachs made the 12-speed Elan hub gear from 1997 to 1999. It had a 339% range with even steps between gears. On the down side, with its shell made of steel, it weighed 7 lbs. (3 kg), and it cost $475.

For the 1998 model year, Sachs made a final run at the U.S. mountain bike market. Mannesman Steel, the corporate owner, reportedly said that they would either get a significant market share, or they would sell the Sachs bicycle component business. OEM sales evaporated with rumors of a sale, as had happened with SunTour a decade before.

There were two potential buyers. One was a group of Sachs executives, led by John Neugent, head of Sachs-USA. SRAM was the other, and reportedly they paid 30 million dollars for Sachs.

French Huret was absorbed by German Sachs, and German Sachs was absorbed by American SRAM. When Mavic gave up in 2001, a century of French derailleur innovation and manufacture came to an end.

Above: Fig. 14.67. The 1996 Sugino VIC rear derailleur was a novel design that came a decade too late.

### Rohloff

Rohloff is a small family-owned company. In the early 1990s, Rohloff made chains for Campagnolo. In 1997, Campagnolo decided to start making their own chains, and Bernhard Rohloff completed the design of the 14-speed Rohloff Speedhub. It has a range of 526%, equivalent to a low gear of 20 in. and a high gear of 105 in. There are uniform 13.6% steps between gears.

The Speedhub uses three sets of planetary gears: two sets provide seven speeds, and a third set provides half-steps between those seven speeds. All of the planetaries run on needle bearings. Cheap it isn't: by 2009 the retail price was around $1,500, due in

Above: Fig.14.65. The 1997 Sachs Elan 12-speed hub, which was abandoned by SRAM after its takeover of Sachs.

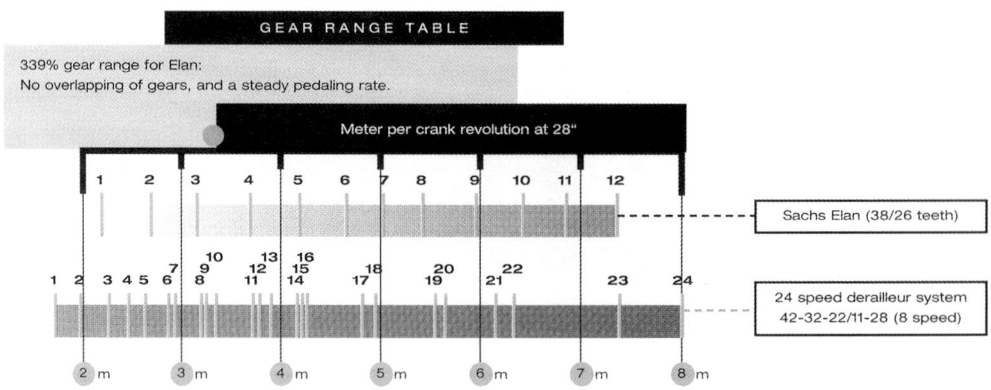


Right: Fig.14.66. Comparison of the Elan's 339% range with even steps between gears and a typical 24-speed derailleur gear-train with uneven steps, overlaps, and a shift sequence that precludes using the intermediate gears.

part to the declining dollar exchange rate.

The Speedhub set a new quality benchmark for multi-speed hub gears. Rohloff claims efficiencies between 96% and 98%, depending on the gear selected. Chester Kyle and I, in our hub gear efficiency test (see Chapter 15), did not do quite that well, but still very good by hub gear standards. The weight was just under 3¾ lbs. (1.7 kg).

### Sugino

Sugino Cycle Industries was a major Japanese crankset maker in the 1970s and 80s. Sugino supplied cranksets for Maeda SunTour's gruppos, and they fell on hard times when SunTour failed. Sugino made quality cranksets for American makers such as Ritchey and White Industries. Sugino lost money in the late 1990s, and they shut down their Chinese factory and reorganized. In 2008, Sugino was still offering a full line of conventional cranksets.

The 1996 Sugino VIC rear derailleur was an interesting design that combined features of the old Huret Allvit with SunTour's slant parallelogram design. It came with either thumb shifters or twistgrip shifters. The rear derailleur was designed to index-shift on 6-, 7-, and 8-speed freewheels. It shifted very nicely, but no major bike manufacturer would specify such an unknown brand of derailleur.

### SR-SunTour

In 1995, five Taiwanese companies made bicycle gear-train components for lower-priced derailleur bicycles. This market wanted less expensive components that looked like Shimano. There was little design innovation. SR-SunTour had some designers from the old Japanese Maeda SunTour, and they were more innovative than the others.

Daisuke Kobayashi and Hideo Hashizume, owners of Sakae Ringyo (SR), sold their company to Mori Industries in 1989. At the end of 1994, when Mori Industries decided to get out of the bicycle business, the previous owners arranged a management buyout of the SR factory in Taiwan and the SunTour name.

At this point, the old Japanese Maeda-SunTour disappeared and the Taiwanese company started using the SunTour name for better brand recognition. They maintained offices in Japan for Re-

Below: Fig. 14.68. The 14-speed Rohloff Speedhub, introduced in 1998, is the most capable and efficient hub gear to date.

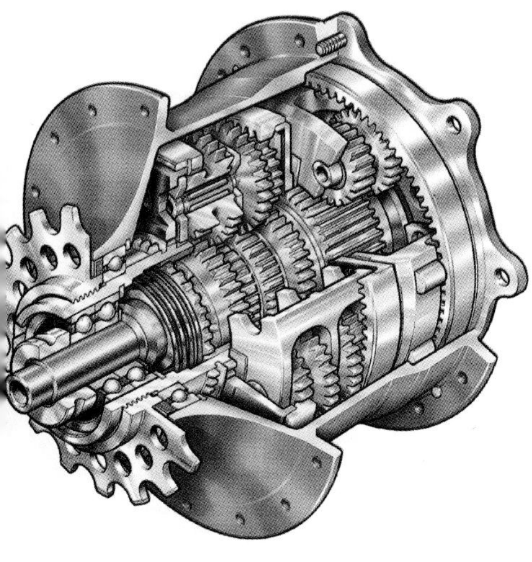

Above: Fig. 14.69. The 1997 SR-SunTour SX05 rear derailleur was a modified version of the S1 design.

Top right: Fig. 14.70. SR-SunTour's 1995 Ergotwist shifter competed with SRAM's Grip-Shift.

Bottom right: Fig. 14.71. The 2003 SR-SunTour front derailleur had a double inner cage plate to improve shifting to the outer chainring.

search & Design and Japanese sales. They called the products SR SunTour to distinguish them from the old Maeda SunTour.

SR SunTour maintained an engineering and design staff, which developed innovative gear-train products.

Above: Fig. 14.72. The 2007 SR-SunTour V-Boxx had 9 speeds and weighed 12 pounds (5.5 kg), which was fine with the downhill crowd.

Below: Fig. 14.73. The SR-SunTour Duro rear derailleur, introduced in 2002, had a swing arm (like the old Huret Duopar) and a slant parallelogram. It could handle an 11-34 freewheel and a total capacity of 45 teeth. It was compatible with both Shimano and SRAM shifters.

SR SunTour closed its U.S. office in early 1995, by which time their share of the OEM market in the U.S. was insignificant. SR SunTour continued to sell components in the rest of the world, where the markets were less demanding.

SR SunTour owned the name SunTour, but they did not own SunTour's patents, intellectual property, or the rights to any of the pre-1995 designs. Mori sold the rest of SunTour's facilities separately. Various people tried to buy the SunTour tooling for the Superbe Pro components, but Mori sold the tooling for scrap.

In 1995, the SR SunTour gruppos were called XR-150 and XR-50. They were plug-and-play compatible with Shimano's inexpensive Alivio and Acera gruppos. SR SunTour called their easy-shifting sprockets and chainwheels PowerFlo. They also made 6- and 7-speed screw-on freewheels.

SR SunTour reopened a U.S. sales office in early 1996. By 1997, the company had 250 employees. They made inexpensive gear-train gruppos consisting of cranksets, screw-on and cassette freewheels, front and rear derailleurs, and twistgrip shifters. They also made suspension forks. Their best gruppo was called Giga. Sales grew from 12 million dollars in 1994 to 27 million dollars in 1997.

In 1999, SR SunTour reintroduced the Superbe crankset as the first component of a high-quality gruppo.

By 2004, SR SunTour's main products were suspension forks. However, they also made a full line of twistgrip shifters, cranksets, cassette sprockets, and front and rear derailleurs.

In 2007, SR SunTour introduced the V-Boxx, a 9-speed bottom bracket gear for the downhill racers. It had all of the century-old problems. It was heavy, expensive, and it required a made-to-order bicycle frame. However, it had a 610% gear range and with 22-tooth sprockets front and rear and a tight chain, it got rid of derailleurs and chain suck.

Right and below: Figs. 14.74 and 14.75. Introduced in 2005, the SunRace Sturmey-Archer S80 8-speed hub gear finally made Sturmey-Archer competitive again.

### SunRace Sturmey-Archer

The SunRace Roots Enterprise Company made a complete line of gear-train components: derailleurs, shifters, cranksets, hubs, cassettes, freewheels, and chains. They supplied the low-priced market with Shimano-compatible products.

SunRace bought Sturmey-Archer in December 2000 for about 3 million dollars, and moved all of the machinery to Taiwan. They made a full line of 3-speed, 5-speed and 8-speed hub gears with all of the optional features.

In 2005, they introduced the S80 8-speed hub gear with a 305% range. There were five 13% steps in the middle and 28% steps to the Low and High gears. It was available with drum brakes, roller brakes, or disk brakes.

Their 2008 top-of-the-line derailleur gear-train gruppo was called DRiVEn. It was 10-speed and it included combined brake shift levers.

### Long Yih

In 2006, the Long Yih Industry Company made inexpensive derailleurs, shifters, cassettes, and freewheels. They called their indexed components Epoch and their non-indexed components Motion.

### Falcon

Falcon Cycle Parts Co. Ltd. had factories in Taiwan and China. They were the world's largest maker of coaster brakes. The 1999 Falcon AD-30 rear derailleur shifted gears automatically. The jockey pulley drove a centrifugal governor that shifted gears to maintain a steady pedaling cadence. In 2008, they made a complete line of inexpensive gear-train components—derailleurs, shifters, cranksets, hubs, cassettes, freewheels, and chains. They also made brakes and brake levers. Their indexed derailleurs were called Novel.

### Other Countries

Over the past 60 years, bicycle production for the mass-market moved from the U.S., Britain, and Europe to Japan, then to Taiwan, and then to China. The driving force in each case was lower production costs due to lower labor costs. As China moved up in the price range of bicycle exports, there was a demand for better quality Chinese components. In response, most of the major component companies opened joint-venture factories in China.

The recent Interbike shows had an area for Chinese, Indian, and Indonesian component makers, hoping to make contacts with low-end bicycle makers. The booths are of-

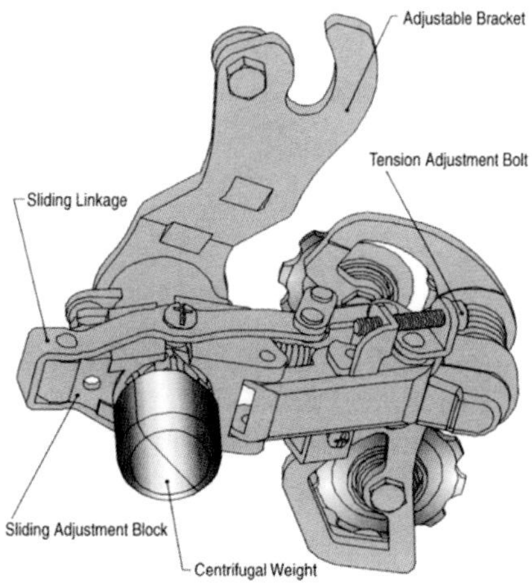

Above: Fig. 14.76. The 1999 Falcon automatically shifting AD-30 rear derailleur maintained a steady pedaling cadence.

Left: Fig. 14.77. The 2008 SunRace DRiVEn rear derailleur for 9-speed operation.

Right: Fig. 14.78. Tiso, an Italian manufacturer, shows that hope still springs eternal. Introduced in 2004, their $300 rear derailleur came in two models: Sereo 130 was for Campagnolo geartrains and Altore 366 for Shimano.

ten empty, because Interbike caters to bike shops, and few U.S. bike shop owners are in the market for 25-cent derailleurs. I usually stop to look at what is being offered. To date, the third-world component makers have been copying the lowest-priced Taiwanese makers, who copied Shimano from the 1980s.

## Status of the Bicycle Gearing Market

Four major bicycle makers—Trek, Giant, Specialized, and Cannondale—supply the bulk of the U.S. market for quality bicycles. Quality bicycles are sold by independent bicycle stores.

I took the 2008 catalogs of these four major bicycle suppliers and counted which road bike models and which mountain bike models use which makes of derailleurs. I did this to get a feel for how the three major suppliers of derailleurs, Shimano, SRAM , and Campagnolo, are faring in the 2008 market.

You can properly complain that I should have counted numbers of bicycles and not numbers of bicycle models. However, the bicycle companies are very careful not to release any data that could reveal market share. I did a similar count twenty years ago. That count had the same problem but it showed quite clearly that Shimano dominated the 1988 market and that SunTour and everyone else had an insignificant share.

This time, I counted expensive bicycles. The mountain bikes used either Shimano XTR, Deore XT, or Deore LX, or SRAM X-O, X-9, or X-7. The road bikes used either Shimano Dura-Ace, Ultegra, or 105, or SRAM Red, Force, or Rival. Just one road bicycle, the Cannondale SuperSix Ultimate, came equipped with Campagnolo Record.

From this count, I draw the following conclusions concerning the status of the major derailleur manufacturers:

- Shimano was still the dominant supplier. They equipped twice as many expensive bicycles as SRAM. They provided an even large percentage of front derailleurs and of cassettes.
- SRAM provided significant competition to Shimano. Their derailleurs and shifters equipped about half as many models as Shimano's. Although I say derailleurs, it was mostly just *rear* derailleurs. No one specified SRAM's mountain bike front derailleurs. In front, everyone used Shimano's derailleurs.
- Campagnolo had only minimal presence in the American road bike market, and Campagnolo doesn't make mountain bike components.
- Finally, a significant and growing number of city and comfort bicycles used multi-speed hub gears—not just inexpensive 3-speed hubs. The industry's fastest growing segment was that of flat-handlebar bikes with hub gears.

Right: Fig. 14.79. The 2008 Sampson Stratics rear derailleur is not a copy of anybody else's, though it is Shimano 10-speed compatible. Sampson also offers their own complete dual-control gruppos at price points well below those of the more established brands.

## More New Stuff

The following are additional observations from the most recent Interbike shows:

Trek and Cannondale continue to boycott the show, to spend their money on private dealer shows.

Since 2009, there ha been an increase of electric bikes (E-bikes. Most major bicycle makers now include electric bikes in their catalogs. Perhaps E-bikes are a trend, but most small independent bike shops can't afford to stock and promote them, and in consequence most are sold at "big-box" stores.

The trend is to diversity, with three major categories; Road, Mountain, and City.

Drop-handlebar road bikes had a small resurgence in 2009, which turned down in 2010. Today's road bikes, especially the carbon-fiber models, are ugly, expensive, harsh-riding, and suitable only for racing. Fixed –gear bikes found a tiny niche market.

Mountain bikes are still a major part of the market. There use is defined by the travel of their suspension systems. The public now realizes that a full-suspension mountain bike with knobby tires isn't particularly pleasant for commuting or shopping. The new fad is for "29-inchers". This is the 700C road bike rim diameter with a wide knobby tire. Many manufacturers now use them on their larger mountain bikes.

City bikes made up the majority of the market in 2010 and beyond. This broad category is defined by low prices, flat handlebars and medium-width tires. These bike satisfy the needs for most short-distance commuters, shoppers, and recreational riders.

## Cycling Magazines

*Cycling Science* ceased publication in 1996. *Velo News* is the magazine for bike racers, but they also publish technical articles, mostly by Lennard Zinn.

*Adventure Cycling* is the house organ of Adventure Cycling (the organization that grew out of Bikecentennial). Their publication caters to bicycle touring enthusiasts, and they have until recently published regular technical articles by John Schubert (one-time editor of *Bicycling*, *Bike Tech*, and *Bicycle Guide*) that are worth reading.

The trade publication *Bicycle Retailer and Industry News* caters to the bike trade but contains very little technical information beyond manufacturers' press releases.

The field of magazines for readers interested in bicycle history and technology is pretty barren these days, except for some "zines"—small-circulation, non-commercial special-interest periodicals.

Grant Petersen, who owns Rivendell Bicycle Works, published *Rivendell Reader* quarterly from 1995 to 2008. Although it certainly promoted Rivendell's lugged-steel bicycles and related products, it also contained a wealth of technical, historical, and general-interest articles. It ceased publication early 2009. Some of these materials can now be found online at the Rivendell web site, *www.rivbike.com*.

Gabe Konrad published 20 issues of *Aeolius Butterfly* from 1994 to 1997. He then published 12 issues of *On the Wheel* from 1998 to 2000. Both magazines were devoted to classic bicycles and racing in the period between about 1920 and 1970.

Jan Heine started *Vintage Bicycle Quarterly* in 2002, which was renamed simply *Bicycle Quarterly* in 2008. His primary interest is classic French touring bicycles from the 1940s to the 1970s, with lots of Daniel Rebour illustrations. Over time Heine has included more and more articles of lasting technical interest. This edition of *The Dancing Chain* contains several illustrations and derailleur tests that were first published in *(Vintage) Bicycle Quarterly*.

Members of the American veteran cyclists organization The Wheelmen receive their quarterly history magazine, *The Wheelmen*. They also publish a quarterly *Wheelmen Newsletter*.

The British Veteran-Cycle Club publishes *The Boneshaker*, their quarterly history magazine, and *News & Views*, their bi-monthly membership newsletter. The emphasis is on historic bicycles, especially British bicycle history.

Two new consumer magazines catering to the urban cyclist were introduced in 2008 and 2009. *Momentum*, though published in Canada, and *Bicycle Times*, published by the same people who publish the mountain bike magazine *Dirt Rag*.

Right: Fig. 14.80. Initially started as a "zine" for vintage bicycle enthusiasts, *Bicycle Quarterly* is a well-produced publication covering serious technical issues that are no longer covered by the conventional cycling monthlies.

# 15. How Derailleurs Work

THE PRECEDING 14 chapters have described more than a century of efforts by hundreds of inventors to devise mechanisms to give a ⅛-horsepower (0.08 kW) human cyclist the ability to pedal a two-wheeled vehicle over almost any terrain. Over that century, thousands of patents were issued on bicycle gear-trains. Many of these patents represented years of work by the inventors.

This chapter is not about history but about the technical nitty-gritty that is so critical to the smooth operation of today's bicycle gear-trains. It also explains terms that I may have used and not explained before. Here you will also find a lot of information about shift levers, front and rear derailleurs, freewheels, chainwheels, and chains. If you are not interested in technical minutiae, you can just skip this chapter.

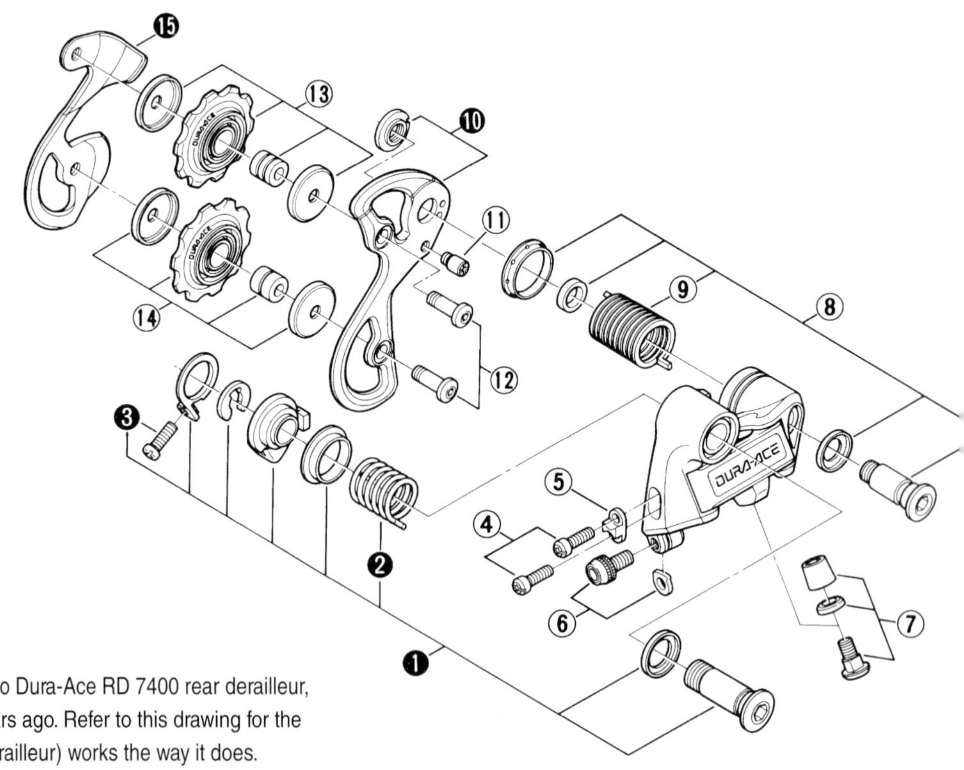


Right: Fig. 15.1. Exploded view of the 1985 Shimano Dura-Ace RD 7400 rear derailleur, a design that's as up-to-date today as it was 25 years ago. Refer to this drawing for the explanation of why it (or any other good modern derailleur) works the way it does.

We have now arrived at an elegant compromise in bicycle gearing. Today's derailleur bicycles have a freewheel with five to ten sprockets (although anything more than six does not contribute very much), a crankset with two or three chainwheels, a chain with a pitch of ½ inch, a front derailleur to move the chain between the chainwheels, and a rear derailleur to move the chain between the sprockets.

As a mechanical engineer who has studied bicycle gear-trains for 35 years, I feel that we are quite close to the optimum. I predict that bicycles (two wheels are also the optimum) will look pretty much the same. I would not be surprised if a bicycle 50 years from now had something that looks very much like today's derailleur gear-train. If these predictions are correct, then it is worthwhile to summarize what is found in a well-engineered bicycle gear-train.

## Rear Derailleurs

I first wrote this section in 1999, and nothing significant has changed in rear derailleur design in the intervening ten years. The cosmetics may be a little different, but the basic elements are the same. I used the 1985 Shimano Dura-Ace RD-7400 rear derailleur for my explanation. An exploded view of a 2009 Shimano, Campagnolo, or SRAM rear derailleur would show the same four essential design elements that are needed to provide crisp, positive indexed shifting.

Shimano did not invent any of the four major elements: Prével d'Arley developed the two pulleys in 1909; Ghiggini, JIC, or Nivex developed the parallelogram in 1937 or 1938; Simplex patented the two spring-loaded pivots in 1944; and SunTour patented the slant parallelogram in 1964.

If you are seriously interested in how and why rear derailleurs work, then follow me through this explanation of the operation of a modern derailleur. The numbers refer to the exploded view in Fig. 15.1 on the preceding page.

- The lower pivot (8) encloses the P-spring (9) that rotates the cage clockwise to wrap up chain.
- The upper pivot (1) encloses the B-spring (2), which swings the whole derailleur body clockwise to wrap up chain.
- The two springs work in concert to position the jockey pulley (13).
- The jockey pulley pushes the chain between the sprockets.
- The tension pulley (14) wraps up the excess chain.
- The cable comes through the cable adjusting barrel (6) and is anchored on the cable fixing bolt (7).
- As the cable is pulled, the parallelogram moves inward, causing the jockey pulley to push the chain onto the larger sprockets. The jockey pulley also moves downward to keep a constant distance from the larger sprockets.
- The cage plates (10 and 15) hold the two pulleys.
- The outer cage plate has a raised guide to help push the chain onto the larger sprockets.
- The cage pivot is offset in front of the jockey pulley, so that the jockey pulley moves downward as the cage rotates to unwind chain when it shift onto the larger sprockets.
- The limit screws (4) limit the total lateral movement of the jockey pulley.

Single-pivot rear derailleurs have only the B-spring, and they do not shift as well as double-pivot derailleurs. They sometimes have an angle-adjustment screw that engages the claw of the derailleur hanger and adjusts the mounting angle.

Double-pivot derailleurs often have a screw that varies the spring-tension in the top pivot. Turning this screw inward tightens the B-spring and rotates the derailleur backward, shortening the chain gap. The B-spring tension adjustment (3) is a patented Shimano feature.

All of the key dimensions and the tensions of the two springs were optimized by a Shimano computer program in the 1980s to provide a nearly constant chain gap in all gears. The chain gap was chosen to provide "early" shifting, which is the key to crisp, positive indexed shifting.

### Rear Derailleur Chain Gap

Chain gap is the distance between the jockey pulley and the sprocket, measured along the chain (see Fig. 15.2). Chain gap is the key design parameter, and determines the shifting performance of rear derailleurs. It took decades for the derailleur makers to figure this out.

Derailleurs with long chain gaps shift late. Late shifting is a design defect that requires the rider to push the shift lever too far to force the derailleur to shift, and then pull the lever back to center the chain. Derailleurs with short chain gaps shift early. Early shifting is a desirable feature in a non-indexed rear derailleur, and a requirement for proper indexed shifting.

The optimum chain gap depends on the design of the chain and the freewheel sprockets. The 1940s Cyclo Standard rear derailleur recommended a mounting position that gave a chain gap of about 2¾ in. (70 mm). It was designed for "stiff" ⅛-in. (3 mm) wide chains and pointed sprocket teeth.

By 1985, the optimum chain gap was about 2 in. (51 mm) for conventional 3/16-in. (5 mm) wide chains and flat-topped sprockets. With narrow flexible bushingless chains, the optimum chain gap is about 1½ inches (40 mm). If the chain gap is below 1½ in. (40 mm), the jockey pulley almost touches the freewheel sprocket and, depending on derailleur design, the chain might have difficulty derailing from the initial sprocket. If the chain gap is more than 2½ in. (63 mm), the derailleur will be late shifting.

By 2004, Shimano's HyperGlide sprockets and their copies had provided shifting "off-ramps" and "on-ramps" and chains were even narrower and more flexible. This permitted smaller chain gaps. By 2004, virtually everyone had copied HyperGlide sprockets.

The goal in designing an indexed shifting rear derailleur is to provide a chain gap that causes early shifting in all gears. The derailleur should maintain this chain gap even though front shifts changes the amount of slack chain and moves the tension pulley.

Left: Fig. 15.2. Author Frank Berto testing a Campagnolo Nuovo Record on his derailleur testing machine during his time as Engineering Editor for *Bicycling Magazine*.

### Early Versus Late Shifting

In the 1970s, I built a derailleur testing machine to measure how derailleurs shifted. I tested more than one hundred rear derailleurs for *Bicycling*. I became convinced that early shifting was desirable. A late-shifting rear derailleur took more lever force, and it would sometimes shift two gears instead of one. Early-shifting derailleurs changed gears with about ¼ turn of the crank. Late-shifting derailleurs took a ½ turn or more.

This section explains rear derailleur design. When you understand early and late shifting and chain gap, you understand why some derailleurs shift crisply and others shift poorly. Refer to Fig. 15.3. for the explanation. The spacing between sprockets on a 7-speed freewheel is about 5 mm (0.2 in.). Imagine that the chain's journey from the initial sprocket to the next sprocket is marked with millimeter posts 1 mm apart. The jockey pulley is pushing the chain from one sprocket to the next one. At some point on the trip between two sprockets, the chain disengages from the starting sprocket. A bit later, the chain settles onto the destination sprocket. In today's easy shifting world, these two actions occur almost simultaneously at the "shift point."

With an early-shifting derailleur/sprocket/chain combination, the chain derails from the starting sprocket to the destination sprocket before it passes the 4 mm post. The shift is completed before the jockey pulley gets to the 5 mm post, which is the centered position under the destination sprocket. After the chain shifts, the indexing shift lever pulls the jockey pulley a bit further to center the chain. Early-shifting systems require less lever force because the chain does not have to bend as much, and there is less sliding friction between the chain and the sprocket.

With a late-shifting derailleur, the jockey pulley has to push the chain past the 5 mm post. The chain does not derail onto the next sprocket until the jockey pulley gets to the 6-mm post, or even to the 7 mm post. After the shift, the chain will not be centered under the destination sprocket, so it will grind away until the jockey pulley is moved back by 1 or 2 mm to center it.

In the old days, before 1985 and the introduction of Shimano Indexed Shifting (SIS), experienced cyclists had learned the shifting characteristics of their derailleurs, and automatically made the necessary adjustments after each shift. Beginning cyclists either rode along with a noisy, uncentered chain, or they tweaked the shifter to move the derailleur in one direction or the other until the grinding noises ceased. Many old derailleurs required different amounts of overshifting on different gears. Shimano's 1985 SIS was a complete system, and it was based on early-shifting derailleurs.

In 1987, SunTour and Campagnolo tried to provide indexed shifting using their old late-shifting derailleurs. They provided overshift in the indexed shift levers. This only worked in one direction, and SunTour Accushift and Campagnolo Syncro were hard to keep in adjustment. The rider often had to push the lever past the detent, wait for the shift, and then release the lever. They were basically indexed centering systems.

Shifting from a small sprocket to a larger one is more difficult than it is to shift from a larger to a smaller sprocket, because the larger sprocket acts like stop, similar to a spoke protector. This was particularly true with old (pre-HyperGlide) sprockets without shifting ramps. The outer cage of the derailleur pushed against the chain to help with the shift. The chain had to bend 3° to 6° before it contacted the larger sprocket.

In the pre-HyperGlide days, either the chain side plates or the rivet projections rubbed against the larger sprocket. This created friction, and required higher lever forces to push the chain up onto the larger sprocket. Dropping from a large sprocket to a smaller sprocket was easier, because the

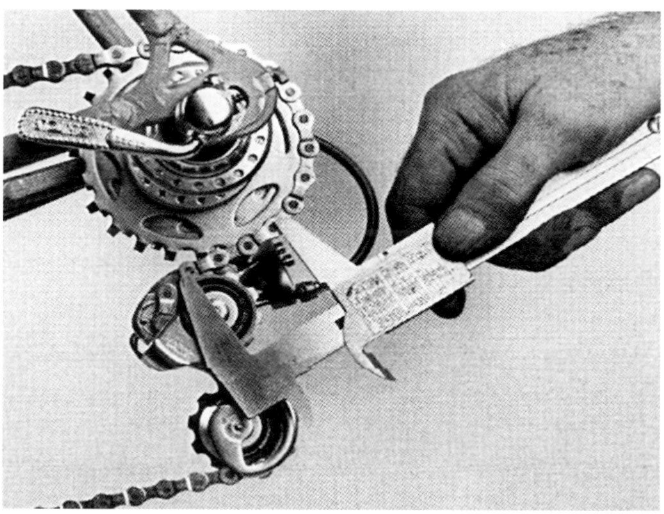

Left: Fig. 15.3. Chain gap is measured along the chain from the centerline of the freewheel to the centerline of the jockey pulley.

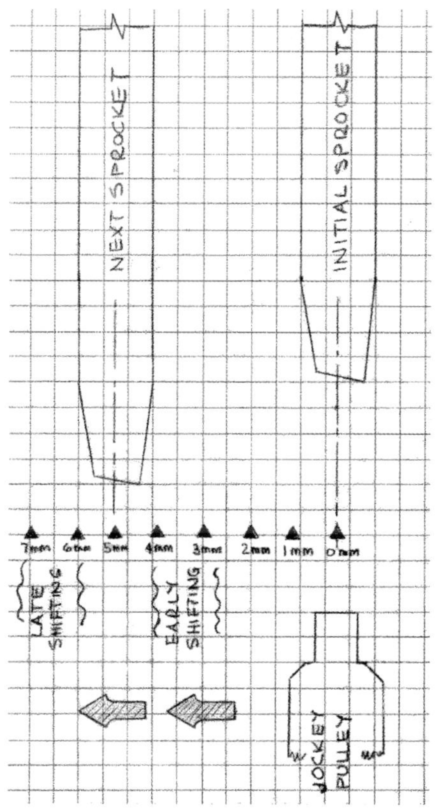


Right: Fig. 15.4. Enlarged diagram of the path of the jockey pulley from one sprocket to the next.

chain could bend freely without running into a sprocket.

With HyperGlide, the chain is pushed into a shifting off-ramp created by thinning the sprockets and cutting away the tooth tops.

## How Rear Derailleurs are Designed

Prior to 1985, rear derailleurs were usually designed by cut-and-try experimentation. Often they were designed by copying the key dimensions of a competitor's derailleur that worked well. In 1985, Shimano used a sophisticated computer design program to simulate proposed designs, and range them through their motions. The designs were fine-tuned and optimized in the computer prototype stage. Chain gap was found to be the key performance criterion.

Two Shimano rear derailleurs came out of this design program: the 1985 Dura-Ace RD-7400 and the 1986 Deore RD-MT-60. Both were optimized to work with the Shimano chains and freewheel sprockets of the mid-1980s. The main difference between the Dura-Ace and the Deore was the longer cage to wrap up more chain on the Deore. There have been minor changes since then, to take advantage of more flexible chains and easier shifting HyperGlide sprockets. Today's chain gaps are a bit shorter, but most of the changes have been cosmetic.

It took about two years for Shimano's competitors to realize that derailleurs with dimensions that are different from those of these two Shimano archetypes would be less than optimum, and would not index shift as well as the Shimano derailleurs did. As I write this, twenty years later, virtually all mass-produced rear derailleurs are based on these two designs.

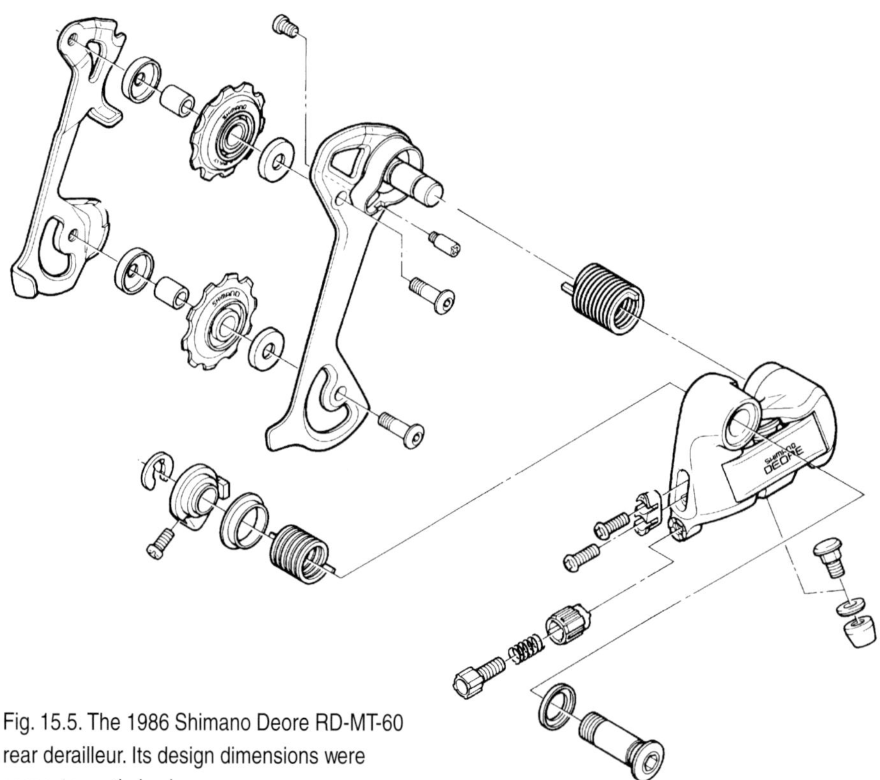

Fig. 15.5. The 1986 Shimano Deore RD-MT-60 rear derailleur. Its design dimensions were computer-optimized.

### Rear Derailleur Design Features

In this section we'll look at the main features that distinguish different rear derailleur designs. These features include the number of pivots, the parallelogram design, the cage design, and the pulley spacing.

### One or Two Spring-Loaded Pivots

Almost all rear derailleurs made since 1946 have had a spring-loaded bottom pivot (P-pivot), which rotated the cage and wrapped up the excess chain. Many early derailleurs had just this one spring-loaded pivot. The Campagnolo Nuovo Record 1020A and the SunTour VGT were classic single-pivot designs. The SunTour VGT had only a single pivot, but it shifted well compared to the competition because of its slant parallelogram.

The Campagnolo Nuovo Record was a fine single-pivot rear derailleur that could never be made to index shift. The top pivot was not spring-loaded. The derailleur could be pulled back to remove the wheel, but chain tension held it against the dropout stop when in use. The cage pivot moved inward and upward a little as it shifted onto the larger sprockets. The jockey pulley moved backward and downward as it shifted onto the larger sprockets. It also moved backward and downward when the rider shifted onto the big chainwheel. The chain gap varied from 50 to 70 mm (1.9 to 2.7 in.) in the various gears. It shifted late in almost every gear. The two holes in the cage allowed adjustment of the spring tension. There were brass bushings in the parallelogram arms. The Nuovo Record was rebuildable. and the parts were available.

Double-pivot derailleurs have two spring-loaded pivots. The spring in the top pivot (B-pivot) rotates the entire derailleur backward. The spring in the bottom pivot (P-pivot) rotates the cage, which winds up the chain and pulls the derailleur forward. The tension of the two springs and the cage geometry are varied to properly position the jockey pulley in the various gears.

The Simplex SLJ 5000T rear derailleur was the classic double-pivot derailleur, a design patented in the 1940s. You will remember the plain hole in the Simplex dropout, described in Chapter 10. It was possible to adjust the spring tension in both the B- and the P-springs by loosening the pivot bolt and turning the body or the cage. Simplex never issued instructions on how to make these adjustments.

## Parallelogram Design

The parallelogram moves the cage with the two pulleys in and out, while keeping the pulleys parallel to the bike's centerline. The distance between the parallelogram pivots varies from 38 to

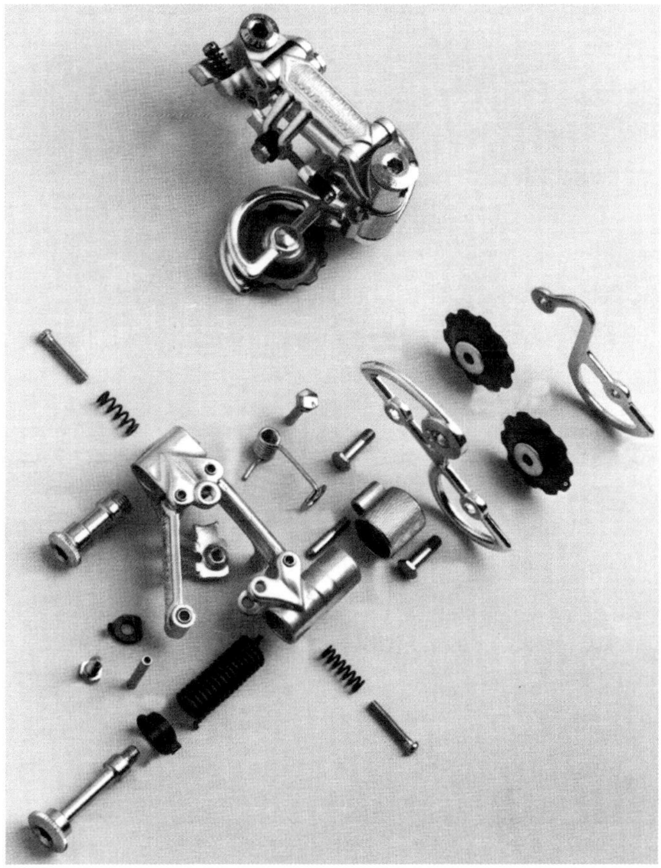

Right: Fig. 15.6. Exploded view of the 1984 Campagnolo Nuovo Record rear derailleur. Its design was based on tradition and evolution dating back to the 1950 Campagnolo Gran Sport.

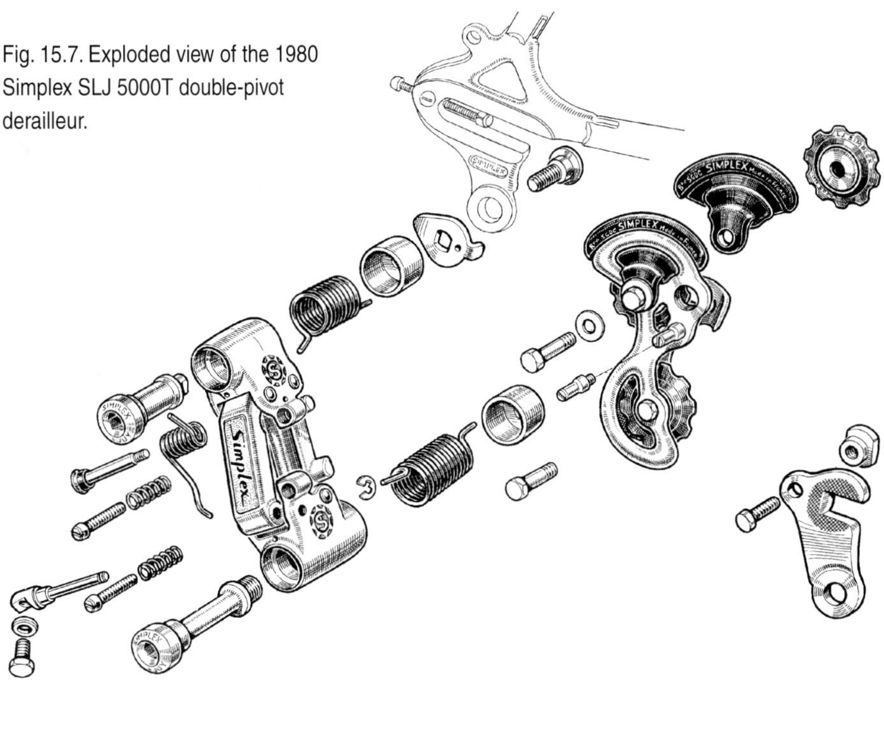


Fig. 15.7. Exploded view of the 1980 Simplex SLJ 5000T double-pivot derailleur.

45 mm (1.5 to 1.75 in.). As the parallelogram swings inward, it also swings upward a little. The length of the arms allows about 1.5 in. (38 mm) of lateral jockey-pulley travel (enough for the widest freewheel), while maintaining a reasonably linear correspondence between cable travel and jockey-pulley travel.

The parallelogram pivots were usually located so that the parallelogram was parallel to the bike's centerline on the middle sprocket. This minimized the amount of upward movement. If you measured very closely, you might find the two arms were not exactly the same length. This pointed the tension pulley inward on the small sprockets and outward on the large sprockets.

The parallelogram needed to be rigid, with adequate bearings. Campagnolo set the standard for rigidity with the Gran Sport in 1951, and today's top-quality derailleurs match their design basis.

### The Slant Parallelogram

SunTour patented the slant parallelogram in 1964. With the parallelogram pivots angled, rather than perpendicular to the axle centerline, the jockey pulley moved down as it moved inward. The slant angle of 30 to 45 degrees gave about 25 mm (1 in.) of downward movement as the jockey pulley moved inward 38 mm (1½ in.).

The parallelogram had to be approximately horizontal. This was accomplished by locating the parallelogram pivots about 25 mm (1 in.) below the top pivot, and by limiting the rotation of the top pivot. The jockey pulley moved back a bit as the parallelogram moved inward.

On the SunTour VGT, the jockey pulley was mounted coaxial with the cage pivot, so that front shifts had no effect on the position of the jockey pulley. The VGT had a big pusher plate on the outer cage, because it was designed to do battle with 34-tooth sprockets.

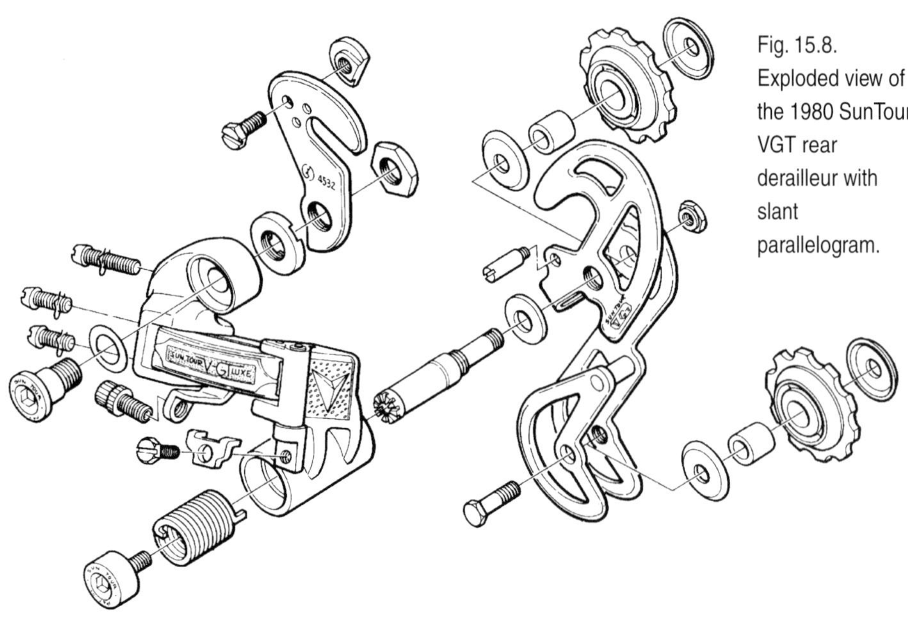


Fig. 15.8. Exploded view of the 1980 SunTour VGT rear derailleur with slant parallelogram.

### Cage Geometry

Cage geometry is a compromise. It can help, but not solve, the problem of providing a constant chain gap. Almost all index-shifting derailleurs have similar cage geometry.

Figs. 15.9 through 15.13 show the locations of the jockey pulley, the tension pulley, and the cage pivot for nine rear derailleurs. These figures allow you to visualize the path of the jockey pulley as the derailleur shifts from the wound-up chain position (little chainwheel and little sprocket), to the unwound position (big chainwheel and big sprocket). The tension pulley moves from the rear position to the front position.

Figure 15.9 shows the cage geometry of the 1951 Campagnolo Gran Sport. The jockey pulley moved back and down as it wrapped up chain. The typical half-step gearing on 1950s racing bikes had a 3- or 4-tooth difference between the chainwheels and about a 12-tooth total difference between the sprockets. The jockey-pulley did a reasonable job following the sprockets.

Fig. 15.10 shows the cage geometry of the 1959 Huret Allvit. It located the jockey pulley coaxial with the cage pivot. The parallelogram of the Allvit moved the jockey pulley down as it moved inward. Front shifts had no effect on the Allvit. It was a brilliant design, though a bit flimsy.

Figure 15.11 shows the cage geometry of the 1963 Campagnolo Record. The Nuovo Record and all other Campagnolo racing derailleurs until the 1990s used very similar geometry. The pivot was moved forward and up. As it wrapped up chain, the jockey pulley moved down, and it did not move backward as much.

This matched the crossover gearing on 1960s racing bikes, which had a 10-tooth difference between the chainwheels and about an 8-tooth total difference between sprockets. The chain gap did not change as much during front shifts. Campagnolo changed cage geometry in the 1950s to match the switch from small-step to crossover gearing.

Fig. 15.12 shows the cage geometries (left to right) of the 1985 Shimano Dura-Ace, the 1987 SunTour Superbe Pro, and the 1988 Huret New Success racing rear derailleurs. They were all similar to the 1963 Campagnolo Record. There is little room for improvement, and today's derailleurs have not changed very much. In fact, the 1954 Simplex Juy 543 had similar cage geometry.

Fig. 15.13 shows the cage geometries of the 1998 Shimano 105SC-GS and the 1998 Shimano Nexave wide-range rear derailleurs. They are nearly identical, except for the distance between the two pulleys, which establishes the amount of chain that can be wrapped up. The 105 had a wrap-up capacity of 36 teeth. The Nexave had a wrap-up capacity of 43 teeth.

### Pulley Spacing

In Fig. 15.13, the distance "A" between the two pulleys should be wide enough to wrap up the excess chain. "A" is about 50 mm (2 in.) for narrow-range rear derailleurs, which wrap up 6 in. of chain, and about 90 mm (3½ in.) for wide-range rear derailleurs, which wrap up 280 mm (11 in.) of chain.

### Rear Derailleur Capacity

Rear derailleurs have two capacity numbers. The largest rear sprocket is self-explanatory. The wrap-up capacity is the sum of the chainwheel difference and the sprocket difference.

- Example: 52/30 chainwheels and 14–28 freewheel sprockets.
  Chainwheel difference: 52 – 30 = 22 teeth.
  Sprocket difference: 28 – 14 = 14 teeth.
  Consequently, the rear derailleur must have a wrap-up capacity of 22 + 14 = 36 teeth.
  Since the chain only wraps around half of the sprocket and the chainring, the amount of chain to be wrapped up in the combination of small sprocket and small chainwheel i.e. the combination 30/14) is: 36 / 2 = 18 teeth.

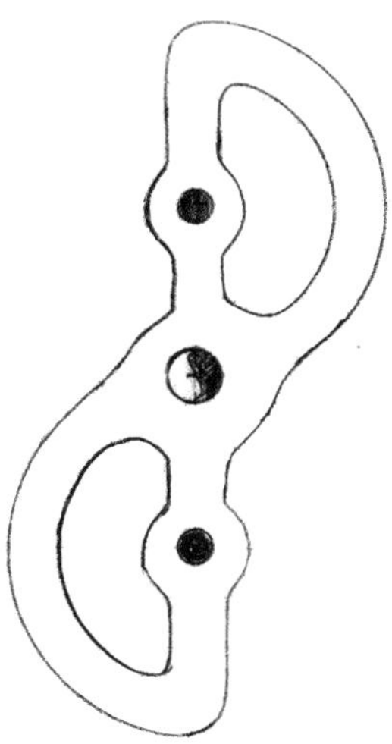

Above: Fig. 15.9. Cage geometry of the 1951 Campagnolo Gran Sport, which set the standard for a decade.

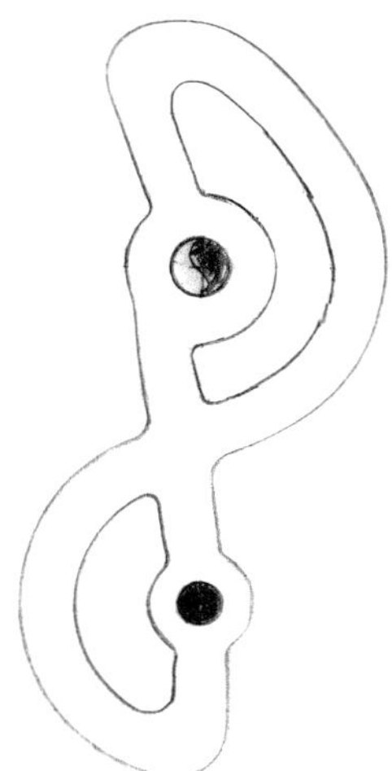

Above: Fig. 15.10. Cage geometry of the 1959 Huret Allvitt, on which the jockey pulley was coaxial with the pivot.

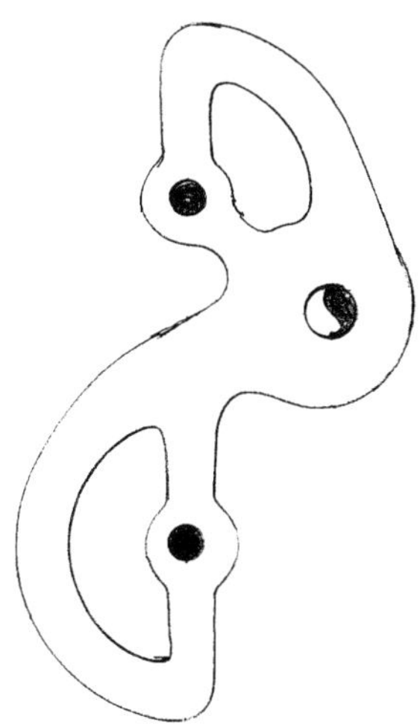

Above: Fig. 15.11. Cage geometry of the 1963 Campagnolo Record, on which the pivot was moved forward and up.

### Offset Distance

In Fig. 15.13, "O", the offset distance between the jockey pulley and the cage pivot, affects the shifting characteristic of the rear derailleur. If the jockey pulley is on the same axis as the cage pivot, ("O"=0), the jockey pulley will not move with front shifts. The old SunTour VGT and the Huret Allvit and Jubilee rear derailleurs had coaxial jockey pulleys.

In general, coaxial jockey pulleys are not used with today's index-shifting rear derailleurs. The 1995 SRAM ESP was an exception. Most index-shifting rear derailleurs locate the jockey pulley 10 to 25 mm (0.4 to 1.0 in.) away from the cage pivot. The angle between the two pulleys and the cage pivot is usually about 90 degrees.

### Chain Wrap

The derailleur should position the jockey pulley high enough so that the chain wraps around the sprocket for nearly 180 degrees, particularly on the small sprockets. With less than about 6 teeth of wrap, the chain might jump forward with high pedaling forces, especially with a worn chain or worn sprockets. The chain should slope downward from the larger sprocket to the jockey pulley, to ensure that the pusher plate clears the adjacent smaller sprocket.

### Cage Pusher Plate Shape

Ideally, the jockey pulley should lead the chain smoothly from sprocket to sprocket. However, shifting to a larger sprocket on a hill imposes extra strain. Most derailleurs have an enlarged outer cage to help push the chain onto the larger sprockets, and to keep the chain from over-shooting when dropping onto the smaller sprockets. This was more critical in the pre-HyperGlide dark ages.

The cage pusher plate of the 1971 to 1974 Campagnolo Gran Turismo was awe-inspiring. This was a derailleur of superlatives: It was the heaviest, it had the stiffest return spring—and it shifted worse than any other wide-range derailleur ever made. They weren't too expensive by Campy standards, and a lot were sold during the bicycle boom. Most of them are still around, almost unused. They make good paperweights.

### Shift Lever Force

Before 1988, every rear derailleur had a spring in the parallelogram to pull the jockey pulley outward back to the small sprockets. A strong spring requires more lever force to shift and more lever friction, or more pronounced detents, to prevent inadvertent shifting.

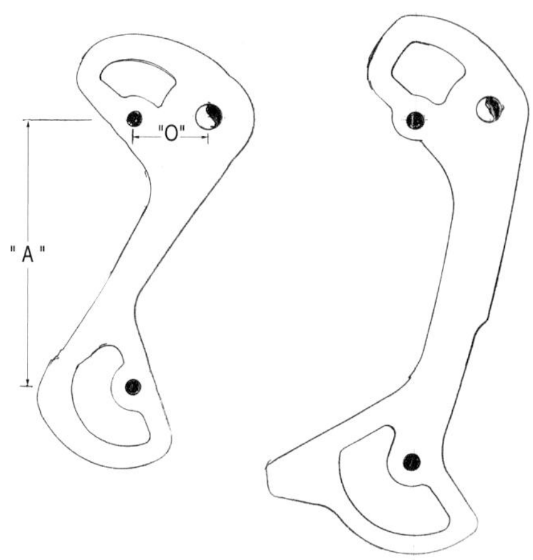


Above: Fig. 15.13. Wide-range rear-derailleur cage geometries: 1998 Shimano 105SC-GS and 1998 Shimano Nexave.

Below: Fig. 15.14. The cage pusher plate of the Campagnolo 1971 Gran Turismo.

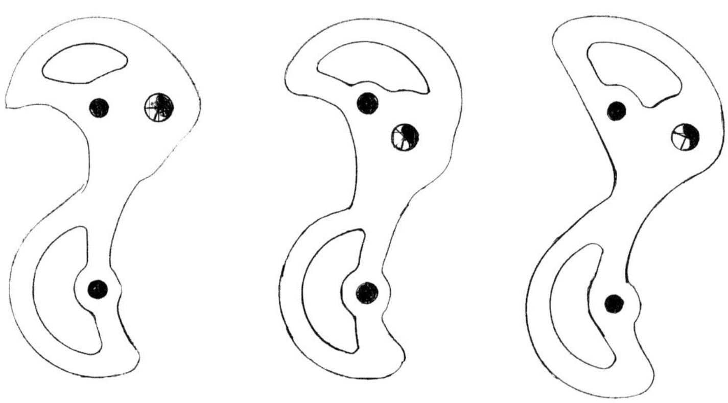

Left: Fig. 15.12. Racing rear-derailleur cage geometries. From left to right: 1985 Shimano Dura-Ace, 1987 SunTour Superbe Pro, and 1988 Huret New Success.

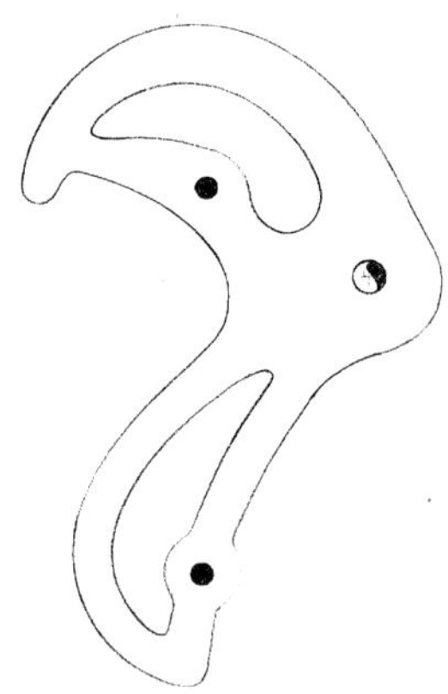

Over the last decade, derailleurs and sprockets have been redesigned for lighter action, and the return springs are less powerful. In 1998, Shimano felt comfortable enough with rear shifting forces that they designed many of their rear derailleurs for reverse action: The spring pulls the jockey pulley inward, and the cable pulls it outward.

### Chain Tension

The spring tensions in the upper and lower pivots and the chain length must be balanced to provide enough tension on the lower run of chain. Too little tension allows "chain slap" (which would destroy the paint on the chainstay), chain "suck" (jamming the chain in the chainwheels), or chain derailment from the chainwheels.

There's an interesting bit of history here: In bygone times, racing cyclists believed that it took less effort to pedal with a slack chain on the lower run. So Huret, Simplex, and Cyclo all made special rear derailleurs that provided adjustable chain tension, and they sold at a premium. These old derailleurs shifted better on the small sprockets with a taut chain, and they had more capacity with a slack chain. Eventually the makers improved their basic designs to avoid the need for adjustable chain tension.

### Stiffness and Longevity

Top-quality rear derailleurs use beefy forgings, sealed bearings, stainless steel pins, brass bushings, and plastic seals to increase stiffness, shifting precision, and longevity. Today's double-pivot, slant-parallelogram derailleurs are rugged and have a minimum number of moving parts. This is an area where you get what you've paid for. The top-of-the-line derailleurs usually last longer than their less expensive brethren.

### Dimensions of the Derailleur Hanger (Dropout)

The derailleur hanger positions the rear derailleur relative to the freewheel sprockets. The location of the hanger directly affects the chain gap and the derailleur capacity. This went through two iterations before arriving at the current standardized design.

- First, there were manufacturer-specific hangers, such as those unique to either Simplex or Huret. These days, this type only concerns restorers of old French bicycles, who have to find matching derailleurs for given derailleur hangers
- Second, there was the Campagnolo "standard," which was, however, not universally used (specifically, not by the French manufacturers). This created a problem with early indexed shifting.
- Finally, Shimano laid down the law in the late 1980s, and their recommendations became the *de facto* industry standards.

With a horizontal dropout (shown in Fig. 15.15), you could alter the chain gap by moving the wheel back or forward. Most of today's quality bicycles use vertical dropouts, while lower-quality bicycles use a separate derailleur hanger plate, inserted between the dropout and the axle nut or quick-release thumbnut. The "drop" was about 25 mm (1 in.) for racing hangers and about (30 mm (1¼ in.) for touring bikes and mountain bikes.

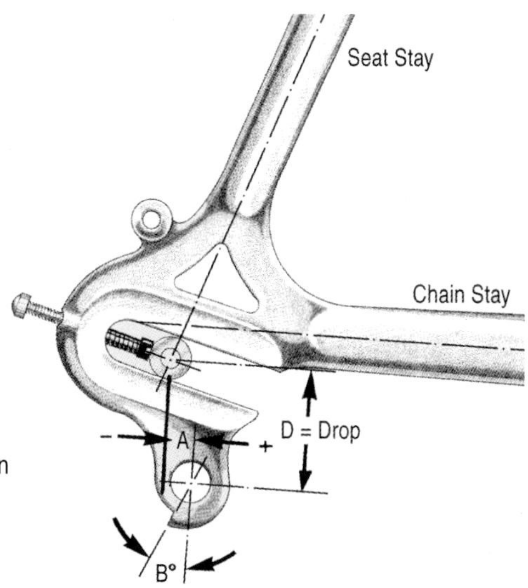


Right: Fig. 15.15. The position of the rear derailleur on the bicycle is determined by the dimensions "A" (axle centerline to derailleur pivot) and "D" (drop), and the claw angle "B."

### Optimum Path for the Jockey Pulley

For good index shifting, the chain gap should be about 40 mm (1½ in.), and the top of the jockey pulley should be slightly lower than the bottom of the sprocket. These two requirements define the optimum path for the jockey pulley as it moves from sprocket to sprocket. About 40 mm (1½ in.) of vertical jockey pulley movement is needed for a wide-range touring rear derailleur, and about 20 mm (¾ in.) for a narrow-range racing rear derailleur.

Rear derailleur design is a mechanical engineering exercise to optimize the tensions of the P- and the B-springs, the location of the parallelogram pivot in relation to the rear axle, the slant and length of the parallelogram, and the key cage-dimensions. It is a complicated exercise, because there are about a dozen variables. The design is successful if the mechanism moves the jockey pulley reasonably close to the optimum path.

It is easy to design a rear derailleur for a single chainwheel. The problems come with multiple chainwheels. Shifting from the largest to the smallest chainwheel on a mountain bike requires the rear derailleur cage to rotate enough to wind up 130 mm (5 in.) of chain. Since the chain is on the same rear sprocket, the chain gap should not change with front shifts.

Today's rear derailleurs look alike because everyone copied Shimano's 1985 and 1986 models. Today's buyers expect derailleurs to have a Shimano look about them.

There are at least five other possible designs. The 1924 Cyclo Touriste, the 1946 Nivex, the 1958 Huret Allvit, the 1964 SunTour Grand Prix, and the 1976 Huret Duopar could all have been modified to provide an optimum jockey pulley path and good index shifting.

## Freewheel Sprocket Design

Early or late shifting depends primarily on the chain gap, but it also depends on the freewheel sprockets. Since 1985, there have been a whole series of improvements in sprocket design, which have greatly reduced the force needed to derail the chain from sprocket to sprocket.

Shimano instigated most of these changes, and the all-inclusive buzzword is HyperGlide. Shimano has never stopped refining sprocket shape, and each refinement has added another buzzword to the Shimano lexicon. I listed most of these changes in Chapters 12 and 13.

HyperGlide was a huge improvement. Use one of your favorite old friction-shifting rear derailleurs with a modern 9-speed HyperGlide cassette and the matching chain, and it will think it has died and gone to heaven. You may need narrower pulleys to match the narrow chain. With that change, it will shift very nicely. It probably would not index with modern indexed levers, but people keep trying.

Try the reverse—using a modern indexed rear derailleur with an old freewheel and chain. It will not be a pleasant experience, because the chain gap is too small, and the sprockets resist shifting. Many of today's cyclists, with less than 15-years of experience, are spoiled rotten. They have never learned how to operate a friction-shifting derailleur.

When you shifted back in the good old days, you eased back on the pedaling force and then pushed the lever an appropriate

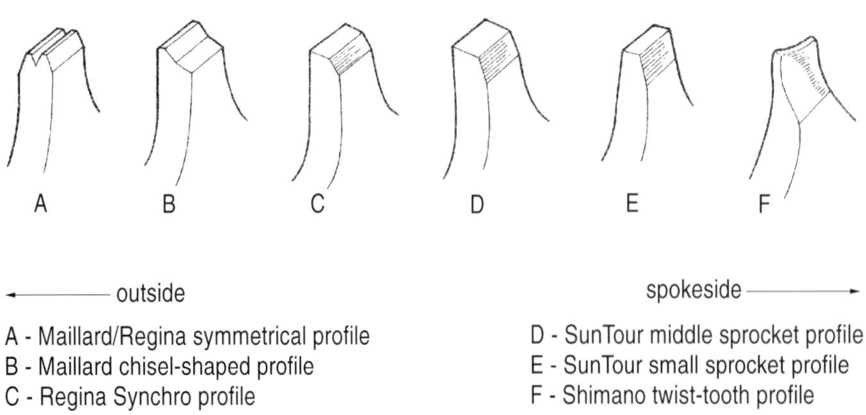


Left: Fig. 15.16. Tooth profiles of various 1980s (pre-HyperGlide) feewheel sprockets.

distance until something happened in the back. During many shifts, the chain would ride on top of the destination sprocket for a bit, and then settle down into engagement. If there was noise in the back, you moved the lever to quiet the complaining. If you were having a bad day, your companion would say, "Grind me a pound while you're at it."

### Pre-HyperGlide Sprockets

Designers knew that there had to be a better tooth shape than the standard pointed top used by industrial chain drives. The changes to sprocket designs started in the 1930s, when derailleur sprockets received flat-topped instead of pointed teeth, so that the chain did not have to climb so high. In the 1950s, Vittoria carved away the front of the teeth, so that their back-pedaling derailleur shifted better.

Several freewheel makers put little ridges on the tips of the teeth. Shimano left out every other tooth on their largest sprocket to aid shifting. By the 1970s, Shimano had sprockets with twisted teeth and sprockets with alternate teeth missing, and SunTour had slanted teeth. Some of these early modifications probably helped, especially when combined with the modern, more flexible chains (which was the feature that helped more than any of those early orthodontic experiments).

### HyperGlide Sprockets

The most significant improvements in rear derailleur design took place before 1990. Since then, the huge improvement in rear shifting has been due to the use of HyperGlide sprockets.

Introduced in 1989, Shimano HyperGlide provided shifting sectors on the sprockets. In these shifting sectors, the height and the

Left: Fig. 15.17. 1997 HyperGlide 9-speed sprockets. Note the cut-down teeth and the sculpting.

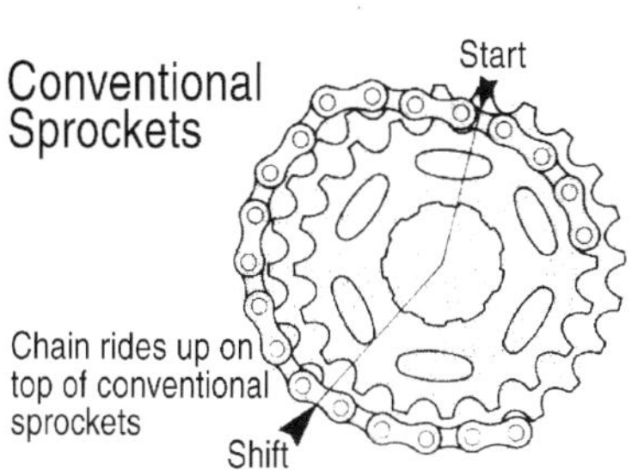


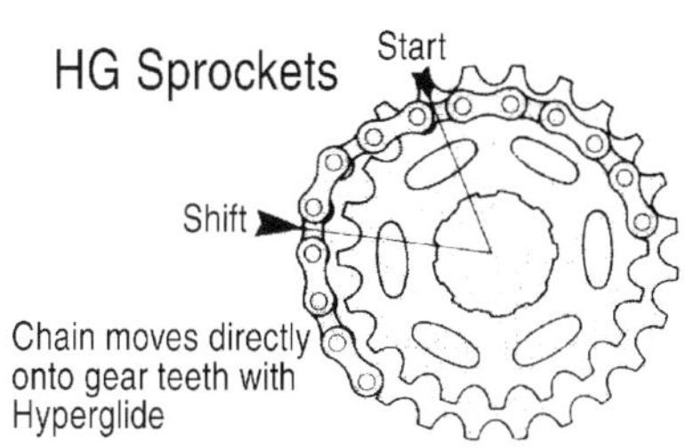


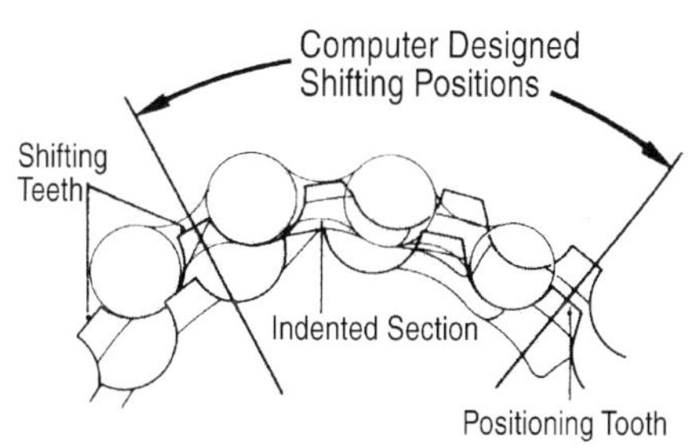


Above and left: Figs. 15.18 and 15.19. HyperGlide sprockets versus conventional sprockets, and detail of chain on sprockets during shift.

width of three or four "release" teeth were revised to provide an "off-ramp" for the chain to slide off the sprocket. On the adjacent sprocket, three or four "pickup" teeth were narrower and lower, to provide an "on-ramp" for the chain to flow gracefully onto the sprocket.

The chain engaged both sprockets at the same time, which was a crucial requirement. The sprockets had to be keyed onto the freewheel, so that the off-ramp on one sprocket was exactly opposite the on-ramp on the next sprocket. That's why you cannot use old freewheels with threaded sprockets or unkeyed splined sprockets on an indexed-shifting bike: Adjacent sprockets are not exactly lined up.

Shimano fine-tuned the concept every other year. Chapters 12 through 14 give the details. The result was that the chain flowed from sprocket to sprocket without having to climb on top of the next sprocket and then drop down again. HyperGlide sprockets shifted almost instantly with low lever forces. Over the years, Shimano's competitors copied HyperGlide.

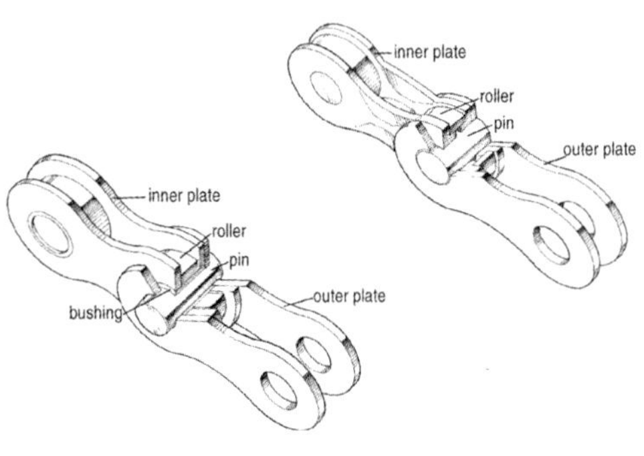


Above: Fig. 15.20. Conventional bush-roller chain (left) compared with bushingless chain (right).

On Shimano's 1997 HyperGlide sprockets, there are four cut-down, or "sculpted," teeth, covering the width of the shifting sector. The shifting sector moves one tooth clockwise as you go to the next larger sprocket. The small sprockets (11 to 18 teeth) have one shifting sector. The larger sprockets have two or three. Both sides of the sprockets are similarly sculpted. See Figs. 15.18 a and b.

My objection as a retro-grouch is that HyperGlide destroyed the ancient and honorable order of "gear freaking," or at least it limited the possibilities. You have to use the sprockets that Shimano provides, and they come in particular prepackaged combinations. In the glory days, a genuine gear freak had a SunTour sprocket board and a Shimano sprocket board, with every sprocket between 11 or 12 and 34 or 38, and you could arrange them in almost any combination. I still have the sprocket boards, and perhaps 200 sprockets, but they are only good for nostalgia.

## Chain Design

Chain design progressed along with HyperGlide, but it was not as critical. You could use a Sedisport chain of the right width with almost anybody's indexed geartrain, and it would shift just fine.

The period after 1983 SIS and before 1989 HyperGlide was more critical for chains. Some of the early indexed SunTour Accushift and Campagnolo Syncro rear derailleurs had wide chain gaps, and they needed stiff chains. This was a band-aid. The real problem was that the old derailleurs were not designed for indexed shifting.

### Chain Flexibility

Flexible chains shift easier because they can bend further with less force to engage the adjacent sprockets. Wide chains are stiffer than narrow chains. Stiff and flexible refer to lateral bending of the chain. All chains are flexible as they go around the sprockets.

The bushingless chain design is inherently more flexible. Sedis introduced the first widely sold bushingless chain, the Sedisport, in 1981. Now all quality narrow chains are bushingless. Look closely at the bushingless chain in Fig. 15.21, and you can see how the inner plates are swaged inward. There is no bushing: The rollers turn on the swages.

### Chain Types

The conventional bush-roller chain was little changed from Hans Renold's 1880 patent. Note the pin extensions. We used to think that they helped shifting.

In pre-HyperGlide times, the chain makers modified the shape of the chain's side plates to shift easier with the available freewheel sprockets. Bulged or chamfered side plates shifted slightly better than conventional chains.

Shimano introduced the Uni-Glide chain in 1976. The outer plates were bulged to the width of the pins, which made engagement and release from the sprockets easier. So much for pin extensions.

The Sedisport chain had bulged inner plates to assist shifting. The flexible bushingless construction made this the best-shifting chain available in the 1980s.

### Chain Width

The nominal width of a chain is the thickness of the sprocket that it matches. The spacing between the inner plates is the sprocket width plus a tad.

Over the century and a quarter since Renold's patent, there has been a continuous trend to narrower chains and thinner, more closely spaced freewheel sprockets. At the time of each transition, the old chain was called "standard," and the new chain was called "narrow."

In the 1890s, 1/4-inch (6 mm) wide chains were made to work with 1/4-inch (6 mm) thick sprockets. Chapter 2 describes the heroic efforts to build 2-sprocket derailleurs and freewheels using these chains.

By the 1910s, 3/16-inch wide (5 mm) chains and sprockets had taken over, and 2-sprocket freewheels were in use.

By the 1920s, the standard chain and sprocket was 1/8-in. (3 mm) wide. This was used on 2-, 3-, and 4-sprocket freewheels. Today, this is the standard chain for single-speed bicycles.

In 1936, Brampton introduced the 3/32-inch chain. This permitted freewheels with 5 sprockets, and later 6 sprockets.

At this point, the law of diminishing returns set in. SunTour introduced the narrow Ultra chain in 1977, which permitted a narrow Ultra-6 freewheel, with 6 sprockets on a 120-mm hub. In 1979, this became the Ultra-7, with 7 sprockets on a 126-mm hub. This was a retrograde step, and the shifting was noticeably worse.

Tolerances became so critical, that chains had to be measured by the outside pin width rather than by the nominal inside width. "Wide" 3/32-inch chains had pins about 8 mm long, and "narrow" 3/32-inch chains had pins about 7.3 mm long.

The total width of the sprockets and spacers on a freewheel (or freehub) can't be too great for two reasons:

- Wheel dish: To avoid excessive wheel dish, spacers on the left side of the hub match the width of the sprockets on the right. This can result in wide over-locknut dimensions of the rear hub. The over-locknut width has grown from 120 mm, to 126 mm, to 130 mm for road bikes, and 135 mm for mountain bikes.

- Chain line and chain bending: Even flexible chains don't like to bend an inch. That's why all of the books tell you not to use the gears that combine the biggest chainwheel with the biggest sprocket, or the smallest chainwheel with the smallest sprocket. The problems get even worse with short chainstays. The practical limit of freewheel width is about 36 mm (1.4 in.).

In 1989, Shimano was able to cram 8 thin sprockets and 7 thinner spacers in 36 mm (1.4 in.). This required a new extra-narrow

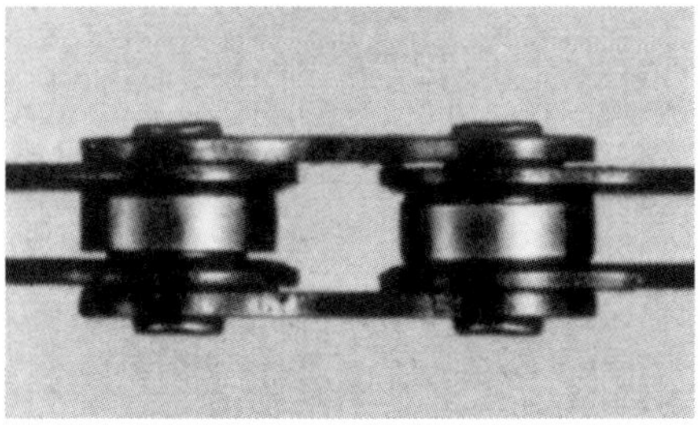

Above: Fig. 15.21. A conventional bush-roller chain.

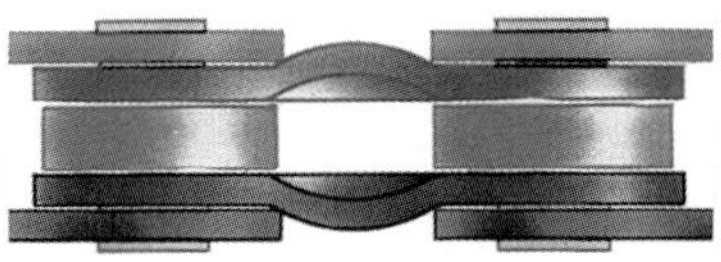

Above: Fig. 15.22. The1981 Sedisport chain was the first bushingless chain.

Above: Fig. 15.23. The 1976 Shimano UniGlide chain.

chain (7.1 mm pin width) and thinner chainwheels and spacers. SunTour, Sachs, and Campagnolo followed a few years later.

Little in the new 8-speed gear-trains was interchangeable with the prior 7-speed gear-trains (except the freehub). In theory, the rider got two extra gears with a double chainwheel, or three extra gears with a triple. In practice, the extra gears were duplicates or triplicates.

However, bicycle buyers assumed that 16- or 24-speed bikes just had to be better than 14-or 21-speed bikes. Soon, 8-sprocket freewheels trickled down to inexpensive bikes.

In 1997, Campagnolo and Shimano were able to cram 9 very thin sprockets and 8 even thinner spacers into the same 36 mm. This required an even narrower chain (6.6 mm pin width) and closer-spaced chainwheels. Of course, little in the 9-speed gear-trains was interchangeable with anything less prestigious.

In 2000, Campagnolo came out with the 10-speed Record gruppo, with 10 very, very thin sprockets and 9 super-thin spacers into the same 36 mm. The super-narrow chain was just 5.9 mm wide, and the pins are flush with the side plates.

Shimano did not respond immediately. In my opinion, they may have felt that the disadvantages outweighed the advantages. I can imagine meetings between the engineers and the marketers. By 2004, Campagnolo's top four gruppos had 10-speeds and triple cranksets. At this point, the Shimano marketing department prevailed, and Dura-Ace became a 10-speed. The CN 7800 super-narrow chain is 5.9 mm wide.

Shimano now has at least five different chain widths. Another dozen companies make quality chains in four or five widths, with dimensions that differ slightly from those used by Shimano. Refer back to the story about the 1880s traveler finding that ½-inch pitch chains were "*tout la même chose*." That is certainly not the case today, and today's world traveler is advised to carry an extra chain, extra pins, and a matching chain tool.

Campagnolo, Shimano, and SRAM all make chains. Not surprisingly, each company recommends that only their chain be used with their cassettes and derailleurs. This is certainly a good idea for 10- and 11-speeds, and probably also for 9-speeds.

At the risk of sounding like a retro-grouch, I don't consider 10- and 11-speed freewheels progress. Tolerances have become just too critical, and there is almost no safety factor. The precision required to align and adjust these 10- and 11-speed gear-trains would do credit to a clock-maker. The chains are expensive, short-lived, and they have to be assembled very carefully.

*Table 15.1. Summary of chain-width developments, 1890–2009*

| Year | Speeds | Name | Nominal Width | Outer Plate Width | Pin Width | Inner Gap |
|---|---|---|---|---|---|---|
| 1890 | 1 | Standard 1-speed chain | ¼ in. | N/A | 14 mm± | 8 mm± |
| 1910 | 1 or 2 | Standard 1-speed chain | 3/16 in. | N/A | 13 mm± | 7 mm± |
| 1920 | 1 to 4 | Standard 1- to 4-speed chain | 1/8 in. | 8.0 mm | 9.4 mm | 3.5 mm |
| 1935 | 5 | Standard 5-speed chain | 3/32 in. | 7.0 mm | 8.0 mm | 2.8 mm |
| 1977 | 5 & 6 | Shimano UniGlide Wide | 3/32 in. | 7.65 mm | 7.95 mm | 2.6 mm |
| 1985 | 6 & 7 | Shimano UniGlide Narrow | 3/32 in. | 7.3 mm | 7.4 mm | 2.6 mm |
| 1989 | 7 & 8 | Shimano InteractiveGlide | N/A | 7.07 mm | 7.15 mm | 2.49 mm |
| 1997 | 9 | Shimano HyperGlide | N/A | 6.70 mm | 6.70 mm | 2.35 mm |
| 2000 | 10 | Campagnolo Record C10 | N/A | 5.94 mm | 6.08 mm | 2.30 mm |
| 2004 | 10 | Shimano Super Narrow HG | N/A | 5.90 mm | 5.90 mm | 2.35 mm |
| 2009 | 11 | Campagnolo Record 11 | N/A | 5.50 mm | 5.50 mm | N/A |

The low-priced Made-in-China bicycle market seems happy with 5-, 6-, or 7-speed freewheels. I will be surprised if 10-speed freewheels catch on with bicycle tourists, or if they do, if the tourists will be happy with their choice in the long run. We shall see.

### Tooth Difference Between Adjacent Sprockets

Freewheels with wide tooth differences between sprockets make it harder to shift. Today's wide-range mountain bike rear derailleurs can index shift over a 6-tooth difference (26- to 32-tooth sprockets). Shimano's 1998 Nexave rear derailleur could shift over an 8-tooth difference (26- to 34-tooth sprockets).

Back in the friction-shifting days, I set up freewheels with 10-tooth differences (24- to 34-tooth), and the best of the old touring derailleurs, like the Duopar, could shift over that difference. Interestingly, the rear derailleur has an easier task shifting over a two-tooth difference than a one-tooth "straight block," but it does not matter very much with HyperGlide sprockets.

## Front Derailleur Design

More than two hundred new gruppos were introduced in the period between 1995 and 2009. Most included a "new" front derailleur that differed only cosmetically from the previous front derailleur and from the competitor's front derailleurs.

In reality, there were few innovations in front derailleur design. In fact, you can substitute a 1960 Campagnolo Record front derailleur for almost any modern front derailleur, and it will friction shift quite adequately. The inner cages of modern front derailleurs have various ridges and bumps in the right positions to help the shifting, particularly with HyperGlide chainwheels.

Racing front derailleurs are designed for either two or three chainwheels. Racing front derailleurs for double-chainwheel cranksets are designed for a maximum difference of about 15 teeth (53/39).

Triple front derailleurs have deeper inner cages. Sport-touring bike triple front derailleurs are designed for a front difference of about 23 teeth (53/39/30). Mountain bike front derailleurs are designed for front differences of about 22 teeth (48/36/26 or 44/32/22).

The cage on a mountain bike front derailleur matches the diameter of the big chainwheels, which are a bit smaller (48 to 44 teeth) than they are on road bikes or sport-touring bikes. Road bike front derailleurs have cages to match the curvature of the larger big chainwheels (48 to 54 teeth).

This is not quite the case anymore, because the manufacturers have created a tolerance problem with index shifting, ultra-narrow chains, and closely spaced, thin chainwheels. This requires the curvature of the cage plates to closely match the chainwheel

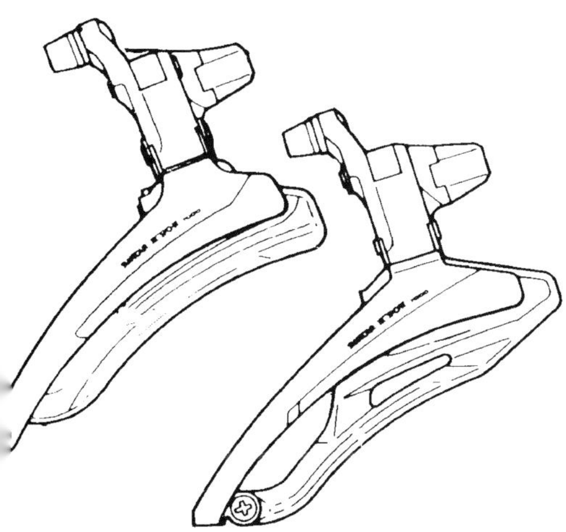

Above: Fig. 15.24. The 1987 SunTour half-step (left) and crossover (right) front derailleurs.

Right: Fig. 15.25. The 1998 Campagnolo Veloce double and triple front derailleurs.

diameter. So now there are different front derailleurs for normal and compact chainwheels, even though the latter are only a few teeth smaller.

For a few years in the 1980s, there were two kinds of front derailleurs for touring triples: Half-step models had shallower inner cages to clear the middle chainrings of half-step triple cranksets (50/45/26). Crossover models had deeper cages to do battle with crossover triple cranksets (50/38/26). Half-step gearing and half-step front derailleurs disappeared in 1988, when mountain bikes took over.

All front derailleurs have a cage that is wider at the rear than it is at the front, so that the chain does not rub on the inside when it runs to the largest sprocket in the rear, or on the outside when it runs to the smallest sprocket. In the friction shifting era, riders learned to "trim" the front derailleur when they heard scraping. Today's top-quality indexed shift levers provide a trim adjustment. Proper adjustment of the limit stops is critical.

Some front derailleur cages move forward, some move upward, and some swing outward during the shift from the small to the big chainwheel. None of these actions make much difference. When I was testing derailleurs for *Bicycling* in the 1980s, I found that rigidity was the key factor: Beefy front derailleurs shifted better than flimsy front derailleurs, and that holds as true today as it did 20 years ago.

## Chainwheel Design

The key improvements in front shifting were not better front derailleurs, but modified chainwheels. Shimano initiated a series of chainwheel changes to assist the shift from the smaller chainwheel to the larger chainwheel. Each change came with a new buzzword. The competitors promptly copied Shimano's designs and provided their own buzzwords.

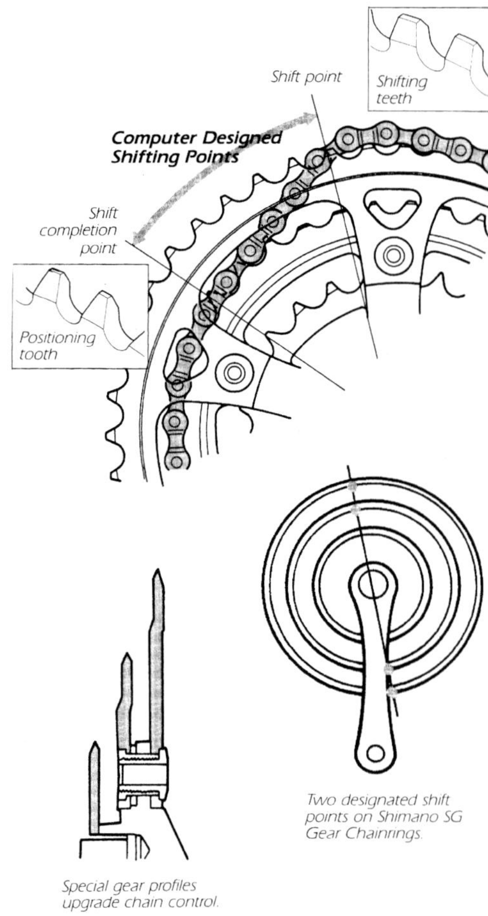


Above: Fig. 15.28. Shimano's 1990 "sculpted" chainwheel teeth.

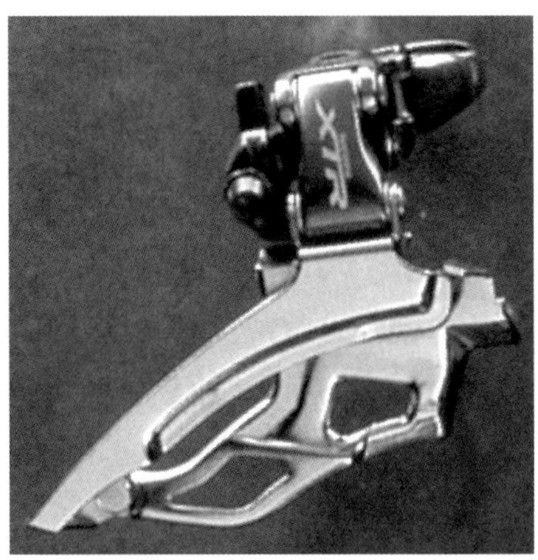


Above: Fig. 15.26. The 2004 Shimano XTR M961 triple front derailleur. Note the smaller radius of the outer cage.

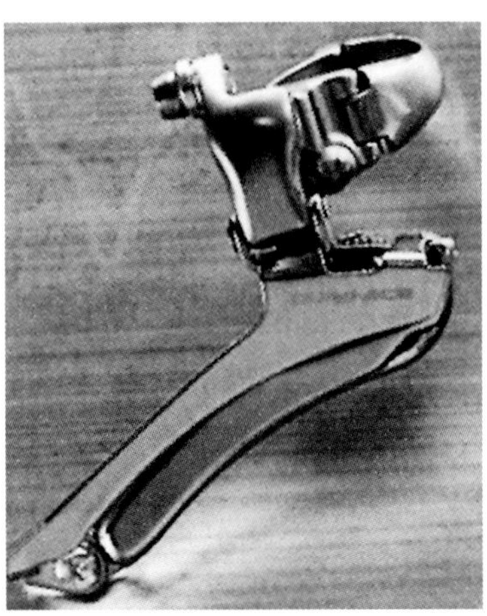

Left: Fig. 15.27. The 2004 Shimano Dura-Ace 7800 double front derailleur. Note the larger radius of the outer cage.

Below: Fig. 15.29. Shimano's SIS front chainwheels.

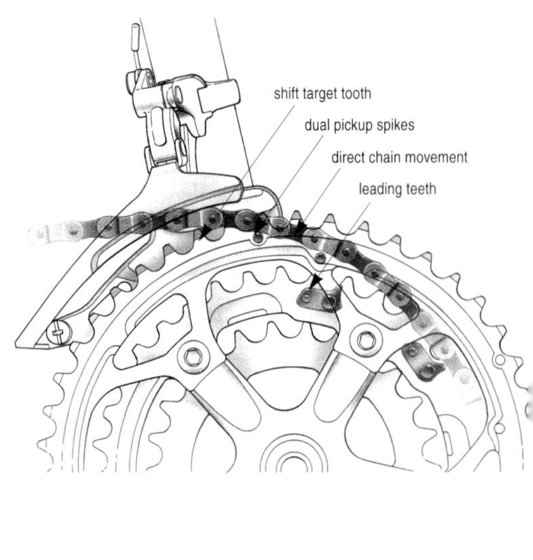

On the 1990 version, the chainwheel teeth were "sculpted" so that front shifts took place on the dead centers, where the chain tension was lowest. Shimano called this SuperGlide. In 1992, Shimano added hooks to the middle chainwheel to help the shift from the little chainwheel. This really provided an intermediate 2-tooth chainwheel between the inner and the middle chainwheels. They called this HyperDrive. In 1993, they added spikes to the outer chainwheel to help the shift from the middle chainwheel. They called this Front SIS. In 1996, they modified the hooks and spikes a bit. They called this Front IG.

The total effect of the improved chainwheels was to make front shifts almost as reliable as rear shifts. The changes required that the chainwheels be installed in the proper orientation so that receiving chainwheel lines up with passing chainwheel.

## Shifters

In the friction shifting days, road bikes used down-tube shifters, bar-end shifters, or stem-mounted shifters. The latter were for inexpensive beginner bikes. Mountain bikes used stem-mounted shifters or handlebar mounted thumb shifters. Indexed shifting required indexed versions.

In 1991, Shimano combined the (road bike) brake lever with the indexed shift lever. They called this STI (Shimano Total Integration). The STI lever is a sophisticated piece of machinery with lots of precision parts. There are no parts lists or repair instructions. Nobody knows how to fix one, but they do not seem to break or wear out.

Campagnolo followed a few years later with Ergopower. Sachs used Campagnolo's Ergopower levers until SRAM acquired Sachs. SRAM introduced its own dual-control sustem called "Double-Tap" in 2008.

Prior to STI, many companies made shifters with detents. They reduced the amount of lever friction required to overcome the tension of the return spring in the derailleur. If the detents were in the right locations, and the cable was properly adjusted, they could provide indexed centering.

Twist-grip shifters have been offered for flat-handlebar bikes, most notably by SRAM, who originally (and currently) offered it for triathlon and time-trial bikes. Most manufacturers offer this type of shifter for use on city and comfort bikes.

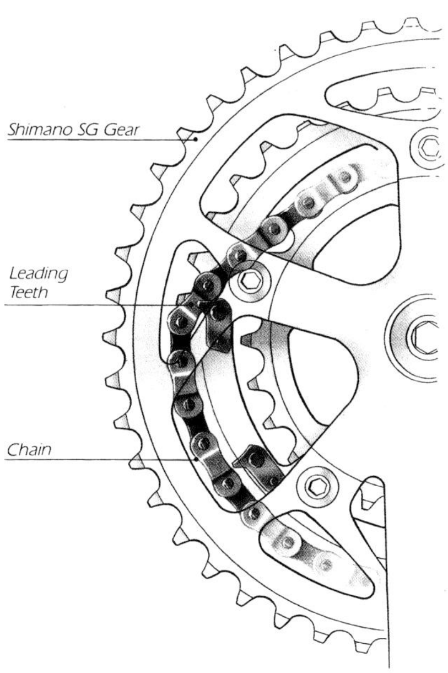


Above: Fig. 15.30. Shimano's 1992 HyperDrive chainwheels.

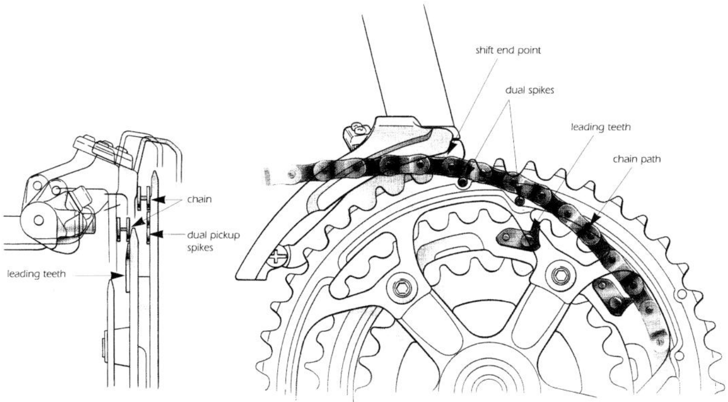


Right: Fig. 15.31. Shimano's 1996 front IG chainwheels.

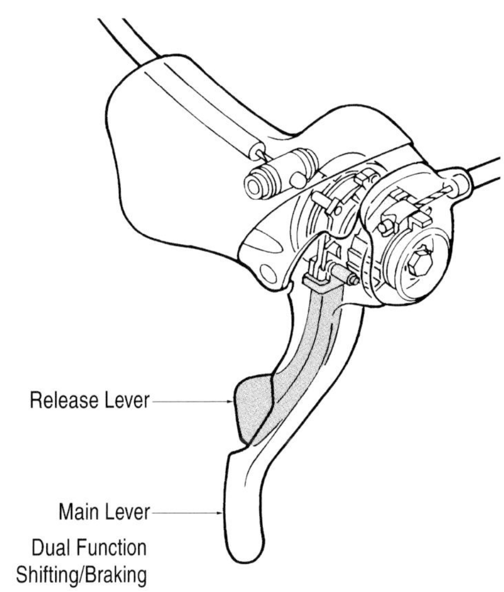


Above: Fig. 15.32. The 1994 Shimano STI (Shimano Total Integration) was the first successful combined brake-shift lever.

### Mountain Bike Shifters

There were more developments with shifters for mountain bikes. For the first few years, everyone used either SunTour or Shimano thumb shifters, mounted on top of the handlebars. SunTour's Power Ratchet had a lighter action. With indexed shifting, Shimano and SunTour both offered the choice of indexed or friction modes with their thumb shifters. In 1990, the levers were replaced by push buttons under the handlebar. One button shifted up and the other shifted down. Shimano called it RapidFire and SunTour called it X-Press.

SunTour's 1990 Command Shifter was essentially a bar-end shifter mounted next to the brake levers. In 2003, Shimano combined the shift lever with the brake lever.

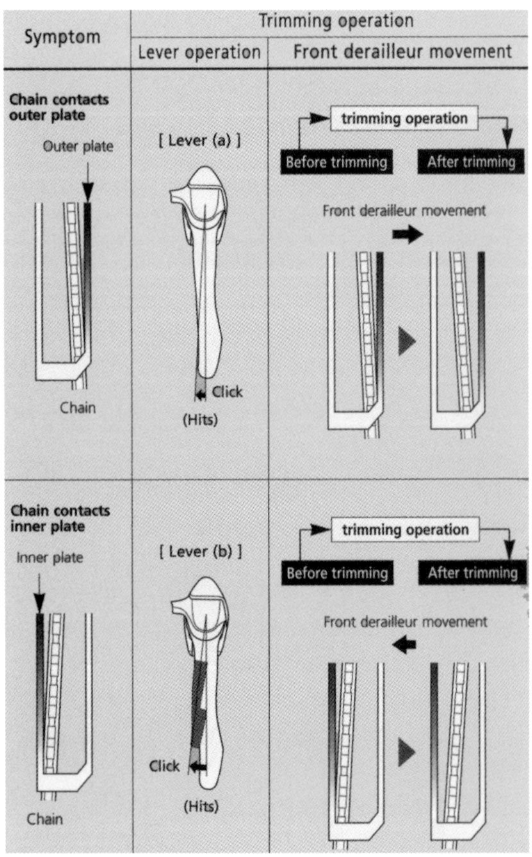


Above: Fig. 15.33. The 1990 Shimano STI lever provided trimming adjustment for all three chainwheels.

### Shift Lever Cable Movement

The entire rear derailleur, indexed shift lever, cable, and casing system has to be designed so that the spacing between the detents on the lever causes a cable movement that positions the jockey pulley directly under each freewheel sprocket.

This is a mechanical design exercise that coordinates the shift lever, the rear derailleur parallelogram, the location of the cable anchor, and the sprocket spacing. The difficult factors are cable stretch, casing compression, and friction. Friction is particularly nasty, because it results in something called hysteresis—a jockey-pulley position that is different for upshifts than it is for downshifts.

There is a "quiet zone" for the jockey-pulley position about 0.8 mm (1/32 in.) wide on either side of the centered position. In the quiet zone, the chain rides smoothly and quietly on the freewheel sprocket. The longer the chain gap the wider the quiet zone. Shimano once used a Centeron pulley on their rear derailleurs, which had a tiny bit of lateral play to allow the chain to center itself. These generalizations do not apply to 9- and 10-speed systems, which have to be precisely adjusted.

Most front and rear derailleurs require about the same amount of cable movement, and you can sometimes mix and match indexed shifters and derailleurs.

SRAM actively promotes their 1:1 shifters and derailleurs, which have about twice as much cable movement as those by other manufacturers.

## Efficiency of Hub Gears Versus Derailleurs

An inter-European gear-train war lasted for 60 years, from the 1900s to the 1960s. The British believed that the chainwheel had to be precisely aligned with the rear sprocket to avoid chain deflection. Consequently, they preferred internal hub gears, which allowed for chain alignment, as well as the use of an enclosed chain case, and presented a neat and tidy appearance.

The hub gear does have advantages: It is light and reliable, and the mechanism is protected from the elements. The chain line is straight (although we've established that that doesn't matter much), and the chain can be protected with a chain case. The derailleur has one major advantage: It was felt to be more efficient than a hub gear.

Derailleurs were more popular in France, Belgium, and Italy from the 1920s to the present. The French said that hub gears were "friction boxes," and that cyclists did not have power to waste overcoming the inherent inefficiency of hub gears.

The bike boom of the 1970s ended the war: The American bike buyers assumed that a 10-speed was at least three times better than a 3-speed and chose accordingly. The same public is applying the same logic to today's 27-speeds and 30-speeds.

Few comparative efficiency tests have been performed on either hub gears or derailleurs. The tests that were done, measured the efficiency of one or the other. Efficiency testing is difficult, because the human power output is small, typically less than ¼ horsepower, and the power losses are less than 10%. No-one performed the rigorous measurements required to determine the actual efficiencies of both devices under the same test conditions. Numbers were bandied about, but if precise tests were performed, the results were never published while the dispute over gearing efficiency raged.

In 1998, Dr. Chester Kyle, Peter Kaufman, and Jim Merz performed a series of rigorous efficiency tests comparing a standard derailleur gear system with two Browning automatic transmissions and a variety of hub gears. The test equipment used a variable-speed motor to drive the crank on the right side. Input power was measured by a strain gauge checked by a spring balance. The rear hub drove the dynamometer flywheel.

A cord was wrapped around the outside of the flywheel, and the cord tension was measured to determine the output power. The frictional losses of the tested drive-trains included the bottom bracket and the rear hub. They also included an extra chain drive to the dynamometer. Figure 15.33 shows the test equipment.

In 2000, Dr. Chester Kyle and I (Frank Berto) measured the efficiency of Sturmey-Archer, Sachs, and Shimano 3-speed and 7-

Above: Fig. 15.34. Rohloff, Sachs, Sturmey-Archer, and Shimano hub gears, and a Shimano 27-speed derailleur gear-train. Not all of the hub gears shown in the picture were tested.

Below: Fig. 15.35. The efficiency comparison test equipment used a variable–speed motor to drive the crank on the right side and a dynamometer to absorb the power.

speed hub gears, and the Rohloff 14-speed hub gear. Using the same test equipment and the same procedures, we also measured the efficiency of a 27-speed Shimano derailleur gear-train and of 4-speed and 12-speed Browning swinging-gate derailleur gear-trains.

After 30 hours of testing and 20,000 data points, we concluded that the French were right—but not by very much: Derailleurs were found to be more efficient than hub gears; however, the differences were smaller than we expected. We concluded that:

- Hub gears were not as bad as described by the old French writers. This was particularly true of pre-World War II hubs that were lubricated with a light oil rather than lubricated-for-life with grease (our test included a 1950s Sachs Torpedo 3-speed that was oil-lubricated). With a 100-watt load, the average efficiency of the Shimano and Sturmey-Archer three-speed hubs in all three gears was 92%. The old Sachs was about 3% better. I attribute the better efficiency to oil lubrication, but it may also have been that this hub was well "broken in." The others were right out of the box. It would have been interesting to test SRAM's current three-speed hub, which is lubricated with grease, like everyone else's.
- The direct-drive middle gear of a three-speed hub is about 95% efficient, not too far from the efficiency of a fixed gear or a derailleur.
- The planetaries used in the other gears lower the efficiency. Low gear is about 93%, and high gear is about 92% efficient. It was these gears that gave three-speed hubs the reputation of being friction boxes.
- Derailleur gear-trains are more efficient than hub gears in non-direct gears.
- 5-speed and 7-speed hub gears use two sets of planetary gearing. The gears that use both planetaries are significantly less efficient than 3-speeds.
- The efficiency of hub gears drops off badly below 100 watts. It rises a bit as the load increases.
- The efficiency of derailleur gears is better at higher power levels. Strong riders and racers could feel and appreciate the derailleur's higher efficiency.
- The straight chain lines, which were so admired by the British writers like F.T. Bidlake, made no difference in efficiency. The efficiency of the "cross-chain" derailleur gears was the same as it was for the aligned gears. This contradicted the accepted wisdom in the bicycle books, yet that is what our tests told us. However, we used modern narrow bushingless chains for the derailleur tests. Old derailleur chains were wider and stiffer. They could not bend sideways very readily, and that might have affected the efficiency of

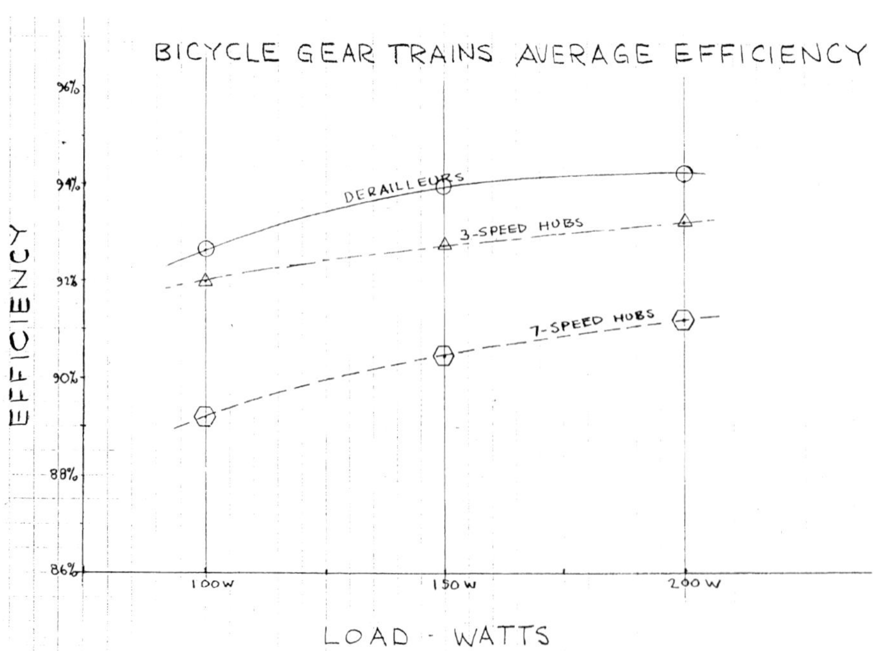


Left: Fig. 15.36. Graph showing the results of the Kyle & Berto efficiency tests.

old derailleur-drives. They certainly made more noise.

- The size of the small sprocket was very significant. Sprockets with at least 15 teeth gave a measurable efficiency gain. The 12-tooth sprocket gave the worst derailleur efficiencies. The big (44-tooth) chainwheel and the small (12-tooth) sprocket were well aligned, yet the efficiency was 92%. The big chainwheel—big sprocket (34 teeth) is the worst cross-chain gear, yet it still had an efficiency of 93%. The efficiency gain between a 12-tooth and a 16-tooth sprocket was about 2%. Early cyclists knew this and generally did not use sprockets smaller than 14 teeth. Hub gears use larger sprockets, which helps their efficiency.

The fact is that the French were right—but the difference was not as much as they thought: Hub gears indeed have more friction than derailleur gears, except in the direct gear. In the indirectly driven gears, the efficiency varied between 85% and 90% for a 7-speed hub.

The 4-speed Shimano Auto D was the best of the hub gears tested. It was 90% efficient in high gear and 95% efficient in low gear, which was the direct drive.

The tests were run at light, medium, and heavy loads, corresponding very roughly to average (1/8 hp), strong (1/4 hp), and elite (1/2 hp) cyclists. The efficiencies go up with the load. The derailleurs peak at about 97% at 500 watt (2/3 hp). The hub gears peak at 93% at 400 watt (1/2 hp). The graph shows the efficiencies that correspond to 100 watt (1/8 hp).

The Rohloff 14-speed Speedhub gear was a special case. Rohloff uses needle-bearing pinions and oil lubrication. Even though it has three sets of planetaries, it was of 88% to 91% efficient at a 100-watt load. This was considerably better than the 7-speed hubs, lending support to Bidlake's assertion that precision manufacture was essential for hub gear efficiency.

To put all of these numbers in perspective, suppose you are an average, 1/8 hp, 160-lb. cyclist. You can ride up a 4% grade in a 31-inch gear with a 96% efficient derailleur gear-train at 5.5 mph (8.8 km/h), at a cadence of 60 RPM.

If you switch to a 7-speed hub gear, and use the same 31-inch low gear, the efficiency is about 90%. Since you are output-limited, your cadence will drop to 56 RPM, and you will slow down to 5.1 mph (8.2 km/h). You be the judge.

## Gear Selection

I started writing about bicycle gearing about thirty years ago. Over the years, I've written more than fifty gearing articles. At one time, I had three different bicycles, with the latest Campagnolo, Shimano, and SunTour gear-trains. I am still proud of my 1975 *Bicycling* article, "Choosing Gears That Suit You." The hardware has changed significantly since then, but the advice is still valid. Selecting gearing to suit you should be a four-step process:

1. Select a high gear to suit your strength, your comfortable cadence, and your riding style.
2. Select a low gear to suit your strength, the hills, and your load.
3. Select chainwheels, sprockets, and derailleurs to provide the selected range, with small steps between gears.
4. Select the component quality that fits your budget and meets the first three criteria.

### Selecting the High Gear

The high gear uses the big chainwheel and the smallest sprocket. A 27-inch-wheel road bike with a 52-tooth chainwheel and a 13-tooth sprocket has a high gear of 52/13x27=108 in.

Our continental friends use the term "development," which is how far the bicycle travels per crank revolution, measured in meters. The development of the above bike is 4 times the wheel circumference, i.e. 339 in. (8.6 m). A convenient rule-of-thumb for calculating development is gear inches divided by 12.5. However, in spite of the metric revolution, most English-speaking cyclists still think in gear inches.

Beginning riders are comfortable with high gears between 60 and 80 in. for level terrain. Racers in top condition use gears of 90 to

100 in. Gears higher than 100 in. are for tail winds, drafting another rider, or for downhill riding. Selecting high gear is a personal matter. In the above example, you could get a higher high gear either by fitting a chainwheel with more than 52 teeth or by using a small sprocket with less than 13 teeth (but, as we found in our tests, larger sprockets are more efficient).

### Selecting the Low Gear

This is surprisingly controversial. There is a macho syndrome that says that a fit cyclist should be able to climb any hill in a gear of, say, 40 in. The other extreme is the mechanical engineer who wants a low enough low gear so that his grandma can pedal up 12% grades. Hence the term "granny gear."

I live in Marin County, in the San Francisco Bay Area, which offers wonderful cycling. My favorite ride goes over White's Hill, which rises 350 ft. in 4/5 of a mile—about an 8% grade. It's not the steepest hill in Marin, but it's pretty tough. Twenty-five years ago, I could climb it in under 6 minutes at better than 8 mph, cranking a 40-inch gear at 70 RPM. In 2004, at age 75, it took me 14 minutes to climb White's Hill at 3.5 mph (5.5 km/h) in a 20-inch gear at 60 RPM. I expect to pedal slower in lower gears and take longer as I grow older.

Tourists with 30 or 40 lbs. of luggage particularly appreciate extra-low gears. One of my early articles gave the following advice for low gear selection:

- Touring in the mountains with a heavy load: 20 in.
- Loaded touring or steep hills: 25 in.
- Cycling in hilly country: 30 in.

For a septuagenarian like myself, I would lower these recommendations by about 5 in. each. Most people reading this book are probably buying their third or fourth bicycle. From experience with their previous bikes, they will have learned their needs for a suitable low gear.

## Bicycle-Type-Specific Gear Selection

Are you buying a gruppo to build the custom bicycle of your dreams, or are you buying a gruppo with a bicycle attached? Either way, you have to decide between a road bike, a mountain bike, or a comfort bike. Road-bike gruppos come as doubles for racing or as triples for sport-touring. If you lust for Campagnolo, you have chosen a road bike geartrain. The following specifications come from the various manufacturers' 2004 catalogs.

### Gearing for Road Bikes

Shimano, SRAM, and Campagnolo all provide very similar options. Their double crankset gruppos

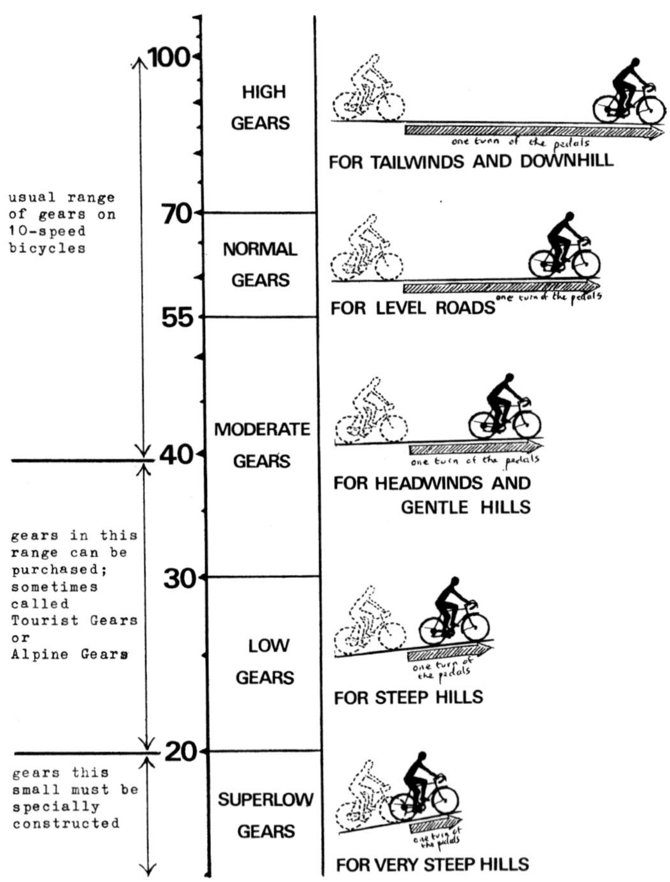


Left: Fig. 15.37. "Gears for easy pedaling." This drawing is taken from contributing author Ron Shepherd's 1975 *Low Gear Bulletin.* Nothing has changed since then.

offer 53-, 52-, or 50-tooth large chain wheels and 42-, 40-, or 39-tooth small chainwheels. The 9-speed or 10-speed cassettes can start with an 11-, 12-, or 13-tooth small sprocket and range up to a large sprocket with 29 teeth for Campagnolo, or 27 teeth for Shimano. That means high gears range between 53 / 11 × 27 = 130 in., and 50 / 13 × 27 = 104 in., and the lowest low gear is 39 / 29 × 27 = 36 in.

Triple cranksets offer the same 53-, 52-, or 50-tooth big chainwheels and 42-, 40-, or 39-tooth (middle) chainwheels, while adding a 30-tooth inner chainwheel. The cassettes are the same, so you have the same range of high gears, while the lowest low gear is 30 / 29 × 27 = 28 in. for Campagnolo, or 30 / 27 × 27 = 30 in. for Shimano. Granny won't be touring on these "sport-touring" bikes.

### Gearing for Mountain Bikes

Campagnolo does not make mountain-bike gruppos. You're dealing with Shimano, although SRAM also provides a few options. You can install Shimano's mountain bike components on road bikes, and you can mix and match shifters, derailleurs, cranksets, and cassettes, as long as everything is 9-speed. That is how to get a lower low gear. Mountain bike cranksets move your feet out a bit, because they have to clear the fat tires.

Shimano makes four high-end mountain-bike gruppos: XTR, Deore XT, Deore LX, and Deore. There are two sets of chainwheels: 44/32/22 and 48/36/26. The large chainwheels are only available with Deore XT.

There are nine different 9-speed cassettes, starting with an 11-tooth or a 12-tooth small sprocket and going up to a 32- or 34-tooth large sprocket.

The high gears range from 48 / 11 × 27 = 118 in. to 44 / 13 × 27 = 91 in. The low gears range from 26 / 32 × 27 = 22 in. to 22 / 34 × 27 = 17 in.

These numbers are all based on 27-inch wheels. Skinny 700C tires or 26-inch fat tires lower the gears by about 3%.

### Gearing for Comfort Bikes

Call them comfort bikes, city bikes, hybrid bikes, trekking bikes, or transportation bikes, we are talking about medium-width tires, flat handlebars, and an upright riding position. They use the same kinds of wide-range gearing. Shimano calls their derailleur gruppos Nexave and their hub gear gruppos Nexus. SRAM is also a factor in this market.

Nexave is available in 8-speed or 9-speed versions, with either double or triple cranksets. The double uses 46/34 chainwheels and the triples use 46/34/24.

The 9-speed gruppos use an 11–34 cassette and the 8-speed gruppos use an 11–33 cassette.

This gives a 113-in. high gear and either a 27-in. low gear with the double crankset or a 19-in. low gear with the triple.

The double crankset versions provide a 4:1 range, which will cover most hills.

The triple crankset versions provide genuine wide-range gearing with a 6:1 range. You will rarely encounter a situation where you will need the highest or the lowest gears. That's good.

### Hub Gears

Shimano, SRAM, and SunRace Sturmey-Archer make 3-, 4-, 5-, 7- and 8-speed hub gears, SRAM also offers a 9-speed (and there are some older 12-speeds still around), while Rohloff makes a 14-speed hub gear. By selecting the appropriate chainwheel and sprocket, you can raise or lower both the high gear and the low gear to match your cycling con-

*Table 15.2. Comparison of hub-gear ranges*

| Make | 3-speed | 4-speed | 5-speed | 7-speed | 8-speed | 9-speed | 14-speed |
|---|---|---|---|---|---|---|---|
| Shimano | 184% | 184% | – | 244% | 307% | – | – |
| SRAM | 186% | – | 251% | 303% | – | 340% | – |
| Sturmey-Archer | 177% | – | 225% | – | 305% | – | – |
| Rohloff | – | – | – | – | – | – | 526% |

ditions. Shimano is making a serious effort to provide electronic control, so that all you will have to do is select fast, medium, or easy pedaling.

## "Gear Freaking"

I've mentioned gear freaking and half-step gearing pretty much in the past tense. This was how mathematically inclined cyclists used to optimize their gear-trains when they only had 10 or 15 gears to play with. They used logarithmic paper to plot out the optimum combination of chainwheels and freewheel sprockets to give the maximum number of usable gears with nice even steps between them.

A "half-step" gearing setup might use a freewheel with 24% steps between the five sprockets, combined with two chainwheels with a 12% difference. That half-step 10-speed would have had ten gears 12% apart. "half-step plus granny" gearing added a small inner chainwheel to provide three extra-low gears for the steepest hills.

A triple crankset with 52-, 46-, and 24-tooth chainwheels, combined with a freewheel with 13, 16, 20, 26, and 34-tooth sprockets was the near-perfect half-step plus granny set-up. Work it out: It provided a 108-in. high gear, a 19-in. low gear, 13 useful gears out of 15, and 13% steps between 10 of the gears. There were lots of other arrangements with fun names. In today's world of 8-, 9-, and 10-sprocket cassettes, gear freaking is almost dead.

Shifting today's 27-speeds is quite simple. You start on the middle chainwheel. When you run out of higher gears, you shift to the big chainwheel and drop back a couple of clicks of the rear derailleur. When you reach a steep hill and run out of lower gears on the middle chainwheel, you shift to the little chainwheel and add a couple of clicks in the rear.

You don't want to plot your gears on logarithmic paper, because you will find that you have only about 15 useful gears out of 27. What's worse, you'll have big steps between many of the gears that you often use. The worst case

Right and below: Figs. 15.38 and 15.39. The 53/39/28-tooth chainwheels and 10-11-12-13-14-15-17-20-26-34 sprockets on Frank Berto's Alex Moulton New Series.

Right: Fig. 15.40. Author Frank Berto, photographed on a blustery day during a 1999 bicycle tour in Holland with his Alex Moulton New Series (before the "granny" chainring was added in the front).

is the 18% step between the 11-tooth and the 13-tooth sprockets.

### Berto's 30-speed Moulton

Even today, gear freaking is not completely dead. With a large supply of Hyperglide sprockets, there is a modern version of Half-Step plus Granny that I call "straight-block plus granny." I bought my Moulton New Series in 1999. Since then, the gear-train has been a work in progress, with annual updates as Shimano provides new options. As the spiritual father of gear freaking, my goal was to use available components to obtain the following:

- The lowest useful low gear.
- The widest spread between high and low gears.
- Small steps between gears.
- An easy shift sequence.

Shimano's 2005 Ultegra triple crankset comes with 52-39-30-tooth chainwheels. However, the 74-mm inner bolt circle accepts a 24-tooth inner chainwheel. I combined the 52-39-24 chainwheels with 10, 11, 12, 13, 14, 15, 17, 20, 26, and 34-tooth Hyperglide sprockets. This uses a Dura-Ace 9-speed cassette (this was before Shimano went to 10-speed) with Moulton's 10-tooth sprocket instead of the retainer and a 135-mm axle. I used a device called an Ericksen Gizmo to provide the 10th click on my 9-speed STI levers. I drilled the rivets out of a MegaRange cassette to get the 20-, 26-, and 34-tooth sprockets.

This may sound like a lot of tinkering to avoid using Shimano's standard cassettes, but it works beautifully. I have 7 closely spaced level-terrain gears on the middle chainwheel. When I get to the 39×10 gear, I shift to the big chainwheel and drop back two clicks on the STI lever to get five more closely spaced gears on the big chainwheel. The straight blocks overlap, which minimizes double shifts. When I get to a hill, I stay on the middle chainwheel, unless it's a really steep hill. Then I use the little chainwheel and the Gizmo. The 30% steps between the lowest gears are just fine on a steep hill.

The current setup has a low of 13 inches (24/34 × 18.5 = 13), a high gear of 96 inches (52/10 × 18.5 = 96), and a 740% spread between high and low gears.

Because of the one-tooth steps between sprockets, the normal gears are closely spaced. The step between the 10- and the 11-tooth sprocket is 10%. The step between the 14- and the 15-tooth sprocket is 7%.

Above and below: Figs. 15.41 and 15.42. Comparison of Shimano's stock cassette (top) with the author's customized one giving the optimum wide-range gearing steps (below).

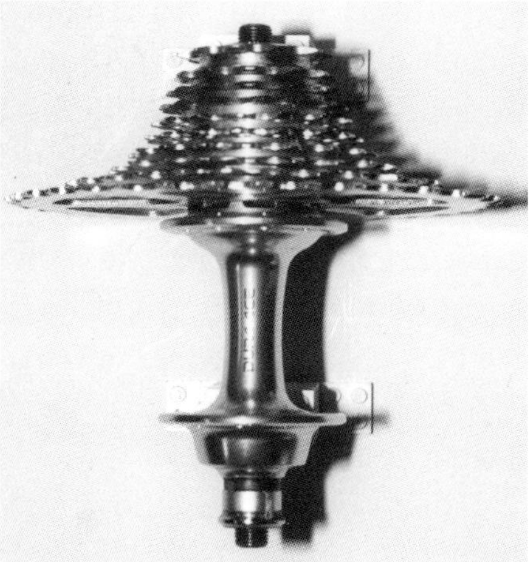

Right: Fig. 15.43. Controls for Frank Berto's Moulton: Shimano Flight Deck (gear and cadence), Avocet 25 (speed and distance), STI brake-shift levers, and Ericksen Gizmo (visible where the gear cable enters the right-hand STI lever) to allow use of a 9-speed STI lever with a 10-sprocket freewheel.

# 16 Nothing New Under the Sun?

THE PRECEDING chapters chronicled the development of bicycle gear-trains from their primitive beginnings to the present level of sophistication. In this chapter, we will look at some of the blind alleys that were explored. Some ideas that were introduced a century ago and failed just won't die. They reincarnate again and again.

By regularly attending bicycle trade shows, I've encountered dozens of these "new" inventions and their inventors. They all share the following characteristics:

- The inventor believes in his invention with true religious fervor.
- Satisfied users provide glowing testimonials.
- The inventor believes that if he can demonstrate the superiority of his invention, buyers will launch him on the road to fame and fortune.

Left: Fig. 16.1. Each Interbike trade show brings "new" familiar products, and not all are meant to revolutionize cycling. This Kangaroo-type children's bike is actually a beautiful modem version of the type described in Chapter 2, see page 36.

The real world is hard on bicycle gear-train inventors. Few of these products survive to the next bicycle show.

The Gay Nineties, the decade from 1890 to 1899, was extraordinarily productive for bicycle designs. Literally thousands of patents were taken out for bicycle improvements. The market became saturated and bicycle prices plummeted. By and large, the best designs survived and the second best designs remain to delight collectors.

Santayana said that those who cannot remember the mistakes

of the past are condemned to repeat them.

Apparently, most of today's inventors haven't read bicycle history. Or perhaps they've read just enough history to copy century-old inventions, and hope that the buyers have not read the same parts of bicycle history. Here are five gear-trains that are reincarnations from the Gay Nineties. Some are legitimate alternatives, but they do not provide significant advantages over today's gear-trains. Judge for yourself.

## Treadle- and Lever-Drives

Treadle- and lever-drives have up-and-down or back-and-forth pedal movement. The idea dates back to the infancy of the bicycle. Kirkpatrick Macmillan's presumed 1842 bicycle (or tricycle) probably used a treadle-drive, but there are no technical details. Certainly the Scottish bicycles built by Gavin Dalzell and Thomas McCall in subsequent decades were driven by treadles connected by rods to crankshafts on the rear wheel.

These early bicycles used treadles because the available chains were unsuitable. By the 1880s, reliable chains were available, and even front-driving high bicycles used chains to lower the pedals or increase the gear. Other safeties with large front wheels used treadle-drives rather than chains to move the rider lower and further back to reduce the risk of "headers." Some early treadle-drive bicycles are shown on pages 35 through 37.

Later lever cycles were more sophisticated. On the 1892 Svea, the pedals were attached to bell cranks. As the pedals moved up and down, the crank arms pivoted back and forth. Chains attached to the crank arms ran around freewheels on either side of the rear hub. Moving the attachment point varied the chain movement (i.e., the gear ratio). The Svea is shown on page 46.

Lever-drive bicycles were popular in France from 1895 to about 1905. The levers of the 1905 Terrot Lévocyclette had 12 notches, giving 12 speeds, with a 10-to-1 gear range. A spring returned each pedal to the top position. The pedals could be pushed alternately or together. The Lévocyclette is shown on page 71.

Lever- and treadle-drive is a broad category, and there have been numerous designs since 1920. I have five different models in my file. The Alenax was one of the better versions.

Lever-drives work on hills at low cadences. They are less effective at normal speeds on the level. The search for faster human-powered vehicles (HPVs) led to considerable testing and analysis of alternatives to conventional circular pedaling. Intermittent drives require the pedal to come up to coupling speed on each stroke before it can transmit additional power. The tests concluded that lever drives are more expensive and less efficient than chain drives.

## Shaft-Drives

Early bicycle chains had problems. Shaft-drive eliminated troublesome greasy chains and gave the bicycle a neat, clean appearance. The French word *acatène* (chainless) was used to describe chainless bicycles. The French company L'Acatène was the major developer of the shaft-drive in the 1890s. There were numerous French (and other) shaft-drive bicycles. Some had two or even three speeds.

Right: Fig. 16.2. The 1985 Alenax Transbar bicycle was almost identical to the Svea, with modern materials and manufacturing. It had four different gears, with a 3-to-1 range. Alenax eventually gave up on this design, but recently they were back with another oddball system.

By 1897, the prices of U.S. chain-driven safety bicycles had fallen dramatically. Columbia saw shaft-drive as a way to provide uniqueness. They introduced the Columbia Chainless with great fanfare. Sales were small, however, because the 1898 shaft-drive model cost $125 (about $6,000 in today's dollars), when the top-of-the-line Columbia chain-drive machine cost only $75.00. The 1899 Overman Victor Chainless used spin-rollers to reduce the cost, but it did not survive either. L'Acatène and Columbia shaft- drives are shown on page 70.

Shaft-drives are an elegant alternative to chains. Shaft drives require precision-cut bevel gears and quality bearings, which makes them much more expensive than a chain drive. The spin-roller was cruder and less efficient. The need to remove the rear wheel reduces precision and makes lubrication of the rear drive more difficult. Shaft drives have a lower mechanical efficiency than chain drives.

## Retro-Direct, or Reverse Pedaling

Retro-directs are pedaled either clockwise or counterclockwise. In most cases, reverse pedaling gave a lower gear. The first patent for a retro-direct was issued to Barberon and Meunier in 1869. Retro-direct appeared anew in France in the Gay Nineties, with models by Hirondelle, Magnat & Debon, and Floquart. Some had as many as three speeds. Retro-direct gearing is practical, reliable, and not too expensive. Hirondelle made bicycles with retro-direct gearing until 1938. Retro-directs are described in Chapters 4 and 5.

I know of no tests of the ergonomic efficiency of reverse pedaling. However, it seems to me that reverse pedaling is like walking backward. It is inherently less comfortable and uses the leg muscles inefficiently. Today's retro-directs often promote reverse pedaling to exercise different leg muscles. Jan Heine has

Right: Fig. 16.3. The 1995 Kariz Shaft Drive System was a recent reincarnation of shaft-drive. It was available with either a single-speed coaster brake or a 3- or 5-speed hub gears. Kariz disappeared after two years.

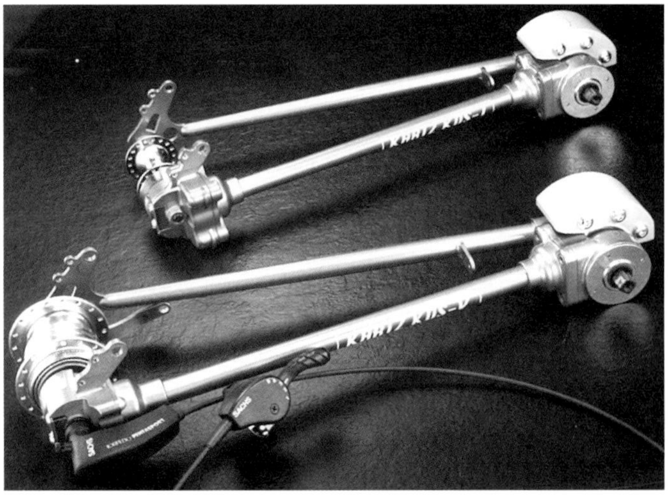

Left: Fig. 16.4. The 2005 Tara No-Chain Drive was a reincarnation of the 1899 Overman Victor Chainless spin-roller drive.

Right: Fig. 16.5. The 1994 MBI Bi-Directional Gear was a single-speed retro-direct kit that was made in Korea. It was available from 1994 to 1999 in two versions—integrated in the chainwheel and in the rear hub respectively.

ridden several retro-directs and reports that he was not able to reach backpedaling speeds in excess of 45 RPM.

## Bottom-Bracket Gears

Shortly after the safety bicycle was invented, variable gears were installed in the bottom bracket, starting in 1870 both in Britain and in France. By the 1900s, front bottom-bracket gears and rear hub gears were similar. They both used planetary gearing to provide two or three speeds. For examples, see 2, 3, 4, and 5.

The 1995 Shimano FG-5S10 was a 5-speed planetary bottom-bracket gear. It was developed for the Japanese shopping/commuter bike market. It failed because in Japan housewives make the bicycle buying decisions, and they resisted the extra cost.

Currently, The Swiss Schlumpf company markets a 2-speed gearbox built into the chainwheel, thus eliminating the need for a front derailleur, at the expense of a slight decrease in efficiency. They are finding a market for use with recumbents.

The size and weight of a gear box depends on the power transmitted and the rate of rotation. For a given horsepower, the faster the rotation, the lower the torque. Since the rear wheel turns about three times faster than the crankshaft (in a "normal" gear), a gear box in the rear hub will be smaller and lighter, and probably more efficient, than a slower turning gearbox in the bottom bracket.

## Expanding Chainwheels

The first expanding chainwheel was the 1896 Linley & Biggs Protean. It expanded from 18 to 19, to 20, to 21 teeth using a 1-inch pitch chain. It shifted by reverse pedaling. See page 54 for an illustration.

The 1903 Paradox expanding chainwheel expanded from 20 to 26 teeth in one-tooth steps, giving seven closely spaced gears. It lasted for a year. The 1903 Mitchell Any-Speed rotated the chainwheel off-center. The eccentricity changed the effective diameter. It provided infinitely variable gears within its range. See page 60 for the Mitchell and Paradox originals.

Left: Fig. 16.6. The Schlumpf Mountain Drive is a heavy-duty wide-range 2-speed bottom-bracket gear with a gear reduction of 1:2.5 (40%). Similar devices, called Speed Drive and High Speed Drive respectively, offer gear ratio increases of 165% and 250% respecively. Shifting is either by kicking the dustcap or by means of an optional handlebar-mounted shift lever. These things are actually being used by some mountain bike racers and recumbent rdiers.

After 70 years, the expanding chainwheel reappeared. The 1973 Deal Drive had 16 speeds, and a 2-to-1 range. It was well built and reliable. However, it was heavy, expensive, and it had a limited gear range. It expired after 2 years. See page 264 for an illustration.

The 1983 Excel Cambiogear had 16 speeds, and a 3-to-1 gear range. It was made of graphite-fiber-reinforced plastic. It lasted a year. See page 243 for an illustration.

Expanding chainwheels must compete with front derailleurs and double or triple cranksets, both of which are lighter, cheaper, and more resistant to mud and dirt.

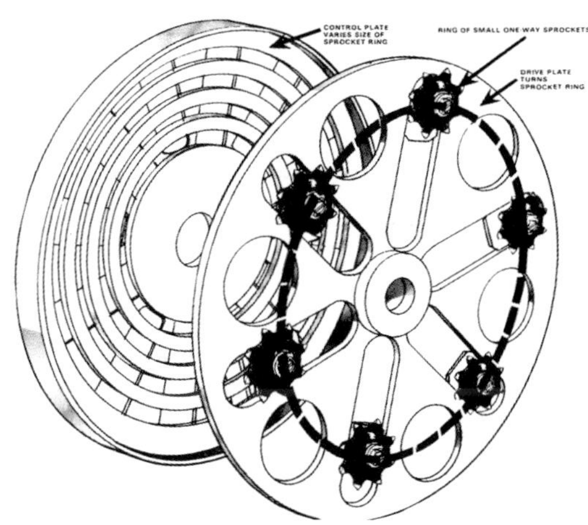


Above: Fig. 16.7. The 1974 Hagan All-Speed Expanding Chainwheel had six sprockets that moved out as the inner spiral turned. The gear range was 2.85 to 1, or 285%.

# Bibliography & Chapter Notes

The most reliable sources for a technical history of the bicycle are old catalogs. In the process of researching and writing *The Dancing Chain*, I reviewed hundreds of old catalogs.

Old magazines, especially the advertisements, were also useful. Two French magazines, *Le Cycliste*, and *Le Cycle*, were major sources of information for *The Dancing Chain*. *Le Cycliste*, and its editor Paul de Vivie (Vélocio), covered the period from the turn of the century to 1930. *Le Cycle*, and its editor and illustrator, Daniel Rebour, covered the period from 1946 to 1974. The British magazine *The Boneshaker*, which started in 1955, contains a wealth of historical material. Finally, I wrote for the American magazine, *Bicycling*, from 1975 to 1991, and I have used many of my old technical articles from that period.

However, old catalogs and old magazines are hard to locate, so historical researchers usually rely on books. The International Cycle History Conferences (ICHC) have been held annually since 1990. The papers that appear in the Proceedings of the ICHC were essential sources for *The Dancing Chain*.

Some books and some ICHC papers cover only material relevant to one chapter in *The Dancing Chain*, and those are listed in the chapter-specific references. The books that are listed in section A below are more general, and cover more than one chapter.

## A. General References

### 1. Papers published in the Proceedings of the International Cycle History Conferences


1.1. *Proceedings of the 1st (1990) International Cycle History Conference*. San Francisco: Van der Plas Publications, 1995.

Nick Clayton. "The Meyer-Guilmet Bicycle, 1869 or 1879?"

Keizo Kobayashi. "The Inventor of the Lallement Pattern."

Hans-Erhard Lessing. "Karl Von Drais' Two-wheeler."

Andrew Millward. "The Genesis of the British Cycle Industry. 1867–1872."

Nicholas Oddy. "Kirkpatrick Macmillan."

1.2. *Proceedings of the 2nd (1991) International Cycle History Conference*. San Francisco: Van der Plas Publications, 1992.

Nadine Besse. "The Hirondelle Bicycles of the Manufacture Française d'Armes et Cycles."

Les Bowerman. "Cycle History Research in Britain."

Nick Clayton. "The Development of the Suspension Wheel."

Raymond Henry. "Vélocio and Multiple Gearing."

Keizo Kobayashi. "French Sources and Bibliography on Cycle History."

—. "The Olivier Brothers."

Hans-Erhard Lessing. "Around Michaux: Myths and Realities."

Augusto Marinoni. "Leonardo da Vinci's Bicycle."

Andrew Millward. "The Birth of the Cycle Industry in England."

Roland Sauvaget. "Cycle History Research in France."

Walter Ulreich. "Austria and Bicycle History."

1.3. *Proceedings of the 3rd (1992) International Cycle History Conference*. San Francisco: Van der Plas Publications, 1993.

Les Bowerman. "Samuel Gompertz and his Addition."

Nick Clayton. "Hans Reynold and the Birth of the Bicycle Chain."

Alastair Dodds. "Kirkpatrick Macmillan."

Raymond Henry. "Vélocio's Crypto Bicycle with Whippet Derailleur."

Jan Holt. "The Svea (1892) and the Itera (1982)."

Keizo Kobayashi. "From Von Drais to Pierre Michaux, 1818–1861."

Hans-Erhard Lessing. "The Bicycle and Science–from Drais until Today."

Wilheim Matthies. "Founded in Crisis: Union Frondenberg."

Andrew Milward. "The Cycle Trade in Birmingham. 1890–1920."

Roland Sauvaget. "Notes on Vernon Blake."

1.4. *Proceedings of the 4th (1993) International Cycle History Conference*. San Francisco: Van der Plas Publications, 1994.

Nadine Besse. "From Small Arms to Bicycles in the St. Étienne Region."

Nicholas Clayton. "Renold Chain Production–1882-1910."
Raymond Henry. "Albert Raymond and the Cyclo Company."
David Herlihy. "Who Invented the Bicycle–Lallement in 1863 or Michaux in 1861?"
Karl Hodges. "Did the Emergence of the Automobile End the Bicycle Boom?"
Hans-Erhard Lessing. "The Front-Wheel Driven Velocipede in Germany."
Roger Street. "The Quest for the Medieval Hobby-Horse Velocipede."

1.5. *Proceedings of the 5th (1994) International Cycle History Conference*. San Francisco: Van der Plas Publications, 1995.
Alastair Dodds and Alex Brown. "The Edinburgh Tricycle."
Ron Shepherd. "The Dancing Chain: One Hundred Years of Derailleur Gears."
Raymond Henry. "Touring Bicycle Technical Trials in France, 1901–1950."
H.D. Higman. "Founding of the Dunlop Tyre Company."
Hans-Erhard Lessing. "Cycling or Roller Skating: The Rise of Personal Mobility."
David L. Patton. "Historical Geography of Cycling Technology."
Albert A. Pope. "Colonel Pope and the Founding of the U.S. Bicycle Industry."
Andrew Ritchie. "Methodological Approach to Velocipedes, Bicycles and Tricycles."

1.6. *Proceedings of the 6th (1995) International Cycle History Conference*. San Francisco: Van der Plas Publications, 1995.
Raymond Henry. "Fifty Years of Development of Simplex Derailleurs."
—. "Daniel Rebour–Illustrative Genius of the Bicycle."
John Stegman and Mike Kramer. "Front Wheel Drive Concepts."

1.7. *Proceedings of the 7th (1996) International Cycle History Conference*. San Francisco: Van der Plas Publications, 1997.
Frank J. Berto. "Rear Derailleur Development since D-Day."
Les Bowerman. Velocipedes in Milan in 1811?"
Nick Clayton. "Who Invented the Penny-Farthing?"
Glen Norcliffe. "Colonel Albert Pope."
Ron Shepherd. "Derailleur Pulley Resistance."

1.8. *Proceedings of the 8th (1997) International Cycle History Conference*. San Francisco: Van der Plas Publications, 1998.
Frank J. Berto. "Who Invented the Mountain Bike?"
Alan Bernstein. "Shimano AX and the Aero Tradition."
Raymond Henry. "Huret: A Notable French Derailleur."
Loren Hufstetler. "The Scottish School of Bicycle Design."
Hans-Erhard Lessing. "The Evidence against Leonardo's Bicycle."
Roger Street. "Denis Johnson's Patent of 1818."

1.9. *Proceedings of the 9th (1998) International Cycle History Conference*. San Francisco: Van der Plas Publications, 1999.
Frank J. Berto. "Sunset for SunTour."
Bruce Epperson. "Failed Colossus: Albert A. Pope & the Pope Manufacturing Co. 1876–1900."

1.10. *Proceedings of the 10th (1999) International Cycle History Conference*. San Francisco: Van der Plas Publications, 2000.
Frank J. Berto. "The Great American Bicycle Boom."
Alastair Dodds. "The Scottish Bicycle Industry before Macmillan."
Bruce Epperson. "After Pope. The Pope Manufacturing Company, 1899–1990."

1.11. *Proceedings of the 11th (2000) International Cycle History Conference*. San Francisco: Van der Plas Publications, 2001.
Frank J. Berto. "History Repeats."
Alastair Dodds. "Dunlop and the Pneumatic Bicycle Tyre."
Bruce Epperson. "How Many Bikes?"
Tony Hadland. "Raleigh UK in the Last Quarter of the $20^{th}$ Century."
Toshio Kajiwara. "Early Japanese Safety Bicycles."
Hans-Erhard Lessing. "What Led to the Invention of the Early Bicycle?"
Andrew Millward. "UK and Japanese Competition in the Inter-War Period."
Hiroshi Nakamura. "History of the Derailleur in Japan."
Ross D. Petty. "The Rise of the Asian Bicycle Business."
Roger Street. "Johnson's 1819 Tour of England."
Takata, Nozomu. "Power-Assisted Bicycles in Japan."

1.12. *Proceedings of the 12th (2001) International Cycle History Conference*. San Francisco: Van der Plas Publications, 2002.

Frank J. Berto. "Hub Gear versus Derailleur Efficiency."
Jacques Graber. 'The Lefèbvre Bicycle."
Chester R. Kyle. "A Brief History of the IHPVA."
Nicholas Oddy. "Cycle Purchases from 1897 to 1899."
Andrew Ritchie. "The Velocipede of Alexandre Lefèbvre."
Roger Street. "The Milan Velocipedi."

1.13. *Proceedings of the 13th (2002) International Cycle History Conference*. San Francisco: Van der Plas Publications, 2003.
Hans-Erhard Lessing. "The Velocipede of 1819 in America."
Hans van Vliet. "The Shimano Dura-Ace Story."

1.14. *Proceedings of the 14th (2003) International Cycle History Conference*. Dickson, ACT, Australia: Canberra Bicycle Museum, 2003.
William Hudson. "Myths and Milestones in Bicycle Development."

1.15. *Proceedings of the 15th (2004) International Cycle History Conference*. San Francisco: Van der Plas Publications, 2005.
Raymond Henri. "Vélocio—Leisure Cyclist, and the Pioneer of Multi-Speed Gearing."

1.17. *Proceedings of the 16th (2005) International Cycle History Conference*. San Francisco: Van der Plas Publications, 2006.
Glen Norcliffe. "The Technical and Social Significance of the Tricycle."

1.18. *Proceedings of the 17th (2006) International Cycle History Conference*. San Francisco: Van der Plas Publications, 2007.
Michael Grützner. "The Chainless Bicycle Craze in Germany Around 1900."

## 2. General Reference Books

*American Bicyclist and Motorcyclist. 100th Anniversary Edition*. New York: Cycling Press, 1979.
Frederick Aldermen. *Bicycling–A History*. Newton Abbot UK: David and Charles (Publishers) Ltd., 1972.
James T. Allen. *Cycles patented in the U.S., from 1789 to 1892*. Washington, D.C: 1892.
Sharon Babaian. *The Most Benevolent Machine—Cycles in Canada*. Ottawa: National Museum of Science and Technology, 1998.
Richard Ballantine. *Richard's 21st Century Bicycle Book*. New York: Overlook Press, 2001.
H.W. Bartleet. *Bartleet's Bicycle Book*. London: Ed. J. Burrow & Co., 1931.
Gregory Houston Bowden. *The Story of the Raleigh Cycle*. London: W.H. Allen, 1975.
Gianni Brera. *The Giant and the File—Tullio Campagnolo*. Vicenza: Campagnolo, 1995.
F.J. Camm. *Every Cyclist's Handbook*. London: George Newnes Ltd., 1936.
Pierre Chany. *La Fabuleuse Histoire du Cyclisme des Origines à 1955*. Paris: Nathan, 1988.
Nick Clayton. *Early Bicycles*. Aylesbury, UK: Shire Publications Ltd., 1994.
Judith Crown and Glen Coleman. *No Hands–The Rise and Fall of the Schwinn Bicycle Company*. New York: Henry Holt and Company, 1996.
*The Data Book*. San Francisco: Van der Plas Publications, 1998.
Fred DeLong. *DeLong's Guide to Bicycles and Bicycling*. Radnor PA: Chilton Book Company, 1974.
Alastair Dodds. *Scottish Bicycles & Tricycles*. Edinburgh: NMS Publishing Ltd., 1999.
Pryor Dodge. *The Bicycle*. Paris: Flamarrion, 1996.
Fermo Galbiati. *Bicycles. Le Biciclette*. San Francisco: Chronicle Books, 1989.
Pablo Fachinetti and Guido P. Rubino. *Campagnolo: 75 Years of Passion*. Boulder, CO: Velo Press, 2008
W. Gronen and W. Lemke. *Geschichte des Radports und des Fahrrades*. Eupen, Belgium: Doepgen Verlag, 1978.
Tony Hadland. *The Sturmey-Archer Story*. Tony Hadland, 1987.
Jan Heine and Jean-Pierre Pradères. *The Golden Age of Handbuilt Bicycles*. Seattle: Vintage Bicycle Press, 2005.
—. *The Competition Bicycle: A Photographic History*. Seattle: Bicycle Quarterly, 2008.
Raymond Henry. *Du Vélocipède au Dérailleur Moderne*. St.-Étienne: Association des Amis du Musée d'Art et d'Industrie de St.-Étienne, 1998.
—. *Paul de Vivie, dit Vélocio: L'évolution du cycle et le cyclotourisme*. St.-Étienne: Association des Amis du Musée d'Art et d'Industrie de St.-Étienne, 2005.
David Herlihy. *Bicycle: The History*. New Haven: Yale University Press, 2004.
William Humber. *Freewheeling–The Story of Bicycling in Canada*. Erin, Ontario: The Boston Mills Press, 1986.
Lionel Joseph. *The First Century of the Bicycle and its Accessories*. Surrey: Lionel Joseph, 1996.

Ian K. Jones. *The Safety Bicycle*. Aylesbury, UK: Shire Publications Ltd., 1986

Ron Kitching. *Everything Cycling*. Harrogate, UK: Ron Kitching, 1979.

—. *A Wheel in Two Worlds*. Harrogate, UK: Ron Kitching, 1993.

Keizo Kobayashi. *Pour Une Bibliographie du Cyclisme 1818–1983*. Paris: Federation Française de Cyclotourisme, 1984.

—. *Cinq Ans de Bibliographie Cycliste. 1984-1988*. Paris: Federation Française de Cyclotourisme, 1991.

Hans-Erhard Lessing. *Automobilität: Karl von Drais und die unglaublichen Anfänge*. Dresden, Germany: Maxime Verlag, 2004.

James McGurn. *On Your Bicycle*. York, UK: Open Road Publishers, 1999.

*New Cyclng Derailleur Collection*. New Cycling (Japan), May 1981.

David B. Perry. *Bike Cult–The Ultimate Guide to Human-Powered Vehicles*. New York: Four Walls Eight Windows, 1995.

Jay Pridmore and Jim Hurd. *The American Bicycle*. Osceola, WI: Motorbooks International, 1995.

—. *Schwinn Bicycles*. Osceola, WI: Motorbooks International, 1996.

Max J. B. Rauck, Gerd Volke, and Felix R. Paturi. *Mit dem Rad durch zwei Jahrhunderte*. Aarau: AT Verlag, 1979.

Daniel Rebour. *La Pratique du Vélo*. Paris: Technique et Vulgaristion, 1949.

—. *Cycle Compétition et Cyclotourisme Sportif.* Paris: Technique et Vulgaristion, 1962.

—. *Cycles de Compétition et Randonneuses*. Paris: Technique et Vulgaristion, 1975.

—. *Le Monde de Daniel Rebour*. Japan: 1976.

Andrew Ritchie. *King of the Road*. London: Wildwood House, 1975.

Derek Roberts. *Cycling History–Myths and Queries*. John Pinkerton, 1991.

—. *This Veteran Business*. Surrey: Derek Roberts, 1992.

—. *The Invention of the Safety Bicycle*. Surrey: Derek Roberts, 1993.

Frank W. Schwinn. *50 Years of Schwinn-Built Bicycles*. Chicago: Arnold, Schwinn & Company, 1945.

Robert A. Smith. *A Social History of the Bicycle*. New York: American Heritage Press, 1972.

Philip Sumner. *Early Bicycles*. London: Hugh Evelyn Ltd., 1996.

Rob van der Plas. *Bicycle Technology*. San Francisco: Bicycle Books, 1991.

R. John Way. *Cycling Manual*. London: Temple Press Books, 1967.

Frank R. Whitt and David G. Wilson. *Bicycling Science: Ergonomics and Mechanics*. 2nd Edition. Cambridge, MA: MIT Press, 1982.

John Woodforde. *The Story of the Bicycle*. New York: Universe Books, 1970.

Wilhelm Wolf. *Fahrrad und Radfahrer.* Leipzig: Otto Spamer Verlag, 1890.

Neil S. Wood. *Evolution of the Bicycle*. Gas City, IN: L-W Book Sales, 1994.

### 3. General Magazine References

*Australian Cycling and Motorcycling* (Australia)
*Australian Cyclist* (Australia)
*Adventure Cyclist* (USA)
*Aeolius Butterfly* (USA)
*American Bicyclist and Motorcyclist (USA)*
*American Wheelmen* (USA)
*The Bicycle* (UK)
*Bicycle Guide* (USA)
*Bicycle Retailer and Industry News* (USA)
*Bicycling* (USA)
*Bicycling News* (UK)
*Bike World* (USA)
*Bike Tech* (USA)
*The Boneshaker* (UK)
*CTC Gazette* (UK)
*Le Cycle* (France)
*Cycling* (UK)
*Cyclist* (UK)
*Cycling Science* (USA)
Le Cycliste (France)
*Le Cyclotouriste* (France)
*The Fat Tire Flyer* (USA)
*Fellowship News* (UK)
*The Moultoneer* (UK)
*Mountain Bike* (USA)
*Mountain Bike Action* (USA)
*The National Cyclists' Union Review* (UK)
*New Cyclng* (Japan)
*News and Views* (UK)
*On the Wheel* (USA)
*L'Officiel du Cycle* (France)
*La Pédale* (France)
*Rivendell Reader* (USA)
*Tour* (Germany)
*Vintage Bicycle Quarterly & Bicycle Quarterly* (USA)
*The Wheelman* (USA)

## B. Chapter-Specific References

There are more than 1,000 illustrations in *The Dancing Chain*. To make the book read smoothly, I have not shown the source of each illustration in the captions. The following references for each chapter show the sources for the illustrations in that chapter.

### Chapter 1. The First Bicycles: 1817–1860


Jean Althuser. *Pierre Michaux and His Sons*. Kenilworth, UK: Jim Willis, 1986.

Les Bowerman. "Samuel Gompertz and his Addition." *1992 ICHC.*

Les Bowerman. "Velocipedes in Milan in 1811." *1996 ICHC.*

—. "Michaux v Lallement." *The Boneshaker.* No. 157, Winter, 2001.

Viscount Bury and G. Lacy Hillier. *Cycling*. London: Longmans Green, 1887.

Nick Clayton. "The Meyer-Guilmet Bicycle. 1869 or 1879?" *1990 ICHC.*

—. "Hans Renold and the Birth of the Bicycle Chain." *1992 ICHC.*

—. "Renold Chain Production–1882-1910." *1993 ICHC.*

Alastair Dodds "Kirkpatrick Macmillan." *1992 ICHC.*

—. "The Scottish Bicycle Industry before Macmillan." *1999 ICHC.*

—. *Scottish Bicycles & Tricycles*. Edinburgh: NMS Publishing Ltd., 1999.

Jacques Graber. "The Lefèbvre Bicycle." *2001 ICHC.*

David Herlihy. "Who Invented the Bicycle–Lallement in 1863 or Michaux in 1861?" *1993 ICHC.*

—. "The First Bicycle–Part One." *Bicycle Guide*, May, 1992.

—. "The First Bicycle–Part Two." *Bicycle Guide,* June, 1992.

—. "Choosing the Strongest Michaux Claim." *The Boneshaker.* No. 157, Winter, 2001.

—. "Choosing the Strongest Lallement Claim." *The Boneshaker.* No. 158, Spring, 2002.

Harry Hewitt Griffin. *Bicycles and Tricycles of the Year 1886*. London: L. Upcott Gill, 1886.

—. *Bicycles and Tricycles of the Year 1889*. London: L. Upcott Gill, 1889.

G. Lacy Hillier. *Cycling*. London: Longmans Green, and Co., 1895.

Keizo Kobayashi. "The Inventor of the Lallement Pattern." *1990 ICHC.*

—. "The Olivier Brothers." *1991 ICHC.*

—. "From Von Drais to Pierre Michaux, 1818–1861." *1992 ICHC.*

Hans-Erhard Lessing. "Karl Von Drais' Two-wheeler." *1990 ICHC.*

—. *Automobilität: Karl von Drais und die unglaublichen Anfänge*. Dresden, Germany: Maxime Verlag, 2004.

—. "The Front-Wheel Driven Velocipede in Germany." *1993 ICHC.*

—. "The Evidence against Leonardo's Bicycle." *1997 ICHC.*

—. "What Led to the Invention of the Early Bicycle?" *2000 ICHC.*

—. "The Velocipede of 1819 in America." *2002 ICHC.*

—. "Evidence Against Leonardo's Bicycle." *The Boneshaker.* No. 145, Winter, 1997.

Augusto Marinoni. "Leonardo da Vinci's Bicycle." *1991 ICHC.*

Nicholas Oddy. "Kirkpatrick Macmillan." *1990 ICHC.*

Albert A. Pope. "Colonel Pope and the Founding of the U.S. Bicycle Industry." *1994 ICHC.*

Charles E. Pratt. *Pierre Lallement and his Bicycle*. Boston: Lallement Memorial Committee, 1992.

Andrew Ritchie. "The Velocipede of Alexandre Lefèbvre." *2001 ICHC.*

Derek Roberts. *The Years of the High Bicycle*. Pinkerton, 1980.

—. *The Years of the High Bicycle*. Pinkerton Reprint, 1980.

Roland Sauvaget. "Michaux v Lallement." *The Boneshaker.* No. 152, Spring, 2000.

Archibald Sharp. *Bicycles and Tricycles*: London: Longmans Green, 1896.

Roger Street."The Quest for the Medieval Hobby-Horse." *1993 ICHC.*

—. "Denis Johnson's Patent of 1818." *1997 ICHC.*

—. 'Johnson's 1819 Tour of England." *2000 ICHC.*

—. "The Milan Velocipedi." *2001 ICHC.*

—. *The Pedestrian Hobby-Horse*. Christchurch, UK: Artesius Publications, 1998.

Aad Streng. "Pierre Lallement, Inventor and Pioneer." *The Boneshaker.* No. 100, Summer, 1986.


### Chapter 2. The Search for Speed: 1861–1890


"Comparative Friction of Chains and Gears on Bicycles." *American Machinist.* September 2, 1897

"The Crypto Dynamic Gear." *The Bicycling World.* February, 1884.

"The H.J. Lawson Myth." *The Boneshaker.* No. 13, Autumn, 1958.

"The Whippet Cycle Syndicate Ltd. with Protean Gear." *The Boneshaker.* No. 15, Spring, 1959.

"The Retro-Direct Gear." *The Boneshaker.* No. 23, Spring, 1961.

"The Paradox Gear." *The Boneshaker.* No. 27, Spring, 1962.

Jacques Seray. *Deux roues: la véritable histoire du vélo.* Millau: Éditions du Rouergue, 1988.

Archibald Sharp. *Bicycles and Tricycles.* London: Longmans Green, 1896.

"The Simpson Lever Chain." *The Boneshaker.* No. 28, Summer, 1962.

Charles Spencer. *Bicycles and Tricycles–Past and Present.* London: Griffith & Farran, 1883.

Michael Stratton and Barrie Trinder. *Industrial England.* London: B. T. Batsford Ltd., 1997.

Henry Sturmey. *Sturmey's Indispensable Handbook to the Safety Bicycle.* London: Iliffe & Son, 1885.

**Chapter 3. The Bicycle Boom: 1891–1899**

W.G. Angell. "The American Star Bicycle." *The Boneshaker.* No. 9, Autumn, 1957.

Edward J. Berry. "The Berry Derailleur Bicycle." *The Boneshaker.* No. 126, Summer, 1991.

Alastair Dodds. "What did Kirkpatrick MacMillan Invent?" *The Boneshaker.* No. 127, Winter. 1991.

Alastair Dodds. "Dunlop and the Pneumatic Bicycle Tyre." *2000 ICHC.*

Bruce Epperson. "Failed Colossus: The Pope Manufacturing Co. 1876–1900." *1999 ICHC.*

Raymond Henry. *Paul de Vivie, dit Vélocio: L'évolution du cycle et le cyclotourisme.* St.-Étienne: Association des Amis du Musée d'Art et d'Industrie de St.-Étienne, 2005.

David Herlihy. "A Thousand Fathers." *Bicycle Guide,* March 1990.

Karl Hodges. "Did the Emergence of the Automobile end the Bicycle Boom?" *1993 ICHC.*

Glen Norcliffe. "Colonel Albert Pope." *1996 ICHC.*

Nicholas Oddy. "Cycle Purchases from 1897 to 1899." *2001 ICHC.*

Andrew Ritchie. "British Opposition to Road Racing." *The Boneshaker.* No. 151, Winter,1999.

Archibald Sharp. *Bicycles & Tricycles.* London: Longmans, Green and Co., 1896.

Ron Shepherd. "Early Australian Bicycle Gears." *The Boneshaker.* No. 141, Summer, 1996.

**Chapter 4. Pre-Derailleurs, Epicyclics and Exotics: 190–1907**

Cycling Staff. *Variable Gears and all about them.* Cycling Penny Handbook No. 2, 1909.

C.L. Freeston. *Cycling in the Alps.* London: Grant Richards, 1900.

Raymond Henry. *Paul de Vivie, dit Vélocio: L'évolution du cycle et le cyclotourisme.* St.-Étienne: Association des Amis du Musée d'Art et d'Industrie de St.-Étienne, 2005.

**Chapter 5. The First Derailleurs: 1908—1919**

Tony Hadland. "The Evolution of Early British Derailleur Gears." Unpublished, download form *http://www. users. globalnet.co.uk/~hadland/old_derail.htm* 2000.

Jan Heine and Jean-Pierre Pradères. *The Golden Age of Handbuilt Bicycles.* Seattle: Vintage Bicycle Press, 2005.

Raymond Henry. *Du Vélocipède au Dérailleur Moderne.* St.-Étienne: Association des Amis du Musée d'Art et d'Industrie de St.-Étienne, 1998.

—. *Paul de Vivie, dit Vélocio: L'évolution du cycle et le cyclotourisme.* St.-Étienne: Association des Amis du Musée d'Art et d'Industrie de St.-Étienne, 2005.

—. "Vélocio and Multiple Gearing." *1991 ICHC.*

—. "Vélocio's Crypto Bicycle with Whippet Derailleur." *1992 ICHC.*

Derek Roberts. "The Invention of the Safety Bicycle." Surrey: Derek Roberts, 1993.

Ron Shepherd. "The Origins of Derailleur Gears." *The Boneshaker.* No. 124, Winter, 1990.

—. "Early Australian Bicycle Gears." *The Boneshaker.* No. 141, Spring, 1996.

—. "Le Premier Derailleur." *The International Veteran Cyclist.* December, 1992.

Paul de Vivie (Vélocio). "Revue des différentes systèmes de polyxcation." *Le Cycliste.* Oct/Nov/Dec, 1919

"The Whippet Cycle Syndicate Ltd." *The Boneshaker.* No. 15, Spring, 1959.

**Chapter 6. Practical Derailleurs: 1920–1929**

Raymond Henry. *Du Vélocipède au Dérailleur Moderne.* St.-Étienne: Association des Amis du Musée d'Art et d'Industrie de St.-Étienne, 1998.

—. *Paul de Vivie, dit Vélocio: L'évolution du cycle et le cyclotourisme*. St.-Étienne: Association des Amis du Musée d'Art et d'Industrie de St.-Étienne, 2005.

—. "Albert Raymond and the Cyclo Company." *1993 ICHC.*

David Herlihy. "The Invention of the Derailleur—Part I." *Bicycle Guide*, March, 1990.

—. "The Invention of the Derailleur—Part II." *Bicycle Guide*, April, 1990,

—. "Albert Raymond and the Cyclo Derailleur." *Bicycle Guide*, January. 1995.

Ron Shepherd. "The Amazing Vernon Blake." *On the Wheel,* March, 1999.

### Chapter 7. The Golden Age: 1930–1939

Riccardo Carnovalini. "La Preistoria del Cambio." *Bicisport*, March, 1990.

*Cycling Manual. 7th Edition*. London: Temple Press Ltd., 1934.

"The Derailleur Gear—Variable Gear by Chain Shift." *The Cyclist,* February 19, 1936.

Fermo Galbiati and Nino Ciravegna. "La bicicletta." *Editrice, Milan*. 1989.

Jan Heine and Jean-Pierre Pradères. *The Golden Age of Handbuilt Bicycles*. Seattle: Vintage Bicycle Press, 2005.

Raymond Henry. *Du Vélocipède au Dérailleur Moderne*. St.-Étienne: Association des Amis du Musée d'Art et d'Industrie de St.-Étienne, 1998.

—. *Paul de Vivie, dit Vélocio: L'évolution du cycle et le cyclotourisme*. St.-Étienne: Association des Amis du Musée d'Art et d'Industrie de St.-Étienne, 2005.

—. "Cambio Vittoria's brief reign." *Bicycle Guide*, February, 1996.

Andrew Millward. "UK and Japanese Competition in the Inter-War Period." *2000 ICHC.*

Peter Nye. *Hearts of Lions—the History of American Bicycle Racing*. New York: Norton, 1988.

H.F. Opperman. *Pedals, politics and people*. Sydney: Haldane, 1977.

Gordon Selby. "History of Super Champion Osgears." *The Boneshaker,* No. 152, Spring, 2000.

—. "Louis Camillis and British Cyclo" (Part I). *The Boneshaker,* No. 165, Summer 2004.

—. "Louis Camillis and British Cyclo" (Part II). *The Boneshaker,* No. 166, Autumn 2004.

Ron Shepherd. "Tullio Campagnolo's First Patent." *On the Wheel*, June, 1999.

F.R. Thompson. "Cycling in the 1930s." *Fellowship News*, No. 122, 1995.

Alexander von Tutschek. "The Constrictor Tyre Company." *The Boneshaker,* No. 146, Spring, 1998.

William F. Wray. *The Kuklos annual and handbook for cyclists and all wayfarers*. London: Burrow & Co., 1939.

Jim and Elisabeth Young. *Bicycle Built for Two*. (Reprint Edn.) Oakland, CA: Cycling Classics, 1996.

### Chapter 8. The Post-War Years: 1945–1954

"Back-pedalling Overhill Derailleur." *Cycling,* August 11, 1948.

Alex Bainbridge. "Bartali and the Cervino Derailleur." *The Boneshaker,* No, 160, Winter, 2002.

*Cycling Book of Maintenance, 1st Edition*. London: Temple Press, 1945.

*Cycling Manual, 20th Edition*. London: Temple Press, 1947.

Jan Heine. "Rear Derailleur Milestones." *Vintage Bicycle Quarterly*, Winter, 2003.

—. "Daniel Rebour—The Life Behind the Illustrations." *Vintage Bicycle Quarterly*, Summer 2004.

*René Herse—New Cycling No. 450. Japan: New Cycling*, 2001.

Ralph Hurne. "When the Campionissimi Rode." *Bicycling*, June, 1991.

H. Ichikawa & Jan Heine. "Campagnolo Gran Sport." *Vintage Bicycle Quarterly*, Winter, 2003.

Alajandro José Nacere. "The Derailleur in the Great Races." *Le Cycle*, January, 1987.

Dave Nuttal. "Campagnolo Corsa and Paris Roubaix Gears." *On the Wheel,* October, 1998.

A.L. Pullen. *The Cycling Handbook*. London: Sir Isaac Pitman & Sons Ltd., 1953.

Daniel Rebour. "Numerous articles on Derailleurs." *Le Cycle,* September 14, 1945 to December 25, 1954.

Richard Yates. "The 1949 Tour de France." *On the Wheel,* December, 1998.

### Chapter 9. Slow Growth: 1955–1964

*Cycling Book of Maintenance*. London: Temple Press Ltd., 1961.

H.E. England. *Cycling Manual. 24th Edition*. London: Temple Press Ltd., 1960.

Liz Fried. *Schwinn Sting-Ray*. Osceola, WI: Motorbooks International, 1997.

Harold Moore. *The Complete Cyclist. 6th Edition.* London: Sir Isaac Pitman & Sons Ltd., 1960.

Jean-Pierre Pradères. "The Life of Daniel Rebour." *Vintage Bicycle Quarterly,* Summer, 2004.

A.L. Pullen. *Cycling Handbook. 3rd Edition.* London: Sir Isaac Pitman & Sons Ltd., 1960.

Daniel Rebour. "Numerous articles on Derailleurs." *Le Cycle,* January 8, 1955 to December, 1964.

### Chapter 10. "The Great American Bike Boom": 1965–1974

Frank J. Berto. "The Great American Bicycle Boom." *1999 ICHC.*

—. "Freewheels, chainwheels, Derailleurs." *Bicycling,* March, 1975.

—. "Choosing Gears to Suit You." *Bicycling,* April, 1975.

—. "First Aid for Alpine Gearing." *Bicycling,* January, 1976.

—. "Getting into Really Low Gears." *Bike World,* August, 1973.

—. "The Sins of Shimano." *Rivendell Reader,* Fall, 1998.

Peter Brown. "The Elswick Hopper Convincible." *The Boneshaker,* No. 164, Spring, 2004.

Sheldon Brown. "Chuck Harris and the First American Derailleur." *Rivendell Reader,* Spring, 1998.

Tony Hadland. *The Moulton Bicycle.* Coventry: Tony Hadland, 1981.

David Herlihy. "Daniel Rebour." *Bicycle Guide,* March, 1995.

Gabe Konrad "Daniel Rebour." *Aeolius Butterfly,* No. 6, August, 1995.

Eleanor Moore. *The Complete Buyer's Guide to Bicycles–1973.* New York: Service Communications Ltd., 1973.

Marc Muller. "Inside the Schwinn Varsity." *Rivendell Reader,* Spring, 1998.

Daniel Rebour. "Numerous articles on Derailleurs." *Le Cycle,* January, 1965 to May, 1974.

Lawrence Teeman. *Consumer Guide 1974 Bicycle Test Reports.* Skokie, IL: Consumer Guide, 1974.

Mark Thiffault. *Bicycle Digest–1973.* Northfield, IL: Digest Books Inc., 1973.

### Chapter 11. The Dawn of Mountain Biking: 1975–1984

Frank J. Berto. *The Birth of Dirt–Origins of Mountain Biking.* San Francisco: Van der Plas Publications, 1998.

—. "All About Freewheels." *Bicycling,* June, 1978.

—. "1981 New York Bicycle Show." *Bicycling,* June, 1981.

—. "Shimano Deore Drivetrain." *Bicycling,* May, 1982.

—."1982 Long Beach Bicycle Show." *Bicycling,* May, 1982.

—. "Campagnolo vs Shimano vs SunTour." *Bicycling,* September, 1982.

—. "1983. New York Bicycle Show." *Bicycling,* June, 1983.

—. "Repack Revisited." *Bicycling,* March, 1984.

—. "Mid-Range Rear Derailleurs." *Bicycling,* April, 1984.

—. "Touring Rear Derailleurs." *Bicycling,* May, 1984.

—. "1984 Long Beach Bicycle Show." *Bicycling,* May, 1984.

Harold Bridge. "History of Derailleur Design." *Rivendell Reader,* January, 1998.

Nicky Crowther. *The Ultimate Mountain Bike Book.* Osceola, WI: Motorbooks International, 1996.

"Gears to suit every pocket." *Cycling,* October 25, 1980.

Tony Hadland. "Raleigh UK in the Last Quarter of the 20th Century." *2000 ICHC.*

—. *"The Spaceframe Moultons."* Coventry: Tony Hadland, 1994.

Raymond Henry "Daniel Rebour." *Bike Culture,* February, 1997.

Charles Kelly and Nick Crane. *Richard's Mountain Bike Book.* New York: Ballantine Books, 1988.

Grant Petersen. "Huret's Jubilee Derailleurs." *Rivendell Reader,* June, 2001.

Rob van der Plas. *The Mountain Bike Book.* San Francisco: Velo-Press/Bicycle Books, 1984.

Grant Wolf. *Mountain Bikers Almanac.* Stateline NV: Grant Wolf, Inc., 1996.

### Chapter 12. The Rise of Shimano: 1985–1994

Frank J. Berto. "Racing Front Derailleurs." *Bicycling,* January, 1986.

—. "Touring and Off-Road Front Derailleurs." *Bicycling,* April, 1986.

—. "1986 Long Beach Bicycle Show." *Bicycling,* May, 1986.

—. "1987 Long Beach Bicycle Show." *Bicycling,* May, 1987.

—. "Indexed Shifting." *Bicycling,* January, 1988.

—. "All About Chains." *Bicycling,* March, 1988.

—. "1988 Long Beach Bicycle Show." *Bicycling,* May, 1988.

—. "1991 Anaheim Bicycle Show." *Bicycle USA,* March, 1991.
—. "1992 Anaheim Bicycle Show." *Bicycle USA,* January, 1992.
—. "1993 Las Vegas Bicycle Show." *Bicycle USA,* March, 1993.
—. "1994 Las Vegas Bicycle Show." *Bicycle USA,* January, 1994.
—. "The Sins of Shimano." *Rivendell Reader,* Fall, 1998.
*History of Shimano*. Shimano Industrial, 1975.
*Profile of Shimano 60th Anniversary*. Shimano Industrial, 1981.
*Shimano 70th Anniversary: The History of Shimano*. Shimano Inc., 1991.

### Chapter 13 & 14. Recent History: 1995–2009

Otto Beaujon. "Recent Raleigh History." *The Boneshaker,* No, 159, Summer, 2002.
Frank J. Berto. "1995 Anaheim Bicycle Show." *Bicycle USA,* January, 1995.
—. "1996 Anaheim Bicycle Show." *Bicycle USA,* January, 1996.
—. "1997 Anaheim Bicycle Show." *Bicycle USA,* March, 1997.
—. "1998 Anaheim Bicycle Show." *Bicycle USA,* March, 1998.
—. "1999 Las Vegas Bicycle Show." *Bicycle USA,* January, 1999.
—. "2004 Las Vegas Bicycle Show." *Australian Cyclist,* January, 2004.
Gerry Moore. "The Fastest Bikes in the World." *The Boneshaker,* No. 142, Winter, 1996.
—. "Today is History." *The Boneshaker,* No. 165, Summer 2004.

### Chapter 15. How Derailleurs Work

Frank J. Berto. "All About Rear Derailleurs." *Bicycling,* January, 1978.
—. "All About Front Derailleurs." *Bicycling,* March, 1978.
—. "All About Shift Levers." *Bicycling,* April, 1978.
—. "All About Bicycle Chains." *Bicycling,* November, 1980.
—. "All About Indexed Shifting Rear Derailleurs." *Cycling Science,* March, 1990.
—. "Hub Gear versus Derailleur Efficiency." *2001 ICHC.*
—. *Bicycling Magazine's Complete Guide to Upgrading Your Bike*. Emmaus PA: Rodale, 1988.
Stuart C. Burgess. "Improving cycling performance with large sprockets." *Department of Mechanical Engineering, Bristol University*, 1998.
Dean Dell'Oro and Matthew Malone. "Bicycle Derailleur losses." Department of Mechanical and Manufacturing Engineering, University of Melbourne, 1995.
Chester Kyle. "The Mechanical Efficiency of Bicycle Transmissions." *Test Report to Browning Research*. September 27, 1998.
—. "Alternative Bicycle Transmissions." *Cycling Science*. September, 1991.
—. and Frank Berto. "The Mechanical Efficiency of Bicycle Derailleur and Hub-Gear Transmissions." *Human Power,* Summer, 2001.
Okajima, Shinpei. "Biomechanics of Shifting Performance: Design of the Shimano New Dura-Ace Shifting System." *Bike Tech*, April, 1985.
—. "When Everything Clicks: Evaluating Index System Derailleurs." *Bike Tech*, June, 1988.
Grant Petersen. "Freewheels & Cassettes." *Rivendell Reader,* September, 2002.
Ron Shepherd. "Derailleur Pulley Resistance." *1996 ICHC.*

### Chapter 16. Nothing New Under the Sun?

*The Data Book: 100 Years of Bicycle Component Design*, Japan, 1983; San Francisco: Cycle Publishing, 1998
Rob van der Plas. *Bicycle Technology*. San Francisco, CA: Bicycle Books, 1991.

# Index

## N



## O



## P



## Q



## R



## S